1994

FLUID, ELECTROLYTE, AND ACID-BASE PHYSIOLOGY

A Problem-Based Approach

Second Edition

FLUID, ELECTROLYTE, AND ACID-BASE PHYSIOLOGY

A Problem-Based Approach

Mitchell L. Halperin, M.D., FRCPC
University of Toronto
Toronto, Ontario

Marc B. Goldstein, M.D., FRCPC
University of Toronto
Toronto, Ontario

W. B. Saunders Company
A Division of Harcourt Brace & Company

Philadelphia, London, Toronto, Montreal, Sydney, Tokyo

W. B. SAUNDERS COMPANY
A Division of Harcourt Brace & Company
The Curtis Center
Independence Square West
Philadelphia, PA 19106

Library of Congress Cataloging-in Publication Data

Fluid, electrolyte, and acid-base physiology: a problem-based approach / [edited by Mitchell L. Halperin, Marc B. Goldstein – 2nd ed.
p. cm
Rev. ed. of: Fluid, electrolyte, and acid-base emergencies. 1988
Includes bibliographical references and index
ISBN 0-7216-5155-0
1. Water-electrolyte imbalances. 2. Acid-base imbalances. 3. Water-electrolyte imbalances—Case studies. 4. Acid-base imbalances—Case studies. I. Halperin, M. L. (MItchell L.) II. Goldstein, Marc B.
RC630.H34 1994
616.3'9—dc20

FLUID, ELECTROLYTE, AND ACID-BASE PHYSIOLOGY
A Problem-Based Approach ISBN 0-7216-5155-0

Printed in the United States of America

Last digit is the print number: 9 8 7 6 5 4 3 2 1

ACKNOWLEDGMENTS

Special thanks to the following who made excellent suggestions to improve this book:
Kathy Dorrington
Kamel S. Kamel
Jean Pierre Mallie
James Oster
Bob Richardson
Adrienne Scheich
Martin Schreiber
Harald Sonnenberg
Bobby Joe Stinebaugh

Preparation of book with great appreciation:
Anne Harmon
Lorne Jacobs
Jolly Mangat

DEDICATION

Brenda and Ellen. At last!

PREFACE

Although only five years have passed since our first book on fluid, electrolyte, and acid-base disorders was published, we were eager to undertake a new edition for several reasons.

- There are many new insights in the field, and virtually every chapter contains new concepts and novel approaches. Therefore, the majority of the book has been totally rewritten.
- Our approach to teaching is continuously in evolution. Because we are committed to the philosophy "learn by understanding," we have provided even more biochemical and physiological bases for the clinical disorders discussed. The first chapter of each section addresses the relevant physiology, and the chapters that follow highlight the basic science in a clinical context.
- We have organized this book to take advantage of the modern trends in the learning process. Each section has a concise statement outlining the concepts needed to understand the clinical material that follows. In each chapter, we start with an introductory case, a series of objectives, and an outline of major principles. The major issues to be discussed in each subsection are highlighted, and each chapter ends with a review section that includes a summary of main points.
- The clinical cases are central to the learning experience, and we have greatly enhanced this component of the book. Because we and our students now focus more on "problem-based learning," we emphasize problem solving throughout the book—either in the form of introductory cases or in the many questions distributed throughout the text. Each clinical chapter ends with several challenging "cases for review." We have also provided a table of contents for all the cases so that one can easily find a particular "challenge."
- In the margins, we have included definitions, clinical pearls, and supplemental information; these items should not distract the reader from the major flow of information. We have placed discussions of the cases and the questions at the end of each chapter so that the reader may have an opportunity to think about the problems posed before seeking the answer.

HOW TO USE THIS BOOK

We have retained a format that allows one to read the book at a level that meets one's needs. For a quick reference when dealing with an emergency situation, we have provided a list of the diagnostic flow charts. These flow charts are invaluable, especially when coupled with the relevant subsection in the text. The index should allow the reader to find the appropriate information quickly.

To gain an overview of a chapter, one should examine both the table of contents and the outline of major principles. Thereafter, one should read the shaded areas in each section to obtain the "bare bones" information of that section. If these areas contain surprising or unfamiliar concepts, one should read the section more closely.

The questions (which are all clearly marked) indicate the areas that require further reading. When time allows, we encourage the reader to study the entire section, from the physiology to the review cases.

One final point merits emphasis. It is no longer adequate to maintain the separation of subjects and subspecialties. Accordingly, another strong theme in this book is integration. We have made an obvious attempt to integrate laterally by including aspects of endocrine, gastrointestinal, respiratory, and cardiovascular physiology. In addition, we have attempted a degree of vertical integration by bringing physiology and biochemistry to the bedside.

Mitchell L. Halperin
Marc B. Goldstein

CONTENTS

Section One
Acid-Base

Section Two
Sodium and Water

Section Three
Potassium

Section Four
Hyperglycemia

LIST OF CASES

Chapter 1
Principles of Acid-Base Physiology

Chapter 2
The Clinical Approach to Acid-Base Disorders

Chapter 3
Metabolic Acidosis

Chapter 3
Metabolic Alkalosis

Chapter 5
Respiratory Acid-Base Disturbances

Chapter 6
Sodium and Water Physiology

Chapter 7
Hyponatremia

LIST OF FLOW CHARTS

SECTION ONE

Acid-Base

1

PRINCIPLES OF ACID-BASE PHYSIOLOGY

Concepts in Acid-Base Physiology:

Objectives

- To provide an understanding of how H^+ may be a threat to the body and how physiologic responses minimize this threat.
- To present a background enabling the recognition of both the production and removal of H^+ by *metabolic processes*.
- To develop the vocabulary required to "speak" the acid-base language.
- To emphasize the magnitude of H^+ production and removal in normal metabolism, thus elucidating the potential for the development of major acid-base disturbances.

Metabolic processes:
A way of examining the overall impact of metabolism by considering only the substrates and products and ignoring all intermediates.

Note:
Square brackets denote concentration.

Outline of Major Principles

1. The $[H^+]$ is very tiny, yet if this concentration rises, H^+ will bind to proteins and change their charge, shape, and possibly function.
2. To quantitate the rate of production or removal of H^+, simply count the valences of substrates and products of metabolic processes.
3. H^+ are produced when certain amino acids in proteins are metabolized. They are also produced if there is insufficient oxygen or insulin.
4. H^+ are produced at a very rapid rate only during hypoxia.
5. Buffering of H^+ occurs initially via the bicarbonate buffer system; this system removes H^+ when HCO_3^- are converted to $CO_2 + H_2O$. A large H^+ load obliges a larger proportion of buffering by intracellular proteins.
6. In people consuming a typical Western diet, the kidney regenerates new HCO_3^-, largely via the excretion of NH_4^+. The kidney must also "reabsorb" all filtered HCO_3^-.

Mole:
The weight in grams of a compound when a given number (6.023×10^{23}) of molecules is present. One nanomole (nmol) is 10^{-9} mole.

pH:
The logarithm of $1/[H^+]$. It rises when the $[H^+]$ falls and vice versa.

HCO_3^- (bicarbonate ions):
The conjugate base of carbonic acid (H_2CO_3) and the major H^+ acceptor in the extracellular fluid. New HCO_3^- are made in the kidney when H^+ are excreted or when ammonium ions (NH_4^+) are formed and excreted.

P_aCO_2:
The partial pressure of CO_2 in arterial blood.

Note:
First-time readers should skip to page 5 and read Parts A and B before returning to this case.

Introductory Case
Lee Wants to Know the Acid Truth

(Case discussed on pages 34–35)

Lee, a 70-kg insulin-dependent diabetic, did not take insulin for the past two days because a GI upset prevented her from eating. The relevant values from a sample of plasma were:

H^+	60 nmol/l (pH 7.22)	P_aco_2	25 mm Hg
HCO_3^-	10 mmol/l	Anion gap	27 mEq/l

The relevant values in the urine on admission were:

NH_4^+	200 mmol/day	HCO_3^-	0 mmol/day

Ketoacid anions: strongly positive

Normal values in plasma:
$[H^+]$ = 40 ± 2 nmol/l
pH = 7.40 ± 0.02
$[HCO_3^-]$ = 25 ± 2 mmol/l
P_aco_2 = 40 ± 2 mm Hg, or 5.3 ± 0.2 kpascal
(1 mm Hg = 0.133 kpascal)
Anion gap = 12 ± 2 mEq/l (excluding K^+)

Anion gap in plasma:
- Calculated as: $[Na^+] - ([Cl^-] + [HCO_3^-])$
- Used to detect new anions in the body (see pages 50–51)

What acid-base abnormality is present?
Were acids (H^+) added?
Have H^+ been buffered? If so, how many?
Have the lungs and the kidneys responded appropriately?
Is the acid-base abnormality a major threat to the body? If so, why?
Is it possible to estimate how quickly more H^+ will accumulate?
Do H^+ kill? If so, how do they do it?

PART A
CHARACTERISTICS OF H^+

H^+ and the Potential Threat to Survival

- The free $[H^+]$ is tiny and must be kept so for survival.
- A very large accumulation of H^+ may kill by binding to proteins in cells and changing their charge, shape, and possibly their function.

The control of the $[H^+]$ in the body is of central importance; this concentration must not be allowed to rise or fall appreciably because H^+ bind avidly to proteins. When bound, they increase the net positive charge of these proteins, thereby changing their shape and possibly their function. The maintenance of the normal very low $[H^+]$ in the face of enormous turnover depends very heavily on *buffers* and on the mechanisms for excretion of CO_2 and nitrogenous wastes. These points will be discussed in the remainder of this chapter.

One should think of H^+ from three perspectives.

1. In relation to the concentrations of other major ions in the *extracellular fluid (ECF)*, the $[H^+]$ is very small. Table 1·1 depicts that the normal plasma $[H^+]$ is 40 ± 2 nmol/l and that deviations from this value (halving or doubling) are clinically very significant; larger changes may become life-threatening.

2. One should examine the $[H^+]$ relative to the "affinity" of H^+ for chemical groups on organic and inorganic compounds in the body. This comparison will provide insights as to whether H^+ will be bound or remain free.

3. It is important to gain a quantitative perspective by examining the relative rates of production and removal of H^+. For example, an enormous number of H^+ are formed and consumed daily (70,000,000 nmol) in comparison with the amount of free H^+ in the body at any one time (close to 4000 nmol); discrepancies between the rate of formation versus removal can result in major changes in $[H^+]$.

Buffers:
Compounds that bind H^+ when the $[H^+]$ rises and release them when the $[H^+]$ falls. Thus, buffers minimize the change in $[H^+]$. They require a high concentration of both the H^+ donor and acceptor in solution.

Extracellular fluid (ECF):
Fluid outside cells.

Intracellular fluid (ICF):
Fluid inside cells.

Table 1·1
RANGE OF $[H^+]$ IN PLASMA IN CLINICAL CONDITIONS

Condition	$[H^+]$ nmol/l	pH	Importance
Acidemia	> 100	< 7.00	Can be lethal
Acidemia	50–80	7.1–7.30	Clinically important
Normal	40 ± 2	7.40 ± 0.02	Normal
Alkalemia	20–36	7.44–7.69	Clinically important
Alkalemia	< 20	> 7.70	Can be lethal

Acidemia and alkalemia:
Acidemia means that the $[H^+]$ in plasma is greater than normal and alkalemia means the opposite.

THE $[H^+]$ RELATIVE TO THE CONCENTRATION OF OTHER IONS

The concentrations of major ions in the ECF (Na^+, K^+, Cl^-, HCO_3^-) are close to a million times higher than that of H^+ (Table 1·2). Nevertheless, this very small $[H^+]$ is important from a biologic perspective, as will be discussed below. It is important to recognize that H^+ are produced in millimolar amounts in biochemical reactions and during the transport of CO_2, yet their concentration must always remain in the nanomolar range (i.e., must always be a millionfold lower than that of H^+ produced or the $[HCO_3^-]$ in the ECF or ICF).

Note:
Normal values for the $[HCO_3^-]$ depend on whether one examines the arterial blood (25 mmol/l) or the venous blood (28 mmol/l).

Table 1·2
NORMAL CONCENTRATIONS OF CATIONS AND ANIONS IN PLASMA

The sum of mEq/l of cations and anions must be equal.

Cations (mEq/l)			Anions (mEq/l)	
Na^+	140		Cl^-	103
K^+	4		HCO_3^-	25
Ca^{2+}	5	(2.5 mmol/l)	Proteins	16
Mg^{2+}	2	(1 mmol/l)	Organic	4
H^+	0.000040	(40 nmol/l)	Other inorganics	3

Acids and bases:
Acids are compounds that are capable of donating a H^+; bases are compounds that are capable of accepting a H^+. When an acid (HA) dissociates, it yields a H^+ and its conjugate base (anion, A^-).

$$HA \longleftrightarrow H^+ + A^-$$

THE $[H^+]$ RELATIVE TO THE AFFINITY OF H^+ FOR THEIR ACCEPTORS

In Certain Locations, H^+ Remain Free and Do Not Bind

- For digestion, a high $[H^+]$ (no buffering) is needed in the lumen of the stomach.

To initiate the digestion of proteins, a very high $[H^+]$ is needed. Accordingly, the anion secreted by the stomach along with H^+ is Cl^- because Cl^- will not bind H^+. Because HCl dissociates completely in aqueous solutions, and there are no major buffers in gastric fluid, H^+ bind avidly when they come in contact with ingested proteins. Binding of H^+ makes the protein much more positively charged and alters its shape so that pepsin can gain access to the sites it will hydrolyze in that protein.

Strength of acids:
Chemists classify an acid as strong or weak on the basis of its dissociation constant, or pK (the pH at which the acid is 50% dissociated); strong acids have a much lower pK. Hydrochloric acid is a very strong acid. This differentiation is of little importance biologically, because at a pH of 7, the dissociation of both weak and strong acids is much greater than 99%.

Questions

(Discussions on page 36)

1·1 Why doesn't the HCl secreted by gastric cells denature proteins in the cell membrane of the stomach?

1·2 What might permit H^+ to bind to the H^+ pump inside cells at a concentration of 0.0001 mmol/l, yet dissociate in the lumen of the stomach at a concentration of 100 mmol/l?

1·3 What is the rationale for stating that only weak acids kill?

Note:
Question 1·2 is for the more curious.

In Cells, H^+ May Be Bound to Proteins or Phosphates

- When *acidosis* is severe, binding of H^+ to proteins is a major way that H^+ are buffered.

The $[H^+]$ in cells is such that proteins have "just the right net charge" to perform essential functions (e.g., enzyme activities). A rise in the $[H^+]$ means that more H^+ bind to proteins; the converse is also true. There are two points to emphasize:

1. the change in function of proteins with a change in charge;
2. the enormous number of binding sites for H^+.

Inorganic phosphate can bind an important quantity of H^+ in some circumstances but not others (see the margin). In the ICF, the ECF, and the urine, the difference in degree of binding of H^+ is due to the $[H^+]$ in these three environments relative to the pK for phosphate. In most conditions, binding of H^+ is lowest in the ECF and highest in the urine, where almost every phosphate ion usually has bound H^+ (Table 1·3). We shall return to this phenomenon when discussing buffering in cells and the excretion of bound H^+ in the urine (titratable acid).

Table 1·3
PHYSIOLOGY OF PHOSPHATE BUFFERS

The pK for inorganic phosphate is close to 6.8 at physiologic ionic strength and temperature. Organic phosphates (e.g., ATP) have too low an affinity for H^+ at the $[H^+]$ in cells to act as physiologic buffers.

Compartment	Total inorganic phosphate	% as $H_2PO_4^-$	Equation
ECF	1 mmol/l	20%	$H^+ + HPO_4^{2-} \longleftrightarrow H_2PO_4^-$
ICF	4–5 mmol/l	33%	$H^+ + HPO_4^{2-} \longleftrightarrow H_2PO_4^-$
Urine	30 mmol/day	90%	$H^+ + HPO_4^{2-} \longleftrightarrow H_2PO_4^-$

Acidosis and alkalosis:
Acidosis is a process that adds acids to the blood or removes bases from the blood; alkalosis is the converse.

Compartmental $[H^+]$:
ECF: 40 nmol/l
ICF: 80–100 nmol/l
Urine: 10 000 nmol/l

Note:
The content of inorganic phosphate is quite low in the ECF and the ICF, so little buffering of H^+ occurs by this buffer in these locations.

HPO_4^{2-} = divalent inorganic phosphate ions.
$H_2PO_4^-$ = monovalent dihydrogen inorganic phosphate ions.

$$\mathbf{pH} = pK + \log\frac{HPO_4^{2-}}{H_2PO_4^-}$$

pH	Compartment	Ratio of $HPO_4^{2-}/H_2PO_4^-$
7.4	ECF	4/1
7.1	ICF	2/1
5.8	Urine	1/10

THE $[H^+]$ RELATIVE TO THE PRODUCTION AND REMOVAL OF H^+ DURING METABOLISM

- Production or removal of H^+ is best understood by examining the "net charge," or *valence*, of substrates and products of a metabolic process—ignore cofactors.
- Two processes indicate net H^+ production in metabolism—accumulation of new anions in the body and the excretion of anions without a H^+ or *NH_4^+*.
- H^+ can be classified as "fast" H^+ or "slow" H^+ depending on their net rate of production.

Valence:
The number of charges a compound or ion bears in solution, expressed in mEq/l. The term "milliequivalent" (mEq) reflects the number of charges or valences; therefore, multiply mmol by the valence to obtain mEq. Valence is especially important for albumin, which has a large valence on each molecule.

NH_4^+:
A cation formed in the kidney during metabolism. When NH_4^+ are excreted, the new HCO_3^- formed along with the NH_4^+ are added to the body.

An enormous number of H^+ are produced and removed during the usual metabolic reactions in the body. When are these H^+ important? We offer the following ways to simplify the approach to understanding the clinical significance of H^+.

Definition of a Metabolic Process

Adenosine triphosphate (ATP): The useful form of energy in cells that enables biochemical work to be performed. The need for regeneration of ATP limits the flux through ATP-producing reactions.

In general, a metabolic process starts with either dietary or stored fuels and ends with ATP or an energy store (glycogen, triglyceride). A metabolic process can span more than one organ. In the example in Figure 1·1, H^+ will accumulate if the ketoacid anions are retained in the body or if they are excreted with a cation other than H^+ or NH_4^+ (e.g., Na^+ or K^+). If part of the pathway generates H^+ and is intimately linked to another part that removes H^+, both parts can be ignored from an acid-base perspective. For example, H^+ are formed when ATP is hydrolyzed to perform biologic work (e.g., reabsorb Na^+). As soon as ATP is regenerated in mitochondria of that cell, H^+ are removed. Hence, the cycle has no net acid-base impact.

$$ATP^{4-} \longleftrightarrow ADP^{3-} + P_i^{2-} + H^+ \quad \text{(pump } Na^+\text{)}$$
$$ADP^{3-} + P_i^{2-} + H^+ \longleftrightarrow ATP^{4-} \quad \text{(mitochondria)}$$

When examining a series of metabolic pathways occurring in a number of organs (Figure 1·1), consider all the substrates and products to determine their acid-base impact. By counting the "net charge," or valence, of substrates and products of the overall series of metabolic reactions, one can deduce the number of H^+ produced or removed.

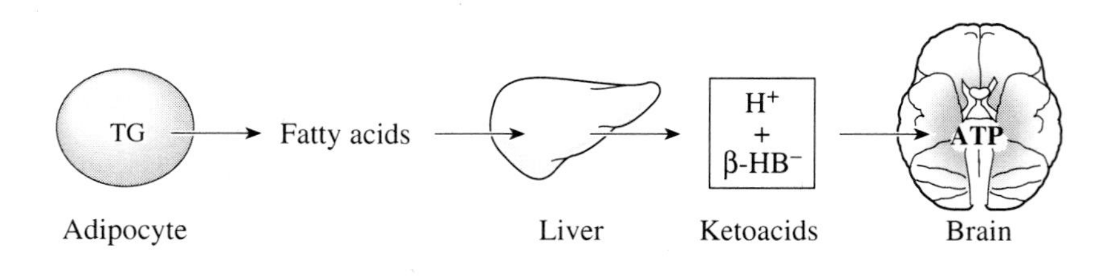

Figure 1·1
Illustration of a metabolic process
The starting point is triglycerides (TG) in adipose tissue and the final product is ATP (and CO_2 + H_2O) in the brain. Ketoacids (H^+ + *β-hydroxybutyrate anion*) are produced in the liver and removed by the brain. This figure is reproduced with the permission of the authors—Halperin and Rolleston, *Clinical Detective Stories* (London: Portland Press, 1993).

Adipocyte: Triglyceride $\longleftrightarrow$ 3 Palmitate$^-$ + 3 H^+ + Glycerol
Liver: 3 Palmitate$^-$ + 3 H^+ + 18 O_2 $\longleftrightarrow$ 12 Ketoacid anions + 12 H^+
Brain: 12 Ketoacid anions + 12 H^+ $\longleftrightarrow$ CO_2 + H_2O + ATP

Overall: Triglyceride yields ATP, CO_2, and H_2O in the brain, but causes no net production or removal of H^+.

β-hydroxybutyrate anion (β-HB⁻): There are two ketoacids formed during the partial oxidation of fatty acids in the liver, acetoacetic acid and β-hydroxybutyric acid (Hβ-HB); the conjugate base of Hβ-HB is β-HB⁻.

Question

(Discussion on page 36)

1·4 In the metabolic process in Figure 1·1, under what circumstances will H^+ accumulate?

Examples of the Generation or Removal of H^+

- H^+ production: Conversion of a neutral compound to anions (usually via oxidation of amino acids) generates most H^+.
- H^+ removal: Conversion of dietary anions to neutral products removes most H^+. In the kidney, generation and excretion of a new cation (NH_4^+) removes H^+.

The following generalizations can be made. When a neutral compound is converted to an anion, or when a cation is converted to a neutral compound, H^+ are produced; conversely, when an anion is converted to a neutral compound, or when a neutral compound is converted to a cation, H^+ are removed. Examples are provided below.

Concept:

1. Identify H^+ production by counting new anions.

(i) All of the following yield H^+ (more net negative charge):

Glucose $\longleftrightarrow$ $Lactate^-$ + H^+ (new anion)

Fatty acid $\longleftrightarrow$ 4 Ketoacid anions + 4 H^+ (new anions)

Cysteine $\longleftrightarrow$ Urea + CO_2 + H_2O + SO_4^{2-} + 2 H^+ (new anion)

$Lysine^+$ $\longleftrightarrow$ Urea + CO_2 + H_2O + H^+ (loss of cation)

(ii) All of the following remove H^+ (more net positive charge):

$Lactate^-$ + H^+ $\longleftrightarrow$ Glucose (anion removed)

$Citrate^{3-}$ + 3 H^+ $\longleftrightarrow$ CO_2 + H_2O (anion removed)

Glutamine $\longleftrightarrow$ Glucose + NH_4^+ + CO_2 + H_2O + HCO_3^- (generation of a new cation)

Note:
Conversion of glutamine to NH_4^+ will have a net yield of HCO_3^- only if NH_4^+ are made as an end product of metabolism (i.e., are excreted in the urine).

(iii) None of the following yields or removes H^+ (no change in net charge):

Glucose $\longleftrightarrow$ Glycogen + CO_2 + H_2O (neutrals to neutrals)

Triglyceride $\longleftrightarrow$ CO_2 + H_2O (neutrals to neutrals)

Alanine $\longleftrightarrow$ Urea + Glucose or CO_2 + H_2O (neutrals to neutrals)

A quantitative description of H^+ balance from metabolism of a typical Western diet is provided in Table 1·4.

Table 1·4 Legend
The following H^+ load is calculated on the basis of the ingestion of 100 g of protein from beefsteak. The net dietary H^+ load is approximately 70 mmol/day, but 40 mmol of H^+ are produced with an anion such as SO_4^{2-} that does not help in their removal (there is virtually no metabolism or excretion of H_2SO_4) so NH_4^+ must be excreted to maintain acid-base balance.

Anion for excretion:
If a H^+ is accompanied by an anion with which it can be excreted in bound form, the acid-base impact of the H^+ is eliminated. For example, $H_2PO_4^-$ is formed from the metabolism of dietary constituents; it yields H^+ and HPO_4^{2-} at a pH of 7.4. Since the kidney secretes H^+, $H_2PO_4^-$ are excreted when the urine pH is less than 6.8 (see page 17).

Table 1·4
DIETARY ACID-BASE IMPACT

Nutrient	Product	H^+ (mmol/day)
Reactions generating H^+		
• Sulfur-containing amino acids - Cysteine/cystine, methionine	H^+	70
• Cationic amino acids - Lysine, arginine, histidine (½)	H^+	140
• Organic phosphates	$HPO_4^{2-} + H^+$	30
Reactions removing H^+		
• Anionic amino acids - Glutamate, aspartate	HCO_3^-	–110
• Organic anions (e.g., citrate^{3-})	HCO_3^-	–60
• Organic phosphate excretion with H^+	$H_2PO_4^-$ excreted	–30
Net total H^+ load to be excreted as NH_4^+		40

Rate of Production of H^+

The production of L-lactic acid via anaerobic glycolysis is the only circumstance in which the rate of production of H^+ is so rapid that it might constitute a serious acid-base threat in minutes to hours. We consider this rate of production to be "fast." Examples of the rate of H^+ production in clinical conditions are provided in Table 1·5.

Table 1·5
RATES OF PRODUCTION AND REMOVAL OF H^+
The total quantity of H+ that can be buffered per day is close to 1000 mmol in a 70-kg person. With very large acid loads, most of the buffering occurs in the ICF.

Event	Rate (mmol/min)	Comments
Production of H^+		
L-Lactic acid (hypoxia)	72	• Rate reflects complete anoxia.
	7.2	• Rate reflects 10% hypoxia.
Ketoacids	1	• Production requires lack of insulin.
Toxic alcohols	< 1	• Poisonous metabolites rather than H^+ are usually the major threat.
Removal of H^+		
Kidney (by excretion of NH_4^+)	0 to 0.2	• Has a lag period. • Metabolic acidosis is needed for rapid rates of excretion.
Metabolism		
L-Lactic acid	4 to 8	• Half by oxidation and half by glucogenesis.
Ketoacids	0.8	• Oxidized primarily in the brain and kidneys.

Clinical Classification of Metabolic Reactions Influencing Acid-Base Balance

An overall approach to the net production of H^+ is provided in Figure 1·2; its basis is that only those H^+ that accumulate provide a H^+ load to the body. The clinical threat of H^+ accumulation is proportional to the magnitude of the H^+ load and the speed with which H^+ accumulate. If H^+ are being formed very rapidly, production of H^+ must be stopped as soon as possible. Therapy with HCO_3^- will at best buy a little time from an acid-base viewpoint. If H^+ are accumulating slowly, there is time for more thorough investigation.

Note:
The rate of H^+ formation is evident from the rate of appearance of anions and the fall in the $[HCO_3^-]$.

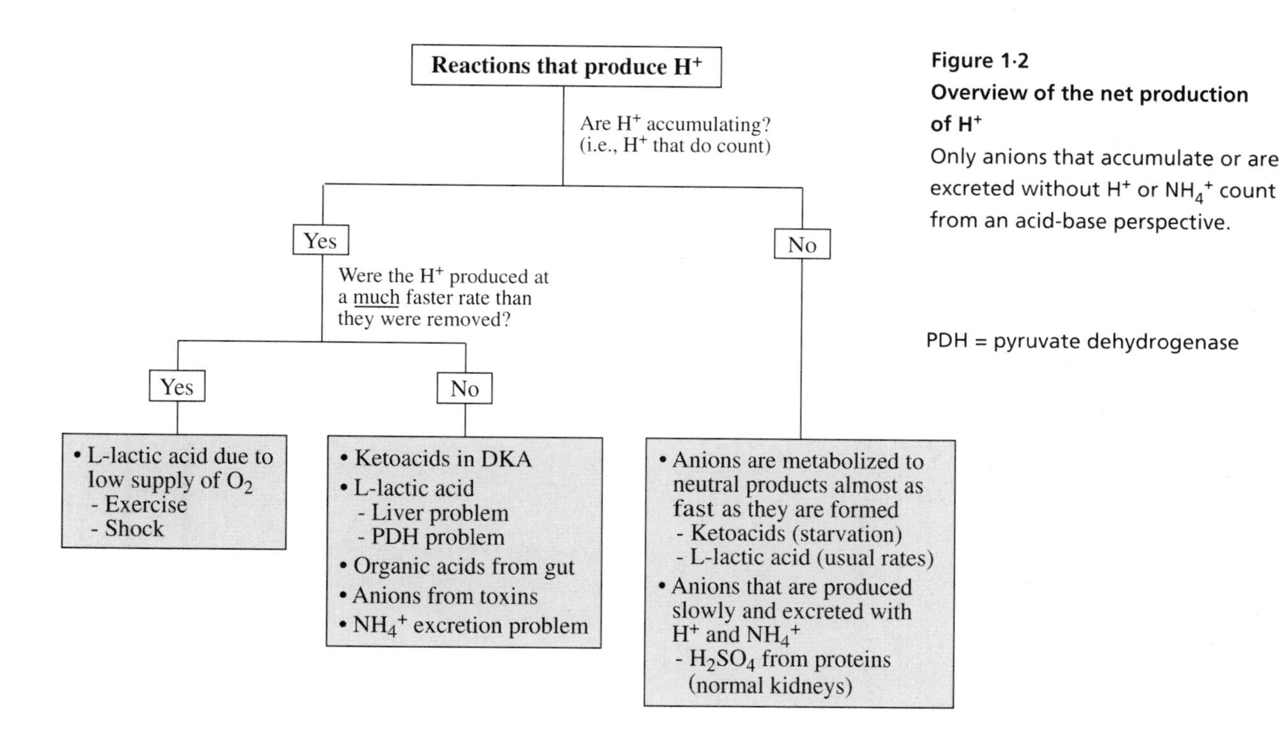

Figure 1·2
Overview of the net production of H^+
Only anions that accumulate or are excreted without H^+ or NH_4^+ count from an acid-base perspective.

PDH = pyruvate dehydrogenase

H^+ THAT DO NOT COUNT

- Since there is a balance between rates of production and removal of H^+ during the normal metabolism of carbohydrates and fats, protons do not accumulate in most of energy metabolism.

Most of the H^+ that do not count are formed along with a partner that aids in their removal. Examples are found in the normal metabolism of carbohydrates and fats and in the metabolism of 13 of the 20 amino acids in proteins. When considering organic phosphates, the anion produced (HPO_4^{2-}) can bind H^+ at most usual [H^+] in the urine so that H^+ do not accumulate in the body.

Carbohydrates

Because all substrates and products of the normal metabolism of carbohydrates have the same "net charge," they do not yield a surplus or deficit of H^+. When glucose is converted to pyruvate or L-lactate anions, H^+ are formed but are then removed when these anions are metabolized to the neutral end products glucose, glycogen, triglycerides, or CO_2 and H_2O (Figure 1·3).

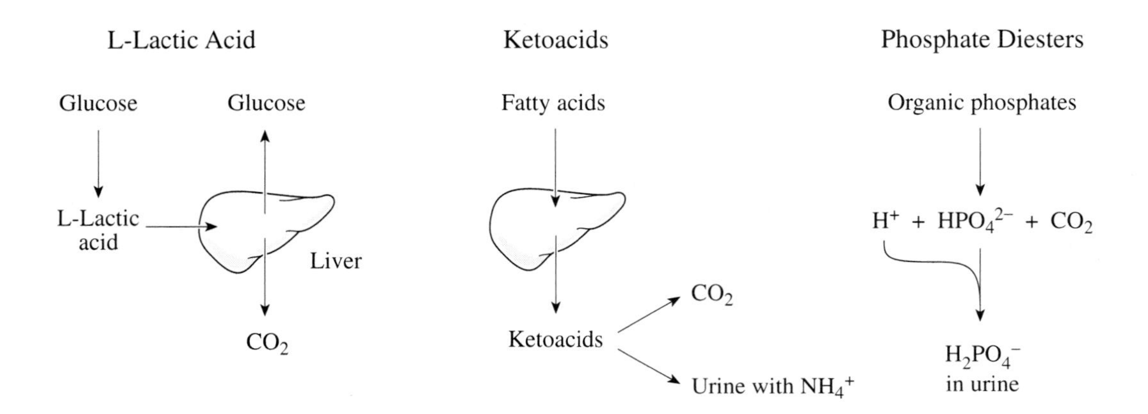

Figure 1·3
H^+ that are removed as fast as they are formed
In each of these examples, an anion is formed but is readily metabolized to a neutral end product or is excreted with a bound H^+ at the [H^+] in urine.

Fats

Despite the fact that fatty acid and ketoacid anions are metabolic intermediates of fat metabolism, neither they nor H^+ accumulate in normal metabolism (Figure 1·3).

Proteins

The majority (13/20) of the amino acids are neutral and their products (urea, glucose, CO_2, and H_2O) are also neutral. Hence, their metabolism does not contribute to net accumulation of H^+.

Anions That Help H^+ Excretion

The major anions of the ICF are organic phosphates. These compounds are primarily phosphate diesters (DNA, RNA, phospholipids, and high-energy phosphates). The products of their metabolism are HPO_4^{2-} + H^+ (Figure 1·3). The divalent phosphate anion (HPO_4^{2-}) is filtered by the kidney. Since the kidney raises the [H^+] in the urine appreciably above that of the plasma, the filtered HPO_4^{2-} are excreted as $H_2PO_4^-$; this process is called "titratable acid excretion." Thus, organic phosphates represent production of H^+ with a partner that aids in their excretion.

H^+ THAT DO COUNT

In the following two sections, the net production of H^+ is considered by examining the properties of the accompanying anions.

> **Key point**
> A gain of anions signals a positive balance for H^+. In this context, we are referring to anions other than HCO_3^-.

Anions That Are Formed Faster Than They Are Metabolized

> • The only time that H^+ are formed very rapidly is during anaerobic metabolism.

Carbohydrates: When oxygen is lacking, carbohydrates cannot be metabolized to the neutral end products CO_2 and H_2O, so the anion L-lactate and a H^+ accumulate (*L-lactic acidosis*). Two general conditions cause L-lactic acidosis: first, when the demand for oxygen is extremely high (e.g., during a sprint or convulsion); second, when the supply of oxygen is very low (e.g., from a low cardiac output; quantitative considerations appear in Chapter 3, pages 99–100).

At times, one of the enzymes involved in the oxidation of L-lactate may limit its oxidation. In this case, there is usually an inborn error of metabolism or a dietary deficiency of a cofactor such as vitamin B_1 (*thiamine*). See Chapter 3, pages 103–04, for discussion.

Fats: When there is a lack of insulin, ketoacid anions and H^+ accumulate (*ketoacidosis*). A relatively small accumulation occurs during chronic fasting. For a large accumulation to occur, β cells of the pancreas must be destroyed (as in diabetic ketoacidosis) or be inhibited (during alcoholic ketoacidosis; see Chapter 3, pages 95–99). In addition, the oxidation of ketoacids must be slower than usual (coma, low GFR; see pages 87–88).

Unusual anions: Bacteria in our GI tract produce a variety of organic acids that are absorbed (D-lactic, butyric, acetic, and propionic acids, among others). Each is metabolized to neutral end products, so there is usually no net H^+ load. Should their production rise dramatically or their metabolism be slower than usual (a liver or kidney problem), H^+ will accumulate.

L-Lactic acidosis:
The accumulation of acid produced from the incomplete oxidation of glucose.

Thiamine:
A vitamin that is required for the activity of the enzyme pyruvate dehydrogenase.

Ketoacidosis:
The accumulation of ketoacids that results from incomplete oxidation of fatty acids and occurs when there is a relative lack of insulin.

Questions

(Discussions on pages 36–38)

1·5 Red blood cells produce 200 mmol of L-lactic acid per day. Why doesn't this production cause severe acidemia?

1·6 When ATP is used to perform biologic work, H^+ are formed. Should these H^+ be considered as part of the quantity that can cause acidosis?

1·7 Acetate is added to the hemodialysis fluid to minimize the net production of H^+. What is the rationale for this maneuver?

1·8 During hemodialysis, β-hydroxybutyric acid is formed from acetate anions and is lost in the dialysis fluid. Does this process result in the net production or removal of H^+?

1·9 What might permit bacteria in the GI tract to overproduce organic acids?

1·10 Will consumption of citrus fruits, which contain a large quantity of citric acid and K^+ citrate, cause an acid or alkali load?

Anions That Cannot Be Metabolized

Sulfate: Amino acids containing the element sulfur (cysteine/cystine and methionine) can be oxidized to yield the terminal anion SO_4^{2-} plus neutral end products (glucose, urea, CO_2 and H_2O). Because the affinity of SO_4^{2-} for H^+ is so low (i.e., SO_4^{2-} has a very low pK), SO_4^{2-} cannot help in removing H^+ by urinary excretion (nor can it help by metabolism). Hence, other ways are needed to remove these H^+ (renal excretion of NH_4^+; see pages 28–33). For each mEq of SO_4^{2-} that accumulates or is excreted without NH_4^+, a H^+ accumulates. Only 20–40 mmol of H^+ are produced as H_2SO_4 each day in people who consume a typical Western diet (obliging a similar excretion of NH_4^+); lower quantities are produced in people who consume a low-protein diet.

Anions from toxins: Metabolism of methanol, ethylene glycol, and toluene will all yield anions that cannot undergo further metabolism (Table 1·6, Figure 1·4). The affinity of these anions for H^+ is also so low that they cannot be excreted with H^+ in the urine. Of greater importance, the production of toxic products, such as formaldehyde from methanol, is a much greater threat to the patient than is the production of H^+ (see Chapter 3, pages 106–08).

Table 1·6
PRODUCTION OF ANIONS THAT CANNOT HELP IN THE REMOVAL OF H^+

Substrate	Anion formed
Methanol	Formate
Ethylene glycol	Glycolate, oxalate
Toluene	Hippurate
Protein	Sulfate

Cations That Are Oxidized to Neutral End Products

Three cationic amino acids (lysine, arginine, histidine) are metabolized to neutral end products in normal metabolism (net negative charge); as a result, H^+ accumulate. Thus, these H^+ also require the excretion of NH_4^+ to prevent the accumulation of protons. The net production of H^+ from protein oxidation is relatively small (Table 1·4, page 10) but constitutes the majority of the usual daily net H^+ production.

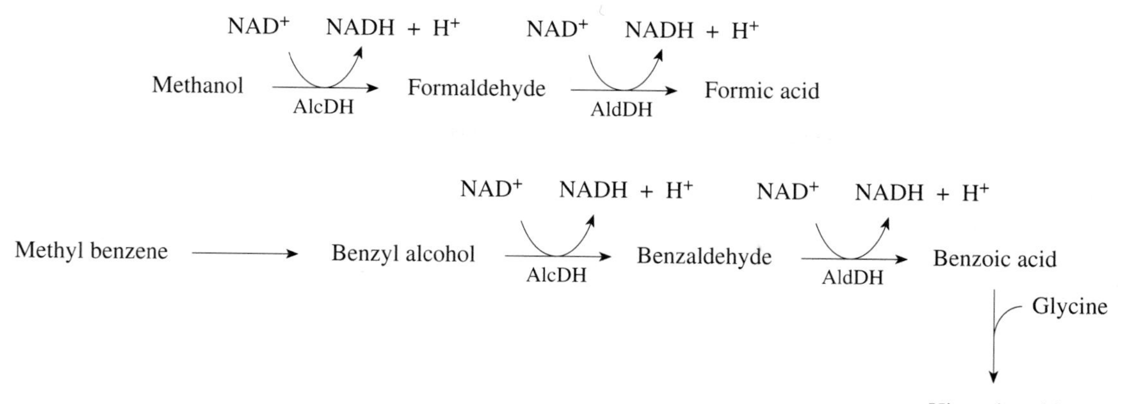

Figure 1·4
Production of acids from alcohols
The initial metabolism of methyl benzene (toluene) takes place in the cytosol of the liver via cytochrome P_{450}; the resulting alcohol and other alcohols such as methanol are acted upon by alcohol dehydrogenase (AlcDH). The aldehydes formed enter mitochondria where aldehyde dehydrogenase (AldDH) converts them to carboxylic acids. In the case of benzoic acid, conjugation with glycine yields hippuric acid.

PART B
DAILY PHYSIOLOGY OF H^+

Figure 1·5
Components of acid-base balance
The body is represented as a HCO_3^- buffer (in the rectangle). There are three components to acid-base balance:
1. A H^+ load is derived from oxidation of fuels.
2. Bicarbonate buffers ultimately buffer this H^+ load. Buffering of H^+ by HCO_3^- leads to production of CO_2, which is removed by the lungs.
3. The kidney regenerates HCO_3^- when glutamine is metabolized to yield HCO_3^- plus NH_4^+. The new HCO_3^- return to the blood when NH_4^+ are excreted in the urine.

There are three major components to the daily turnover of H^+ (Figure 1·5). Each will be considered briefly to demonstrate the overall picture, and then the latter two will be discussed in more detail.

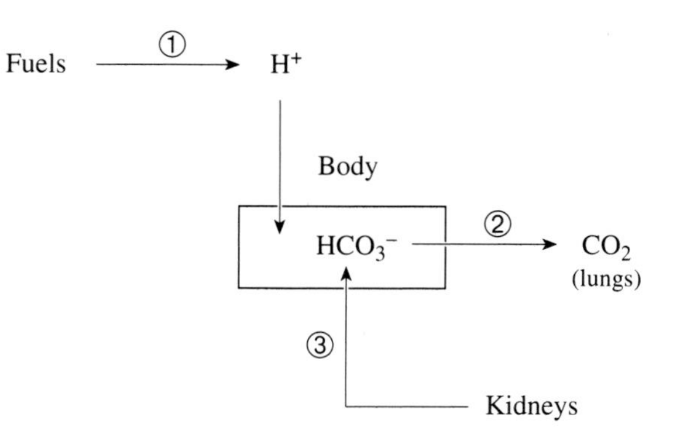

Diet and the Production of H^+

- The majority of net production of H^+ occurs because protein is metabolized.
- Close to 60 mmol of H^+ are removed when fruits and vegetables are metabolized.

An individual eating the usual North American diet is faced with a net daily H^+ load of approximately 1 mmol/kg/day. As listed on page 10, the net production of H^+ is largely from the metabolism of 7 of the 20 amino acids from the proteins of the diet (Table 1·4). Metabolism of certain dietary constituents adds H^+, while metabolism of other constituents removes H^+; examples of the latter group include the organic anions (usually as their K^+ salts) in fruits and vegetables consumed (citrate^{3-}, malate^{2-}, acetate$^-$, etc.). Quantitatively, some 50 to 60 mEq of H^+ are removed each day as a result of this metabolism.

Buffering of H^+

Bicarbonate buffer system (BBS): The major buffer in the ECF and an important buffer in the ICF (providing that the lungs can lower the Pco_2).

The effectiveness of buffering is remarkable. The input of H^+ is more than 1,000,000 nmol per liter of body water, yet the $[H^+]$ is remarkably stable (40 ± 2 nmol/l). Two types of buffers are the proteins in the body and the *bicarbonate buffer system (BBS)*. In a later section (page 20), we shall discuss how the ability of the lungs to lower the $[CO_2]$ (Pco_2) in the body permits the BBS to be the primary means of alleviating a modest H^+ load.

Generation of New HCO_3^- by the Kidney

- The kidneys must generate 1 mmol of new HCO_3^- per kg body weight each day.
- New bicarbonate is generated when NH_4^+ and $H_2PO_4^-$ are excreted.

Only the kidneys excrete acid or alkali on a continuing basis and therefore can regulate the balance of HCO_3^- (called *net acid excretion*) They do so by adjusting the rates of excretion principally of NH_4^+ and HCO_3^-. Each day, 70 mmol of H^+ are derived from the normal oxidative metabolism of dietary constituents and are buffered initially by the BBS. To achieve acid-base *balance*, the kidneys must generate 70 mmol of new HCO_3^- to replace the HCO_3^- consumed by the buffering process (Figure 1·6). Should this process of new HCO_3^- generation fail, the patient will become progressively acidemic, as a result of the continued net negative balance for HCO_3^-.

Net acid excretion (NAE):
This excretion is the same as HCO_3^- balance across the kidney.
NAE = HCO_3^- gain minus HCO_3^- loss via the kidney.
NAE = NH_4^+ plus $H_2PO_4^-$ minus HCO_3^- excreted in the urine.

Balance:
The difference between input and output. Positive balance is net accumulation and negative balance is net loss.

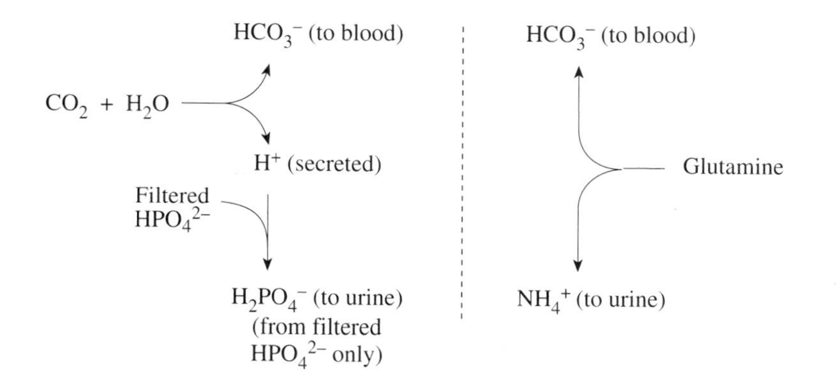

Figure 1·6
Generation of new HCO_3^- in the kidneys
The kidneys generate new HCO_3^- in two ways—by converting CO_2 to H^+ and HCO_3^- and by converting glutamine to NH_4^+ plus HCO_3^-. In the former case (left side of figure), the H^+ must be excreted bound to HPO_4^{2-}, whereas in the latter (right side of figure), NH_4^+ must be excreted in the urine. The net result is that for every mmol of NH_4^+ as well as every mmol of *new $H_2PO_4^-$* excreted, one mmol of new HCO_3^- will be added to the blood.

A More In-depth Look at Buffering of H^+

- Buffers minimize the change in $[H^+]$ when an acid or alkali is added.
- Buffers are a very effective but temporary means of removing H^+ from the body.

The ability of a compound to bind or release H^+ depends on the affinity of that compound for H^+ and the $[H^+]$ in that compartment of the body. The $[H^+]$ when the concentrations of the H^+ donor (acid) and H^+ acceptor (base) are equal is called the "buffer constant" (the pK in negative log terms). At

New $H_2PO_4^-$:
There are two species of inorganic phosphate: divalent (HPO_4^{2-}) and monovalent ($H_2PO_4^-$) ions. To generate new HCO_3^-, HPO_4^{2-} must be converted to $H_2PO_4^-$ via H^+ secretion by the kidney. A $H_2PO_4^-$ that is both filtered and excreted will not contribute to the generation of new HCO_3^-.

Chemistry of buffers:
All buffers depend on a simple equilibrium. Each buffer has its unique dissociation constant (K_d or pK), which determines the range of $[H^+]$ at which the buffer is effective. A buffer is most effective at a $[H^+]$ or pH that is equal to its K_d or pK, i.e., when $[HA] = [A^-]$ (A^- is the conjugate base of the weak acid buffer).

$$H^+ + A^- \longleftrightarrow HA$$

this $[H^+]$, the buffer is best poised to donate or accept H^+ (see Table 1·7 for a quantitative example and Figure 1·7 for a visual representation). The other important factor determining the effectiveness of a given buffer is its concentration in the compartment in which it is to act as a buffer (e.g., inorganic phosphate in the ICF has an appropriate pK, but the concentration is too low to be a physiologically important buffer).

Table 1·7
ILLUSTRATIVE EXAMPLE OF BUFFERING

	At the pK	Far from the pK
Starting point		
Total $HA + A^-$	1000	1000
Ratio of HA to A^-	500/500	800/200
$[H^+]$	$500/500 \times K_{eq}$*	$800/200 \times K_{eq}$
Add 100 new H^+		
Total $HA + A^-$	1000	1000
Ratio of HA to A^-	600/400	900/100
$[H^+]$	$600/400 \times K_{eq}$	$900/100 \times K_{eq}$

Table 1·7 Legend
When 100 new H^+ are added to two buffer solutions poised at different ratios of proton donor (HA) to proton acceptor (A^-), there is a lesser absolute change in free $[H^+]$ when this ratio is closer to 1 ($[H^+]$ is near the pK of that buffer).

* K_{eq} is the equilibrium constant for the dissociation of HA to $A^- + H^+$.

$$H^+ + A^- \longleftrightarrow HA$$
$$K_{eq} = [H^+] \times [A^-]/[HA]$$
$$[H^+] = K_{eq} \times [HA]/[A^-]$$

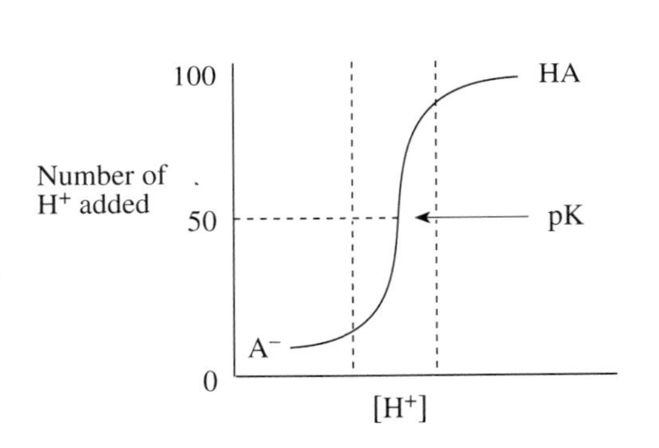

Figure 1·7
Buffer curve describing the buffering of H^+
When H^+ are added, the $[H^+]$ rises quickly until it approaches the pK (between dotted lines) where H^+ can be added with little change in $[H^+]$ (efficient buffering).

Idea:
Instead of considering the location of buffers, classify them as BBS or non-BBS.

The body contains many substances that can bind or release H^+; some of these are effective buffers at $[H^+]$ that are in the physiological range. For buffering to be effective, the quantity of H^+ acceptors obviously must exceed the quantity of H^+ added. The two most important buffer systems are the protein buffer system and the BBS.

The traditional approach to buffering is based on "geography"—ECF vs ICF. It is said that just over half of the H^+ load is buffered in the ICF; however, this approach ignores the type of buffer, its physical and chemical properties, and whether it is possible to change the buffer capacity. These points will be considered below.

PROTEIN BUFFER SYSTEM

- The major non-BBS buffer is protein in the ICF (imidazole group in histidines).
- Recall that when H^+ bind to proteins, the charge, shape, and possibly function may change.

The major buffers of the ICF are histidines (in proteins and in the dipeptides *carnosine and anserine*). The total content of histidines is close to 2400 mmol in a 70-kg individual. Since the pH of the ICF is close to the mean pK of histidine, only half of these histidines (1200 mmol) are potential H^+ acceptors. With a change in ICF pH of 0.3 units (twofold rise in $[H^+]$), about 400 mmol of H^+ can be buffered via histidines (the ratio of H^+ donors to H^+ acceptors in histidine changes from 1200/1200 to 1600/800).

Carnosine and anserine: Compounds that act as a major buffer in skeletal muscle. They contain two amino acids, β-alanine and histidine. The β-alanine permits each compound to have a high pK of its terminal amino group; these compounds can therefore remain with no net charge in cells.

BICARBONATE BUFFER SYSTEM (BBS)

- The BBS is the initial buffer for a H^+ load.
- The BBS is virtually the only buffer of the ECF, but it is also important in the ICF.
- A clinical evaluation of acid-base balance is made by examining the $[H^+]$, $[HCO_3^-]$, and the Pco_2 in plasma.

$$H^+ + HCO_3^- \longleftrightarrow H_2CO_3 \longleftrightarrow H_2O + CO_2$$

There are two ways to consider the BBS—a view related to body geography (ECF vs ICF) and a view related to physiology (BBS is used before proteins for the most part); both will be considered below.

Quantities

- Close to 1000 mmol of H^+ can be buffered, 350 mmol via the BBS in the ECF, 250 mmol via the BBS in the ICF, and 400 mmol via intracellular histidines.

Given the $[HCO_3^-]$ in plasma (25 mmol/l) and the volume of the ECF (15 liters in a 70-kg adult), the ECF can buffer close to 375 mmol of H^+ (see Table 1·8 and the margin). The ICF contains almost the same amount of HCO_3^- as the ECF (close to half the $[HCO_3^-]$ but twice the volume). The critical question is this: What permits preferential buffering of H^+ by the BBS?

Content of HCO_3^- in the ECF: The total content is the $[HCO_3^-]$ × the ECF volume (25 mmol/l × 15 liters = 375 mmol). Each mmol of HCO_3^- can remove 1 mmol of H^+.

Table 1·8
BUFFERS FOR AN ACID LOAD

Values are reported as normal and are approximate for a 70-kg person. The maximum quantity of buffering is close to 1000 mmol. Additional buffering occurs in skeletal muscle during exercise.

Location	Buffers (mmol)			
	HCO_3^-	Protein	Phosphate	Other
ECF	375	<10	<15	0
ICF (muscle)	330	400	<50	CrP*

* Creatine phosphate (CrP) is hydrolyzed to inorganic phosphate, which removes H^+ during anaerobic exercise (see "'Metabolic' Buffering in Skeletal Muscle," page 22).

Physiology: a Reinterpretation

Concept:

2. Buffers work physiologically to keep added H^+ from binding to proteins; instead, H^+ are forced to react with HCO_3^-.

- A function of the BBS is to prevent H^+ from binding to proteins in the ICF.
- The BBS is used first to remove a H^+ load, providing that hyperventilation occurs.
- The key to the operation of the BBS is the control of the P_{CO_2}.

Following an acid load, there is a coordinated operation of the BBS in the ECF and the ICF. It should be appreciated that the fall in P_{CO_2}, which occurs in metabolic acidosis, not only increases the effectiveness of the BBS in the ECF but also prevents a large rise in the "net positive charge" on intracellular proteins following an acid load; the key to understanding this relationship is summarized in Figure 1·8. When an acid is generated, the $[H^+]$ rises, the respiratory center is stimulated, alveolar ventilation rises, and the P_{CO_2} falls. The major effect of this fall in P_{CO_2} differs in the ECF and ICF because the ICF contains a large quantity of non-BBS buffers that are virtually absent in the ECF. Accordingly, with a fall in the P_{CO_2} there will be a decrease in the $[H^+]$ in the ECF and little change in the $[HCO_3^-]$ because the $[HCO_3^-]$ is 10^6-fold larger than the $[H^+]$, yet they are consumed in a 1:1 stoichiometry with CO_2 formation. In contrast, in the ICF, there will be a decrease in the $[HCO_3^-]$ and an increase in the nonprotonated proteins (B°). The reasons for these changes in the ICF are as follows: when the P_{CO_2} fell, the $[H^+]$ began to fall. As soon as this $[H^+]$ fell, the equilibrium of protein buffers was driven from HB^+ to B° plus H^+. Recall that there is a very large quantity of groups in proteins that are capable of binding H^+ relative to the content of HCO_3^- in the ICF. Accordingly, this cycle will continue and lead to the consumption of HCO_3^- and the generation of B° with only a small fall in the $[H^+]$ in the ICF.

The BBS is the first buffer to remove almost all of the H^+ load when the P_{CO_2} falls (Table 1·9). The BBS in the ICF removes about one-third of the H^+ load. In the absence of hyperventilation, close to half of the H^+ are buffered by intracellular proteins. With a much larger acid load, the bulk of

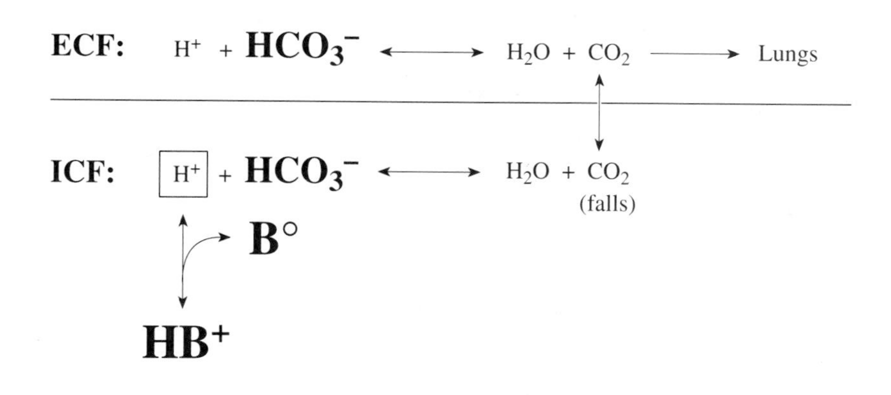

Figure 1·8
Teamwork in buffering
The top of this figure represents events in the ECF, a compartment with only a minor contribution of protein buffers. The main effect of hyperventilation (a lower [CO_2]) in the ECF is to lower the [H^+] in this compartment. In contrast, the lower part of the figure represents events in the ICF. When acid is added, H^+ are buffered by proteins (HB^+ rises and B° falls). When the Pco_2 falls, there is an initial small fall in the [H^+] in the ICF, which leads to *back-titration* of the large content of protein buffers (HB^+) and the subsequent consumption of HCO_3^-. The net effects are to raise the concentration of nonprotonated intracellular proteins (B°) towards normal and to lower the [HB^+] and [HCO_3^-] in this compartment.

Back-titration:
The release of H^+ that were previously bound to a buffer.

the BBS buffers will have already been consumed, so that most additional buffering will occur in the ICF using the protein buffer capacity. This information forms the the basis for the following statements:

1. the volume of distribution of H^+ is larger in more severe metabolic acidosis;
2. compared with a milder degree of metabolic acidosis, a severe degree of metabolic acidosis requires a greater quantity of HCO_3^- to be given to raise the [HCO_3^-] by a given amount.

Table 1·9
BICARBONATE BUFFER SYSTEM: IMPORTANCE OF CO_2 REMOVAL, A QUANTITATIVE EXAMPLE

If the lungs were not present and a patient had a H^+ load that reduced the [HCO_3^-] in the ECF from 25 to 12.5 mmol/l, the Pco_2 would be 455 mm Hg and the resultant [H^+] would be lethal. Because ventilation removes CO_2, and ventilation is stimulated by acidemia, patients with a [HCO_3^-] of 12.5 mmol/l should have a Pco_2 of 27 mm Hg, which produces a [H^+] of 52 nmol/l. Had the Pco_2 been maintained at 40 mm Hg, the [H^+] would be considerably higher (77 vs 52 nmol/l).

Impact of CO_2 removal on the [H^+] in plasma during metabolic acidosis

Condition	[H^+] (nmol/l)	pH	Pco_2 (mm Hg)	HCO_3^- (mmol/l)
Closed system (Pco_2 rises)	871	6.06	455	12.5
No change in Pco_2	77	7.11	40	12.5
Lower Pco_2	52	7.29	27	12.5

Questions

(Discussions on pages 38–39)

1·11 How much would the [HCO_3^-] in the ECF decrease when a load of 70 mmol of H^+ is retained in a 70-kg person who lacks renal generation of new HCO_3^-?

1·12 The following results for a blood sample arrive from the lab: $[H^+]$ = 60 nmol/l, pH = 7.22, Pco_2 = 50 mm Hg, $[HCO_3^-]$ = 32 mmol/l. What do you conclude?

1·13 What would the P_aco_2 be in a normal man if his alveolar ventilation rate decreased by half (assume no change in his metabolic rate)? What must you do to keep his pH ($[H^+]$) from changing?

1·14 Some say that hyperventilation is "maladaptive" because in an experimental situation, there was a fall in the pH of plasma after hyperventilation occurred. How might this phenomenon be interpreted?

"METABOLIC" BUFFERING IN SKELETAL MUSCLE

- During ischemia in skeletal muscle, metabolic reactions take place that result in the formation of HPO_4^{2-}, a new H^+ acceptor.

Buffering during a sprint or a convulsion (explosive, vigorous exercise) differs from buffering in other circumstances of metabolic acidosis, the main issue being the role of creatine phosphate as a precursor for a H^+ acceptor.

During a sprint or convulsion, the following events occur:

1. When ATP breaks down to provide energy for the contraction-relaxation cycle, it produces a divalent phosphate ion (HPO_4^{2-}) and a H^+.

$$ATP^{4-} \longleftrightarrow ADP^{3-} + HPO_4^{2-} + H^+$$

2. The rise in ADP and fall in ATP drives the breakdown of creatine phosphate. This process consumes a H^+ and regenerates ATP without using the divalent phosphate ion.

$$\text{Creatine phosphate}^{2-} + ADP^{3-} + H^+ \longleftrightarrow \text{Creatine} + ATP^{4-}$$

3. Combining the first two reactions will reveal that creatine phosphate is converted to creatine and divalent inorganic phosphate.

$$\text{Creatine phosphate}^{2-} \longleftrightarrow \text{Creatine} + HPO_4^{2-}$$

4. Divalent inorganic phosphate picks up a H^+ avidly at the intracellular $[H^+]$ so that the concentrations of monovalent and divalent phosphates are equal.

$$0.5\ HPO_4^{2-} + 0.5\ H^+ \longleftrightarrow 0.5\ H_2PO_4^-$$

5. Overall: These initial events in a cell provide ATP for a sprint and "parking-spots" (divalent phosphate) for the H^+ (in L-lactic acid) made during anaerobic glycolysis. Thus, more ATP can be generated with less net accumulation of free H^+. With a rise in intracellular $[H^+]$, virtually all the inorganic phosphate becomes monovalent dihydrogen phosphate. The reverse occurs upon recovery.

$$0.5\ HPO_4^{2-} + 0.5\ H^+ \longleftrightarrow 0.5\ H_2PO_4^-$$

Questions

(Discussions on pages 39–40)

1·15 Is it more important to control the $[H^+]$ in the ECF or the ICF?

1·16 What should the pH of the ICF of muscle be in the first 10 seconds of a sprint in a normal person and in a patient with a defect in the hydrolysis of glycogen?

A More In-depth Look at the Kidney and HCO_3^- Balance

- The first component of renal regulation of plasma $[HCO_3^-]$—preventing the loss of the large quantity of filtered HCO_3^-—is primarily the task of the proximal convoluted tubule (PCT).
- The PCT has a high capacity to secrete H^+ but cannot generate steep $[H^+]$ gradients. Components include a Na^+ and a H^+ story.

Concept:
3. The $[HCO_3^-]$ is adjusted because the kidneys can excrete HCO_3^- and/or generate new HCO_3^-.

There are two components of renal generation of HCO_3^-:
1. indirect reabsorption of filtered HCO_3^-, which is achieved primarily by proximal H^+ secretion;
2. generation of new HCO_3^-, which is achieved principally by NH_4^+ production and excretion.

INDIRECT REABSORPTION OF FILTERED HCO_3^-

Quantities

At a normal *glomerular filtration rate* (180 l/day), 4500 mmol of HCO_3^- are filtered, and approximately 85% (4000 mmol/day) are reabsorbed in an indirect fashion by the proximal convoluted tubule (PCT) (Table 1·10).

Glomerular filtration rate (GFR): The volume of plasma filtered at the glomerulus per unit time (usually milliliters per minute, or liters per day).

Table 1·10
QUANTITY OF HCO_3^- INDIRECTLY REABSORBED IN THE NEPHRON

The numbers are estimates for a 70-kg adult based on micropuncture data in rats.

Event	HCO_3^- (mmol/day)
Filtered	4500
Reabsorbed	
- Proximal	4000
- Loop	400
- Distal	100

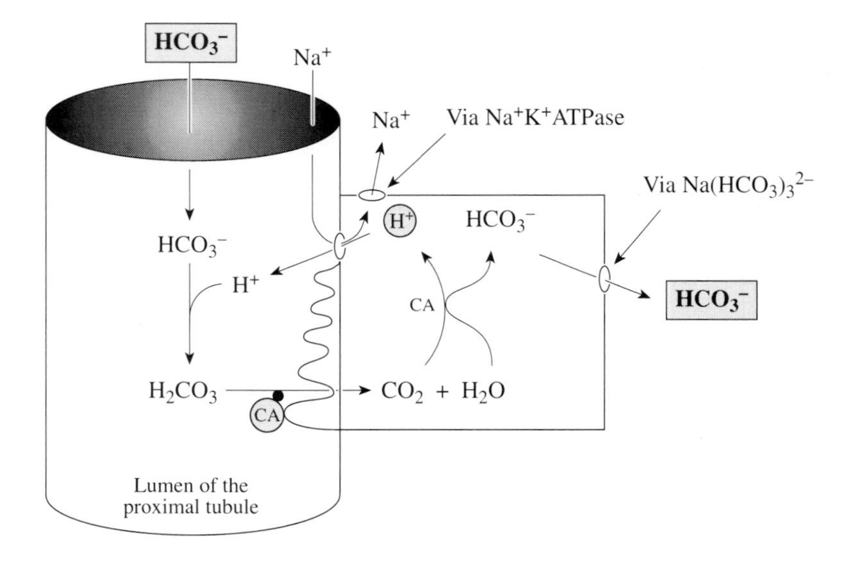

Figure 1·9
H^+ secretion in the PCT
There are two components to H^+ secretion in the PCT: reabsorption of Na^+ and secretion of H^+; they are linked via the Na^+/H^+ antiporter in the luminal membrane. (CA = carbonic anhydrase.)

"Reabsorption" of HCO_3^-:
As shown in Figure 1·9, HCO_3^- are not truly reabsorbed; they disappear from the lumen and reappear in the peritubular blood. We have chosen to call this process "indirect reabsorption."

Physiology

The bulk of filtered HCO_3^- is reabsorbed in an indirect fashion in the PCT as a result of H^+ secretion. Molecules of HCO_3^- disappear from the tubular lumen and reappear in the blood. Because there is neither net HCO_3^- gain or loss, there is no direct acid-base impact of the indirect *reabsorption of HCO_3^-* (Figure 1·9). The events in the PCT result in conservation of both Na^+ and HCO_3^- and can be viewed as two parallel stories—one dealing with the reabsorption of Na^+ and the other dealing with the secretion of H^+.

The Na^+ story: This story has three components: one at the luminal membrane, one in the cell, and another at the basolateral membrane. In the luminal membrane, Na^+ are transported on a special transporter, the Na^+/H^+ antiporter. For every Na^+ reabsorbed, one H^+ must be secreted into the lumen. The intracellular component is the very low $[Na^+]$ inside these cells, which permits this antiporter to reabsorb Na^+. The basolateral story focuses on the $Na^+K^+ATPase$ in the basolateral membrane. This ion transport system provides the driving force for the overall process—maintaining the low $[Na^+]$ in these cells by transporting 3 Na^+ out of the cell in conjunction with the entry of 2 K^+. In general, this component of the process will be driven by the need to reabsorb Na^+.

Note:
A smaller quantity of Na^+ also exit this cell as $Na(HCO_3^-)_3^{2-}$ down their electrochemical gradient.

The H^+ story: This story also has three components: one at the luminal membrane, one in the cell, and a third at the basolateral membrane. On the luminal membrane, there are two unique features: the Na^+/H^+ antiporter described above and luminal carbonic anhydrase. The latter enzyme hydrolyzes carbonic acid that is formed in the lumen to CO_2 and H_2O; the enzyme-catalyzed rate of hydrolysis is virtually as fast as the rate of generation. If this catalysis did not occur, indirect reabsorption of HCO_3^- would be retarded and urinary excretion of HCO_3^- would ensue (see the discussion of Question 1·18).

Inside the cell, there are two important components of the story: a different carbonic anhydrase enzyme species that prevents the accumulation of OH^- (or makes HCO_3^- available inside cells) and a second modifier site on the Na^+/H^+ antiporter for H^+ that activates this transporter. On the basolateral membrane, there is a unique transport system for the exit of HCO_3^- from these cells—a channel that permits a complex of Na^+ and HCO_3^- to exit as an anionic form ($Na(HCO_3)_3^{2-}$).

Regulation of Proximal H^+ Secretion

- The Na^+/H^+ antiporter is a pump with high capacity, but the "leaky" luminal membrane in the PCT prevents the generation of steep $[H^+]$ gradients.
- Regulators include the luminal $[H^+]$, the intracellular $[H^+]$, the stimuli for Na^+ reabsorption, and hormonal influences.

The Na^+/H^+ antiporter in the luminal membrane of PCT cells has a very large capacity for H^+ secretion but it cannot generate steep $[H^+]$ gradients. Therefore, the following regulating influences should be evident.

The filtered load of HCO_3^-: When the plasma $[HCO_3^-]$ is 25 mmol/l and the GFR is 180 liters/day, 4500 mmol of HCO_3^- are filtered and 4000 are indirectly reabsorbed proximally each day (Table 1·10). With metabolic acidosis, lowering the plasma $[HCO_3^-]$ to 10 mmol/l reduces its filtered load to 1800 mmol/day. In this case, H^+ secretion in the PCT will be reduced by more than 50% (there are no other H^+ acceptors of quantitative importance).

Luminal $[H^+]$: If a patient is given a drug that inhibits luminal carbonic anhydrase (e.g., *acetazolamide*), the luminal $[H^+]$ will rise abruptly and bring H^+ secretion to a halt long before 85% of the filtered HCO_3^- are reclaimed (see discussion of Question 1·18).

$[H^+]$ in PCT cells: A rise in the *$[H^+]$ in the ICF* stimulates proximal tubular H^+ secretion here for two reasons: first, the addition of H^+ causes a *substrate effect* on the Na^+/H^+ antiporter, which increases secretion in proportion to the degree of rise in $[H^+]$; second, the binding of H^+ to a separate site on the Na^+/H^+ antiporter causes an additional direct activation of this antiporter. This second site permits amplification of the stimulus of a small rise in $[H^+]$. Contrary to intuition, this activation is not very important during metabolic acidosis because of the small filtered load of HCO_3^-. Nevertheless, a modest increase in Na^+/H^+ antiporter activity can be seen during hypokalemia and with an elevated Pco_2—conditions that may be associated with a higher $[H^+]$ in the ICF. In contrast, when a $NaHCO_3$ load is administered, there is a fall in the $[H^+]$ in the ICF that does not permit an augmented flux through the Na^+/H^+ antiporter despite an increase in the

Acetazolamide:
A drug used for patients with glaucoma. These patients may develop a modest degree of metabolic acidosis from inhibition of luminal carbonic anhydrase.

$[H^+]$ in the ICF:
In hypokalemia, K^+ leave the ICF; to maintain electroneutrality, Na^+ or H^+ enter the ICF, and the $[H^+]$ in the ICF rises.

Substrate effect:
When an enzyme is not saturated with its substrate, its velocity increases with an increase in substrate concentration.

number of H^+ acceptors in the lumen. Thus, there is a prompt excretion of the excess HCO_3^-.

Renal threshold for reabsorption of HCO_3^-:
The "apparent" renal threshold for reabsorption of HCO_3^- represents that maximum [HCO_3^-] in plasma at which proximal H^+ secretion is sufficient to "reabsorb" all the HCO_3^- that were filtered. In truth, there is no real threshold because factors such as ECF volume expansion (which limits Na^+ reabsorption) and a fall in the [H^+] in the ICF independently diminish proximal H^+ secretion.

Avidity for Na^+ reabsorption: With a $NaHCO_3$ load, the increased Na^+ load via ECF volume expansion seems to depress net indirect reabsorption of $NaHCO_3$, contributing to the failure to increase H^+ secretion. In contrast, with a contracted ECF volume, proximal Na^+ reabsorption (and thereby H^+ secretion) tends to rise. The mediator of this effect is probably *angiotensin II*, acting through a phosphorylation/dephosphorylation mechanism on the Na^+/H^+ antiporter.

Angiotensin II:
The active messenger synthesized when renin levels are high. It stimulates the release of aldosterone, is a vasoconstrictor, and increases the indirect reabsorption of $NaHCO_3$ in the PCT.

Minor factors: Hypercalcemia and low parathyroid hormone levels tend to augment proximal H^+ secretion. Perhaps the mechanism involves cyclic AMP and phosphorylation of the Na^+/H^+ antiporter by protein kinases; phosphorylation inhibits the antiporter.

Proximal Renal Tubular Acidosis (pRTA)

- The failure of proximal H^+ secretion is called "pRTA."

Distal H^+ secretion in proximal renal tubular acidosis (pRTA):
In pRTA, distal H^+ secretion is normal but overwhelmed by the large distal HCO_3^- delivery (it is a low-capacity system).

The failure of proximal renal H^+ secretion results in:

- a fall in the plasma [HCO_3^-], caused by renal excretion of HCO_3^-;
- inability to increase the plasma [HCO_3^-] with administration of HCO_3^- because of low proximal HCO_3^- reabsorption (i.e., HCO_3^- excretion rather than retention);
- alkaline urine (pH > 7.0) in the presence of acidemia when the plasma [HCO_3^-] is below its normal concentration.

Questions

(Discussions on pages 40–41)

1·17 What other role might the Na^+/H^+ antiporter have in the PCT?

1·18 How do carbonic anhydrase inhibitors compromise the indirect reabsorption of HCO_3^-?

1·19 If the patient with pRTA develops diarrhea and the [HCO_3^-] in the ECF drops 2 mmol/l below its usual value, how will the urine pH and [HCO_3^-] change?

1·20 How would you establish whether or not a patient with a low [HCO_3^-] and diarrhea had pRTA?

Loop of Henle H^+ Secretion

Because 10–15% of filtered HCO_3^- (500 mmol) leave the PCT and approximately 100 mmol enter the distal convoluted tubule, close to 400 mmol are removed either in the pars recta of the PCT or in the thick ascending limb of the loop of Henle. H^+ secretion is via a Na^+/H^+ antiporter.

GENERATION OF NEW HCO_3^-

There are two processes by which new HCO_3^- are generated, the excretion of titratable acid and the excretion of NH_4^+ (*net acid excretion*). Common to both processes is the secretion of H^+. In the following section, secretion of H^+ in the distal nephron will be considered, followed by a discussion of the excretion of NH_4^+.

Net acid excretion (NAE):

- Calculate the urine NAE rate by adding the rates of excretion of NH_4^+ and $H_2PO_4^-$ and subtracting that of HCO_3^-. NAE regenerates the HCO_3^- consumed by the H^+ load from the diet.
- NH_4^+ are quantitatively the most important component of NAE in the absence of bicarbonaturia.

Distal Nephron H^+ Secretion

- H^+ATPases have a low capacity, but can generate steep $[H^+]$ gradients because the luminal membrane has tight junctions.
- Luminal H^+ acceptors (NH_3) are needed to continue pumping H^+.

Quantities: Distal nephron H^+ secretion is required to reabsorb 100 mmol of HCO_3^- and promote the net secretion of close to 20 mmol of NH_4^+. Most of the HPO_4^{2-} were titrated in earlier nephron segments.

H^+ pumps: H^+ pumps are located primarily in the mitochondria-rich intercalated cells; those involved in secretion of H^+ are located in the luminal membranes of α-intercalated cells (Figure 1·10). There are two major H^+ pumps in this segment, a *H^+ATPase* (in contrast to the Na^+/H^+ antiporter of the proximal nephron) and a *H^+/K^+ATPase*. The H^+ATPase helps to reab-

H^+ATPase:
The major H^+ pump in the distal nephron. It is important for NH_4^+ excretion.

H^+/K^+ATPase:
A pump of low capacity that seems to be important primarily for the reabsorption of K^+ in states of K^+ depletion and hypokalemia.

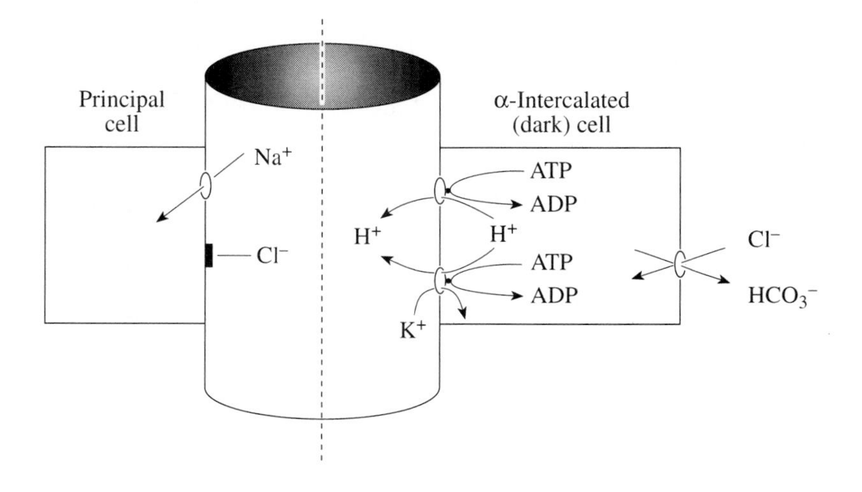

Figure 1·10
H^+ secretion in the distal nephron
The main H^+ pump is a H^+ATPase located in the luminal membrane and in secretory vesicles (precursors or stores of H^+ pumps) inside α-intercalated cells. H^+ secretion generates a steep $[H^+]$ gradient because the luminal membrane is relatively impermeable to H^+. The principal cell is important for K^+ secretion.

HCO_3^- transport:
There is a second type of intercalated cell, the β-intercalated cell. It is a 180° reversed cell in that it secretes H^+ into the peritubular capillaries via a $H^+ATPase$ and secretes HCO_3^- in exchange for luminal Cl^- (the Cl^-/HCO_3^- exchanger). The function of these cells is not really clear; some believe that they are important in excreting a load of HCO_3^-.

sorb any remaining filtered HCO_3^- and also permits an increased renal excretion of NH_4^+. In the absence of a H^+ acceptor in the lumen, however, the free $[H^+]$ in the tubular fluid rises quickly (but is only 0.1 mmol/l), and H^+ secretion by these cells ceases when a gradient limit for H^+ secretion has been reached (the urine pH cannot be lowered below 4.0). With an excess of H^+ acceptors, the $[H^+]$ in the lumen does not rise as markedly because of the limited capacity of the $H^+ATPase$ in the distal nephron.

Excretion of NH_4^+

- One can equate renal NH_4^+ excretion with HCO_3^- generation on a 1:1 basis.

From a practical point of view, for every NH_4^+ excreted, one new HCO_3^- is added to the body (Figure 1·11); knowing this relationship serves the needs of the clinician.

One can consider the NH_4^+ system in the kidney from two perspectives: first, from those factors leading to a high $[NH_3]$ in the renal medullary interstitium; second, from those determining transfer of this NH_3 to the lumen of the collecting duct.

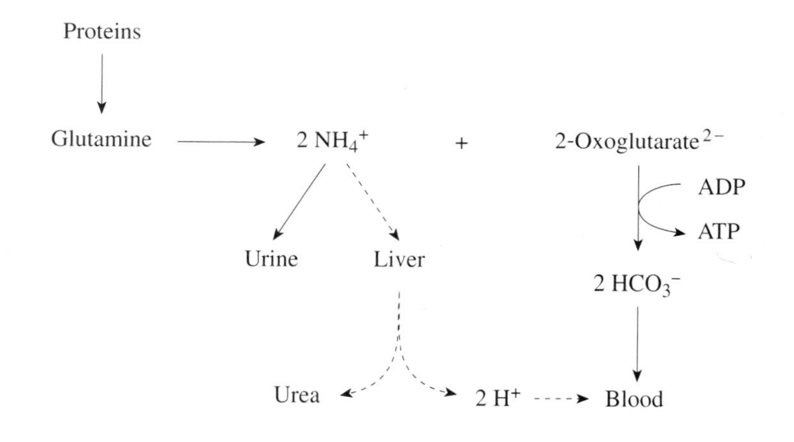

Figure 1·11
Biochemistry of new HCO_3^- synthesis
Glutamine, a neutral compound, is converted to the cation NH_4^+ plus the anion 2-oxoglutarate^{2-} (TCA cycle intermediate) in PCT cells. Further metabolism of 2-oxoglutarate^{2-} results in the formation of HCO_3^- and requires that ATP be formed. If the NH_4^+ are not excreted in the urine, they will be delivered to the liver (dashed line) where subsequent metabolism will yield H^+ plus urea; thus, there is no net HCO_3^- gain if NH_4^+ are not excreted (dashed lines).

Generating a high $[NH_3]$ in the renal medullary interstitium: There are four components to this story (Figure 1·12):

1. **Production of NH_4^+:**
   The metabolic process of NH_4^+ excretion begins with proteins from dietary or endogenous sources (Figure 1·13). These proteins release the amino acid glutamine, which, when metabolized in PCT cells, yields NH_4^+ and HCO_3^-. Proteins, when broken down, also yield the amino acids that generate H^+ (see page 10), but, because glutamine is very abundant in most proteins (and it can be made in the liver and muscle), its supply exceeds that of the amino acids that generate H^+.

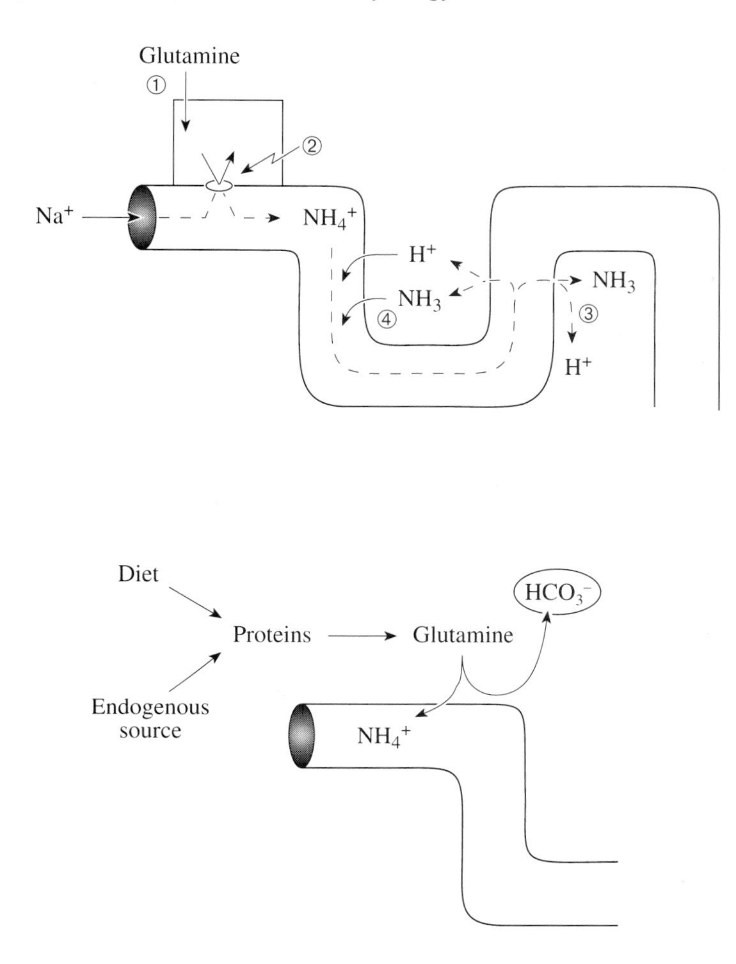

Figure 1·12
Generation of a high $[NH_3]$ in the medullary interstitium
Four steps are required: (1) production of NH_4^+ in PCT cells; (2) secretion of NH_4^+ into the lumen of the PCT via the Na^+/H^+ antiporter (replacing H^+); (3) reabsorption of NH_4^+ in the thick ascending limb of the loop of Henle via the Na^+, K^+, 2 Cl^- cotransporter (NH_4^+ replace K^+); and (4) secretion of NH_4^+ into the descending limb of the loop of Henle together with the operation of a countercurrent system. All of these events lead to the trapping of NH_3 in the medulla.

Figure 1·13
Supply of glutamine to PCT cells
Glutamine is derived from proteins of dietary or endogenous origin.

The metabolism of glutamine to glucose or CO_2 proceeds via the *TCA cycle*, so ATP must be generated when NH_4^+ are formed. The generation of ATP has several implications for *ammoniagenesis* (Figure 1·14). First, ATP cannot be stored; it is produced in response to demand. In the kidney, the major ATP-utilizing function is Na^+ reabsorption. Second, the rate of ammoniagenesis can also be limited if other fuels are present to regenerate ATP. Third, ammoniagenesis is located in PCT cells since they have the largest rate of turnover of ATP. Fourth, acidosis may occur in some situations when very high rates of ammoniagenesis do not confer an advantage to the host (i.e., during acute L-lactic acidosis of exercise and in ketoacidosis of fasting). These aspects are explored further in Questions 1·21–1·25.

The specific pathway involved in the production of NH_4^+ requires that glutamine enter mitochondria of PCT cells. This entry process is probably the major barrier that prevents high rates of conversion of glutamine to NH_4^+ by the enzyme phosphate-dependent glutaminase (PDG) in mitochondria. Two stimuli for the entry of glutamine into mitochondria and its subsequent metabolism via PDG are chronic metabolic acidosis and hypokalemia. Although other pathways exist to generate NH_4^+, they do not appear to be quantitatively important.

TCA cycle:
The pathway in mitochondria that results in the oxidation of acetyl-CoA and the production of the precursors to regenerate ATP.

Hypokalemia and ammoniagenesis:
The most likely reason for the stimulation of ammoniagenesis is the entry of H^+ into PCT cells when K^+ exit; the converse may apply in hyperkalemia.

Other regulators of ammoniagenesis:
Alkalemia, a low GFR, and a low supply of glutamine or a high supply of fat-derived fuels may also inhibit production of NH_4^+.

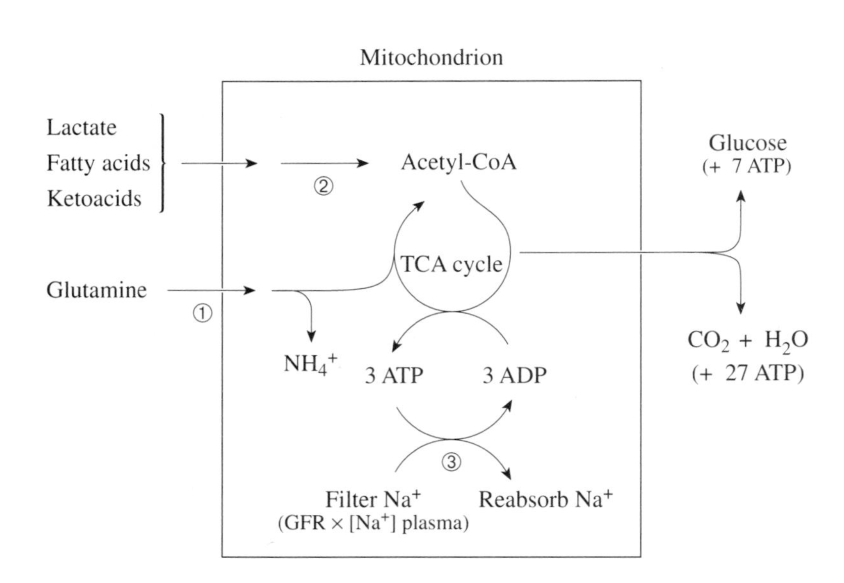

Figure 1·14
Control of ammoniagenesis by ATP turnover and fuel selection
There are three major constraints on the production of NH_4^+: (1) the amount of glutamine that enters the mitochondrion; (2) the "competition" between glutamine and other fuels to generate the quantity of ATP needed at that moment; (3) the turnover of ATP to perform biological work (Na^+ reabsorption, which reflects the GFR).

Questions

(Discussions on pages 41–42)

1·21 Why is the production of NH_4^+ low in renal insufficiency?

1·22 In the ketoacidosis of chronic fasting, the rate of excretion of NH_4^+ is lower than in chronic metabolic acidosis caused by ingestion of HCl. What teleologic advantages might this lower rate of excretion provide and what mechanisms might be involved?

1·23 There is a lag period between the net addition of acids and induction of ammoniagenesis. Why might this time be advantageous to a sprinter?

1·24 Why might ammoniagenesis be low in a patient with hyperkalemia?

1·25 Why is Na^+ acetate added to *total parenteral nutrition (TPN)* solutions? Answer from the perspective of NH_4^+ excretion.

Total parenteral nutrition (TPN): A form of therapy in which patients who are unable to obtain adequate nutrients from their GI tract are given intravenous feeding.

2. **Secretion of NH_4^+ into the lumen of the proximal nephron:**
   There are two theoretical ways NH_4^+ can enter the lumen of the PCT, by non-ionic diffusion or via a transporter. In the former, the uncharged NH_3 readily crosses the membrane down its difference in concentration. For this movement to occur, NH_3 must be permeable and the concentration difference favorable. Recently, it was demonstrated that these criteria were not always met, so the general feeling is that NH_4^+ enter the lumen by replacing H^+ on the Na^+/H^+ antiporter.

3. **Reabsorption of NH_4^+ in the loop of Henle:**
   To have a very high $[NH_3]$ in the renal medullary interstitium, NH_4^+ are reabsorbed in the thick ascending limb of the loop of Henle (the single effect). Reabsorption occurs via the Na^+, K^+, 2 Cl^- cotransporter, with

NH_4^+ taking the place of K^+. This replacement has two implications for regulation of NH_4^+ excretion. First, as more NH_4^+ are reabsorbed, the $[NH_3]$ rises in the medullary interstitium (see point 4 below). A second implication concerns K^+ in the lumen of the thick ascending limb. As the $[K^+]$ rises (it is higher in hyperkalemia), fewer NH_4^+ are reabsorbed because of competition between NH_4^+ and K^+. As a result, the $[NH_3]$ in the renal medullary interstitium falls (the NH_4^+ not reabsorbed in the loop are delivered to the distal convoluted tubule and are probably reabsorbed there, not raising the medullary interstitial $[NH_3]$). Accordingly, with hyperkalemia, the rate of NH_4^+ excretion will be lower, but the $[H^+]$ in the urine will be high because H^+ secretion is normal and there is less H^+ acceptor present.

Clinical pearl:
A low rate of excretion of NH_4^+ that is secondary to hyperkalemia (from low production of NH_4^+ or low reabsorption in the loop of Henle) will be corrected by lowering the $[K^+]$. In contrast, if the low rate of excretion of NH_4^+ is caused by a process that reduces the lumen negative potential difference in the CCD (see page 333), an action that also leads to hyperkalemia, correction of the hyperkalemia will not reverse the low rate of excretion.

4. **Recycling of NH_4^+ in the loop of Henle:**
The NH_4^+ reabsorbed in the loop of Henle are also secreted into the early descending limb of the loop of Henle. Given the very low medullary blood flow, NH_3 is concentrated in the medullary interstitium. This high $[NH_3]$ is essential for excretion of large quantities of NH_4^+.

Clinical pearl:
Patients who have diseases that damage their renal medulla may have low rates of excretion of NH_4^+.

Transfer of NH_3 to the lumen of the collecting duct: To excrete NH_4^+, the lumen of the collecting duct must "attract" NH_3. This action is achieved by the presence of a low $[NH_3]$ in the lumen consequent to the secretion of H^+ by collecting duct cells (Figure 1·15). The H^+ATPase produces a high $[H^+]$ in the lumen, which in turn lowers the $[NH_3]$ in the lumen by driving the equation below to the right.

$$H^+ + NH_3 \longleftrightarrow NH_4^+$$

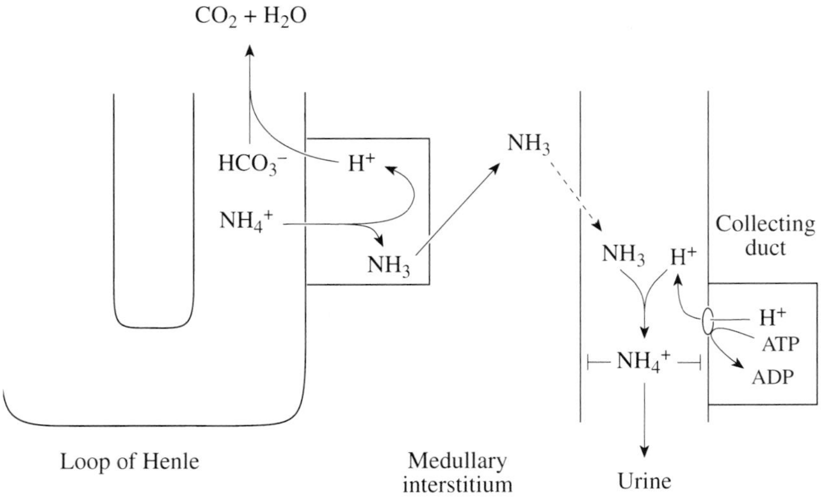

Figure 1·15
Transfer of NH_3 to the final urine
The major feature is active secretion of H^+ by collecting duct cells, which raises the $[H^+]$ and lowers the $[NH_3]$ in the luminal fluid. NH_3 then diffuses (dashed line) from its high concentration (medullary interstitium) to where its concentration is low (lumen of the collecting duct). This process is called "diffusion trapping" or "non-ionic diffusion."

HCO_3^- and the NH_4^+ Story

There are two HCO_3^- stories.

New HCO_3^-: All new HCO_3^- are produced in the PCT when glutamine is converted to NH_4^+ + HCO_3^-. These HCO_3^- are added to the renal venous blood (Figure 1·13, page 29).

Indirect reabsorption of HCO_3^- in later nephron segments: Approximately 500 mmol of the filtered HCO_3^- are not reabsorbed in the PCT. Most of them were reabsorbed by H^+ secretion in the loop of Henle (Table 1·10, page 23). There is a second pathway to reabsorb some of these HCO_3^- using NH_4^+ generated and secreted in the PCT (Figure 1·16). In this process, $NaHCO_3$ is filtered. The Na^+ are reabsorbed via the Na^+/H^+ antiporter in the PCT, but NH_4^+, not H^+, are the countertransporting cations in this case, so that NH_4^+ and HCO_3^- remain in the luminal fluid of the PCT. In the loop of Henle, several events occur. NH_4^+ are reabsorbed and then split into NH_3 and H^+ in the cells of the thick ascending limb. The NH_3 exits via the basolateral membrane, and the H^+ are secreted into the lumen. These secreted H^+ react with luminal HCO_3^- so that filtered HCO_3^- disappear from the lumen, but HCO_3^- do not return to the blood. Carbonic acid (or CO_2 plus H_2O) is delivered to the cortical distal nephron. In α-intercalated cells of the collecting duct, CO_2 is converted to H^+ and HCO_3^-; when the H^+ are secreted, the $[H^+]$ in the luminal fluid increases, and, as a result, the $[NH_3]$ in this fluid decreases. Because of the large concentration difference for NH_3, it diffuses from the medullary interstitium to the luminal fluid, where it is trapped as NH_4^+. The HCO_3^- exit via the basolateral membrane and enter the blood, thereby completing the "tricky" indirect reabsorption of filtered HCO_3^- using the NH_4^+ system (Figure 1·16).

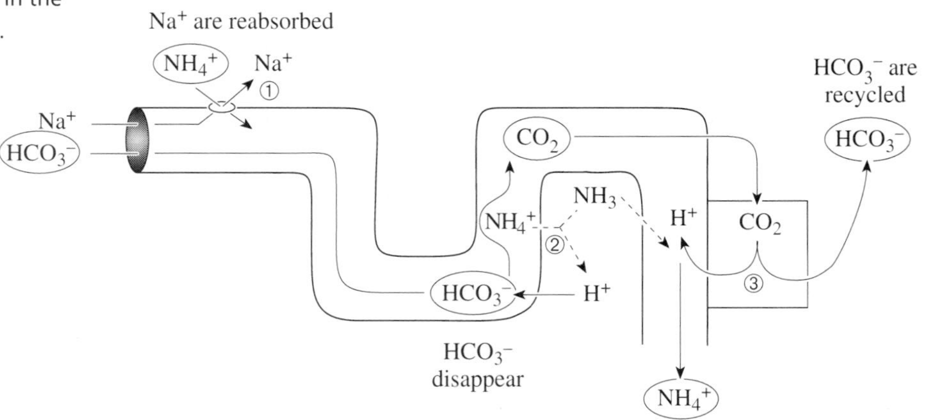

Figure 1·16
The indirect reabsorption of HCO_3^- using the transport of NH_4^+
$NaHCO_3$ is reabsorbed in three separate steps: (1) Na^+ are reabsorbed and NH_4^+ are secreted in the PCT; (2) HCO_3^- disappear in the loop of Henle as a result of H^+ secretion, but there is no addition of HCO_3^- to the blood; (3) HCO_3^- are finally indirectly reabsorbed once H^+ are secreted in the collecting duct.

NH_4^+ made in the PCT end up in the final urine via an indirect route.

Distal renal tubular acidosis: Renal generation of bicarbonate (excretion of NH_4^+) can fail for two major reasons:

1. failure to raise the [NH_3] in the renal medullary interstitium (medullary disease or impaired ammoniagenesis);
2. failure to trap NH_3 in the lumen of the collecting duct because of impaired secretion of H^+ in these nephron segments.

Whatever the basis, a low rate of excretion of NH_4^+ or renal generation of new HCO_3^- is known as "distal renal tubular acidosis (dRTA)." As one can appreciate, there are many possible pathogenic mechanisms to explain the presence of dRTA; the specific basis must be established in each case. Nevertheless, the hallmark is a relatively low rate of excretion of NH_4^+ in the urine. The various causes of dRTA are described in Chapter 3.

PART C
REVIEW

Discussion of Introductory Case
Lee Wants to Know the Acid Truth

(Case presented on page 4)

What acid-base abnormality is present?

Lee has acidemia ([H^+] = 60 nmol/l) and a plasma [HCO_3^-] that is low (10 mmol/l); therefore, metabolic acidosis is present.

Were acids (H^+) added?

The fact that acids have accumulated is reflected by two observations: the fall in plasma [HCO_3^-] and the increase in the anion gap in plasma (accumulation of the anion that accompanied the H^+—in this case, undoubtedly ketoacid anions).

Have H^+ been buffered? If so, how many?

Note:
There is no fixed percentage of buffering of a H^+ load in the ECF vs the ICF. At smaller H^+ loads, most buffering is by the BBS. With very large H^+ loads, all new buffering is via proteins because there are too few HCO_3^- remaining to buffer the H^+ load. In the introductory case, we assumed that 40% of the H^+ were buffered in the ECF at this given [H^+].

The fact that H^+ have been buffered is indicated by the fall in [HCO_3^-] of 15 mmol/l and the rise in the anion gap in plasma of 15 mEq/l. Despite the large input of H^+, the [H^+] has only increased by 0.000020 mmol/l. The H^+ have been buffered by HCO_3^- in the ECF, which has lost 15 mmol of HCO_3^- per liter (some H^+ have been buffered by HCO_3^- in the ICF as well). Assuming an ECF volume of 15 liters, then the H^+ load in the ECF was 225 mmol (15 liters × 15 mmol/l). Because close to 40% of the H^+ load was buffered in the ECF at this [H^+], the total H^+ load was 563 mmol.

Have the lungs and the kidneys responded appropriately?

The lungs have responded to the acidemia because the P_aco_2 has fallen to 25 mm Hg (normal P_aco_2 is 40 mm Hg). The kidneys have also responded, as evidenced by the increased excretion of NH_4^+ from the normal 40 mmol/day to 200 mmol/day.

Is the acid-base abnormality a major threat to the body? If so, why?

This acid-base abnormality is a potential threat to Lee's life. The H^+ load has already consumed just over half of the total buffer capacity. Whereas the overall H^+ production in ketoacidosis is close to 1 mmol/min, the rate of accumulation of ketoacids depends heavily on their rate of removal by oxidation. Therefore, ketoacids can accumulate at an appreciable rate if the rate of ketoacid anion metabolism declines. Such a decline would occur if the patient's GFR is low (the need for ATP to reabsorb Na^+ would be reduced) or if the patient develops coma (the need for ATP to maintain cerebral activity would be reduced). Hence, ketoacidosis may become a major threat if Lee becomes comatose and has prerenal failure (a low GFR).

Is it possible to estimate how quickly more H^+ will accumulate?

Lee is currently regenerating 200 mmol of HCO_3^- per day via the excretion of NH_4^+. If the net ketoacid production increases much more, H^+ will accumulate (production is 1500 mmol/day, and metabolism is 1200 mmol/day via normal brain and kidneys). One can estimate how many H^+ will accumulate by quantitating the number of new anions that appear in Lee over a period of time. These anions may remain in the body (rise in the plasma anion gap × total body water) or be excreted in the urine without H^+ or NH_4^+ (urine ($[Na^+] + [K^+] - [Cl^-]$) × urine flow rate).

Clinical pearl:
One can calculate how many H^+ are accumulating by counting new anions in the body plus anions excreted without H^+ or NH_4^+.

Do H^+ kill? If so, how do they do it?

If H^+ kill, they do so by binding to intracellular proteins and altering their charge and function. We are not sure that H^+ actually kill most patients with metabolic acidosis. Our rationale is that people who sprint vigorously can have very low pH values, but all survive.

Summary of Main Points

- Count the number of charges on substrates and products of the reactions to see if H^+ are produced or removed in the metabolic process.
- H^+ accumulate when anions are formed and cannot be metabolized or when they are metabolized more slowly than they are produced.
- H^+ accumulate during the metabolism of proteins because the number of sulfur-containing amino acids (methionine, cysteine/cystine × 2) plus cationic amino acids (lysine, arginine, histidine) exceeds the number of anionic amino acids (glutamate and aspartate).
- Incomplete metabolism of carbohydrates (usually from hypoxia) or of fats (from a lack of insulin) causes H^+ to accumulate.
- "Fast" H^+ are only produced during hypoxia.
- H^+ accumulate when alcohols (or neutral compounds that are converted to alcohols) are oxidized to anions in the body. Of greater importance, metabolic intermediates of this metabolism may accumulate, and they are very toxic to humans.

Discussion of Questions

1·1 Why doesn't the HCl secreted by gastric cells denature proteins in the cell membrane of the stomach?

The [H^+] is so high in gastric fluid that proteins in cell membranes will be denatured if exposed to this [H^+]. The defence mechanisms include a protective mucus barrier along the luminal lining of the stomach and, located below this lining, a relatively alkaline solution that contains HCO_3^-. Upon secretion, H^+ are "channelled" quickly through this mucus barrier. Hence, H^+ in the lumen are kept away from the gastric mucosa.

1·2 What might permit H^+ to bind to the H^+ pump inside cells at a concentration of 0.0001 mmol/l, yet dissociate in the lumen of the stomach at a concentration of 100 mmol/l?

This question is vexing. The chemical group that binds H^+ in the cell must have its pK changed by local influences (structural changes), but this change is probably not enough to alter the pK so drastically. We envision other forces operating. Perhaps the binding of K^+ in the lumen helps "repel" the positively charged proton from the carrier. This theory might provide a rationale for having a special type of H^+ pump in the stomach, a H^+/K^+ antiporter.

1·3 What is the rationale for stating that only weak acids kill?

The major consideration from a clinical perspective is the size of the H^+ load. Since strong acids are 99.999% dissociated and weak acids are close to 99% dissociated at the [H^+] in the body, the strength of acids is not an important determinant of the number of H^+ released. Of greater importance is the quantity of acid produced; weak acids (L-lactic acid, ketoacids) are produced at a much higher rate than strong acids (HCl and H_2SO_4). Hence, there will be a larger total load of H^+ with weak acids.

1·4 In the metabolic process in Figure 1·1, under what circumstances will H^+ accumulate?

If the rate of production of ketoacids equals their rate of oxidation, then H^+ do not accumulate. Nevertheless, if the rate of ketogenesis increases or the rate of oxidation of ketoacids declines, H^+ will accumulate (see Chapter 3 for details).

1·5 Red blood cells produce 200 mmol of L-lactic acid per day. Why doesn't this production cause severe acidemia?

Although red blood cells produce 200 mmol of L-lactic acid per day, acidemia does not usually result because the hepatic metabolism of the L-lactate anions to neutral end products consumes all the H^+ that were produced. As long as L-lactate metabolism occurs as quickly as this acid is formed, there is no net production of H^+.

1·6 When ATP is used to perform biologic work, H^+ are formed. Should these H^+ be considered as part of the quantity that can cause acidosis?

This question and answer illustrate the difference between net and absolute rates of production of H^+; the former is very small (but important) and the latter is enormous (and not so important). In more detail, as soon as ATP is hydrolyzed to perform biologic work, it is regenerated by metabolic reactions. Hence, there is a large turnover of H^+ but no net production of H^+. Because the quantity of ATP in tissues is quite small, there would be only a minor change in the maximum rate of H^+ accumulation even if all the ATP in cells was converted to ADP. Therefore, the answer to the question is no.

$$ATP^{4-} \longrightarrow ADP^{3-} + P_i^{2-} + H^+ + \text{work}$$

$$ADP^{3-} + P_i^{2-} + H^+ \longrightarrow \text{ATP (regeneration)}$$

1·7 Acetate is added to the hemodialysis fluid to minimize the net production of H^+. What is the rationale for this maneuver?

Acetate anion as its Na^+ salt is added to the dialysis fluid so that H^+ can be removed by metabolic pathways that yield CO_2 as a final product (equation below).

$$\text{Acetate anion} + 2\,O_2 + H^+ \longrightarrow 2\,CO_2 + 2\,H_2O$$

1·8 During hemodialysis, β-hydroxybutyric acid is formed from acetate anions and is lost in the dialysis fluid. Does this process result in the net production or removal of H^+?

Two molecules of acetate anion (Na^+ salt) are converted to one molecule of β-hydroxybutyrate anion in the liver. In this process, the product has less anionic charge than its substrates, so H^+ are removed despite the fact that a ketoacid (β-hydroxybutyric acid) is the product (equation below).

$$2\,\text{Acetate anion} + 2\,H^+ \longrightarrow \beta\text{-hydroxybutyrate anion} + H^+$$

Note:
In contrast to acetate anions, conversion of neutral triglycerides to β-hydroxybutyrate anions does result in the net production of H^+ (see pages 85–94).

1·9 What might permit bacteria in the GI tract to overproduce organic acids?

The normal digestive process, with largely neutral foodstuff, feces, and absorbed fuels does not pose a net acid load. However, changes in the normal bacterial content of the intestines (e.g., from new flora produced by the actions of antibiotics or from stasis, which permits more bacterial growth) can potentiate the release of an unusual quantity of organic acids into the body. The production of organic acids might become limited if these bacteria lack a carbohydrate fuel; when the intestines contain more bacteria, eating foods that contain carbohydrates can suddenly increase the rate of production of these acids (see Case 3·5, pages 128–29 and 133–34).

$$\text{Carbohydrate} \longrightarrow \text{Organic anions} + H^+$$

1·10 Will consumption of citrus fruits, which contain a large quantity of citric acid and K^+ citrate, cause an acid or alkali load?

Very early on, before citrate anions are metabolized, there is an initial H^+ load because citric acid dissociates into H^+ and citrate anions (Citric acid $\longrightarrow$ Citrate^{3-} + 3 H^+). Later, when all citrate anions are removed by metabolism to neutral end products, there will be a net alkali load because some citrate anions are added to the body with K^+ and not H^+ (Citrate^{3-} + 3 H^+ + 4.5 O_2 $\longrightarrow$ 6 CO_2 + 4 H_2O).

1·11 How much would the $[HCO_3^-]$ in the ECF decrease when a load of 70 mmol of H^+ is retained in a 70-kg person who lacks renal generation of new HCO_3^-?

The kidneys normally regenerate 70 mmol of HCO_3^- per day. If the body gains 70 mmol H^+ per day, and just over half are buffered in cells, 30 mmol of H^+ would be buffered in the ECF by HCO_3^-. Since the ECF volume is close to 15 liters, the $[HCO_3^-]$ will fall by close to 2 mmol/l each day while the BBS is still a major buffer system operating in the body.

1·12 The following results for a blood sample arrive from the lab: $[H^+]$ = 60 nmol/l, pH = 7.22, Pco_2 = 50 mm Hg, $[HCO_3^-]$ = 32 mmol/l. What do you conclude?

Henderson equation:

$$[H^+] = \frac{24 \times P_{CO_2}}{HCO_3^-}$$

$$60 \neq \frac{24 \times 50}{32}$$

The values are inconsistent with the Henderson equation. An analytical error may have occurred, measurements may have been made on two different samples of blood, or there may have been a problem with the temperature or ionic strength of the sample. Repeat the analysis and then decide.

1·13 What would the P_aco_2 be in a normal man if his alveolar ventilation rate decreased by half (assume no change in his metabolic rate)?

If the rate of formation of CO_2 by the body is kept constant, halving the alveolar ventilation will temporarily slow the rate of excretion of CO_2. This decreased rate of excretion will cause the Pco_2 in each liter of alveolar air to increase so that all the CO_2 produced will be excreted and a new steady state will exist (compare with clearance of creatinine—when the GFR is halved, the serum creatinine doubles so the daily creatinine production is excreted with half the GFR).

The alveolar ventilation during hypoventilation will fall from 5 to 2.5 liters/min in this example.

Values:

- CO_2 production = 10 mmol/min at rest.
- Alveolar ventilation at rest is 5 liters/min.

Therefore, each liter of alveolar air contains 2 mmol CO_2 (1 liter of alveolar air has (40/760) × 1000 ml CO_2; at standard temperature and pressure, 22.4 ml = 1 mmol).

CO_2 removal = $[CO_2]$ in alveolar air × volume of alveolar air exhaled/min

- Normal: 10 mmol/min = 2 mmol/l × 5 liters/min
- Hypoventilation: 10 mmol/min = 4 mmol/l × 2.5 liters/min

Therefore, halving the alveolar ventilation will lead to a doubling of the P_aco_2 to 80 mm Hg.

What must you do to keep his pH ($[H^+]$) from changing?

In order to keep the $[H^+]$ normal at a P_aco_2 of 80 mm Hg, the $[HCO_3^-]$ must be increased twofold to 48 mmol/l.

1·14 Some say that hyperventilation is "maladaptive" because in an experimental situation, there was a fall in the pH of plasma after hyperventilation occurred. How might this phenomenon be interpreted?

The key question is this: What is the purpose of hyperventilation—to change the $[H^+]$ in the ECF or to influence events in the ICF (return the valence of proteins towards normal)?

We seem preoccupied with events in the ECF because we measure samples from this compartment. As judged from the ability of subjects to perform vigorous exercise with a very high $[H^+]$ in the ECF (the sprint), there is little cause to worry if the $[H^+]$ rises in the ECF.

In our opinion, the crucial compartment is the ICF. When the tissue Pco_2 falls, the BBS equation is driven to the right and H^+ are removed from proteins and transferred to HCO_3^- (Figure 1·8, page 21). Thus, there can be a major advantage of a lower Pco_2 because now proteins will have a valence that is closer to normal and perhaps their charge, shape, and functions will return to normal.

Regarding the question of a maladaptive response with a lower Pco_2, if, as a result of a more normal ICF environment after hyperventilation, there is additional net loss of some HCO_3^-, the $[H^+]$ in the ECF might not fall and may even rise a bit. This response is the "price to pay" for a better environment in cells. Therefore "maladaptive or not" depends on where you stand on this issue.

1·15 Is it more important to control the $[H^+]$ in the ECF or the ICF?

It is more important to control the $[H^+]$ in the ICF.

$[H^+]$ in the ECF: There are three major ways in which a high $[H^+]$ in the ECF can compromise cellular function and limit survival:

1. By virtue of the high $[H^+]$, hormone receptors on the cell surface can be protonated and thereby compromise signal transduction and cellular function. Examples include the lower affinity for the binding of insulin and of adrenaline to their receptors in severe acidemia; binding of adrenaline to its receptor could compromise myocardial function.
2. It is also possible that protonation of other cell membrane proteins can compromise function.
3. A high $[H^+]$ in the ECF could bring about a shift of K^+ from the ICF to the ECF. In this setting, acute hyperkalemia could induce a cardiac arrhythmia.

In spite of the theoretical arguments above, during a sprint, the $[H^+]$ rises markedly in the ECF, yet cardiac performance is close to normal.

$[H^+]$ in the ICF: It is more likely that the detrimental effect of a high $[H^+]$ may be related to titration of intracellular proteins. When more H^+ are bound to these proteins, they will have a more positive charge, which could alter their shape and function. Therefore, whereas the traditional clinical view in the assessment of the acid-base status has focused on the assessment of the $[H^+]$ and $[HCO_3^-]$ in the ECF, a broader view would consider the impact of H^+ not only on the titration of the BBS, but also (and maybe more importantly) on the non-BBS buffers (intracellular proteins for the most part). Therefore, it seems to us that it is more important to defend the $[H^+]$ in the ICF.

1·16 What should the pH of the ICF of muscle be in the first 10 seconds of a sprint in a normal person and in a patient with a defect in the hydrolysis of glycogen?

Vigorous muscle activity requires a large supply of ATP, which is provided by the metabolism of glucose and glycogen. Because there is insufficient O_2 to regenerate the needed ATP, anaerobic metabolism proceeds (glucose $\longrightarrow$ L-lactic acid). The $[H^+]$ in the ICF rises markedly in a sprint in a normal person because of the huge production of L-lactic acid.

In contrast, if only a small quantity of glycogen in muscle cells can be broken down during a sprint (which is the case in the patient with a defect in hydrolysis of glycogen), little L-lactic acid will be formed during the sprint. Exercise induces the hydrolysis of creatine phosphate, thereby producing HPO_4^{2-}, a H^+ acceptor that will cause the muscle cell to become more alkaline. Thus, normally, the impact of the excess local L-lactic acid production in muscle during vigorous exercise is diminished somewhat by the generation of an additional H^+ buffer, HPO_4^{2-}. Parenthetically, the content of creatine phosphate is very high in skeletal muscle cells (20 mmol/kg).

Therefore, in the normal person, the $[H^+]$ in the ICF will be increased during vigorous exercise, but in a patient with a defect in glycogenolysis, the $[H^+]$ in the ICF will be decreased from the increased HPO_4^{2-} coupled with the low H^+ load.

1·17 What other role might the Na^+/H^+ antiporter have in the PCT?

The Na^+/H^+ antiporter helps in the excretion of NH_4^+. NH_4^+ replace H^+ on the Na^+/H^+ antiporter in PCT cells so that NH_4^+ can be added to the lumen. If too many NH_4^+ are secreted, they can be reabsorbed via the diffusion-trapping mechanism.

1·18 How do carbonic anhydrase inhibitors (CAI) compromise the indirect reabsorption of HCO_3^-?

CAI inhibit carbonic anhydrase in the lumen of the PCT. In so doing, H_2CO_3 builds up in the lumen. As a result of the carbonic acid dissociation, the $[H^+]$ in luminal fluid increases and prevents further net secretion of H^+ (which would reabsorb more of the filtered HCO_3^-).

1·19 If the patient with pRTA develops diarrhea and the $[HCO_3^-]$ in the ECF drops 2 mmol/l below its usual value, how will the urine pH and $[HCO_3^-]$ change?

Under some circumstances (e.g., after gastric HCl secretion), the patient with pRTA has alkaline urine because the filtered load of HCO_3^-

exceeds the reduced capacity of the proximal H^+ secretory process to reclaim it. Thus, HCO_3^- are delivered to the distal convoluted tubule. At this point, some (but not all) are reabsorbed, since distal H^+ secretion is a low-capacity system. The rest of the HCO_3^- that were not reabsorbed are excreted and the urine becomes alkaline. If some process (e.g., loss of $NaHCO_3$ via the GI tract) lowers the filtered load of HCO_3^- so that it does not exceed the proximal H^+ secretory capacity, most of the HCO_3^- are then reabsorbed indirectly in the PCT (some HCO_3^- always leave the PCT since the membrane is leaky and steep $[H^+]$ gradients cannot be generated by the proximal nephron; i.e., the pH of the luminal fluid at the end of the PCT is in the mid-6 range). Under these circumstances, a small enough quantity of HCO_3^- are delivered to the distal tubule so that distal H^+ secretion is sufficient to reabsorb these HCO_3^-; the urine is rendered free of HCO_3^-. The remaining H^+ secretion will lower the urine pH to well below 6; how low the urine pH falls is determined by the quantity of NH_3 available to react with secreted H^+ (see the margin).

Note:
The urine pH is generally low in patients with proximal RTA. This low pH permits them to excrete NH_4^+ for daily acid-base balance.

1·20 How would you establish whether or not a patient with a low $[HCO_3^-]$ and diarrhea had pRTA?

Administer a load of $NaHCO_3$. If the patient continues to reabsorb HCO_3^- normally, a small quantity of HCO_3^- will be excreted until the plasma $[HCO_3^-]$ approaches 25 mmol/l. In contrast, with pRTA, bicarbonaturia will be prominent at concentrations of HCO_3^- in plasma that are distinctly subnormal.

1·21 Why is the production of NH_4^+ low in renal insufficiency?

The concentration of ADP, the precursor for ATP, is present in very tiny amounts in cells. Thus, no organ or organelle can generate more ATP than it consumes to perform biological work (hydrolysis of ATP yields ADP). Since renal work is largely the reabsorption of filtered Na^+, a low GFR means fewer filtered Na^+, less work, and less ATP turnover. Hence, in renal insufficiency, less fuel (glutamine) can be oxidized and fewer NH_4^+ will be produced.

1·22 In the ketoacidosis of chronic fasting, the rate of excretion of NH_4^+ is lower than in chronic metabolic acidosis caused by ingestion of HCl. What teleologic advantages might this lower rate of excretion provide and what mechanisms might be involved?

Excretion of more NH_4^+ during fasting requires a source of nitrogen. This source is body protein. Hence, excessive catabolism of lean body mass could occur and, from a teleologic viewpoint, might not be advantageous during fasting. With respect to mechanisms, the lower GFR and the oxidation of fat-derived fuels (ketoacids) by PCT cells could limit the rate of production of NH_4^+ because there would be less ATP to regenerate from the metabolism of glutamine (see the discussion of Question 1·21).

1·23 There is a lag period between the net addition of acids and induction of ammoniagenesis. Why might this time be advantageous to a sprinter?

During sprinting, L-lactic acid is formed. Over the subsequent period of rest (minutes to hours) all this L-lactic acid will be removed by metabolism. Hence, there is no need for extra HCO_3^- to be made by the kidney.

1·24 Why might ammoniagenesis be low in a patient with hyperkalemia?

Hyperkalemia resulting from a K^+ load can cause K^+ to enter cells. For electroneutrality, some H^+ will exit. This cell alkalinization will lower NH_4^+ production. Hyperkalemia may also lower the rate of excretion of NH_4^+ by mechanisms independent of ammoniagenesis. In addition, K^+ and NH_4^+ compete for reabsorption in the thick ascending limb of the loop of Henle. As a result, there is less NH_3 in the medullary interstitium (Figure 1·17).

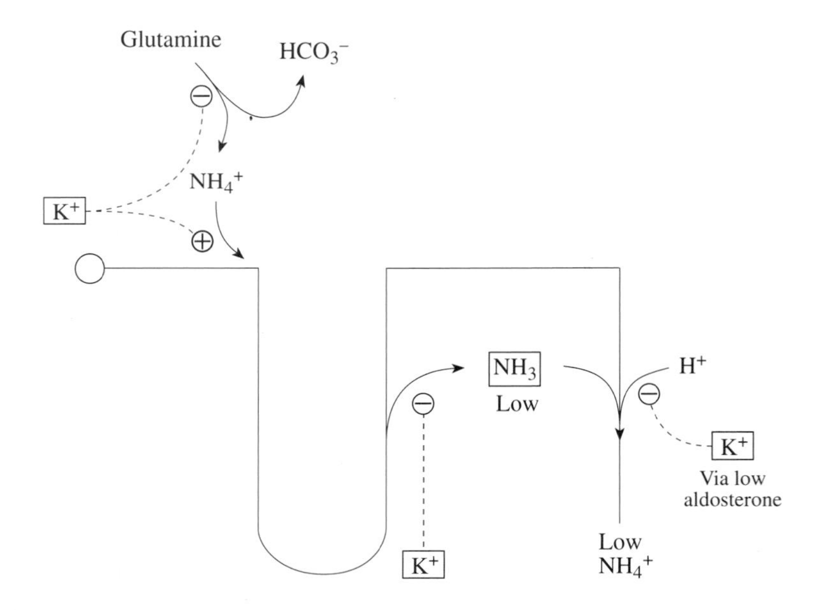

Figure 1·17
Influence of hyperkalemia on the excretion of NH_4^+
In the PCT, hyperkalemia has two effects: it decreases the production of NH_4^+ but enhances their secretion so that there is little reduction in the quantity of NH_4^+ leaving the luminal fluid in the PCT. In the loop of Henle, hyperkalemia causes a higher $[K^+]$ in the luminal fluid and thereby decreases the reabsorption of NH_4^+. This action leads to a low $[NH_3]$ in the medullary interstitium. Both the low medullary $[NH_3]$ and the low distal H^+ secretion (results of the low aldosterone if that was the cause of the hyperkalemia) lead to a low excretion of NH_4^+.

1·25 Why is Na^+ acetate added to TPN solutions? Answer from the perspective of NH_4^+ excretion.

Patients on TPN develop metabolic acidosis with a low rate of renal excretion of NH_4^+. A possible explanation for the low rate of NH_4^+ excretion (and hence the acidemia) is that perhaps fuels provided during TPN (other than glutamine) regenerate ATP, but not NH_4^+, in cells of the PCT. When the acetate anion added to the TPN is metabolized to CO_2 and H_2O, the result is consumption of H^+ (or production of HCO_3^-) and correction of the acidemia.

2

THE CLINICAL APPROACH TO ACID-BASE DISORDERS

Objectives

- To provide the background so that the four primary acid-base disturbances can be identified if present.
- To illustrate the expected physiologic responses to a primary acid-base disorder and thereby enable the recognition of mixed acid-base disorders.
- To provide a group of tools including several "gaps" to assist in the diagnosis of individual acid-base disturbances.

Outline of Major Principles

1. The parameters in plasma that describe the patient's acid-base status are summarized in the margin.
2. The *Henderson equation* relates three parameters; calculate the third knowing the other two. This equation also helps in identifying errors in measurement.
3. There are four primary acid-base disorders:
   - Metabolic acidosis: increased $[H^+]$ and reduced $[HCO_3^-]$ in plasma;
   - Metabolic alkalosis: reduced $[H^+]$ and increased $[HCO_3^-]$ in plasma;
   - Respiratory acidosis: increased $[H^+]$ and P_aco_2 in plasma;
   - Respiratory alkalosis: reduced $[H^+]$ and P_aco_2 in plasma.
4. Specific physiologic responses occur as the result of acid-base disorders. In the ECF, these responses return the plasma $[H^+]$ toward normal, but not into the normal range. It is more instructive to consider their impact on events in the ICF.
5. The clinical and laboratory data must be evaluated in concert to reach an acid-base diagnosis.
6. Several acid-base disorders may coexist in a patient.

Normal values in plasma:
pH: 7.40 ± 0.02
$[H^+]$: 40 ± 2 nmol/l
P_aco_2: 40 ± 2 mm Hg
$[HCO_3^-]$: 25 ± 2 mmol/l
Anion gap: 12 ± 2 mEq/l (excluding K^+)
$([Na^+] - ([Cl^-] + [HCO_3^-]))$

Henderson equation:

$$[H^+] = \frac{24 \times Pco_2}{[HCO_3^-]}$$

Abbreviations:
ECF = extracellular fluid.
ICF = intracellular fluid.
Pco_2 = partial pressure of CO_2.
P_aco_2 = Pco_2 in arterial blood.

Introductory Case
Lee's Acidosis Is Mixed Up

(Case discussed on page 60)

When Lee, an insulin-dependent diabetic, developed abdominal pain and diarrhea, her food intake was curtailed and the insulin dose was reduced. Because the diarrhea persisted, Lee saw a physician, and the following results were obtained from blood tests.

Na^+	140 mmol/l	H^+	60 nmol/l (pH = 7.22)
K^+	5.0 mmol/l	P_aco_2	25 mm Hg
Cl^-	103 mmol/l	HCO_3^-	10 mmol/l

What is the acid-base diagnosis?
Is there a primary respiratory acid-base disorder?

PART A
BACKGROUND

Methods to Interconvert pH and $[H^+]$

At times, values may be reported in pH units; at others, the $[H^+]$ is utilized. Therefore, it is important to know how to interconvert these units. Methods for this interconversion are provided below.

RULE OF THUMB #1:
DROP THE 7 AND DECIMAL POINT

There is a near-linear relationship between the $[H^+]$ and the pH in the range of pH 7.28 to 7.55. At a pH of 7.40, the $[H^+]$ is 40 nmol/l (note that if you drop the 7 and the decimal point you have the $[H^+]$ of 40 nmol/l). Next, after dropping the 7 and the decimal point, take the difference of the number from 40. Add that value to 40 if the pH is less than 7.40, and subtract it if the pH is greater than 7.40; the result is the $[H^+]$.

Reference:
Kassirer, J. P., and H. L. Bleich. 1965. *N. Engl. J. Med.* 272:1067.

pH	Drop 7 and decimal point	Difference from 40	$[H^+]$ nmol/l
7.40	40	0	40
7.38	38	2	42
7.42	42	2	38

THE 0.1 pH CHANGE RULE

For every 0.1 unit increase in pH, multiply the $[H^+]$ by 0.8. Given that a pH of 7.00 equals a $[H^+]$ of 100 nmol/l, a rise in pH of 0.1 (pH 7.10) equals 0.8 $\times$ 100, or a $[H^+]$ of 80 nmol/l. For values less than 7.00, divide by 0.8 (or multiply by 1.25). Intermediate values are calculated by interpolation.

Reference:
Fagan, T. J. 1973. *N. Engl. J. Med.* 288:915.

pH	Conversion factor	$[H^+]$ nmol/l
6.90	100×1.25	125
7.00	100	100
7.10	100×0.8	80
7.20	$100 \times 0.8 \times 0.8$	64

LOG TABLE

A log table (Table 2·1) is provided to aid in interconverting the $[H^+]$ and pH.

Table 2·1

INTERCONVERSION OF pH AND $[H^+]$

pH	$[H^+]$	pH	$[H^+]$	pH	$[H^+]$	pH	$[H^+]$
.01	9772	.26	5495	.51	3090	.76	1738
.02	9550	.27	5370	.52	3020	.77	1698
.03	9333	.28	5248	.53	2951	.79	1660
.04	9120	.29	5129	.54	2884	.79	1622
.05	8913	.30	5012	.55	2818	.80	1585
.06	8710	.31	4898	.56	2754	.81	1549
.07	8511	.32	4786	.57	2692	.82	1514
.08	8318	.33	4677	.58	2630	.83	1479
.09	8128	.34	4571	.59	2570	.84	1445
.10	7943	.35	4467	.60	2512	.85	1413
.11	7762	.36	4365	.61	2455	.86	1380
.12	7586	.37	4266	.62	2399	.87	1349
.13	7413	.38	4169	.63	2344	.88	1318
.14	7244	.39	4074	.64	2291	.89	1288
.15	7079	.40	3981	.65	2239	.90	1259
.16	6918	.41	3890	.66	2188	.91	1230
.17	6761	.42	3802	.67	2138	.92	1202
.18	6607	.43	3715	.68	2089	.93	1175
.19	6457	.44	3631	.69	2042	.94	1148
.20	6310	.45	3548	.70	1995	.95	1122
.21	6166	.46	3467	.71	1950	.96	1096
.22	6026	.47	3388	.72	1905	.97	1072
.23	5888	.48	3311	.73	1862	.98	1047
.24	5754	.49	3236	.74	1820	.99	1023
.25	5623	.50	3162	.75	1778	1.00	1000

Examples

pH	$[H^+]$ nmol/l
7.01	97.7
7.00	100
6.90	125.9

Making an Initial Acid-Base Diagnosis by Examining Parameters in Plasma

- Integrate clinical and laboratory pictures.
- Examine all four parameters in plasma ($[H^+]$, $[HCO_3^-]$, P_aco_2, anion gap).

pH vs $[H^+]$:
We prefer to think in terms of the $[H^+]$ rather than the pH, but the principles are the same: a low $[H^+]$ is a high pH, and vice versa.

In making an acid-base diagnosis, there are two points to stress. First, one must integrate the clinical picture and the laboratory values to make a proper diagnosis. For example, finding acidemia, a high P_aco_2, and an elevated $[HCO_3^-]$ does not indicate that chronic respiratory acidosis is present in a patient who does not have a chronic problem with ventilation; in this case, more than one acid-base disturbance is likely to be present. Second, from the laboratory perspective, we recommend that four parameters in plasma be examined, the pH or $[H^+]$, the P_aco_2, the $[HCO_3^-]$, and the anion gap; we usually start with the plasma $[H^+]$, as illustrated in Figure 2·1.

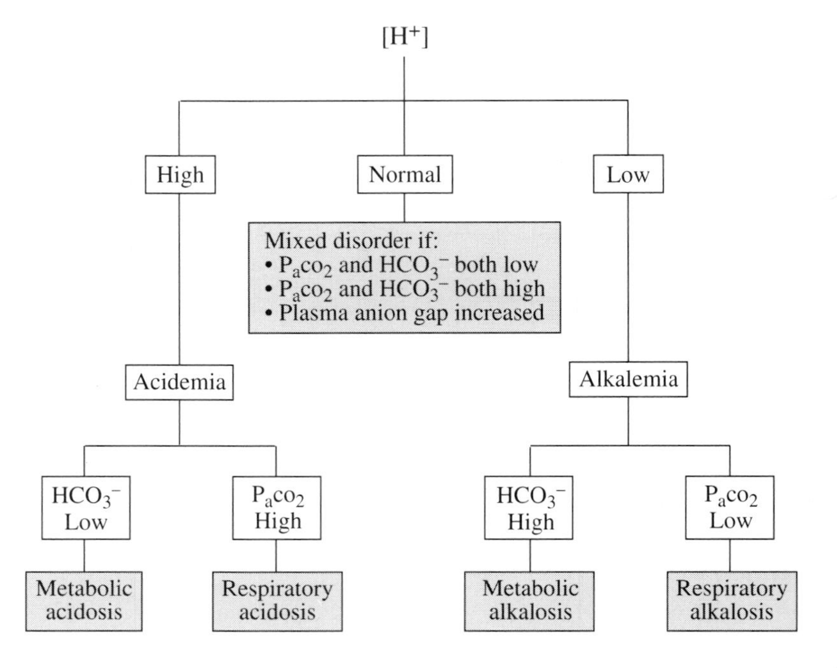

Figure 2·1
Initial diagnosis of acid-base disorders
Start with the plasma $[H^+]$. The information to be interpreted to make a final diagnosis is indicated in open boxes; the final diagnoses are shown in the shaded boxes (see the text for details).

P_ao_2:
The P_ao_2 has no direct application with respect to the acid-base analysis; however, calculating the alveolar-arterial (A-a) O_2 difference provides additional important information (see pages 193–95 for more discussion).

HIGH $[H^+]$

If the $[H^+]$ is increased, the patient has acidemia, and there are two potential causes—metabolic or respiratory acidosis.

- **Metabolic acidosis** is the result of a process that lowers the $[HCO_3^-]$ and raises the $[H^+]$; the expected adjustment is a low P_aco_2.
- **Respiratory acidosis** is characterized by an increased P_aco_2 and $[H^+]$; the expected adjustment is an increased $[HCO_3^-]$. This adjustment is minimal in acute respiratory acidosis and larger in chronic respiratory acidosis.

LOW $[H^+]$

If the $[H^+]$ is low, the patient has alkalemia. Again, there are two potential causes—metabolic or respiratory alkalosis.

- **Metabolic alkalosis** is the result of a process that raises the $[HCO_3^-]$ and lowers the $[H^+]$; the expected adjustment is a rise in the P_aco_2, which is usually modest because of respiratory stimulation from the resultant hypoxia. Nevertheless, if the patient is receiving O_2, and hypoxia is prevented, hypoventilation can be more profound.
- **Respiratory alkalosis** is characterized by an unexpectedly low P_aco_2, which results in a lower $[H^+]$. The expected physiological response is a

reduction in plasma [HCO_3^-]. As in respiratory acidosis, this response is modest in acute disorders and more significant in chronic disorders.

NORMAL [H^+]

A normal [H^+] implies either no acid-base disorder or the presence of two acid-base disorders—one tending to raise the [H^+] and the other tending to depress the [H^+].

Question

(Discussion on page 61)

2·1 A patient has diabetic ketoacidosis and the following lab data: pH = 7.10, P_aco_2 = 30 mm Hg, [HCO_3^-] = 13 mmol/l, anion gap = 25 mEq/l. What do you conclude?

Tests Used in Making Acid-Base Diagnoses

Most hospital-based biochemistry labs do not routinely perform all the tests needed to make a definitive acid-base diagnosis. In this section we shall provide the rationale for some tests that can supply additional information at the bedside (Table 2·2).

Table 2·2
TOOLS USED AT THE BEDSIDE TO MAKE ACID-BASE DIAGNOSES

The tests described help in establishing the basis of the metabolic acidosis.

Test	Major function assessed	Disadvantages
• Anion gap in plasma	• Accumulation of acids other than HCl and H_2CO_3.	• Several lab measurements, so errors are possible. • [Albumin] must be known.
• Osmolal gap in plasma	• Presence of alcohols in plasma.	• Osmolality method may cause a problem.
• Urine net charge	• [NH_4^+] in urine.	• Interpretation is confounded by the excretion of anions other than Cl^-.
• Osmolal gap in urine	• [NH_4^+] in urine.	• May not know [urea] or [glucose] in urine.
• Urine Pco_2	• Distal H^+ secretion.	• Not quantitative. • Many quibbles with this test.

Caution regarding calculation of the osmolal gap:
Beware of rare circumstances where unusual cations (e.g., myeloma proteins) may be present in plasma.

Note:
The rate of increase in plasma anion gap and the rate of decrease in plasma $[HCO_3^-]$ provide insights into the rate of H^+ accumulation.

Anion gap in plasma:
$[Na^+] - [Cl^-] - [HCO_3^-]$
- The normal value is 12 ± 2 mEq/l.
- Expect close to a 1:1 reciprocal change in anion gap and $[HCO_3^-]$.
- The anion gap changes with blood pH, but this change is small (0.5 mEq/l for a 0.1 unit change in pH).
- An increased anion gap may be the only clue that metabolic acidosis is present in a mixed acid-base disorder.

Distal RTA:
A renal disorder characterized by a low rate of excretion of NH_4^+.

In almost every case, the tests are performed in patients with metabolic acidosis. The specific questions to be addressed are as follows:

1. Is the net production of acids occurring at an unusually rapid rate?
   Measuring the *anion gap in plasma* and the net charge in urine may indicate the presence of new anions added to the blood or urine (see the margin).
2. Have alcohols accumulated in the body?
   Measuring the osmolal gap in plasma will indicate whether or not many uncharged particles are present in plasma.
3. In a patient with metabolic acidosis, what quantity of NH_4^+ is being excreted?
   To determine if the disease renal tubular acidosis (*distal RTA*) is present, two indirect tests are performed in tandem: the urine net charge is used to estimate NH_4^+ excreted with Cl^-, and the urine osmolal gap is used to detect the excretion of NH_4^+ with all anions.
4. Is H^+ secretion by the distal nephron reduced?
   To determine the basis of the lesion in a patient who has a low rate of excretion of NH_4^+, measure the Pco_2 in alkaline urine.

THE ANION GAP IN PLASMA

- The anion gap ($[Na^+] - [Cl^-] - [HCO_3^-]$), a calculation for diagnostic convenience:
  - normally equals 12 ± 2 mEq/l;
  - indicates the quantity of added acids (the fall in $[HCO_3^-]$ equals the rise in anion gap);
  - is useful in following the patient's response to therapy;
  - helps in detecting laboratory errors, cationic proteins, and mixed acid-base disorders.

Figure 2·2
The anion gap in plasma
(a) In this portion of the figure, the concentrations of ions in plasma are depicted. The difference between the $[Na^+]$ and the sum of $[Cl^-]$ and $[HCO_3^-]$ is the anion gap, shown as the shaded area between the columns (A^-).

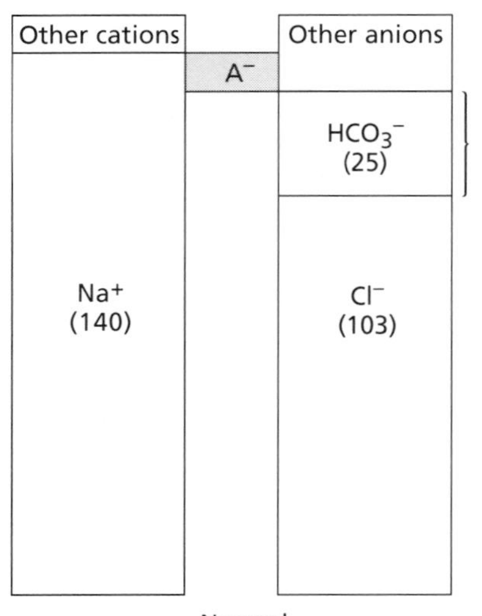

In every solution, the number of positive charges on cations must equal the number of negative charges on anions. If measurements indicate an imbalance, either the measurements are wrong, or not all the ionized materials are identified.

Although physiological fluids may contain many anions and cations, only a few are present in significant amounts, and the concentrations of others usually undergo only minor changes. Hence, by measuring the concentrations of only a few cations and anions, and by making assumptions about the other ions normally found in plasma, it is possible to obtain a rough measure of whether large amounts of unsuspected anions or cations are present. The term "plasma anion gap" is used to signify the difference between the $[Na^+]$ and the sum of $[Cl^-] + [HCO_3^-]$. This shortfall in the number of anions is due to the fact that a significant number of the anions are unmeasured (e.g., albumin; note the quantity of unspecified anions in

Table 1·2, page 6). The concentrations of the cations K^+, Ca^{2+}, and Mg^{2+} are usually almost constant (minor variations are life-threatening). The concentration of albumin, the only quantitatively important protein that is routinely present, is usually assumed to be constant (although some clinical situations, such as cirrhosis of the liver or nephrotic syndrome, lead to significant changes). The concentrations of other unmeasured anions—primarily SO_4^{2-} and HPO_4^{2-}—do not change appreciably except in renal insufficiency. Hence, subtraction of the measured values for $[Cl^-]$ and $[HCO_3^-]$ from that for $[Na^+]$ will normally yield a value of 12 ± 2 mEq/l, the normal anion gap. Values for the plasma anion gap that are significantly larger than normal indicate the presence of and approximate total concentration of one or more abnormal unmeasured anions (Figure 2·2).

Figure 2·2, continued

(b) When an acid such as lactic acid is added, the $[HCO_3^-]$ will fall, and the HCO_3^- will be replaced with an anion such as lactate anion (L^-). The bracketed area represents the normal $[HCO_3^-]$.

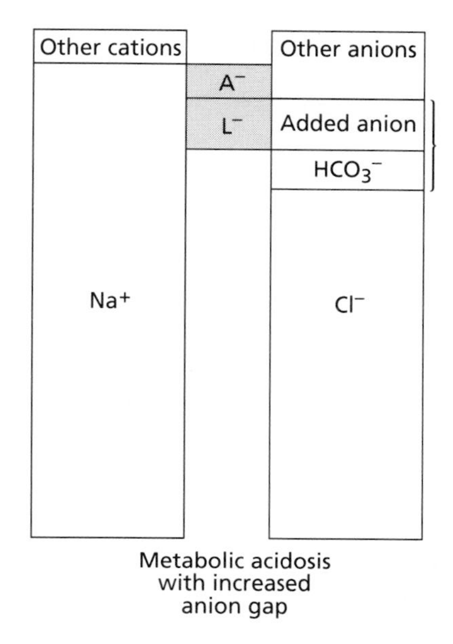

An Example

When lactic acid dissociates into H^+ and lactate anions in the ECF, H^+ are buffered by HCO_3^-, leaving lactate anions as the "footprint" of the lactic acid added to the ECF. These events are depicted below, with values before and after the addition of 10 mmol of lactic acid (HL) to each liter of ECF.

Plasma (mEq/l)	$[Na^+]$	$[Cl^-]$	$[HCO_3^-]$	Anion gap
Normal	140	103	25	12
+ 10 mmol/l lactic acid	140	103	15 = 25–10	22 = 12 + 10

The patient begins with a normal plasma anion gap of 12 mEq/l (140 - [103 + 25]). The protons accompanying 10 mmol of lactate anion per liter of ECF react with HCO_3^-; the result is a decline in plasma $[HCO_3^-]$ to 15 mmol/l and an increase in the plasma anion gap to 22 mmol/l. This value, best thought of as the increment over normal (i.e., 12 + 10), forces you to think of the normal plasma anion gap and to ask whether the patient has any reason not to have a normal anion gap (see the discussions of Questions 2·2 and 2·3). Thus, the increase in the plasma anion gap reflects, in a semi-quantitative fashion, the presence of the anion of the organic acid (in this case, 10 mmol of lactate per liter of ECF).

If, in the above example, the increase in the anion gap were the same 10 mEq/l, but the concentration of lactate in plasma were only 5 mmol/l, lactic acidosis would not be the sole cause of the metabolic acidosis; the patient must have also accumulated other acids and their unmeasured anions (e.g., ketoacid anions).

(c) Note that with a loss of $NaHCO_3$, the $[HCO_3^-]$ will fall, but no new anions will be added.

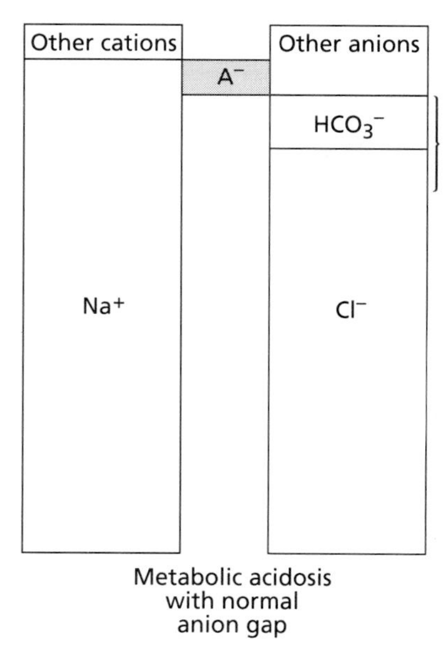

Questions

(Discussions on page 61)

2·2 If the concentration of albumin in plasma is half of normal, what adjustments should be made when interpreting the plasma anion gap?

2·3 Patients with multiple myeloma may have a protein in plasma that bears a net positive charge. What is the impact of this protein upon the value of the plasma anion gap?

2·4 Are there other reasons for having a low value for the plasma anion gap in patients with multiple myeloma?

THE OSMOLAL GAP IN PLASMA

Rule:
Doubling the $[Na^+]$ closely approximates the number of ions in plasma.

- The osmolal gap in plasma is used to reveal alcohols in blood.
- This gap is the measured osmolality minus the calculated osmolality ($2[Na^+]$ + [Glucose] (mmol/l) + [Urea] (mmol/l)).

Clinical note:
If there is a laboratory error with the $[Na^+]$ (i.e., hyperlipidemia), the formula for the osmolal gap will have an error introduced.

The osmolal gap is a useful means of detecting the presence of uncharged molecules in the plasma. The osmotic pressure of a solution is determined by the concentration of dissolved particles in the water; a protein molecule, a molecule of glucose, and a Na^+ make virtually equal contributions to the osmotic pressure. Because the numbers of anions and cations must be equal in any solution, and the concentrations of K^+, Ca^{2+}, and Mg^{2+} in plasma are low and approximately constant, the value of $2 \times [Na^+]$ will account for the osmotic pressure of normal anions plus cations in plasma. Glucose and urea are the two major non-ionized molecules in plasma that are likely to change in concentration. Hence, the calculated osmolality is $2 \times [Na^+]$ + [glucose] + [urea] (all measurements in mmol/l; see Table 2·3 for conversion of concentrations in mg/dl to mmol/l). The difference between the measured and the calculated osmolality is the osmolal gap. A high plasma osmolal gap indicates the presence of an unmeasured compound; since the unmeasured compound is not charged, it is probably an alcohol (see the margin for more information).

Notes on Table 2·3:
To convert mg/dl to mmol/l, multiply mg/dl by 10, then divide by molecular weight.

In certain circumstances, molality is more useful (e.g., urine osmolality), but in others, knowing the weight is valuable (e.g., what weight of protein was oxidized).

Table 2·3
CONVERSION BETWEEN mg/dl AND mmol/l

Constituent	Molecular weight	Sample concentrations	
		mg/dl	mmol/l
Glucose	180	90	5
Urea	60	30	5
Urea nitrogen (2 × 14)	28	14	5

THE URINE NET CHARGE

- The urine net charge is usually used to detect NH_4^+ excreted with Cl^- in the urine.
- Sometimes the urine net charge reveals the excretion of unusual anions.

Since most hospital biochemistry laboratories do not routinely measure the $[NH_4^+]$ in urine, it is useful to have an indirect way to estimate this concentration (Figure 2·3). In normal urine, the major cations are Na^+, K^+, and NH_4^+, and the major anions are Cl^- and HCO_3^-. The principle is that NH_4^+ are usually excreted along with Cl^-. Therefore, if NH_4^+ are plentiful in that urine, there will be a much greater quantity of Cl^- than the measured cations Na^+ plus K^+, so that "electrical" room will remain for NH_4^+. On the other hand, if the sum of Na^+ and K^+ is greater than Cl^-, there will be no "electrical" room for NH_4^+ unless there are large amounts of unmeasured anions in that urine.

$[Cl^-] > [Na^+] + [K^+]$ = High $[NH_4^+]$
$[Cl^-] < [Na^+] + [K^+]$ = Either a low $[NH_4^+]$ or excretion of NH_4^+ with an anion other than Cl^-

Assumptions Required to Interpret the Urine Net Charge

Urine net charge:
Measures NH_4^+ excreted with Cl^-.

1. No cations other than Na^+, K^+, and NH_4^+ are quantitatively important.
   - This statement is true unless intake is very unusual.
2. There are no anions in the urine other than Cl^-.
   - This statement is not true, but the usual anions (phosphate, sulfate, organic anions) do not vary appreciably in most cases.
   - Alkaline urine contains HCO_3^-, but we are not looking for NH_4^+ in alkaline urine. A urine pH that is less than 6 eliminates bicarbonaturia.
   - In some cases of metabolic acidosis (ketoacidosis, glue sniffing), the urine contains a large quantity of anions other than Cl^-. In these cases, the urine osmolal gap must be examined if you want a quick estimate of NH_4^+ excretion.

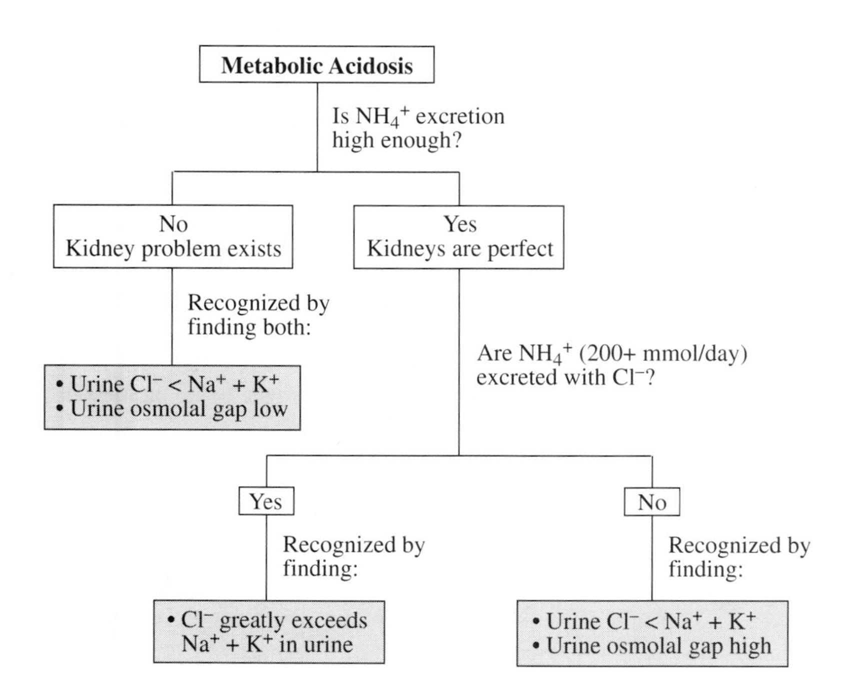

Figure 2·3
Sequence of steps in evaluating the rate of excretion of NH_4^+
These tests are performed in a patient with metabolic acidosis. The expected value for NH_4^+ in the urine is more than 200 mmol/day with normal kidneys and less than 40 mmol/day when the kidneys are the sole cause of the acidosis. Remember that these tests provide rough estimates, not precise values.

Estimate of 24-hour urine:

- Creatinine excretion is relatively constant over the 24-hour period.
- Normal individuals excrete 200 μmol (20 mg) of creatinine per kg body weight in 24 hours.
- Divide the $[NH_4^+]$ in urine by the concentration of creatinine in urine and multiply by the 24-hour value for creatinine (see the discussion of Question 2·5 for an example).

Precautions for the Urine Net Charge

1. One obtains only a rough estimate of the $[NH_4^+]$ in the urine (see the margin).
2. The volume of urine must be known to estimate NH_4^+ excretion. Since most adults excrete 80 mEq of anions other than Cl^- per day, if the urine volume is close to 1 liter in a 24-hour period, the following formula applies:

$$NH_4^+ \text{ (mEq/day)} = 80 + (Cl^- - Na^+ - K^+), \text{ all in mEq/day}$$

THE OSMOLAL GAP IN URINE

- The osmolal gap in the urine is used to detect NH_4^+.

In a patient with chronic metabolic acidosis and normal renal function, one expects to find more than 200 mmol of NH_4^+ excreted each day. Under most circumstances where this excretion is high (e.g., loss of $NaHCO_3$ in diarrhea fluid), the NH_4^+ are excreted with Cl^- and the urine net charge is very negative. Now consider a second circumstance where there is overproduction of acids, but the anion produced with H^+ is not retained because it is excreted in large quantities in the urine (see the discussion of Case 3·2, pages 127 and 130–31). For instance, the following values (all in mEq/l) might be measured:

$$[Na^+] = 50, [K^+] = 50, [Cl^-] = 25, [NH_4^+] = 200, A^- = 275$$

Note that the sum of the concentrations of Na^+ plus K^+ exceeds that of Cl^-, yet the $[NH_4^+]$ is high. In this circumstance, the urine net charge does not reflect NH_4^+ excretion; a calculation of the osmolal gap in the urine, however, would provide a more accurate estimate of the $[NH_4^+]$ in the urine. More data are required, though—namely, the concentrations of glucose and urea, the major organic molecules that might be present in the urine.

$$[\text{Glucose}] = 0 \text{ mmol/l}, [\text{Urea}] = 250 \text{ mmol/l}, \text{Osmolality} = 850 \text{ mosm/kg } H_2O$$

In this example, the measured urine osmolality (850) exceeds the calculated osmolality—urea (250) + glucose (0) + 2 × ($[Na^+]$ + $[K^+]$) (200)—by 400 mosmoles. Because the osmolal gap contains NH_4^+ and equal number of anions, the quantity of NH_4^+ (if accompanied by a monovalent anion) would be half the difference of measured and calculated osmolalities, or 200 mmol/l (see the margin).

Notes:

The formula to use to calculate $[NH_4^+]$ in this case is 0.5 (measured minus calculated osmolality) where the calculated osmolality = [Urea] + [Glucose] + 2($[Na^+]$ + $[K^+]$), all in mmol/l (for mg/dl values, see Table 2·3).

One other point is obvious: in this case, the number of NH_4^+ (plus Na^+ and K^+) greatly exceeds the $[Cl^-]$, so there are many unmeasured anions in the urine and the sum of their concentrations is close to 275 mEq/l. The unmeasured anions one might generally encounter are ketoacid anions, drug metabolites, or hippurate (in toluene "intoxication").

THE URINE pH

If the $[H^+]$ or the $[NH_3]$ is high, the $[NH_4^+]$ will rise. Therefore, with a separate measure of two of these parameters (NH_4^+, H^+), you can deduce the third (NH_3). This calculation will become important when examining the urine in patients who have a low excretion of NH_4^+ during metabolic acidosis.

Questions

(Discussions on page 62)

2·5 How can one use the urine creatinine concentration to estimate the rate of excretion of NH_4^+?

$$NH_3 + H^+ \longleftrightarrow NH_4^+$$

2·6 A patient has a bladder infection with bacteria that release the enzyme urease. The enzyme urease catalyzes the following reaction:

$$\text{Urea} \longrightarrow 2\ NH_4^+ + 2\ HCO_3^-$$

Which test would you select to determine how many of the NH_4^+ excreted were of renal origin: (a) direct assay of NH_4^+ in urine; (b) urine osmolal gap; (c) urine net charge?

THE URINE Pco_2, A TEST REFLECTING SECRETION OF H^+ BY COLLECTING DUCTS

- Once you know that NH_4^+ excretion is low, the urine Pco_2 can help you decide if the cause is low distal H^+ secretion.

Another test used to determine the probable cause of a low excretion of NH_4^+ is the urine Pco_2. In a very alkaline urine, the secretion of H^+ by the collecting duct leads to the formation of H_2CO_3. Since there is no luminal carbonic anhydrase here, H_2CO_3 is slowly dehydrated to CO_2 in the medulla; the result is an increased Pco_2 in the renal medulla and bladder. Patients with a defect in secretion of H^+ have a urine Pco_2 close to that of their blood (Figure 2·4).

Notes on the urine Pco_2 test:
The urine is collected in a bottle with a small surface area relative to volume to minimize CO_2 loss. No oil is necessary if the Pco_2 in the urine is measured promptly. The sample is aspirated from the bottom of the bottle into a sealed syringe. The pH and Pco_2 are measured anaerobically.

Because this test requires the administration of sufficient $NaHCO_3$ to render the urine frankly alkaline, you can rule out proximal RTA (a disorder in which the urine pH rises long before the $[HCO_3^-]$ in plasma becomes normal).

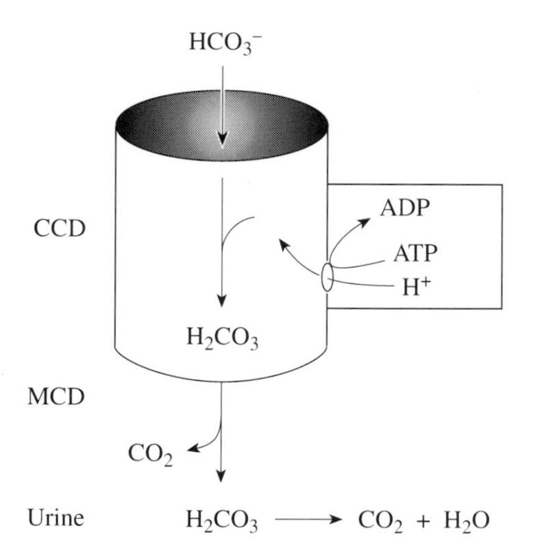

Figure 2·4
The basis of an increased Pco_2 in alkaline urine
Because the urine is alkaline, virtually the only H^+ acceptor is HCO_3^-. Any H^+ secretion in the cortical collecting duct (CCD) leads to the formation of H_2CO_3. Since there is no carbonic anhydrase present in the lumen of this nephron segment, the H_2CO_3 dehydrates slowly in the medullary collecting duct (MCD) and the urine; the result is an increase in the urine Pco_2. If H^+ secretion in the CCD and MCD is absent, the urine Pco_2 will be close to that of the blood.

The test is valid only when the urine pH exceeds 7.40 or the urine [HCO_3^-] is greater than 50 mmol/l. These levels can usually be achieved by giving an oral load of $NaHCO_3$ (0.5 to 2 mmol/kg body weight) on the morning of the test. The patient's K^+ deficit should be corrected prior to giving HCO_3^-.

The patient with normal secretion of H^+ from collecting ducts should have a urine Pco_2 that exceeds 70 mm Hg. This test also enables the identification of patients with a gradient limit to H^+ secretion (back-diffusion of H^+, e.g., resulting from the use of amphotericin B); these patients may have a low urine NH_4^+ excretion but have a normal urine Pco_2.

Question

(Discussion pages 62–63)

2·7 In the following examples of urine excreted by a patient with chronic metabolic acidosis, what is the rate of excretion of NH_4^+, and what might the urine Pco_2 be?

Osmolality mosm/kg H_2O	Na^+ mmol/l	K^+ mmol/l	Glucose mmol/l	Urea mmol/l
600	50	50	0	100
600	50	50	0	350

PART B
IDENTIFYING ACID-BASE DISORDERS

Expected Responses to Primary Acid-Base Disorders

It has been empirically observed that when a patient has one of the four primary acid-base disturbances, a predictable response occurs to return the $[H^+]$ toward normal (Table 2·4). Only in chronic respiratory alkalosis may the $[H^+]$ actually return to the normal range as a result of the expected response.

Mixed acid-base disorders:
When faced with an acid-base disorder, one should ensure that the expected response has occurred; its absence indicates the presence of a second primary acid-base disturbance.

Table 2·4
EXPECTED RESPONSES TO PRIMARY ACID-BASE DISORDERS

Disorder	Response
Metabolic Acidosis	• For every mmol/l fall in plasma $[HCO_3^-]$ from 25, the P_aco_2 should fall by 1 mm Hg from 40, or use the "*Rule of Thumb #2.*"
Metabolic Alkalosis	• For every mmol/l rise in plasma $[HCO_3^-]$ from 25, the P_aco_2 should rise by 0.7 mm Hg from 40, or use the "Rule of Thumb #2."
Respiratory Acidosis	
• Acute	• For every mm Hg rise in P_aco_2 from 40, the plasma $[H^+]$ should rise by 0.8 nmol/l from 40. • Alternatively, for every twofold increase in P_aco_2, the plasma $[HCO_3^-]$ should increase by 2.5 mmol/l.
• Chronic	• For every mm Hg rise in P_aco_2 from 40, the $[H^+]$ should rise by 0.3 nmol/l, or the plasma $[HCO_3^-]$ should rise by 0.3 mmol/l from 25.
Respiratory Alkalosis	
• Acute	• For every mm Hg fall in P_aco_2 from 40, the plasma $[H^+]$ should fall by 0.8 nmol/l from 40.
• Chronic	• For every mm Hg fall in P_aco_2 from 40, the plasma $[H^+]$ should fall by 0.2 nmol/l, or the plasma $[HCO_3^-]$ should fall by 0.5 mmol/l from 25.

Rule of Thumb #2:
This rule helps in identifying if the expected change of ventilation is present.

- Drop the 7 and the decimal point.
- The number remaining is the expected P_aco_2 in patients with metabolic acidosis or metabolic alkalosis (e.g., if the plasma pH is 7.30, the expected P_aco_2 is 30 mm Hg).

How to Recognize Mixed Acid-Base Disorders

1. **Examine the Paco2 in metabolic acidosis and alkalosis to identify the presence of a respiratory acid-base disturbance.**
   If the P_aco_2 is much higher than expected (refer to Table 2·4), there is a coexistent respiratory acidosis. If the fall of the P_aco_2 is much lower, respiratory alkalosis is also present.

2. **Compare the increase in anion gap with the fall in [HCO3–] in plasma.**
   In metabolic acidosis of the increased anion gap type, the fall in $[HCO_3^-]$ is generally equal to the rise in the plasma anion gap. If the rise in the plasma anion gap substantially exceeds the fall in $[HCO_3^-]$, there must be an additional source of HCO_3^- (e.g., coexistent metabolic

Note:
Be sure to rule out other factors, such as a low concentration of albumin, that lead to a low value for the plasma anion gap (page 50).

alkalosis). On the other hand, if the fall in [HCO_3^-] markedly exceeds the increase in plasma anion gap, there must be two processes contributing to the fall in [HCO_3^-]: a normal anion gap type and an increased anion gap type of metabolic acidosis.

3. **Integrate the clinical picture and lab tests with the acid-base disturbances.**
4. **Assess the metabolic adjustment in respiratory acid-base disorders, differentiating between acute and chronic disorders on clinical grounds.**
   (a) With all the respiratory disturbances, if the plasma [HCO_3^-] is unexpectedly high, there is a coexistent metabolic alkalosis; if the plasma [HCO_3^-] is lower than expected, there is also a metabolic acidosis.
   (b) In acute respiratory acidosis or alkalosis, there should be only a slight change in [HCO_3^-].
   (c) In chronic respiratory acidosis or alkalosis, the slope of [H^+] vs P_aco_2 is much flatter because of a change in the [HCO_3^-] in plasma (see Table 2·4 for expected changes).

Strong ion difference:
Every decade or so, a new way to analyze acid-base disorders is proposed. The impetus for these approaches is that the traditional measures described in this chapter do not permit a total explanation for all the pathophysiology (these newer ways also do not provide a more comprehensive analysis). The current vogue in some centers is to use the strong ion difference (SID) proposed by Stewart. It is based on the sound chemical principles of electroneutrality, dissociation, and reliance on independent variables, such as the SID and the Pco_2. Chemically, its unfortunate but necessary weakness is that it assigns an arbitrary and fixed valence to all proteins. The bottom line in our opinion is that it offers no major advantage, and it has some disadvantages; radical changes in the clinical approach described in this chapter are therefore not indicated. We have stressed that the laboratory values interpreted by any approach are just one step in an acid-base analysis. To make rational decisions, one must correlate these values with the clinical picture and a knowledge of the underlying biochemistry and physiology.

Question

(Discussion page 63)

2·8 Consider the following values in a patient who complains of diarrhea and vomiting.

Na^+	140	mmol/l	H^+	40	nmol/l
K^+	2.3	mmol/l	P_aco_2	40	mm Hg
Cl^-	103	mmol/l	Anion gap	12	mEq/l
HCO_3^-	25	mmol/l			

What are the acid-base disorders?

Guidelines for the Diagnosis of Mixed Disorders

1. Clearly, a correct analysis of the laboratory results presumes that the data are accurate. There are two ways to detect laboratory errors. The first is to calculate the plasma anion gap. If it is very low or negative, there is probably an error in one of the electrolyte values, unless the patient has multiple myeloma or hypoalbuminemia. The second way to evaluate the laboratory results is to insert the [H^+], Pco_2, and [HCO_3^-] (as reflected in the plasma electrolyte values) into the Henderson equation. If there is a substantial error (> 10%), one of the three parameters is incorrect; if the discrepancy is great enough to change the diagnosis, the tests should be repeated and the error identified.

2. Calculate the plasma anion gap; if it is increased by more than 5 mEq/l from the expected value, the patient probably has metabolic acidosis.

3. Compare the magnitude of the fall in plasma [HCO_3^-] with the increase in plasma anion gap. These changes should be similar in magnitude. If the change in [HCO_3^-] differs from the change in the plasma anion gap by more than 5 mEq/l, a mixed disturbance is present. A rise in the plasma anion gap that is less than the fall in plasma [HCO_3^-] suggests that a component of the metabolic acidosis involves loss of $NaHCO_3$ or renal tubular acidosis. On the other hand, an increase in the plasma anion gap that is much greater than the fall in [HCO_3^-] suggests that there is a coexistent metabolic alkalosis (i.e., an additional source of HCO_3^-).

Delta [HCO_3^-]/Delta anion gap: Some people rely on the ratio of these changes when defining mixed disorders, but we do not think this practice adds anything of value. Instead, we prefer to consider the difference in the magnitude of these changes and, even then, not to rely too heavily on the 1:1 relationship.

4. In metabolic acidosis or alkalosis, look for the expected change in P_aco_2. If the P_aco_2 is significantly lower, the patient has coexistent respiratory alkalosis; if higher, respiratory acidosis.

5. In respiratory acid-base disturbances, one must distinguish between acute and chronic (more than 3–4 days) conditions on clinical grounds, since different degrees of physiologic compensation are present. Table 2·4 summarizes the expected changes in [HCO_3^-] for a given acute or chronic change in P_aco_2.

Questions

(Discussions on pages 63–65)

2·9 A 23-year-old woman with rheumatoid arthritis increased her dose of salicylates because of a flare-up. She then developed epigastric pain and vomited frequently for two days. She went to the local hospital where the following blood results were obtained:

H^+	20 nmol/l (pH = 7.70)	P_aco_2	25 mm Hg
Anion gap	17 mEq/l		

What is (are) her acid-base disorder(s)?

2·10 A 50-year-old woman underwent intestinal bypass for morbid obesity. Because she was having 10–15 watery stools per day, she was treated with tincture of opium and was found somnolent and somewhat hypotensive the next morning. Plasma values were:

Na^+	130 mmol/l	H^+	96 nmol/l (pH = 7.02)
K^+	3.2 mmol/l	P_aco_2	40 mm Hg
Cl^-	102 mmol/l	HCO_3^-	10 mmol/l
Albumin	40 g/l		

What is (are) her acid-base disorder(s)?
What treatment would you consider for her very high [H^+] in plasma?

PART C REVIEW

Discussion of Introductory Case Lee's Acidosis Is Mixed Up

(Case presented on page 45)

Increase in anion gap:
$[Na^+] - [Cl^-] - [HCO_3^-] - 12$
(The expected value for the normal anion gap is 12 because the concentration of albumin is normal.)

What is the acid-base diagnosis?

The parameters necessary to make an acid-base diagnosis are the $[H^+]$ or pH, $[HCO_3^-]$, P_aco_2, and the anion gap ($[Na^+] - [Cl^-] - [HCO_3^-]$) in plasma. Given the high $[H^+]$ and low $[HCO_3^-]$, the major acid-base diagnosis is metabolic acidosis with an increased anion gap. Since the *increase in anion gap* is 15 mEq/l and the decrease in $[HCO_3^-]$ from its expected value of 25 mmol/l is also 15 mEq/l, the basis of the metabolic acidosis is added acids, most likely diabetic ketoacidosis.

Is there a primary respiratory acid-base disorder?

Since the fall in P_aco_2 (15 mm Hg from 40), equals the fall in $[HCO_3^-]$ (15 mmol/l from the normal value of 25), there is no primary respiratory acid-base disorder.

Summary of Main Points

- To identify acid-base disturbances, evaluate the $[H^+]$, $[HCO_3^-]$, P_aco_2, and the anion gap and correlate them with the clinical picture. The reliability of the data should be confirmed using the Henderson equation.
- Each primary acid-base disorder is characterized by a specific physiological response, which should be evaluated. Absence of the expected response signals the coexistence of a second acid-base disorder.
- In metabolic acidosis, additional clues concerning the basis of the acidosis can be obtained from the plasma anion gap, the urine net charge, the plasma and urine osmolal gaps, and the urine Pco_2.
- In all acid-base disturbances, diagnosis of the presence of the acid-base disorder is only the first step; one must then establish the underlying basis of the specific acid-base disorder and then draw up a plan of management.

Discussion of Questions

2·1 A patient has diabetic ketoacidosis and the following lab data: pH = 7.10, P_aco_2 = 30 mm Hg, $[HCO_3^-]$ = 13 mmol/l, anion gap = 25 mEq/l. What do you conclude?

When the numbers are inserted in the Henderson equation, it appears that at least one of the three parameters is in error (see the margin). When faced with these laboratory results, be cautious; repeat the blood tests to clarify the basis of the discrepancy.

$[H^+] = 24 \times Pco_2/[HCO_3^-]$
$80 \neq 24 \times 30/13$
$80 \neq 56$

Whatever the scenario, the patient has metabolic acidosis with an increased plasma anion gap and presumably diabetic ketoacidosis. This diagnosis should be confirmed with the appropriate tests (serum ketones, blood glucose) and initial treatment should be instituted.

2·2 If the concentration of albumin in plasma is half of normal, what adjustments should be made when interpreting the plasma anion gap?

When correcting for a change in the concentration of albumin, use a value of 16 for the normal plasma anion gap to include all the major positive charges in solution (K^+ of 4 mmol/l). Therefore, if the concentration of albumin is 20 g/l, or 2 g/dl (half the normal concentration), the expected value for the plasma anion gap should be reduced to 8 mEq/l.

2·3 Patients with multiple myeloma may have a protein in plasma that bears a net positive charge. What is the impact of this protein upon the value of the plasma anion gap?

With a lysine-rich or arginine-rich protein in plasma (IgG myeloma), this paraprotein carries a net positive charge, and, if high enough, it can actually render the value for the plasma anion gap negative because these "unmeasured" positive charges are associated with "measured" Cl^-.

2·4 Are there other reasons for having a low value for the plasma anion gap in patients with multiple myeloma?

Although hypercalcemia may be seen with multiple myeloma, hypercalcemia generally is not severe enough to have a significant impact on the plasma anion gap. Patients may have an unexpectedly low plasma anion gap because of hypoalbuminemia, quirks of the laboratory methods, or actual laboratory errors. The plasma $[Cl^-]$ will be overestimated in patients with halide intoxication (bromide or iodide may elevate the reported value for Cl^- depending on the method used to measure Cl^-) and hyperlipidemia (techniques for measurement depend on turbidity). Similarly, a simple error in the measurement of $[Na^+]$, $[Cl^-]$, or $[HCO_3^-]$ will make the calculation of the plasma anion gap invalid.

2·5 How can one use the urine creatinine concentration to estimate the rate of excretion of NH_4^+?

Creatinine is excreted at a relatively constant rate throughout the day; its rate of excretion in an adult male is close to 0.2 mmol/day/kg body weight (20 mg/day/kg body weight). Thus, estimating the proportion of the 24-hour urine contained in a sample is possible if the concentration of creatinine and the urine volume are known. In fact, the rate of excretion of NH_4^+ can be calculated without knowing the urine volume.

- $[NH_4^+]$ = 40 mmol/l, [Creatinine] = 10 mmol/l
- $[NH_4^+]$/[Creatinine] = 4 mmol/mmol
- A 70-kg person excretes 14 mmol of creatinine per day.

Therefore, NH_4^+ excretion is 56 mmol/day if there is no diurnal variation in NH_4^+ excretion (see the margin).

Calculation:
Multiplying 4 mmol of NH_4^+ per millimole of creatinine by 14 mmol of creatinine per day equals 56 mmol of NH_4^+ per day.

2·6 A patient has a bladder infection with bacteria that release the enzyme urease. The enzyme urease catalyzes the following reaction:

$$\textbf{Urea} \longrightarrow 2\,NH_4^+ + 2\,HCO_3^-$$

Which test would you select to best reflect how much of the NH_4^+ excreted was of renal origin: (a) direct assay of NH_4^+ in urine; (b) urine osmolal gap; (c) urine net charge?

The urine contains NH_4^+ from two sources, the kidney and urea via urease.

(a) Total NH_4^+ assay cannot determine how much was from each source so it is not the best test to use.

(b) Urine osmolal gap just estimates total NH_4^+, so it is not the best test to use.

(c) Urine net charge does not reveal NH_4^+ if its salt is HCO_3^-. Therefore, if renal NH_4^+ were excreted with Cl^-, the urine net charge would provide the only real estimate of the renal component of NH_4^+ excretion (see the margin).

Note:
Humans do not excrete NH_4^+ with HCO_3^-, but alligators do. This excretion provides a way for alligators to excrete more water at the same urine osmolality (alligators do not have a loop of Henle, and although urea has one particle, four particles are excreted when these ions are excreted as 2 NH_4^+ and 2 HCO_3^-).

2·7 In the following examples of urine excreted by a patient with chronic metabolic acidosis, what is the rate of excretion of NH_4^+, and what might the urine Pco_2 be?

Osmolality mosm/kg H_2O	Na^+ mmol/l	K^+ mmol/l	Glucose mmol/l	Urea mmol/l
600	50	50	0	100
600	50	50	0	350

To obtain the calculated urine osmolality, double ($[Na^+]$ + $[K^+]$) and add the concentration of urea. The $[NH_4^+]$ is half the difference of the measured and calculated urine osmolality. In the first example, the urine osmolal gap suggests that the $[NH_4^+]$ is high (150 mmol/l). Since a high distal H^+ secretion is needed to have a high rate of excretion of NH_4^+, the urine Pco_2 should be high in alkaline urine (following a $NaHCO_3$ load).

Conversely, in the second example, the estimated concentration of NH_4^+ is low (25 mmol/l). In this case, a low Pco_2 in alkaline urine would suggest a defect in H^+ secretion, and a high Pco_2 in alkaline urine would suggest a very low $[NH_3]$ in the medulla (either impaired ammoniagenesis or impaired function of the loop of Henle).

2·8 Consider the following values in a patient who complains of diarrhea and vomiting.

Na^+	140	mmol/l	H^+	40	nmol/l
K^+	2.3	mmol/l	P_aco_2	40	mm Hg
Cl^-	103	mmol/l	Anion gap	12	mEq/l
HCO_3^-	25	mmol/l			

What are the acid-base disorders?

On the surface, most of the laboratory data do not suggest an acid-base disturbance, but the hypokalemia must be explained. It is most likely that the patient has a mixed metabolic acidosis (HCO_3^- loss due to diarrhea) and metabolic alkalosis (HCO_3^- gain due to vomiting) with the hypokalemia associated with excessive renal loss of K^+. The acid-base status, which results from the combination of these two disorders, is normal when judged solely by the four parameters in plasma (pH, P_aco_2, $[HCO_3^-]$, and anion gap). The physical exam was not normal; the ECF volume was contracted. In therapy, the ECF should be reexpanded with an isotonic solution containing K^+ (40–60 mmol/l). Half-normal saline plus KCl would be a good starting solution.

2·9 A 23-year-old woman with rheumatoid arthritis increased her dose of salicylates because of a flare-up. She then developed epigastric pain and vomited frequently for two days. She went to the local hospital where the following blood results were obtained:

H^+	20 nmol/l (pH = 7.70)	P_aco_2	25 mm Hg
Anion gap	17 mEq/l		

What is (are) her acid-base disorder(s)?

The pH of her blood is very alkalemic ($[H^+]$ = 20 nmol/l, pH = 7.70). Her $[HCO_3^-]$ can be calculated from the Henderson equation:

$$[HCO_3^-] = 24 \times P_aco_2/[H^+] = 30 \text{ mmol/l}$$

Thus, she has metabolic alkalosis (low $[H^+]$ and elevated plasma $[HCO_3^-]$). However, her P_aco_2 is low (the expected response during metabolic alkalosis is hypoventilation to return the $[H^+]$ in cells toward normal), and she is hyperventilating; therefore, she has a second primary acid-base disorder—respiratory alkalosis—which is why she is so alkalemic. Her metabolic alkalosis was secondary to vomiting and HCl loss (see Figure 4·1, page 150), and her respiratory alkalosis was secondary to salicylate intoxication. In addition, the small rise in her plasma anion gap suggests a third acid-base disorder, metabolic acidosis resulting from added acids (salicylic acid and the more negative valence of albumin, most likely).

2·10 A 50-year-old woman underwent intestinal bypass for morbid obesity. Because she was having 10–15 watery stools per day, she was treated with tincture of opium and was found somnolent and somewhat hypotensive the next morning. Plasma values were:

Na^+	130	mmol/l	H^+	96	nmol/l (pH = 7.02)
K^+	3.2	mmol/l	P_aco_2	40	mm Hg
Cl^-	102	mmol/l	HCO_3^-	10	mmol/l
Albumin	40	g/l			

What is (are) her acid-base disorder(s)?

The patient is very acidemic ($[H^+]$ = 96 nmol/l, pH = 7.02). The $[HCO_3^-]$ is low (10 mmol/l); therefore, she has metabolic acidosis. The plasma anion gap (18 mEq/l) is increased by about 6 mEq/l because her albumin level is normal; however, the $[HCO_3^-]$ has fallen by 15 mmol/l (from 25 mmol/l). Thus, the fall in $[HCO_3^-]$ exceeds the increase in the plasma anion gap and indicates two components to the metabolic acidosis: part is due to the accumulation of an organic acid, as reflected by the increase in the plasma anion gap (presumably D-lactic acidosis, L-lactic acidosis, or ketoacidosis), and part is due to $NaHCO_3$ loss in the diarrhea.

The patient's P_aco_2 is 40 mm Hg, which is higher than expected during metabolic acidosis with a plasma $[HCO_3^-]$ of 10 mmol/l (the P_aco_2 should be 40 – 15 = 25 mm Hg). Thus, this patient also has a respiratory acidosis.

Therefore, this patient has three acid-base disturbances:

1. metabolic acidosis resulting from $NaHCO_3$ loss (diarrhea);
2. D-lactic acidosis (abnormal bowel flora and GI motility suppression), L-lactic acidosis (hypotension), or ketoacidosis (starvation);
3. respiratory acidosis (suppression of ventilation).

What treatment would you consider for her very high $[H^+]$ in plasma?

The treatment is determined by the underlying causes for the acid-base disturbances. Presumably the respiratory acidosis is due to the central nervous system suppression by the narcotic. Therefore, treatment with naloxone (a morphine antagonist) would be an appropriate first step. One could also stimulate the patient (verbally and physically) to breathe. If the patient does not respond to the naloxone, or if her condition deteriorates, mechanical ventilation will give the quickest control of the acidemia. Reducing her P_aco_2 to 25 mm Hg will lower her plasma $[H^+]$ to 24 × 25/10, or 60 nmol/l (pH = 7.22).

The patient is very acidemic and has lost some $NaHCO_3$; therefore, one could also give $NaHCO_3$ to alleviate the severe acidemia, but the danger of more severe hypokalemia makes this option unattractive.

Time is required to slow the rate of production of D-lactic acid. Do not give food (carbohydrate) by mouth because the bacteria may make more D-lactic acid and other toxic metabolites. Restoring the ECF volume and giving thiamine, if indicated, could alleviate L-lactic acidosis. Giving glucose will correct ketoacidosis if the patient is hypoglycemic; otherwise, reexpansion of the ECF volume should do the trick.

3

METABOLIC ACIDOSIS

PART C
SPECIFIC DISORDERS 85

PART D
REVIEW 125

Objectives

- To provide a diagnostic classification of *metabolic acidosis* based on:
  - acid accumulation;
  - HCO_3^- loss;
  - failure of the kidneys to generate new HCO_3^-.
- To explain the roles of hyperventilation in metabolic acidosis.
  - Hyperventilation not only defends the plasma $[H^+]$ but also reduces the binding of H^+ on ICF proteins.
- To emphasize the critical importance of disturbances in the plasma $[K^+]$ in the genesis of metabolic acidosis and in the response to therapy.
- To elucidate the pathogenesis of the various forms of lactic acidosis, ketoacidosis, and the acidosis associated with certain intoxications.
- To provide a therapeutic approach to metabolic acidosis based on physiologic priorities and to consider the controversy concerning therapy with $NaHCO_3$.

Metabolic acidosis:
A process that tends to lower the $[HCO_3^-]$ and to increase the $[H^+]$ in plasma. Metabolic acidosis is defined in terms of the bicarbonate buffer system (BBS) in plasma.

Abbreviations:
AcAc = acetoacetate.
ACE = angiotensin-converting enzyme.
AG = anion gap.
AKA = alcoholic ketoacidosis.
ASA = acetylsalicylic acid.
BBS = bicarbonate buffer system.
CA = carbonic anhydrase.
CAI = carbonic anhydrase inhibitors.
DKA = diabetic ketoacidosis.
ECF = extracellular fluid.
GFR = glomerular filtration rate.
β-HB$^-$ = β-hydroxybutyrate.
Hβ-HB = β-hydroxybutyric acid.
HSL = hormone sensitive lipase.
ICF = intracellular fluid.
IDDM = insulin-dependent diabetes mellitus.
NIDDM = noninsulin-dependent diabetes mellitus.
P_aco_2 = partial pressure of CO_2 in arterial blood.
Pco_2 = partial pressure of CO_2.
RTA = renal tubular acidosis.

Threats to life:
- Severe acidosis
- Poisonous products
- K^+ changes

Outline of Major Principles

1. Metabolic acidosis occurs with acid gain (other than H_2CO_3) or loss of HCO_3^- plus Na^+ and/or K^+. Both result in a rise in the $[H^+]$ and a fall in the $[HCO_3^-]$ in plasma.
2. Metabolic acidosis often occurs as a complication of catastrophic illness (shock, sepsis) and adds to the seriousness of the clinical setting.
3. Because metabolic acidosis is not a primary diagnosis, the underlying cause must be sought; specific therapy may be life-saving (e.g., insulin for diabetic ketoacidosis, ethanol for methanol intoxication).
4. The rate of H^+ input may be very high (hypoxia) or normal (renal failure). In the latter case, the plasma $[HCO_3^-]$ declines slowly but progressively; thus, on admission to the hospital, both types of acidemia might be equally severe.
5. The accumulation of new anions in the plasma or the urine indicates an overproduction of acids. The plasma anion gap provides a useful clue to determine the basis of the metabolic acidosis. It is elevated in most patients whose metabolic acidosis results from overproduction of acids.
6. The impact of metabolic acidosis may depend upon the quantity of H^+ bound to intracellular proteins. This quantity is minimized by a reduction in the tissue Pco_2. Normally, there is a predictable decline in P_aco_2 for a given degree of metabolic acidosis.

7. Profound derangements in plasma [K^+] may accompany either metabolic acidosis or its therapy. At times, the abnormal level of K^+ may pose a greater threat than the acidemia.
8. The therapeutic role of $NaHCO_3$ in metabolic acidosis varies; for example, its use is important when the degree of acidemia is very severe, but it may endanger patients who have a severe degree of hypokalemia along with metabolic acidosis. The decision whether or not to give $NaHCO_3$ must be individualized.

Introductory Case
To Make a Diagnosis, Step on the Gas

(Case discussed on pages 125–26)

The following results in plasma were obtained in Lee, a diabetic patient who presented with weakness.

Na^+	mmol/l	140	H^+	nmol/l	144	(pH 6.84)
K^+	mmol/l	1.8	P_aCO_2	mm Hg	30	
Cl^-	mmol/l	125	HCO_3^-	mmol/l	5	

What is (are) your acid-base diagnosis(es)?
What would the plasma [H^+] be if the respiratory response were appropriate?
What are the likely causes for the metabolic acidosis in Lee?
What is the significance of the hypokalemia?
Must all patients with metabolic acidosis have a high plasma [H^+]?
Can a patient have a low plasma [HCO_3^-] and not have metabolic acidosis?
Is it possible to have a persistently alkaline urine (i.e., no renal HCO_3^- generation) and maintain acid-base balance?

PART A BACKGROUND

Development of Metabolic Acidosis

The bicarbonate buffer system (BBS) equation shown in Figure 3·1 reveals two major ways that one can have a high [H^+] and a low [HCO_3^-] (metabolic acidosis). First, the addition of H^+ (acid) will consume HCO_3^- and drive this equilibrium to the right. Conversely, if there is a loss of HCO_3^- (i.e., when $NaHCO_3$ is excreted in the urine or is *lost indirectly*), the [H^+] will be increased along with the decline in the [HCO_3^-]. In this case, the BBS equilibrium is shifted to the left. Making this distinction between *acid accumulation* and loss of $NaHCO_3$ is the first step to take in the differential diagnosis of metabolic acidosis (Table 3·1).

Indirect loss of HCO_3^-:
When an acid is produced and dissociates, if its H^+ is titrated by HCO_3^- and its anion is excreted with a Na^+ or a K^+, an indirect loss of HCO_3^- has occurred (see Figure 3·3, page 73).

Acid accumulation:
Addition of acids to the body faster than they can be removed. H^+ are produced when anions without a cation such as Na^+ or K^+ are made from a neutral substance.

$$N^0 \longrightarrow A^- + H^+$$

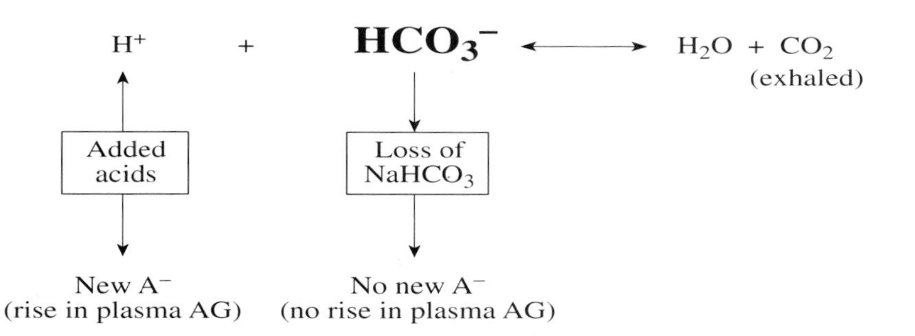

Figure 3·1
Basis of metabolic acidosis
Metabolic acidosis (a rise in the [H^+] and a fall in the [HCO_3^-]) is recognized by examining the BBS in plasma. When acids are added and H^+ are retained, "new anions" appear in the plasma or in the urine. Failure to find new anions suggests that there is a loss of $NaHCO_3$. (AG = anion gap.)

Table 3·1
OVERVIEW OF THE ETIOLOGY OF METABOLIC ACIDOSIS

Each disorder is discussed in more detail in later sections of this chapter.

Overproduction of acids

- Retention of anions in the plasma (increased plasma anion gap)
  - L-Lactic acidosis (L-lactic acid)
  - Ketoacidosis (largely *β-hydroxybutyric acid*)
  - Overproduction of organic acids in the GI tract (D-lactic acidosis)
  - Conversion of alcohols (methanol, ethylene glycol) to acids and poisonous aldehydes
- Excretion of anions in the urine (no increase in the plasma anion gap)
  - Ketoacidosis and impaired renal reabsorption of β-HB^-
  - Inhalation of toluene (hippurate)

Actual bicarbonate loss (normal plasma anion gap)

- Direct loss of $NaHCO_3$
  - Gastrointestinal tract (e.g., diarrhea, ileus, fistula or T-tube drainage, villous adenoma, ileal conduit combined with delivery of Cl^- from urine)
  - Urinary tract (e.g., proximal renal tubular acidosis, use of carbonic anhydrase inhibitors)
- Indirect loss of $NaHCO_3$
  - Failure of renal *generation of new bicarbonate* (low NH_4^+ excretion)
    - Low production of NH_4^+ (e.g., renal failure (low GFR), hyperkalemia)
    - Low transfer of NH_4^+ to the urine (e.g., medullary interstitial disease, low distal net H^+ secretion)

β-Hydroxybutyric acid:
One of the so-called ketoacids produced in the liver when levels of insulin are low. (β-HB^- = β-hydroxybutyrate anion.)

Generation of new bicarbonate:
The usual diet generates approximately 1 mmol of H^+ per kilogram of body weight. The normal kidney regenerates 1 mmol of "new" HCO_3^- per kilogram of body weight ; failure to regenerate this new HCO_3^- results in metabolic acidosis.

Use of the Anion Gap in Plasma to Detect the Net Addition of Acids

- In patients with a metabolic acidosis associated with an increase in the plasma anion gap, there are two possible reasons for the metabolic acidosis:
  1. overproduction of an organic acid;
  2. renal failure (low GFR).

OVERPRODUCTION OF ORGANIC ACIDS

Salicylate:
Acetylsalicylic acid (ASA), the active moiety of aspirin, causes harm to the body via direct toxicity to cells rather than injury from acidosis. ASA stimulates the respiratory center and disturbs a variety of metabolic processes.

Patients may produce organic acids as a result of the excess activity of a normal metabolic pathway. For example, there is an exceedingly rapid rate of L-lactic acid production during hypoxia and a modest rate of production of ketoacids during states with a relative deficiency of insulin (Table 3·2). In addition, the metabolism of an ingested substance (methyl alcohol, ethylene glycol) may lead to a moderate rate of acid production (Table 3·2). In *salicylate* intoxication, the usual problem is respiratory alkalosis, but overproduction of acids may be a problem, especially in children.

Table 3·2
RATES OF PRODUCTION AND REMOVAL OF H^+

The total quantity of H^+ that can be buffered is close to 1000 mmol in a 70-kg person. With very large acid loads, most of the buffering occurs in the ICF.

Production of H^+	Rate (mmol/min)	Comments
L-Lactic acid (Hypoxia)	72 7.2	• Rate reflects complete anoxia. • Rate reflects 10% hypoxia.
Ketoacids	1	• Production requires lack of insulin.
Toxic alcohols	< 1	• Poisonous metabolites rather than H^+ are usually the major threat.
Removal of H^+		
Kidney (by excretion of NH_4^+)	0 to 0.2	• Has a lag period. • Metabolic acidosis is needed for rapid rates of excretion.
Metabolism		
- L-Lactic acid	4 to 8	• Half by oxidation and half by glucogenesis.
- Ketoacids	0.8	• Oxidized primarily in the brain (⅔) and kidneys (⅓).

RENAL FAILURE

During renal failure, metabolic acidosis is usually accompanied by an increase in the plasma anion gap. This condition is the one exception to the rule that an increase in the plasma anion gap signals an overproduction of

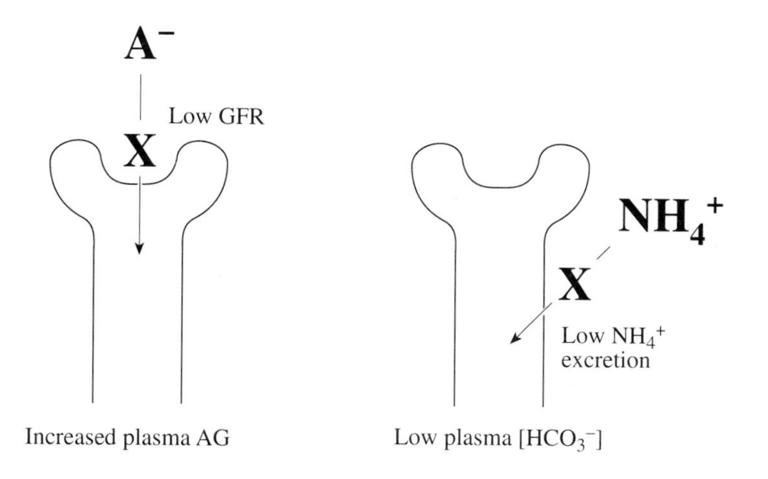

Figure 3·2
Basis of high plasma anion gap and acidosis in renal failure
The basis of the increased plasma anion gap (AG) is the low GFR with reduced excretion of anions such as phosphate or sulfate (left side). The acidosis is due to a low rate of excretion of NH_4^+ (right side).

acids. As illustrated in Figure 3·2, the cause of the rise in the plasma anion gap is a decrease in the GFR, but the cause of the acidosis is a tubular problem—a low rate of excretion of NH_4^+.

Note:
In renal failure, there is no obvious relationship between the plasma anion gap and the [HCO_3^-]. In some patients, the degree of rise in the plasma anion gap is much less than the fall in the plasma [HCO_3^-]; the converse is also true.

METABOLIC ACIDOSIS AND A NORMAL PLASMA ANION GAP

When metabolic acidosis is not associated with an increase in the plasma anion gap, it is due to a direct or indirect loss of $NaHCO_3$. Direct HCO_3^- loss occurs either via the gastrointestinal tract or in the urine. Indirect loss of HCO_3^- occurs as follows: an organic acid is produced and dissociates into an organic anion plus a H^+; when more organic anions than H^+ or NH_4^+ are excreted in the urine, there is a net gain of H^+ (or loss of HCO_3^-) in the body. Two major subgroups can be identified:

1. A large number of organic anions are filtered and not reabsorbed so that their excretion exceeds the quantity of NH_4^+ that can be excreted.
2. Only a modest number of anions are filtered and escape reabsorption, but this quantity exceeds the amount of NH_4^+ excreted because there is a major reduction in the rate of excretion of NH_4^+; consequently, the kidneys are unable to generate enough "new" HCO_3^- (Figure 3·3).

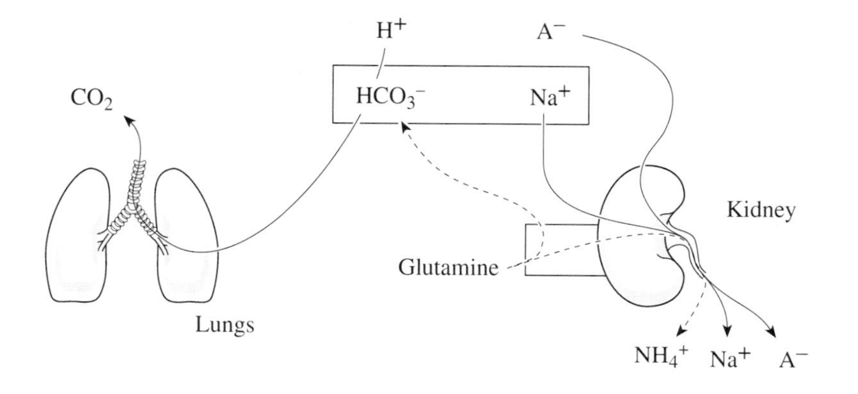

Figure 3·3
Indirect loss of $NaHCO_3$
The body is represented by a rectangle and contains Na^+ and HCO_3^- for simplicity. An acid (H^+ + A^-) is produced. The H^+ is titrated by a HCO_3^- and the A^- is excreted with a Na^+. The net result is loss of Na^+ and HCO_3^- involving the lungs and the kidneys. As shown in the dashed line, there is no net loss of HCO_3^- if the anion is excreted with NH_4^+ rather than Na^+.

Because the generation of "new" HCO_3^- is largely the result of the urinary excretion of NH_4^+, tests to measure the rate of excretion of this cation are important (see Chapter 2, pages 52–54).

Respiratory Response During Metabolic Acidosis

- Acidemia stimulates ventilation and lowers the P_aco_2.
- In metabolic acidosis, there is an empiric, predictable relationship between the fall in the P_aco_2 and the degree of acidemia.

$$H^+ + HCO_3^- \leftrightarrow H_2CO_3 \leftrightarrow H_2O + CO_2$$

Acidemia is a potent stimulus to the respiratory center. In metabolic acidosis, hyperventilation (lowering the P_aco_2) displaces the equilibrium of the BBS equation downwards and lowers the $[H^+]$ (see the margin). As discussed in Chapter 1, the net effect of this fall in P_aco_2 is to minimize binding of H^+ to intracellular proteins (page 20). The relationship between the P_aco_2 and the fall in $[HCO_3^-]$ is predictable from empiric data. Two ways to remember this relationship are as follows:

1. **Rule of thumb:** The P_aco_2 should equal the value after the "7." in the pH (see page 57).
2. **Relationship between P_aco_2 and $[HCO_3^-]$:** As the $[HCO_3^-]$ in the ECF falls, the P_aco_2 also falls; the slope of the line is approximately 1, which means that the P_aco_2 falls 1 mm Hg from 40 mm Hg for every 1 mmol/l fall in plasma $[HCO_3^-]$ from 25 mmol/l.

Expected P_aco_2:
If the P_aco_2 is significantly lower than predicted by the relationship between P_aco_2 and $[HCO_3^-]$, the patient has another primary stimulus to the respiratory center in addition to the acidemia (respiratory alkalosis). Similarly, if the P_aco_2 is significantly higher than predicted, there is a compromised ability to ventilate in response to normal stimuli (respiratory acidosis).

DETRIMENTAL EFFECT OF AN INADEQUATE DEGREE OF HYPERVENTILATION

- When the $[HCO_3^-]$ is very low, small changes in the P_aco_2 or the $[HCO_3^-]$ will result in large changes in the $[H^+]$.

In patients who are incapable of appropriately lowering the arterial Pco_2 during metabolic acidosis, the degree of rise in $[H^+]$ is much more extensive (see the discussion of Question 3·1 and Tables 3·3 and 3·4). Although the $[H^+]$ and $[HCO_3^-]$ in the ECF are the variables that are clinically evident, the impact of hyperventilation on the defense of the net charge on intracellular proteins must also be kept in mind whenever a patient with metabolic acidosis is subjected to any procedure that may interfere with ventilation (see the discussions of Questions 3·2 and 3·3).

Table 3·3
PLASMA [H^+] IN PATIENTS WHO HAVE PROGRESSIVE METABOLIC ACIDOSIS WITH AND WITHOUT AN APPROPRIATE DEGREE OF HYPERVENTILATION

Patients without hyperventilation in this table are assumed to have a P_aco_2 of 40 mm Hg.

	Status with appropriate hyperventilation			Status without hyperventilation		
[HCO_3^-] (mmol/l)	P_aco_2 (mm Hg)	[H^+] (nmol/l)	pH	P_aco_2 (mm Hg)	[H^+] (nmol/l)	pH
20	35	42	7.38	40	48	7.32
15	30	50	7.30	40	64	7.19
10	25	60	7.22	40	96	7.02
5	20	96	7.02	40	191	6.72

Table 3·4
IMPACT OF SMALL CHANGES IN THE [HCO_3^-] OR P_aCO_2 ON THE ACID-BASE STATUS OF THE PATIENT WITH A PLASMA [HCO_3^-] OF 7 mmol/l

A small increase in P_aco_2 in patient B or a small fall in [HCO_3^-] in patient C converts a modest degree of acidemia into a severe one.

Patient	Condition	[H^+] nmol/l	pH	P_aco_2 (mm Hg)	[HCO_3^-] (mmol/l)
A	Stable metabolic acidosis	72	7.13	20	7
B	Small reduction in hyperventilation	102	6.99	30	7
C	Further fall in plasma [HCO_3^-]	96	7.02	20	5

Questions

(Discussions on pages 135–36)

3·1 Which patient has a primary respiratory acid-base disorder? How should each patient be managed from an acid-base point of view?

Patient	[H^+] nmol/l	pH	P_aco_2 (mm Hg)	[HCO_3^-] (mmol/l)
A	64	7.20	20	8
B	120	6.90	40	8
C	30	7.50	10	8

3·2 Does the arterial or venous Pco_2 best reflect the degree of protonation of intracellular proteins during metabolic acidosis?

3·3 How can reexpansion of the ECF volume affect the Pco_2 in vital organs?

Diagnostic Approach Metabolic Acidosis

Figure 3·4
Diagnostic approach to metabolic acidosis
Metabolic acidosis is present when the [H^+] in plasma is higher than expected and the [HCO_3^-] in plasma is lower than expected. The first step is to assess the respiratory response to acidosis and then define the basis of the acidosis—a gain of acid vs a loss of HCO_3^-. The final diagnoses are shown in the shaded boxes. (AG = plasma anion gap.)

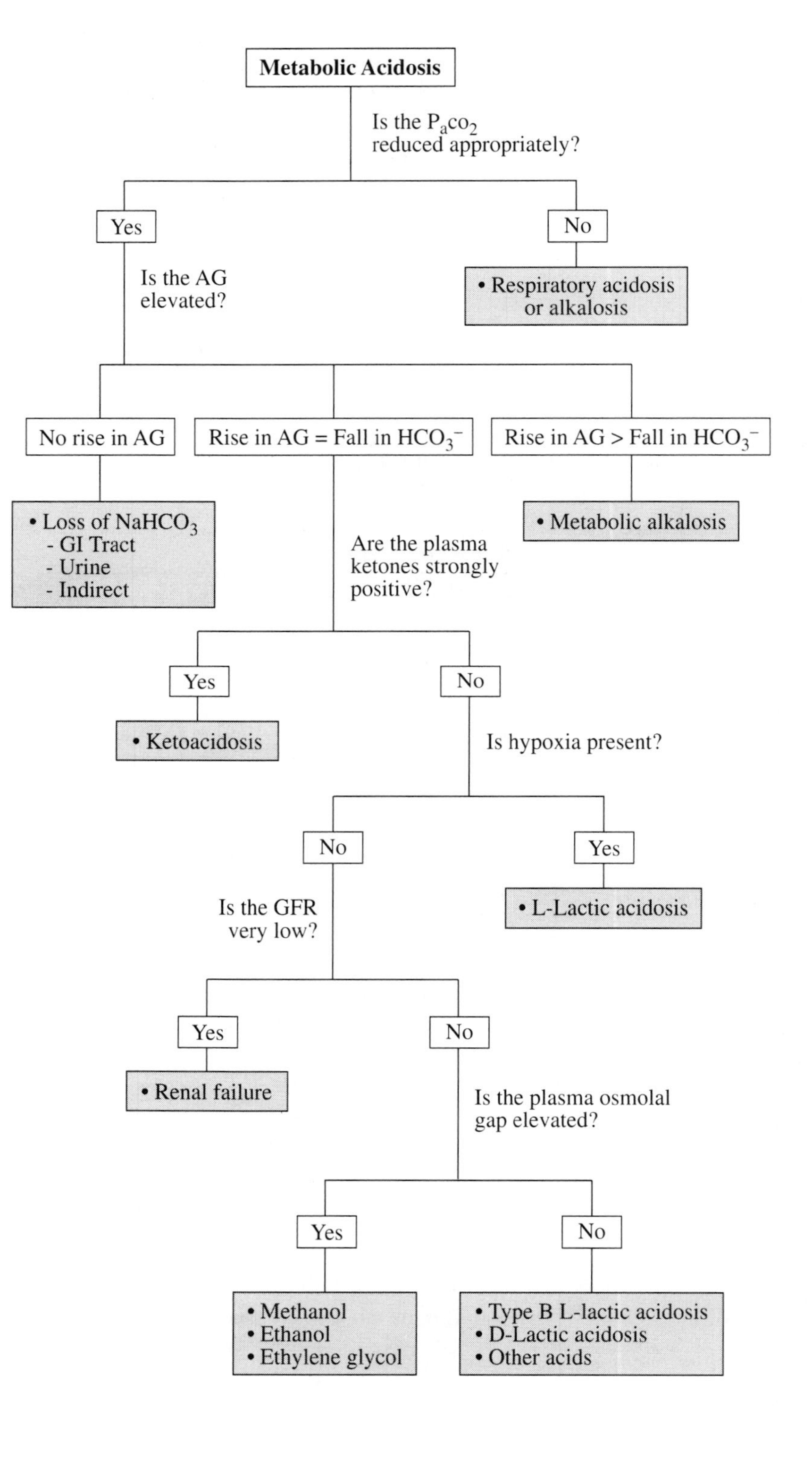

The overall diagnostic approach to the patient with metabolic acidosis is outlined in Figure 3·4 . There are four steps to take:

1. **Confirm that metabolic acidosis is present.**
The presence of metabolic acidosis is confirmed by finding a higher $[H^+]$ and a lower $[HCO_3^-]$ in plasma than expected. Alternatively, metabolic acidosis is present in virtually all cases when the plasma anion gap is unexpectedly high.

2. **Has the ventilatory system responded appropriately?**
In patients with an elevated $[H^+]$, a fall in the tissue Pco_2 is necessary to minimize binding of H^+ to intracellular proteins. If the P_aco_2 does not fall by the appropriate amount (defined on page 57), a coexistent respiratory acid-base disorder is also present.

3. **Does the patient have metabolic acidosis and no increase in the plasma anion gap?**
In patients with metabolic acidosis and no increase in the plasma anion gap, begin by ruling out inapparent accumulation of organic acids because another condition (e.g., hypoalbuminemia) obscured the expected rise in the plasma anion gap.
The two major diagnostic possibilities are the loss of $NaHCO_3$ and the overproduction of acids with anions that do not appear in plasma; either these anions were excreted without H^+ or NH_4^+ or the plasma anion gap was underestimated. If the kidneys are not the cause of the metabolic acidosis, the rate of excretion of NH_4^+ should be high (> 200 mmol/day in a 70-kg adult). The *urine net charge* can be used to estimate the $[NH_4^+]$ in urine (pages 52–54). A nonrenal basis of metabolic acidosis (e.g., diarrhea) is suggested when the urine NH_4^+ excretion is > 200 mmol/day. In this case, the urine $[Cl^-]$ will exceed the urine $[Na^+]$ + $[K^+]$. In contrast, with a low urine $[NH_4^+]$, suspect distal RTA (urine $[Na^+]$ + $[K^+]$ exceeds $[Cl^-]$). Rarely, NH_4^+ may be in the urine in conjunction with an anion other than Cl^- (β-HB^-, hippurate anion). If a patient excretes β-HB^- with NH_4^+, the marked ketonuria will result in ketoacidosis and no increase in the plasma anion gap. In this case, a high rate of excretion of NH_4^+ may be revealed by calculating the *urine osmolal gap* (page 54).

4. **Has the plasma anion gap risen appropriately?**
If the rise in the plasma anion gap is approximately equal to the fall in $[HCO_3^-]$ in plasma, the patient has a gain of acids or renal failure. Now one must detect the basis of the added acids: the presence of ketoacids, hypoxia, or a very low GFR. In the absence of these findings, suspect the presence of toxic products from unusual alcohols (an increased plasma osmolal gap). If the plasma osmolal gap is normal, the most likely diagnoses are a low rate of removal of L-lactic acid (usually a liver problem) and the accumulation of D-lactic acid (a GI problem).

The detailed approach to each category will be provided later.

Urine net charge:
Detects NH_4^+ + Cl^-.

Urine osmolal gap:
Detects NH_4^+ + any anion.

Clinical pearls:
- Always suspect methanol or ethylene glycol poisoning.
- In a patient with metabolic acidosis, an increased plasma anion gap, and a normal ECF volume, be extremely suspicious of methanol or ethylene glycol poisoning especially if the GFR, GI tract, and the liver are normal.

PART B
TREATMENT OF METABOLIC ACIDOSIS

The therapeutic decisions about patients with metabolic acidosis revolve around the following issues:

1. What emergency measures are required?
2. How can the threats to life be avoided?
3. What are the options for treating the acidosis per se?
4. How should one deal with an abnormal [K^+] in plasma?

Emergency Measures

Before the biochemical results are available, measures to ensure a proper airway, adequate circulation, and O_2 delivery must be pursued vigorously. These measures will not be discussed further.

Avoiding Threats to Life

There are three critical reasons for making a specific diagnosis:

1. It is important to determine the rate of H^+ production, which may be so high that the most effective means of arresting it is to increase the delivery of O_2 (e.g., L-lactic acidosis caused by a low cardiac output; see Table 3·2, page 72).

2. The cause of the metabolic acidosis may pose a serious but independent threat to the patient (e.g., methanol overdose). Its specific therapy (ethanol administration) is the most important therapeutic measure.

3. In certain types of metabolic acidosis that are associated with hypokalemia (low distal tubular H^+ secretion, diarrhea), K^+ replacement may be necessary before or along with administration of $NaHCO_3$ in order to avoid serious cardiac arrhythmias or respiratory failure (the HCO_3^- administered might promote entry of K^+ into cells).

STOP H^+ PRODUCTION

Arresting H^+ production is critical in conditions with a very rapid rate of H^+ production (Figure 3·5). This rate can be 72 mmol/min in L-lactic acidosis from anoxia. Because the rate of production of H^+ is much lower in diabetic ketoacidosis and methanol overdose (1 mmol/min), it is less urgent to stop the production of H^+ in these situations (Table 3·2). Instead, other measures can be life-saving: delivery of oxygen in L-lactic acidosis (so that ATP can be regenerated), stopping the production of toxins by ethanol administration in methyl alcohol intoxication, and possibly gastric lavage in certain intoxications.

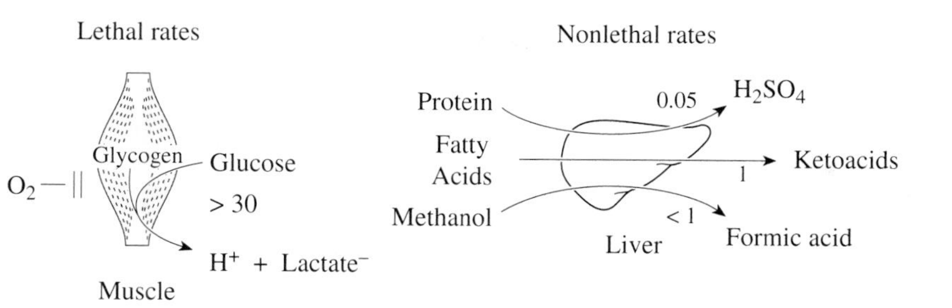

Figure 3·5
The production of acids
The numbers next to the arrows represent the rate of production of H^+ in mmol/min.

LOWER THE NUMBER OF H^+ BOUND TO PROTEINS BY LOWERING THE VENOUS Pco_2

There are two therapeutic options for rapidly lowering the quantity of H^+ bound to intracellular proteins. First, ensure an adequate degree of hyperventilation. Second, increase the rate of blood flow to vital organs (see discussion of Questions 3·2 and 3·3). These options are most useful in coexistent respiratory and metabolic acidosis and are the initial treatments of choice.

INCREASE ENDOGENOUS HCO_3^- FORMATION

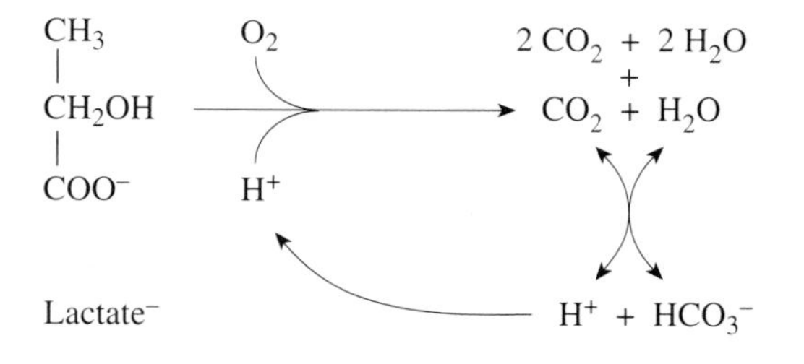

Figure 3·6
Oxidation of organic anions and the generation of HCO_3^-
Oxidation of an organic anion to a neutral end product consumes a H^+ and yields a HCO_3^-. One CO_2 plus one H_2O is converted to a H^+ and a HCO_3^- and leads to the net yield of a HCO_3^-.

Increasing the formation of endogenous HCO_3^- is a therapeutic option only in patients with metabolic acidosis and an increased plasma anion gap (Figure 3·6). The only emergency measure of value is to increase the metabolism of the circulating organic anions (renal new HCO_3^- formation can only occur at a rate of 0.3 mmol/min with perfectly adapted kidneys). Net removal of L- or D-lactate anions and β-HB^- by metabolism requires that the rate of their removal exceed their ongoing production.

Treatment of the H^+ Load

To assess the need for $NaHCO_3$ therapy, use the plasma $[HCO_3^-]$ instead of the pH to avoid being misled by an unusually low P_aco_2 (see the margin). The major dangers of this therapy are summarized in Figure 3·7.

Quantities:
Consider two cases of metabolic acidosis; in both the $[HCO_3^-] = 1$ mmol/l, but the P_aco_2 changes from 5 to 20 mm Hg. The $[H^+]$, as calculated via the Henderson equation, yields a nonlethal value in the first example (120 nmol/l, pH 6.9), but a lethal value in the second example (480 nmol/l, pH 6.3).

$$H^+ = 24 \times Pco_2/[HCO_3^-]$$
$$= 24 \times 5/1$$
$$= 24 \times 20/1$$

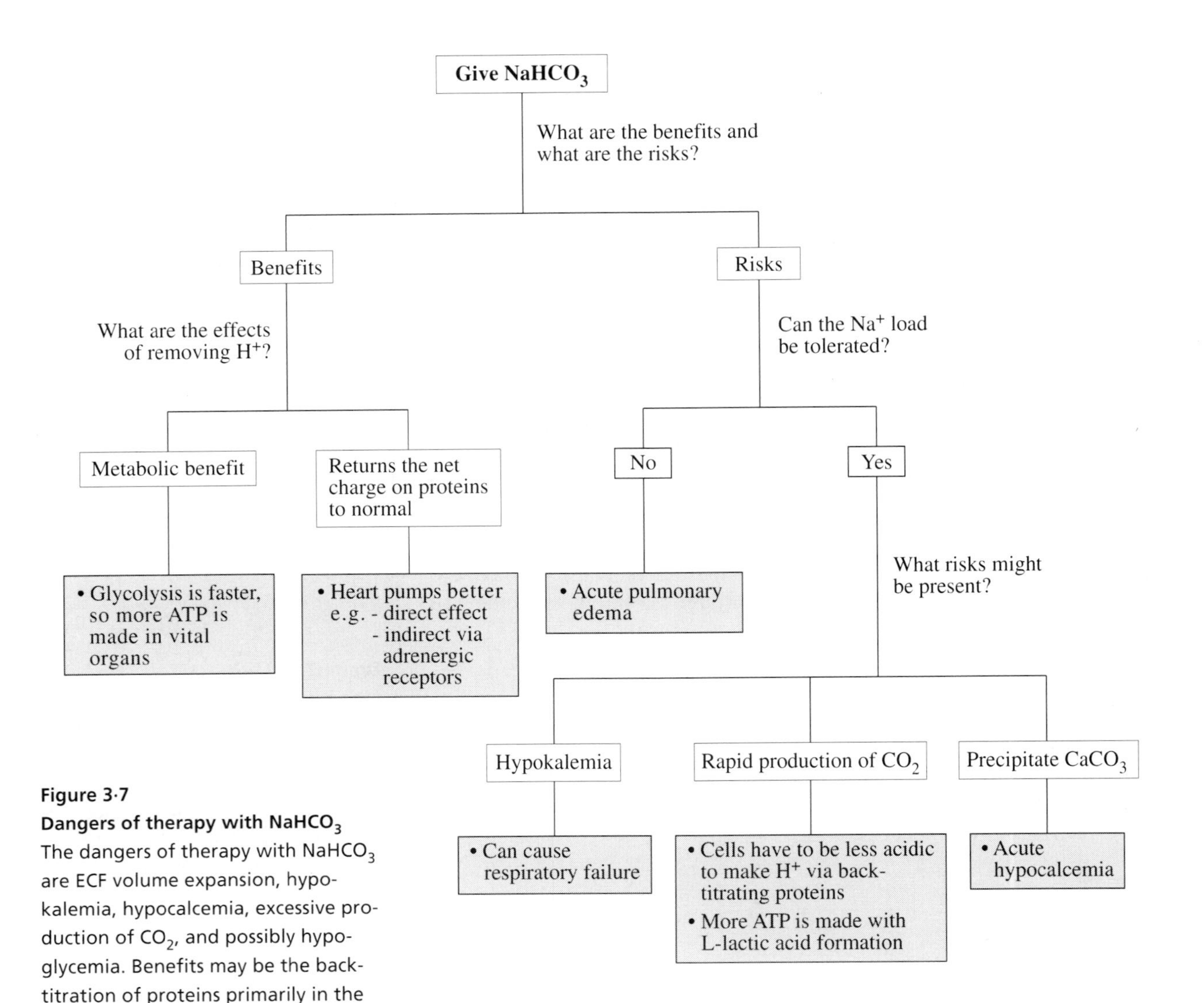

Figure 3·7
Dangers of therapy with $NaHCO_3$
The dangers of therapy with $NaHCO_3$ are ECF volume expansion, hypokalemia, hypocalcemia, excessive production of CO_2, and possibly hypoglycemia. Benefits may be the back-titration of proteins primarily in the ICF and the generation of more ATP because of a fall in the $[H^+]$ in the ICF.

THE USE OF $NaHCO_3$ IN L-LACTIC ACIDOSIS

Treatment with $NaHCO_3$ is a controversial area. Some strongly advocate the use of $NaHCO_3$, but others consider it harmful. The following comments reflect our view.

The Database Is Weak

Humans: L-Lactic acidosis is an acid-base "cough"! It is a common finding that results from a heterogeneous group of disorders (Table 3·12, page 100). Even the subgroup of Type A L-lactic acidosis is not homogeneous; there are many different causes for inadequate delivery of O_2 to tissues, as well as varying degrees of tissue hypoxia. Further, delivery and demand do not remain constant from minute to minute. Hence, evaluating modes of therapy is fraught with hazard and the conclusions drawn are not really justified.

Note:
- Type A lactic acidosis is the hypoxic type.
- Type B lactic acidosis is not due to hypoxia.

Animals: The basis of L-lactic acidosis in animals differs markedly from that seen in clinical situations. For example, in dogs breathing air with a low Po_2, the cardiac output is as high as possible and the heart is normal. What $NaHCO_3$ might or might not do in this setting cannot be translated reliably to the bedside where most cases with L-lactic acidosis are caused by a failing, diseased heart.

Stoichiometry

One mmol of HCO_3^- is needed to titrate each millimole of H^+ produced or each millimole of H^+ bound to intracellular proteins. The following statements illustrate some of the problems.

- The quantity of H^+ produced in each subtype of metabolic acidosis is not constant and may vary considerably (see Table 3·2).
- The number of H^+ bound to intracellular proteins rises with increased severity of the acidosis. For example, when acidosis is very severe, several hundred millimoles of $NaHCO_3$ are needed to raise the $[HCO_3^-]$ in the ECF even 1 to 2 mmol/l (see the example in the margin).

Quantitative example:
If you want to raise the $[HCO_3^-]$ so that the $[H^+]$ is halved, the $[HCO_3^-]$ must be doubled if the Pco_2 does not change. In a patient with a $[HCO_3^-]$ of 2 mmol/l, a $[H^+]$ of 125 nmol/l, a pH of 6.9, one need only raise the $[HCO_3^-]$ to 4 mmol/l to halve the $[H^+]$ (62.5 nmol/l, pH 7.20; assume the Pco_2 remained close to 10 mm Hg). To raise the $[HCO_3^-]$ 2 mmol/l in this setting, more than 200 mmol of $NaHCO_3$ may have to be given (the content of HCO_3^- in the ECF expands by close to 30 mmol).

Basics of $NaHCO_3$ Therapy

1. **Treatment of patients with a normal plasma anion gap and a defect in NH_4^+ excretion (moderate degree of acidosis):**
   Because these patients have no renal or metabolic source of HCO_3^-, they must receive exogenous HCO_3^- if their acidemia is to be corrected. When calculating the amount of $NaHCO_3$ to administer to a patient with a moderate degree of metabolic acidosis, assume a volume of distribution of 50% of total body weight. This figure is derived from experimental data and reflects the fact that 60% of buffering occurs in the ICF with a modest degree of metabolic acidosis. This assumption is not valid for a more severe or a less severe degree of metabolic acidosis (illustrated in the calculation in the margin) because much more than 70 mmol of $NaHCO_3$ would be needed to raise the plasma $[HCO_3^-]$ by 2 mmol/l from 2 to 4 mmol/l (0.5×70 kg $\times$ 2 mmol/l).

2. **Treatment of patients with a severe degree of metabolic acidosis:**
   In these patients, the initial therapy with $NaHCO_3$ should remove the patient from immediate danger. It should raise the plasma $[HCO_3^-]$ to close to 5 mmol/l if the P_aco_2 is less than 20 mm Hg. With a higher P_aco_2, the alveolar ventilation should be increased by means of intubation and ventilation.

   The rapidity with which the $NaHCO_3$ should given is determined by the severity of the acidemia, the rate of H^+ production, and the K^+ and cardiac status of the patient. The tonicity of the $NaHCO_3$ administered should be determined primarily by the patient's tonicity.

   The administration of $NaHCO_3$ in an acidemic patient is a CO_2-producing process. If the patient is being artificially ventilated, the P_aco_2 will rise if the alveolar ventilation is not increased appropriately.

Clinical note:
At times, the Na^+ load will limit how much $NaHCO_3$ can be given. In these settings, diuretics are rarely helpful and alternate measures are needed.

Therapeutic Options in Patients with Metabolic Acidosis, Renal Failure, and ECF Volume Expansion

Use of $NaHCO_3$ in a patient with metabolic and respiratory acidosis: Liberalize the goal of a plasma $[HCO_3^-]$ of 5 mmol/l in a patient with chronic lung disease because this patient's ability to hyperventilate during metabolic acidosis might be severely compromised.

Gastric HCO_3^- generation: Each day the stomach usually generates 150 mmol of HCO_3^-, which can be mobilized as a therapeutic tool in certain challenging patients. One can insert a nasogastric tube, stimulate gastric acid secretion with pentagastrin (provided that the patient has not received H_2-receptor blockers and is not achlorhydric), and remove significant amounts of HCl (Figure 4·1, page 150). One must ensure that the nasogastric tube is well situated to remove most of the acid. The periodic instillation of antacids down the tube helps in preventing complications secondary to excess HCl secretion. Should sufficient Na^+ be removed via this route, some $NaHCO_3$ could be administered intravenously.

Phlebotomy and dialysis: If a patient with a severe degree of metabolic acidosis also has pulmonary edema, an additional therapeutic maneuver is a phlebotomy to permit the administration of $NaHCO_3$ (acidemia may also impair cardiac function). The phlebotomized blood should be packed and the cells returned to the patient. Early dialysis with a HCO_3^- bath should be planned.

Ventilation: One should ventilate patients to lower their P_aco_2 if it is unduly high. This treatment can influence the acid-base state much faster than administering $NaHCO_3$. In patients with pulmonary edema, ventilation is also beneficial for the pulmonary edema (provided the patient can tolerate positive end-expiratory pressure, which might decrease the cardiac output).

Guidelines for $NaHCO_3$ Therapy

We propose the following guidelines for the use of $NaHCO_3$. The issues concerning the use of $NaHCO_3$ in specific diagnostic categories will be considered in Part C of this chapter.

Ketoacidosis: In ketoacidosis, the rate of H^+ production is slow and $NaHCO_3$ therapy may carry the risk of provoking severe hypokalemia; therefore, $NaHCO_3$ should be avoided in most cases. Cases in which one should consider $NaHCO_3$ (along with K^+ if significant K^+ depletion exists) are as follows:

1. when hyperkalemia is severe despite insulin therapy;
2. in very severe acidemia ($[HCO_3^-] < 5$ mmol/l) to raise the plasma $[HCO_3^-]$ close to twofold;
3. when acidemia worsens despite insulin therapy (perhaps insulin resistance is a result of acidemia).

Type A L-lactic acidosis: In Type A L-lactic acidosis, the primary efforts should be directed at improving delivery of O_2. $NaHCO_3$ should be used in the following instances, and the $[HCO_3^-]$ need not be increased above 8 mmol/l.

1. In states of low cardiac output, raising the cardiac output will have a larger impact on the pH of the ICF via a reduction in tissue Pco_2 than will therapy with $NaHCO_3$ (see the discussions of Questions 3·2 and 3·3).
2. In cases with low alveolar ventilation, plan to increase ventilation to lower the tissue Pco_2.

Question

(Discussion on page 136)

3·4 Some recommend that a 50:50 mixture of $NaHCO_3$ and Na_2CO_3 (carbicarb) be used as a source of alkali. If a patient has L-lactic acidosis and is producing 12 mmol of L-lactic acid per minute, how much less CO_2 (expressed in percentage form) will be produced by titrating the H^+ produced with carbicarb instead of $NaHCO_3$? (Assume 12 mmol of O_2 are consumed each minute by the body.)

K^+ and Metabolic Acidosis

- One must avoid a severe degree of hypokalemia when $NaHCO_3$ is given to a patient with a severe degree of metabolic acidosis.

The principal cation in the ICF is K^+. There is normally a very large electrochemical gradient for K^+ across cell membranes by virtue of the Na^+K^+ATPase in cells, the selective permeability of cell membranes to Na^+ (very low) and K^+ (high), and the fact that most intracellular anions are macromolecular and do not cross cell membranes. This K^+_{in}/K^+_{out} ratio is largely reflected by the *resting membrane potential*, with hyperkalemia diminishing the magnitude of this negative voltage and hypokalemia increasing it. Cells are more excitable during hyperkalemia and less excitable during hypokalemia. Notwithstanding, there is a tendency for cardiac arrhythmias in both cases. The challenge in therapy of metabolic acidosis is that correction of the acidemia will be associated with movement of K^+ into the ICF as H^+ move in the opposite direction.

Resting membrane potential =

$$-61 \times \log\frac{[K^+]_{in}}{[K^+]_{out}}$$

HYPOKALEMIA

There are two general features leading to a K^+ deficit and/or hypokalemia in a patient with metabolic acidosis: altered release of K^+ from cells, and increased renal excretion of K^+ (Table 3·5). A severe degree of hypokalemia has two major negative consequences, cardiac arrhythmias and respiratory failure from muscle weakness. In either case, the aim of therapy is to infuse K^+ quickly (Table 3·6). We recommend the following treatment:

1. Give K^+ rapidly in the case of an arrhythmia or respiratory arrest. Also, when a patient with a very severe degree of metabolic acidosis has hypokalemia and/or a deficit of K^+, promptly administer K^+ with HCO_3^- (see the margin).
2. Give K^+ more slowly (0.5 to 1 mmol/min) in the absence of important ECG changes or respiratory failure.

Aggressive intravenous K^+ therapy:

1. Calculate the increment between the current plasma $[K^+]$ and 3.0 mmol/l; multiply this value by the plasma volume (plasma volume is close to 20% of the ECF volume) and give this amount over 1 minute via a central line.
2. Reduce the infusion rate to 1 mmol/min and recheck the ECG and plasma $[K^+]$ in 5–10 minutes.
3. Repeat steps 1 and 2 if the $[K^+]$ remains well below 2.5 mmol/l.

Table 3·5
POTASSIUM DEPLETION AND METABOLIC ACIDOSIS

In all three settings, there is a deficit of K^+. In diabetic ketoacidosis (DKA), hyperkalemia is usually present; be wary if the diabetic patient has normokalemia or hypokalemia.

Disorder	Basis of K^+ depletion
Distal RTA (low H^+ secretion type)	Renal K^+ loss
Loss of $NaHCO_3$ from the GI tract	GI K^+ loss, renal K^+ loss
DKA	Renal K^+ loss (osmotic diuresis)

Table 3·6
PRINCIPLES OF THERAPY IN PATIENTS WITH ACIDEMIA AND K^+ DEPLETION

Principle	Comments
Use oral route for K^+ whenever possible.	• Giving large amounts of K^+ orally may prevent the IV problems.
Use several IV sites.	• Using different sites allows the dissociation of K^+ vs HCO_3^- infusion rates and permits more aggressive administration of K^+ by peripheral vein.
Use a cardiac monitor.	• A monitor will enable early detection of arrhythmias.
Ensure adequate K^+ output via the urine for continued infusion of K^+.	• If the patient has renal failure, give K^+ more cautiously.

HYPERKALEMIA

In the patient with hyperkalemia in the presence of important ECG abnormalities (Table 3·7), administer Ca^{2+} to minimize the electrical disturbance and give HCO_3^- and insulin to shift K^+ into cells (see Chapter 11 for details).

Table 3·7
HYPERKALEMIA AND METABOLIC ACIDOSIS

Although some patients with metabolic acidosis have hyperkalemia and a total body surfeit of K^+, others have a deficit of K^+. Obviously, treatment differs in the long run, but not necessarily in the acute situation if the ECG is very abnormal.

Cause	Total body K^+	Comment
Renal failure	Increased	Excretion of K^+ is low.
Low aldosterone	Increased	Administer aldosterone and assess bioactivity by the degree of kaliuresis.
DKA	Decreased	The shift to ECF reflects the low level of insulin, not the acidosis.

PART C
SPECIFIC DISORDERS

Ketoacidosis

- The basis of ketoacidosis is relative insulin deficiency.

In order to understand why ketoacidosis develops and why it might be so severe, one must evaluate in a quantitative fashion the rates of production and removal of *ketoacids*. Production of ketoacids occurs in the liver if there is a lack of insulin and/or a resistance to its actions (Table 3·8).

Ketoacids:

- The most abundant ketoacid, β-HB$^-$, is really a hydroxy-acid.

$$CH_3-\underset{\displaystyle H}{\overset{\displaystyle OH}{C}}-CH_2-COO^- + H^+$$

- Acetoacetate (AcAc) is the only real ketoacid.

$$CH_3-\underset{\displaystyle O}{\overset{}{C}}-CH_2-COO^- + H^+$$

- Acetone is made from AcAc; it is not an acid.

$$CH_3-\underset{\displaystyle O}{C}-CH_3$$

Table 3·8
CAUSES OF KETOACIDOSIS

Ketoacidosis with normal β-cell function (i.e., physiologically low release of insulin):
- hypoglycemia
- inhibition of β cells (α-adrenergics)
- excessive lipolysis
Ketoacidosis with abnormal β-cell function:
- insulin-dependent diabetes mellitus
- pancreatic destruction

PRODUCTION OF KETOACIDS

Insulin acts at two major sites to influence the rate of ketogenesis, an extrahepatic site (adipocyte) and an intrahepatic site.

Extrahepatic Effects

Low levels of insulin combined with high levels of hormones whose actions oppose insulin (e.g., adrenaline, ACTH) lead to *activation of hormone sensitive lipase* (HSL) in adipocytes and the release of larger quantities of fatty acids (Figure 3·8).

Activation of HSL:
High levels of hormones such as adrenaline are more important than the lack of insulin in this regard.

Intrahepatic Effects

Low levels of insulin and, more importantly, elevated levels of glucagon lead to a fall in the level of malonyl-CoA, the key intermediate in the hepatocyte. This fall permits fatty acids to enter mitochondria where they are oxidized to acetyl-CoA. Acetyl-CoA has three possible fates:

1. oxidation in the TCA cycle to yield ATP (inhibited by the high ATP levels that result from fatty acid oxidation);
2. reconversion to fatty acids (inhibited by a lack of insulin);
3. conversion to ketoacids (Figure 3·9).

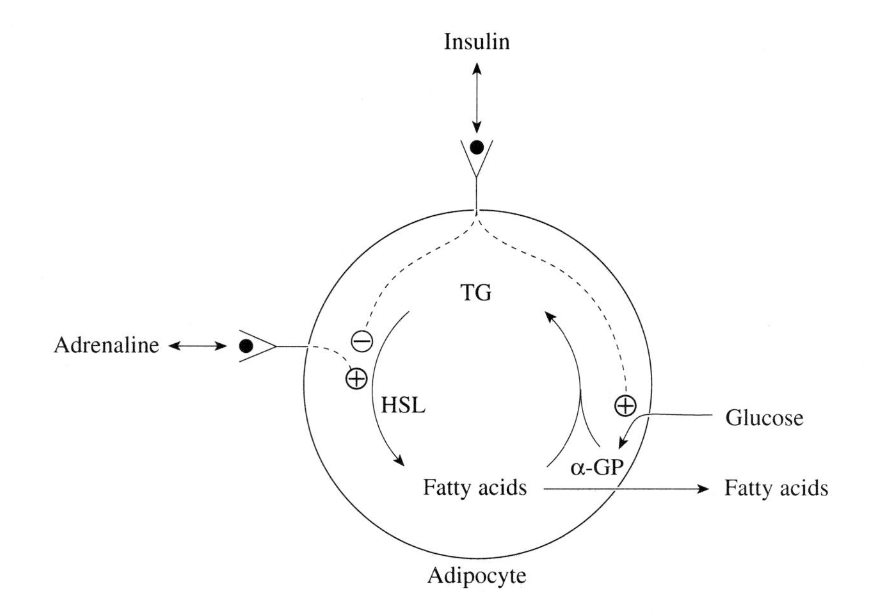

Figure 3·8
Release of fatty acids from adipocytes
When levels of hormones such as adrenaline are high, HSL is activated and fatty acids are released. If insulin levels are low, HSL is more active and reesterification of fatty acids is low because less α-glycerol phosphate (α-GP) is available. (TG = *triglyceride*; ⊕ = stimulated; ⊖ = inhibited.)

Triglycerides:
The major storage form of fat. Three fatty acids are each linked by an ester bond to one of the hydroxyl groups of α-glycerol phosphate.

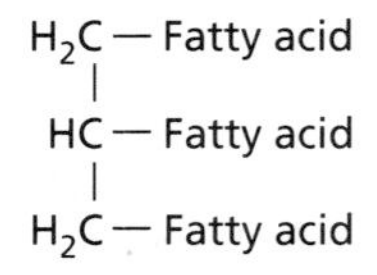

When more acetyl-CoA is produced than can be oxidized to regenerate ATP and when fatty acid synthesis is blocked, the only major fate left for acetyl-CoA is to be converted to ketoacids, a pathway driven by a high level of acetyl-CoA (Figure 3·9).

Control of the Production of Ketoacids

There are two major features in the control of ketogenesis:

1. Low levels of insulin give permission for a high rate of production of ketoacids by activating HSL, which provides more fatty acids for the liver, promotes fatty acyl entry into hepatic mitochondria, and also inhibits fatty acid synthesis.

2. Once fat-derived fuels are "selected," the rate of turnover of ATP in hepatocytes sets the upper limit on ketogenesis, a pathway that must regenerate ATP. The key features in understanding the constraints on ketogenesis by the turnover of ATP are as follows:
   - The pathway must generate ATP and consumes ADP in the process (Figure 3·9).
   - Cells contain a very tiny amount of ADP, the precursor of ATP. Hence, for ketogenesis to proceed, ADP must be reformed—via hydrolysis of ATP when hepatic work is performed (i.e., biosynthesis or ion pumping).

Counter-insulin hormones:
Hormones with metabolic effects that oppose those of insuline. Examples include glucagon, adrenaline, and glucocorticoids.

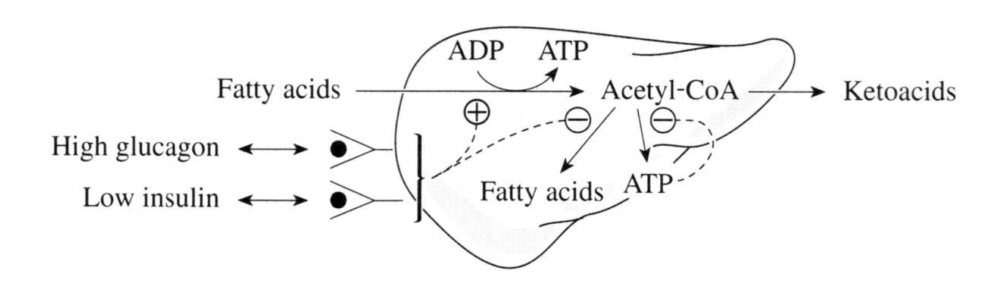

Figure 3·9
Control of ketoacid production in the liver
The hormonal setting of low levels of insulin and high levels of glucagon, via intrahepatic signals (low level of malonyl-CoA), promotes the entry of fatty acids into mitochondria and the formation of intramitochondrial acetyl-CoA. The rate of regeneration of ATP limits the ultimate fate of acetyl-CoA (i.e., it is converted to ketoacids at a controlled rate).

- The rate of turnover of ATP in hepatocytes permits the generation of only 1.3 mmol of ketoacids per minute at the usual rates of O_2 consumption in the liver (see the margin).

Ketogenesis and hepatic O_2 consumption:

- Hepatic blood flow is 1 liter/min.
- Each liter of portal blood contains 6 mmol of O_2 and 0.6 mmol of fatty acid.
- The liver extracts 2 mmol of O_2/min (33% of delivery).
- Stoichiometry:
  6 O_2 needed per C_{16} fatty acid $\longrightarrow$ 4 ketoacids. Therefore, with 2 mmol of O_2 consumed per minute, only 1.3 mmol of ketoacids can be formed per minute; this formation requires that 0.3 mmol fatty acids be extracted per minute, a difficult task because fatty acids are sparingly soluble in water.

Bottom line:
Because hepatic O_2 consumption rarely rises appreciably, ketogenesis is a rather slow way to generate H^+. This rate will be even slower if other fuels are used to regenerate ATP in the liver (e.g., the conversion of amino acids to glucose).

Hepatocytes use O_2 primarily during biosynthesis and to pump Na^+ via the Na^+K^+ATPase.

REMOVAL OF KETOACIDS

Oxidation of Ketoacids

Two organs—the brain and the kidneys—are primarily involved in the oxidation of ketoacids (Figure 3·10).

Brain: The brain can oxidize 750 mmol of ketoacids per day, half the quantity of ketoacids produced when ketogenesis is most rapid. The following influence the oxidation of ketoacids:
- The brain will oxidize ketoacids preferentially if their levels are high because the products of their metabolism (NADH, acetyl-CoA) inhibit the key step in the oxidation of glucose (pyruvate dehydrogenase, Figure 12·1, page 430).
- If the utilization of ATP declines in the brain (less ion pumping in the CNS), fewer ketoacids can be oxidized. In the presence of coma, anaesthetics, or sedation, the brain consumes less O_2 (and utilizes less ATP) . Sedation might be important during the generation of alcoholic ketoacidosis because alcohol may act as a depressant of metabolism in the brain.

Kidney: The kidneys remove about 350–400 mmol of ketoacids per day. If renal work (largely the reabsorption of Na^+) is normal, the kidneys will oxidize 250 mmol ketoacids per day. Since more ketoacids are filtered than reabsorbed, close to 150 mmol of ketoacid anions will be excreted per day during the ketoacidosis of fasting. Because virtually all of these anions are excreted along with NH_4^+ (major) and H^+ (minor), acid-base balance results (Figure 3·11). Much lower renal removal of β-HB^- and H^+ occurs if the filtered load of Na^+ declines (from prerenal failure secondary to loss of Na^+ in the glucose-induced osmotic diuresis) because the rates of both NH_4^+ production and β-HB^- oxidation are both reduced.

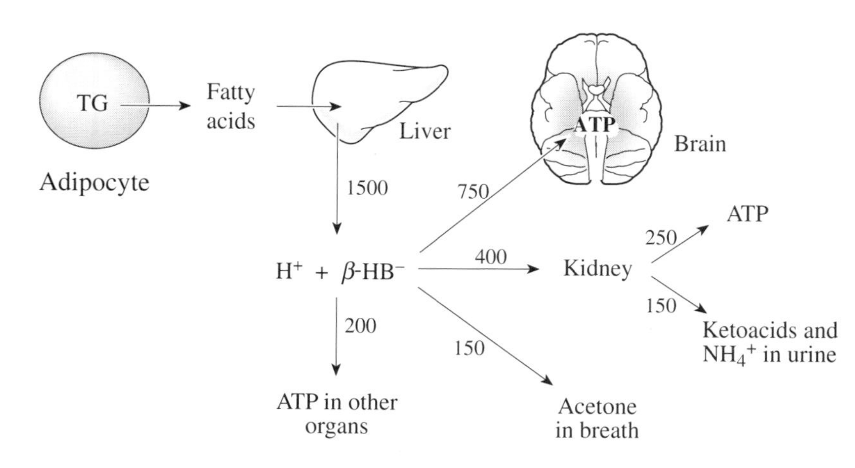

Figure 3·10
Production and removal of ketoacids, an overview
Ketoacids are produced in the liver at a maximum rate of 1500 mmol/day in the absence of insulin; half are oxidized in the brain, and one-fourth are removed by the kidney. The numbers beside the arrows refer to mmol/day and are approximate values. The pathways leading to ATP generation will be limited by the rate of ATP utilization in the organ concerned. Reproduced with permission from *Clinical Detective Stories.*

Figure 3·11
Excretion of β-HB$^-$ + NH_4^+ has no net acid-base effect
The liver produces H^+ with β-HB$^-$. H^+ are removed after reacting with HCO_3^- to form CO_2 + H_2O; the result is a deficit of HCO_3^-. When glutamine is metabolized in the kidney, NH_4^+ and HCO_3^- are produced. If NH_4^+ are excreted, HCO_3^- are added to the body and balance for H^+ and HCO_3^- is restored. To the degree that β-HB$^-$ are excreted with Na^+ or K^+, a deficit of HCO_3^-, Na^+, and K^+ may occur.

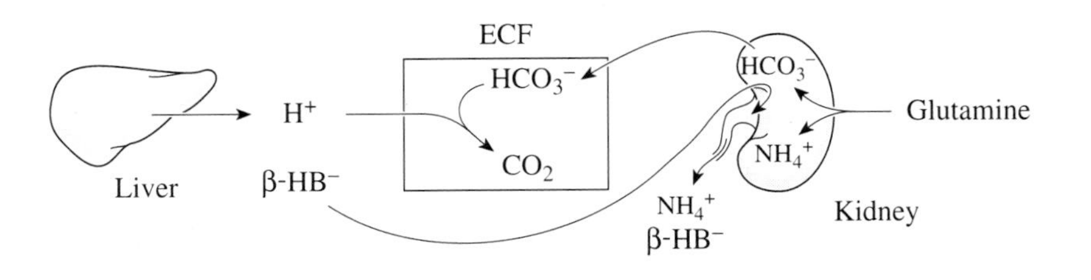

Other organs: The intestinal tract will oxidize ketoacids. If digestion and absorption are proceeding, this utilization can be appreciable (200–300 mmol/day). Notwithstanding, absorption is low in fasting and during DKA.

Skeletal muscle does not seem to oxidize ketoacids if fatty acid levels are high.

Conversion of Ketoacids to Acetone

Note:
High levels of NADH occur most commonly when the supply of O_2 is low or during the oxidation of ethanol.

When the level of acetoacetate is high, this ketoacid is converted spontaneously to acetone and CO_2. High levels of acetoacetate require both a high total ketoacid level and a low level of NADH, as illustrated in the equations below:

$$\text{Acetoacetate}^- + H^+ + \text{NADH} \longleftrightarrow \beta\text{-HB}^- + \text{NAD}^+$$
$$\text{Acetoacetate}^- + H^+ \longrightarrow \text{Acetone} + CO_2$$

Clinical pearl:
In ketoacidosis, production of acetone not only leads to an important diagnostic clue (acetone in the breath) but is also a H^+-removing process. Therefore, a high NADH/NAD$^+$ provides the following four negative effects:
1. less acetone on the breath;
2. fewer H^+ removed with acetone formation;
3. false negative test for ketoacids (see page 97);
4. overproduction of lactic acid.

CAUSES OF A RELATIVE DEFICIENCY OF INSULIN

The causes of relative deficiency of insulin are listed in Tables 3·8 and 3·9. Two groups are evident, those with normal β cells of the pancreas that either lack a stimulus or are inhibited and those with damage to these β cells (diabetes mellitus). Diabetic and alcoholic ketoacidosis are discussed in detail in the next section because of their clinical importance.

Questions

(Discussions on page 137)

3·5 Why is the rate of production of ketoacids so much lower than that of L-lactic acid if both are regulated by the rate of turnover of ATP?

3·6 The rates of ketoacid production and removal are usually equal in a person who lacks insulin. Why is this equality beneficial?

3·7 What makes DKA severe in degree compared with the ketoacidosis of chronic fasting?

Table 3·9
ETIOLOGIC CLASSIFICATION OF KETOACIDOSIS

For details, see the text.

Cause	Special Features	Treatment
Insulin Deficiency with Normal β Cells		
• Hypoglycemia.	• If fasting is the cause, then: - $[HCO_3^-]$ is not < 18 mmol/l; - anion gap is not > 19 mEq/l; - plasma concentration of glucose is 3–4 mmol/l.	• The intake of glucose cures hypoglycemia and ketoacidosis.
• Liver problem, e.g., glycogen storage disease or a defect in gluconeogenesis (GNG).	• Hypoglycemia can be marked. • Plasma $[HCO_3^-]$ can be < 18 mmol/l. • If there is a defect in GNG, the patient may also have L-lactic acidosis.	• Glucose eliminates ketoacidosis. • Provide special therapy for the underlying disease.
• Inhibited insulin release by α-adrenergics; an example is vomiting with marked ECF volume contraction in the alcoholic.	• The ECF volume is very low. • Ketoacidosis may be severe. • Mixed acid-base disorders, K^+ depletion, and phosphate depletion will be present.	• Give NaCl to restore ECF volume. • Give KCl to replace K^+ but do not give insulin. • Give B vitamins. • Give glucose only if the patient is hypoglycemic.
β Cell Destruction (Diabetes Mellitus)		
• Insulin-dependent diabetes mellitus (IDDM).	• Patient has severe hyperglycemia and ketoacidosis, a low ECF volume, K^+ depletion, and hyperkalemia.	• Give NaCl, and insulin. KCl later • Hypokalemia will become a threat in two hours. • Glucose will be needed in six hours. • Treat precipitating factors.
Other Causes		
• Excessive lipolysis:		
- after exercise	• Although fatty acid mobilization is high, oxidation rate slows.	• Because no real danger exists, no special treatment is needed.
- salicylate overdose.	• Excess salicylates activate hepatic lipase and cause hyperventilation, CNS toxicity, and K^+ depletion.	• Remove salicylates by promoting excretion plus GI lavage; dialysis may be necessary. • Replace K^+ deficit.

Diabetic Ketoacidosis

CLINICAL FEATURES

Although diabetic ketoacidosis (DKA) often occurs in the previously diagnosed insulin-dependent diabetic (often with a precipitating event), it may be the initial mode of presentation in young patients with *insulin-dependent diabetes mellitus* (IDDM).

Insulin-dependent diabetes mellitus (IDDM):
A form of diabetes mellitus that occurs most commonly in young, thin patients. Those with IDDM are prone to ketoacidosis.

Noninsulin-dependent diabetes mellitus (NIDDM):
A form of diabetes mellitus that is most common in older, obese patients; ketoacidosis is rare. Complications focus on long-term disorders.

Hormonal Events

Look for the cause of the lack of insulin and the high levels of counter-insulin hormones. Failure of a patient with IDDM to take insulin is the most common reason for lack of insulin; an associated illness (e.g., infection, stress, or even pancreatitis) can lead to elevated concentrations of hormones such as adrenaline and glucocorticoids.

Consequences of Hyperglycemia

The major complaints are polyuria, which is secondary to the osmotic diuresis, and thirst, which accompanies contraction of the ECF volume. Because lean body mass is catabolized during *gluconeogenesis*, there is excessive weight loss and a loss of the sense of well-being.

Gluconeogenesis:
The synthesis of glucose in the liver (or kidney) from compounds not derived directly from glucose—predominantly, the conversion of proteins to glucose.

The severity of hyperglycemia is influenced mainly by the degree of contraction of the ECF volume (a low GFR impairs the excretion of glucose) and to a lesser degree by the quantity of glucose ingested (see Chapter 12).

Signs of the Ketoacidosis

The major signs of the ketoacidosis are ECF volume contraction, the smell of acetone on the breath, and an extreme degree of hyperventilation (*Kussmaul's respirations*).

Kussmaul's respirations:
Deep and rapid breathing.

Symptoms Related to Specific Organs

The most important signs and symptoms occur in the CNS. As DKA becomes more severe, confusion and even coma may develop. Other symptoms and signs that can be attributed to hyperglycemia include problems with vision from swelling of the lens of the eye, nausea and vomiting from poor gastric emptying, and symptoms such as abdominal pain secondary to hyperlipidemia. There may also be complications related to longstanding diabetes mellitus (autonomic or peripheral neuropathy, visual problems, nephropathy, and atherosclerosis).

Natural History

The signs and symptoms of DKA develop very slowly in the initial 12–24 hours after stopping the administration of insulin. This time course can be accelerated if there are high levels of counter-insulin hormones present. When hyperglycemia and acidosis become more prominent, a vicious cycle develops. As the patient starts to become confused, cerebral metabolism of ketoacids declines and the degree of ketoacidosis suddenly becomes even more severe. The GFR and renal work also decline as a result of a lesser filtered load of Na^+, so fewer ketoacids are oxidized and excreted by the kidney. Taken together, these factors largely account for the near-terminal accelerated phase of DKA.

CHANGES IN BODY COMPOSITION

The features of hyperglycemia and ketoacidosis have been discussed above. As a consequence of the osmotic diuresis, there are major changes in Na^+, water, and K^+.

Sodium

As a result of the osmotic diuresis, there is a major loss of Na^+ in the urine, a loss that exceeds any reasonable intake of Na^+. Accordingly, a major feature of DKA is a significant degree of ECF volume contraction. This aspect may dominate the clinical picture. Deficits of Na^+ are close to 5–10 mmol/kg body weight unless renal failure is present (Table 3·10).

Hyponatremia

Hyponatremia reflects the Na^+:H_2O ratio in the ECF; it means that Na^+ was lost and/or water was gained in the ECF (see Chapter 6). Hyponatremia may be present for four major reasons:

1. Hyperglycemia induces the movement of water to the ECF from the ICF of cells that require insulin for glucose transport (osmoles restricted to the ECF attract water from the ICF; see Chapter 12).
2. Na^+ are lost in the urine in the osmotic diuresis and may be excreted with β-HB^-.
3. Water is ingested (because of thirst) while ADH is present.
4. There may be a laboratory error secondary to hyperlipidemia if certain techniques are used to measure the [Na^+] (see page 293). Although these patients are hyponatremic, they are still hyperosmolar because of the hyperglycemia.

Summary:
1. Osmotic shift
2. Loss of Na^+ in the urine
3. Thirst plus ADH
4. Lab error

Table 3·10
DEFICITS IN DIABETIC KETOACIDOSIS

The deficits represent typical values in an adult with DKA. As a result of the osmotic diuresis, anticipate a further renal excretion of at least 100 mmol of Na^+.

Substance	Typical deficit	Therapy
Na^+	5–10 mmol/kg	Give 1–2 liters isotonic saline plus > 2 liters ½ isotonic saline.
K^+	5–10 mmol/kg	Add 20–40 mmol/l KCl to the IV once [K^+] < 5.0 mmol/l.
Water	3 liters from ECF and 3 liters from ICF	Give free water, especially if the patient is hypernatremic.
HCO_3^-	2–3 mmol/kg	Give $NaHCO_3$ initially only if there is life-threatening acidemia ([H^+] > 100 nmol/l) and later only if the excretion of NH_4^+ is low.
Phosphate	0.5 mmol/kg	A deficit is not life-threatening, but therapy is advisable (6 mmol/hr).

Note:
On admission, a patient with diabetes in poor control has a compromised ability to excrete K^+, even though aldosterone is present and Na^+ are being excreted; these patients have a unexpectedly low TTKG (see Chapter 9, pages 340–42).

Potassium

The plasma [K^+] is usually somewhat elevated (5.3–5.7 mmol/l) but there is a large overall total body deficit of K^+ from prior renal loss early during the osmotic diuresis (Table 3·10). Hyperkalemia is due to insulin deficiency, which causes K^+ to shift out of cells; hyperkalemia may be aggravated by tissue catabolism. Some patients may be hypokalemic if they had large prior loss of K^+ (vomiting or a prolonged osmotic diuresis with high aldosterone bioactivity; see the margin).

DIAGNOSIS OF DKA

The diagnosis of DKA is usually not a difficult one to establish. It should be ruled out in all patients with metabolic acidosis and an increase in the plasma anion gap. Hyperglycemia and ketonemia (positive qualitative test for acetoacetate in a serum dilution of 1:8) are sufficient criteria in a patient likely to have IDDM. The fall in plasma [HCO_3^-] should initially approximate the increase in the plasma anion gap, but this equality is a mirage (Table 3·11).

Pitfalls in the Laboratory Diagnosis of DKA

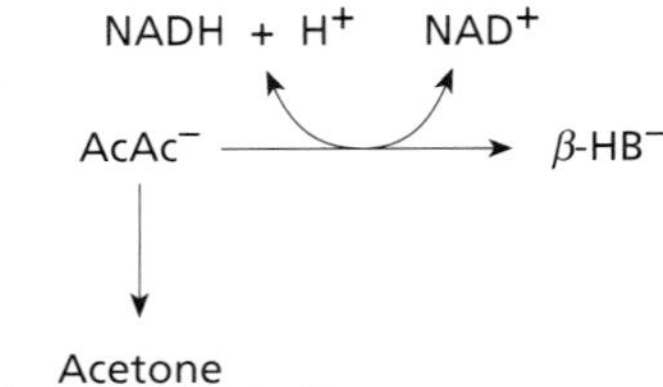

β-Hydroxybutyric acidosis but little acetoacetic acidosis: The β-HB^- and acetoacetate ($AcAc^-$) are in equilibrium because of high activity of the enzyme β-hydroxybutyrate dehydrogenase in mitochondria. The quick screening test for ketoacids (nitroprusside reaction) detects only $AcAc^-$ and acetone. Therefore, if the patient has NADH accumulation in mitochondria (e.g., in hypoxia or during alcohol metabolism; Figure 3·12, page 95), the equilibrium of this equation is displaced to the right and [$AcAc^-$] falls. Because this test may yield only a weakly positive serum ketone result, it is possible to underestimate the degree of ketoacidosis. If hyperglycemia and glycosuria are present without ketonemia or with only moderate ketonemia, suspect coexistent ketoacidosis and L-lactic acidosis. This suspicion is supported by a strongly positive test for ketones in the urine. Enzymatic determinations for β-HB^- and L-lactate anions in blood confirm that diagnosis.

A plasma anion gap that is not increased enough: An unexpectedly low plasma anion gap may result from hypoalbuminemia, an unusual degree of ketonuria, or therapy with NaCl.

1. Hypoalbuminemia resulting from diabetic glomerulosclerosis is common in longstanding diabetes mellitus and can obscure the expected increase in the plasma anion gap. Thus, if a patient has hyperglycemia, ketonemia, metabolic acidosis, ketonuria, glycosuria, and proteinuria, do not be deterred from the diagnosis of DKA by the normal or only slightly elevated level of the plasma anion gap.
2. If the patient has impaired proximal tubular reabsorption of ketoacid anions, the plasma anion gap may not be very increased in the presence of ketoacidosis. The clue to the diagnosis is metabolic acidosis and ketonuria. Suspect DKA if an analysis of the urine electrolytes reveals an unusually high urine osmolal gap and a positive urine net charge

($Na^+ + K^+ > Cl^-$; see pages 52–54). To confirm the diagnosis, quantitate β-HB^- excretion.

3. As shown in Table 3·11, even though the fall in [HCO_3^-] equals the rise in the plasma anion gap in a patient with DKA, there is still a large indirect loss of $NaHCO_3$ (Figure 3·3, page 73). This latter loss is somewhat occult because of the marked difference in the ECF volume (15 liters in the normal adult and close to 12 liters in the patient presenting with a severe degree of DKA). This difference becomes obvious when one considers the content instead of the concentration of HCO_3^- in the ECF, and it is unmasked when saline is infused and the ECF volume is restored; this therapy lowers both the concentration of HCO_3^- and the anion gap in plasma.

Table 3·11
CHANGES IN [HCO_3^-] AND ANION GAP DURING TREATMENT OF DKA

The content of ketone body anions (KB^-) in the ECF is their concentration multiplied by the ECF volume. The sum of the contents of HCO_3^- plus KB^- is lower than normal in DKA (content is only 300 mmol vs 375 mmol), even though the rise in [KB^-] equals the fall in [HCO_3^-]. Despite the conversion of KB^- to HCO_3^-, there is not an equivalent rise in the [HCO_3^-] in the ECF. This inequality reflects three processes: (1) the infusion of HCO_3^--free solution to reexpand the ECF volume; (2) the entry of H^+ in excess of KB^- into the ECF from the ICF, where they were buffered on ICF proteins; and (3) continued excretion of KB^- with Na^+ and K^+. With increased renal new HCO_3^- generation (excretion of NH_4^+), the HCO_3^- deficit will be repaired.

Condition	ECF Volume (liters)	Conc. ECF (mmol/l) HCO_3^-	KB^-	Content ECF (mmol) HCO_3^-	KB^-	Sum
Normal	15	25	0	375	0	375
Admission	12	10	15	120	180	300
Early treatment	15	12	5	180	75	255
Later treatment	15	18	1	270	15	285
Recovery	15	25	0	375	0	375

Note:
"Early treatment" refers to the expected response in the first 4–6 hours after insulin action, whereas "later treatment" would be 12–15 hours after the initial therapy.

TREATMENT OF THE PATIENT WITH DKA

Before dealing with the details of treatment, the clinician should recognize the deficits present (Table 3·10). The therapeutic approach to DKA involves attention to four major issues. First, one must reexpand the ECF volume; second, the rate of H^+ production must be diminished; third, the deficit of K^+ must be replaced, but timing is critical; fourth, one must always look for an underlying event that precipitated DKA or was a complication of it.

Reexpand ECF Volume

If the patient is in impending shock (systolic blood pressure < 90 mm Hg with tachycardia), use isotonic saline at 1000 ml/30 minutes until systolic blood pressure > 100 mm Hg or for the first 2–3 liters. If the patient is severely acidemic ([HCO_3^-] < 5 mmol/l), give some of the Na^+ as isotonic $NaHCO_3$ instead of NaCl until the plasma [HCO_3^-] is in the 5–6 mmol/l range. Once the patient is hemodynamically stable (systolic blood pressure > 100 mm Hg), use 1/2 isotonic NaCl at 0.5–1 liters/hr, depending on the remaining degree of ECF volume depletion.

In the first several hours of treatment, the fall in blood glucose will be largely due to dilution (reexpansion of the ECF volume) and renal excretion (from the rise in GFR)—actions that are not caused by insulin (Table 12·3, page 434); the rate of fall is close to 100 mg/dl/hr (5.5 mmol/l/hr). Glucose should be added to the infusion once the blood sugar reaches 250 mg/dl (12–15 mmol/l).

Clinical pearls:

- The fall in glucose early in therapy reflects the actions of IV saline rather than insulin.
- Do not permit too great a fall in glycemia. Give glucose once the blood sugar approaches 250 mg/dl (12–15 mmol/l).

Stop Ketoacid Production

Regular (crystalline) insulin is needed to stop the formation of ketoacids. An initial bolus of 5–10 units should be given intravenously, and a continuous infusion of 0.1 units/kg body weight/hr (in normal saline) should be started. Even though the lipolytic rate declines promptly, there is a lag period of several hours before there is a net decline in the degree of ketoacidosis. Hence, expect little rise in the [HCO_3^-] in plasma in the first several hours, despite the actions of insulin. The plasma anion gap should return to normal in 8–10 hours (the time required for ketoacid anions to disappear).

K^+ excretion:

Early in therapy, the excretion of K^+ is trivial. Consider the following example: the [K^+] in urine is 30 mmol/l and the urine flow rate is 2 ml/min. In two hours (0.24 liters excreted), K^+ excretion is only 7.2 mmol—a tiny amount vs the quantity of K^+ infused.

K^+ Status

The major parameter to guide K^+ replacement therapy is the plasma [K^+]. Patients with DKA usually are quite severely K^+-depleted; therefore, if the plasma [K^+] < 5 mmol/l, add 20 mmol KCl to each liter of infusion once insulin is given. If the plasma [K^+] < 4 mmol/l, use $NaHCO_3$ with caution (HCO_3^- cause K^+ to enter into the ICF).

If the plasma [K^+] > 5 mmol/l, wait at least one hour before initiating K^+ replacement. If the plasma [K^+] < 3.5 mmol/l, the patient is profoundly K^+-depleted and special precautions should be taken (aggressive replacement of K^+, cardiac monitoring, and attention to alveolar ventilation).

Phosphate and K^+:

In the first six hours of therapy, it is also reasonable to use some K^+ with phosphate (up to 6 mmol/hr) because these patients are also phosphate-depleted. Nevertheless, it takes a considerable period of time before anabolism (synthesis of RNA, etc.) will occur in a patient during a catabolic state, even after insulin is administered. Hence, it is too optimistic to think that cells will incorporate much of the administered phosphate (the deficit is close to 100 mmol, yet only 50 mmol are needed in first 12 hours).

Underlying Illness

Always look for the factors (such as infection or myocardial infarction) responsible for initiating this metabolic emergency, and consider the events secondary to DKA, such as thrombotic complications or possibly aspiration

Alcoholic Ketoacidosis (AKA)

Clinical pearls:
- When diagnosing alcoholic ketoacidosis, look for the cause of the inhibited release of insulin from β cells; IDDM or NIDDM are not necessarily present.
- The concentration of glucose in plasma is usually lower in AKA than in DKA.

Some subjects who consume ethanol in large amounts develop ketoacidosis (Figures 3·12 and 3·13). The following features are required to develop this metabolic picture:

1. **Lack of insulin:**
   A lack of insulin may be due to IDDM, but this disease need not be present. Although destruction of β cells by pancreatitis is also possible, the most important cause of lack of insulin is inhibition of its release from β cells by an intense adrenergic response (Table 3·8, page 85). Hence, an extreme degree of ECF volume contraction from excessive vomiting is almost always part of the clinical picture of AKA.

2. **Synthesis of acetyl-CoA in hepatic mitochondria:**
   The usual precursor of large quantities of acetyl-CoA in hepatic mitochondria is fatty acyl-CoA (acetyl-CoA itself prevents rapid synthesis via pyruvate dehydrogenase, so glucose cannot be a precursor of appreciable quantities of ketoacids). A considerable lag period is required to activate this step (via low malonyl-CoA levels). Notwithstanding, if ethanol is oxidized, acetyl-CoA is generated in hepatic mitochondria and a lag period is not required (Figure 3·12).

V_{max}:
The maximum rate catalyzed by an enzyme.

3. **Conversion of ethanol to Hβ-HB:**
   The oxidation of ethanol occurs in the liver, for the most part, and is catalyzed by the enzyme alcohol dehydrogenase. The affinity of ethanol for its dehydrogenase is high so that it functions at its $\underline{V}_{max}$ with a concentration of ethanol of just several millimoles per liter. The limit on ethanol metabolism is via the rate of removal of the other product, NADH (see Figure 3·12). The following quantitative example could illustrate why β-HB^- is the predominant product of ethanol metabolism when insulin levels are low.
   - Hepatic oxygen consumption is 2 mmol/min, or 3000 mmol/day (18000 mmol of ATP produced daily).

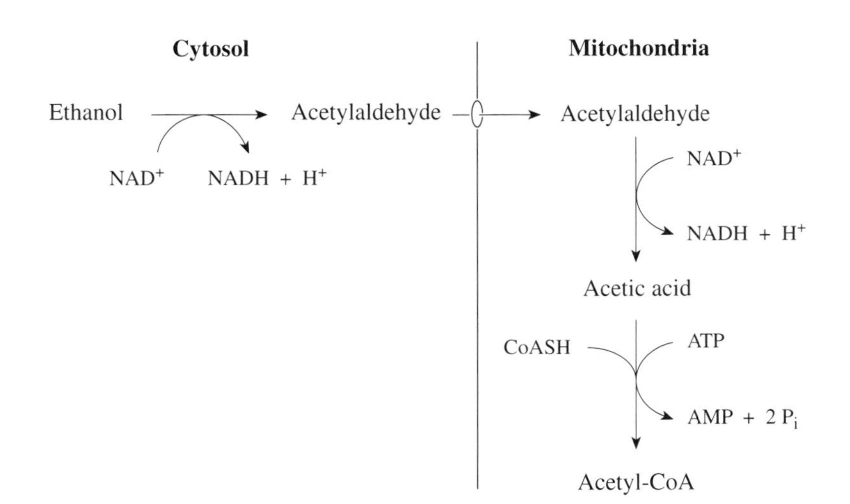

Figure 3·12
Synthesis of acetyl-CoA from ethanol
The conversion of ethanol to acetyl-CoA generates 2 (NADH + H^+) (equivalent to 6 ATP) and uses the equivalent of two ATP bonds when acetyl-CoA is formed.

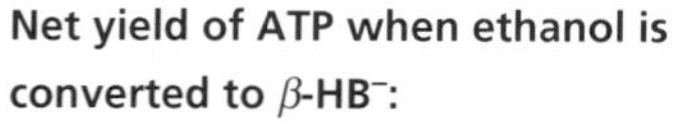

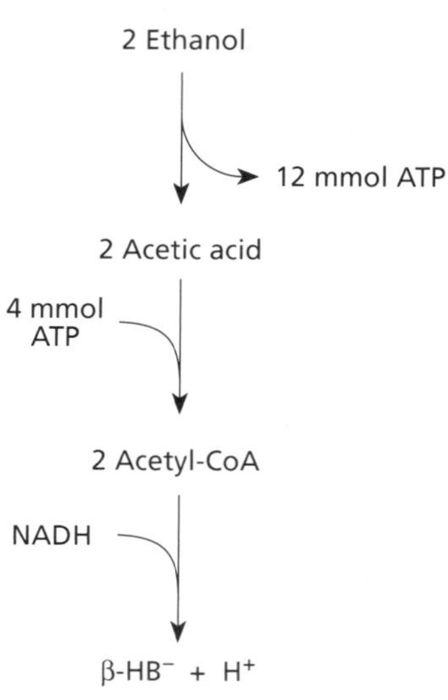

- When ethanol is oxidized in the TCA cycle, 16 mmol of ATP are produced per millimole of ethanol oxidized. Therefore, only 113 mmol (5 g) of ethanol could be metabolized by this route in a day.
- When acetic acid is the product, 2 NADH (6 mmol of ATP) are produced per millimole of ethanol consumed. Thus, this pathway could remove 3000 mmol (138 g) of ethanol per day.
- When β-HB$^-$ is the product of ethanol metabolism, 0.5 mmol of NADH is consumed for each millimole of ethanol converted to β-HB$^-$. Thus, the net yield is 2.5 mmol of ATP formed per millimole of ethanol consumed in this pathway. Now 7200 mmol (close to 350 g) of ethanol can be consumed per day, and this value is close to the rate of ethanol metabolism in vivo. Thus, AKA is commonly seen in states where ethanol is present in abundant amounts and insulin levels are low.

4. **Low oxidation of ketoacids:**
Both suppression of CNS metabolism (intoxication) and a low GFR (prerenal failure) decrease the rate of oxidation and excretion of ketoacids.

5. **Low conversion of ketoacids to acetone:**
The high NADH/NAD$^+$ ratio in the liver causes a lower concentration of AcAc$^-$ relative to β-HB$^-$. This low level of AcAc$^-$ leads to decreased synthesis of acetone. Hence, the quick test for ketoacids, which mea-

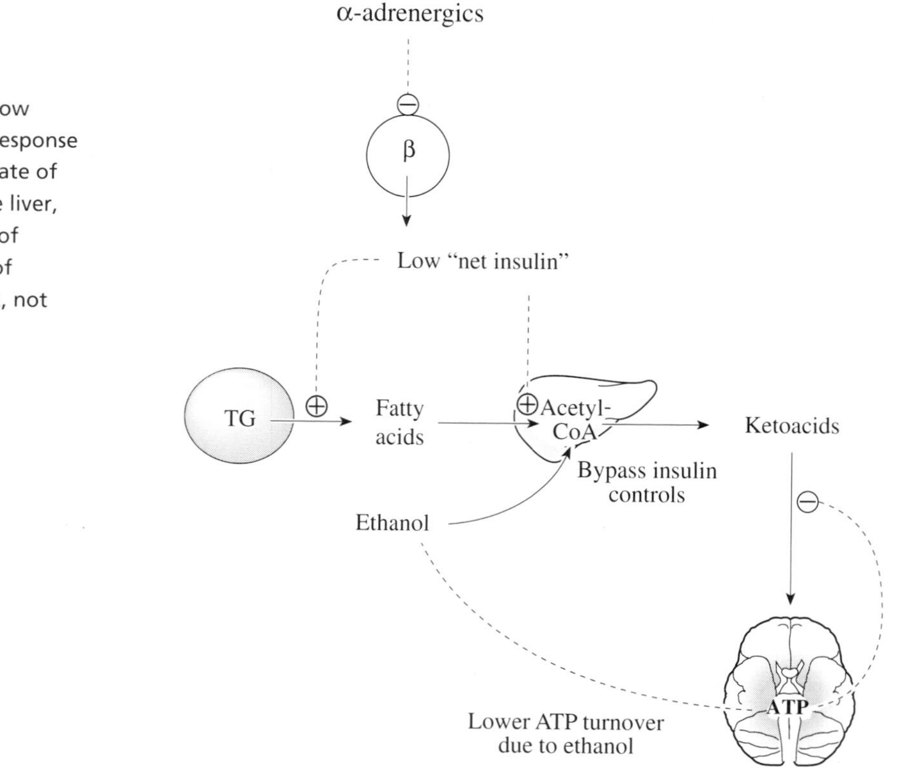

Figure 3·13
Pathophysiology of alcoholic ketoacidosis
The important features are a low level of insulin (α-adrenergic response to a low ECF volume), a high rate of formation of acetyl-CoA in the liver, and a lower rate of oxidation of ketoacids in the brain (effect of ethanol) and kidneys (low GFR, not shown).

sures $AcAc^-$ and acetone, might be unduly low given the degree of ketoacidosis (β-hydroxybutyric acidosis, see the margin).

6. **Other features:**
   The impact of the metabolism of ethanol on the the concentration of glucose will be considered in Chapter 12 (see also Figure 3·15, page 103).

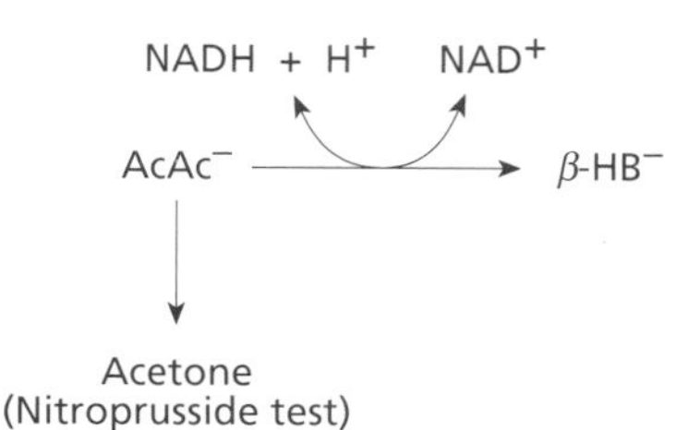

Questions

(Discussions on pages 137–38)

3·8 In hepatic mitochondria, acetyl-CoA is formed from ethanol and from fatty acids at about the same rate. Why are these rates so similar?

3·9 Must ethanol levels be elevated on admission for metabolism of ethanol to be an important cause of ketoacidosis?

DIAGNOSIS OF AKA

The major clinical features include the recent ingestion of ethanol together with a relative lack of insulin from a marked contraction of the ECF volume; this contraction, similar to shock, is often due to excessive vomiting related to alcoholic gastritis or possibly pancreatitis.

Biochemically, ketoacidosis is present by definition, but the patient might not be acidemic because of the metabolic alkalosis consequent to vomiting. The key to the diagnosis of this *mixed acid-base disturbance* is the discovery of an unusually large increase in the plasma anion gap. A note of caution must be inserted here because *hypoalbuminemia* may make the rise in the plasma anion gap less apparent (for every 10 g/l decline in the concentration of albumin in plasma, the anion gap will fall by 4 mEq/l). Given the high propensity to aspiration pneumonitis and the fact that alcohol withdrawal may stimulate ventilation, it is not uncommon to see a respiratory acid-base disorder as well.

Mixed acid-base disturbance:
- When ketoacids are added, H^+ are added to the body. The "footprint" of the added acid is the β-HB^- or a rise in the plasma anion gap.
- With vomiting, Cl^- are lost and HCO_3^- are added to the body (Figure 4·1, page 150).
- Overall, HCO_3^- are lost and added, so the $[HCO_3^-]$ may not change, but the plasma anion gap will be higher.

Hypoalbuminemia:
A low level of albumin in plasma that may be due to liver damage or poor nutrition.

Several other biochemical features merit emphasis. From an acid-base perspective, there may be *L-lactic acidosis* if the metabolism of ethanol raises the $NADH/NAD^+$ in the cytosol of hepatocytes; the lower $[AcAc^-]$ that results may give a false negative or weakly positive test for ketoacids in plasma. L-lactic acidosis may be present for a number of other reasons besides ethanol metabolism; two of the most common causes are hypoxia in tissues because of the very low ECF volume, and muscular contraction (delirium, tremors, or convulsions). The most important cause, however, is thiamine deficiency, which may cause permanent cerebral damage if it is not recognized and treated (Figure 3·14).

Causes of L-lactic acidosis related to ethanol intake:
1. Raised $NADH/NAD^+$
2. Thiamine deficiency
3. Low ECF volume and hypoxia
4. Muscular contraction

From the perspective of K^+, the insulin deficiency tends to raise the $[K^+]$ in plasma, but the excessive vomiting leads to a large kaliuresis. Therefore, the K^+ deficit is usually much larger in AKA than in DKA, and it is not uncommon to see a plasma $[K^+]$ in the 3–4 mmol/l range (an alarming K^+ deficit is present). In parallel, there is a large deficit of phosphate, as in DKA. Finally, the blood glucose level can be low, normal, or high, depend-

Insulin release in AKA:
In the setting of AKA, excessive vomiting leads to ECF volume contraction, an α-adrenergic response, and an inhibited release of insulin from β cells of the pancreas. Reexpansion of the ECF volume will remove this inhibition.

ing primarily on the quantity of glucose ingested. Ethanol levels in plasma can be measured directly or inferred from the plasma osmolal gap. An important point to keep in mind is that other alcohols (methanol, ethylene glycol) might also be present (see pages 106–08).

TREATMENT OF THE PATIENT WITH AKA

There are several life-threatening components to the clinical picture of AKA, and each requires urgent attention.

ECF Volume Reexpansion

Rapid infusion of saline that is isotonic to the patient will take care of this potential danger. The first liter should be given as quickly as possible (30 minutes) and, depending on the blood pressure, the next liter can be given over a period of 30 or 60 minutes. Anticipate that perhaps one more liter of isotonic fluid will be needed; renal losses should be small.

K^+ Deficit

Once the ECF volume is reexpanded, insulin will be released and K^+ will move rapidly from the ECF to the ICF. Accordingly, replacement of part of the K^+ deficit should begin early in treatment, but the actual amount given will depend on the plasma $[K^+]$.

Nutritional Therapy

Thiamine: It is critical that B vitamins be added to the first IV solution to replace a potential thiamine deficit. This addition will permit aerobic oxidation of glucose in the brain once ketoacids disappear (Figure 3·14).

Phosphate: Although phosphate depletion is quite marked, it will take time before anabolic reactions occur subsequent to the actions of insulin. Hence, as in DKA, one must delay replacing the majority of the phosphate deficit.

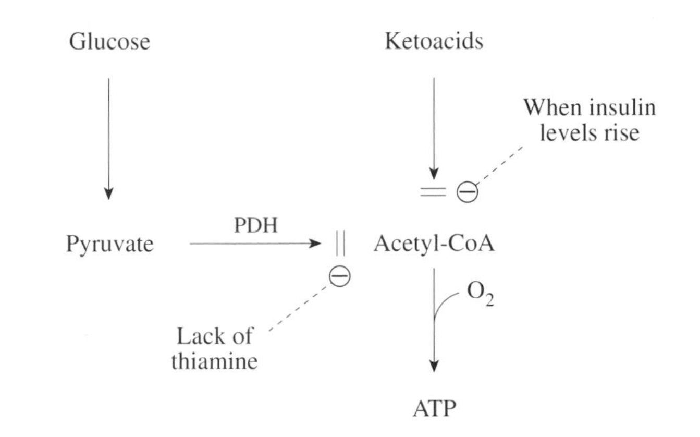

Figure 3·14
Treatment of AKA with thiamine
Thiamine is a cofactor for PDH. Without this B vitamin, oxidation of ketoacids, but not glucose, can provide the ATP needed by brain cells. Infusion of glucose and the subsequent release of insulin could lead to local overproduction of L-lactic acid and a local deficit of ATP with subsequent harm to brain cells.

Glucose: Glucose is needed only if hypoglycemia is present. In this situation, just give enough glucose to achieve normal values because metabolism of glucose will be slow in this setting (see the margin). Further, the advantage of avoiding very high levels of insulin once normovolemia is achieved is the prevention of a sudden fall in glycemia.

Insulin: Insulin need not be given initially unless there is a severe degree of hyperkalemia or acidemia. Otherwise, delay giving insulin, especially in the patient with hypokalemia.

Dose of glucose:
Assume that the concentration of glucose must rise by 3 mmol/l in a 70-kg patient with a volume of distribution of glucose that is close to 20 liters.
- 3 mmol/l × 20 liters = 60 mmol.
- 60 mmol of glucose = 10.8 g of glucose.
- Give 200 ml D_5W, or 20 ml $D_{50}W$.

L-Lactic Acidosis

BACKGROUND

Although both overproduction and underutilization of L-lactic acid result in L-lactic acidosis, their quantitative dimensions may differ by an order of magnitude.

Accumulation of L-lactic acid:
- There are two reasons for L-lactic acid to accumulate, overproduction and underutilization.
- Production of L-lactic acid can occur very quickly, but utilization of L-lactic acid occurs slowly.

Overproduction of L-Lactic Acid

Net production of L-lactic acid occurs when the body must regenerate ATP without oxygen. The bottom line is that 1 H^+ is produced per ATP regenerated from glucose. Quantitatively, because a patient at rest needs to regenerate 72 mmol of ATP per minute, as much as 72 mmol of H^+ can be produced per minute during anoxia (see the margin).

$$2\ \text{ATP} \longrightarrow 2\,(\text{ADP} + \text{P}_i) + \text{Biologic work}$$
$$\text{Glucose} + 2\,(\text{ADP} + \text{P}_i) \longrightarrow 2\,\text{H}^+ + \text{L-Lactate}^- + 2\ \text{ATP}$$

Underutilization of L-Lactic Acid

There are two major ways to remove L-lactic acid: oxidation or conversion to glucose. To oxidize 1 mmol of L-lactate anion, 3 mmol of oxygen must be consumed (18 mmol of ATP must be formed). Hence, if all organs could be "persuaded" to oxidize L-lactate anion to yield 100% of their requirement to regenerate ATP, only 4 mmol of L-lactate anion could be oxidized per minute at rest (since 72 mmol of ATP are used per minute).

$$\text{L-Lactate}^- + \text{H}^+ + 3\ \text{O}_2 + 3\,(\text{ADP} + \text{P}_i) \longrightarrow 18\ \text{ATP} + 3\ \text{CO}_2 + 3\ \text{H}_2\text{O}$$

Stoichiometry of ATP and O_2:
- The ratio of phosphorus to oxygen is 3:1.
- Because there are two atoms of O in O_2, 6 ATP can be produced per O_2 (3 ATP/0.5 O_2).
- Consumption of O_2 at rest is close to 12 mmol/min. Therefore, the amount of ATP needed per minute is 12 × 6, or 72 mmol/min.

The other major metabolic fate for L-lactate anion is glucogenesis, a pathway that occurs in the liver and in the kidney cortex. Applying stoichiometry and a quantitative analysis, the maximum rate of glucogenesis is close to 4 mmol/min in either the liver or kidney cortex. This calculation is determined by the rate at which the liver and the kidneys consume oxygen (2 mmol/min in each organ) and their requirements for ATP in this process.

$$\text{L-Lactate}^- + \text{H}^+ + 3\ \text{ATP} \longrightarrow 0.5\ \text{glucose}$$

Note:
See the margin of page 141 for extra definitions of glucogenesis, gluconeogenesis, and glucopaleogenesis.

Question

(Discussion on page 138)

3·10 An anoxic limb needs to regenerate 18 mmol of ATP per minute (25% of the ATP needed in the body) via anaerobic glycolysis. If the rest of the body was "persuaded" to oxidize L-lactate anions (+ H^+) to regenerate all needed ATP (54 mmol/min, equivalent to the utilization of 9 mmol of O_2), would L-lactic acid accumulate?

CLINICAL PICTURE

Pathogenesis

Causes of L-lactic acidosis:

- The cause of Type A is usually hypoxia (acute problem).
- The cause of Type B is usually a liver problem (can be chronic).
- The treatment depends on the cause; in Type A, one must stop H^+ production (increase O_2 delivery to tissues).

L-Lactic acidosis, a common cause of metabolic acidosis, often occurs in a life-threatening situation (Table 3·12). The most common cause of L-lactic acidosis is relative hypoxia (when the O_2 demand exceeds the O_2 supply—Type A L-lactic acidosis). Hypoxia may be due to hypoxemia, hypotension, or impaired blood supply to an organ. There are a number of causes of L-lactic acidosis in which hypoxia does not play a major role (Type B L-lactic acidosis); almost all of the patients with Type B L-lactic acidosis have liver problems.

Table 3·12
CAUSES OF L-LACTIC ACIDOSIS

Type A: Deficit of oxygen

- Lung problem (low $P_{a}O_2$)
- Circulatory problem (poor delivery of O_2)
- Hemoglobin problem (low capacity of blood to carry O_2)

Type B: Compromised metabolism of L-lactate without hypoxia

- Excessive formation of L-lactic acid (increased glycolysis as a result of low ATP, e.g., from inhibitors of mitochondrial generation of ATP, such as cyanide, or from the presence of agents that uncouple oxidative phosphorylation)
- Insufficient utilization of L-lactic acid
  - PDH problem (from a deficiency of thiamine or an inborn error)
  - Increased availability of other fuels (fatty acids)
  - Low flux through the ATP generation system (less biological work)
- Decreased conversion of L-lactate to glucose, a liver problem
  - Destruction or replacement of cells in the liver (see the margin)
  - Defect in glucogenesis (from an inborn error or from inhibitors of glucogenesis, such as drugs, ethanol, tryptophan)

L-Lactic acidosis and malignancy:

- Malignant cells produce more L-lactic acid than do normal cells under aerobic conditions (mechanism unknown). More L-lactic acid will be produced if the tumor outgrows its blood supply or if there is a nutritional deficit (thiamine).
- Low removal of L-lactate is usually a liver problem resulting from replacement of mass or inhibition of gluconeogenesis by tumor products.
- Drugs used for therapy can aggravate L-lactic acidosis. Even adding $NaHCO_3$ can increase the production of L-lactic acid.

Quantitative Aspects

In Type A L-lactic acidosis, the rate of H^+ production is close to 72 mmol/min with total body anoxia. Hypoxia rather than anoxia is usually the case; thus, survival for more than a few minutes depends entirely on delivering more O_2 to hypoxic tissues. To develop L-lactic acidosis, a supply of glucose is needed. The major endogenous source of glucose is hepatic glycogen, but the glycogen in skeletal muscle can be converted to L-lactic acid if there is a specific stimulus for glycogenolysis in muscle (e.g., a sprint).

Question

(Discussion on pages 138–39)

3·11 If a patient has hypoxia but little glycogen in the liver, will L-lactic acidosis develop? If not, what changes would you expect to find in the concentration of metabolites in blood?

Diagnosis

The diagnosis of Type A L-lactic acidosis must be made quickly because of the high rate of H^+ production or, more likely, the failure to generate ATP quickly enough. This form of L-lactic acidosis is easily established on the basis of the clinical setting of low O_2 content (severe anemia or cyanosis) or much more commonly on very poor delivery of O_2 to tissues (shock). There may also be symptoms of ischemia in one region of the body. However, for a severe degree of L-lactic acidosis, the patient must also be hypotensive because:

1. with local arterial inflow obstruction, the reduction in O_2 delivery is accompanied by a reduction in the delivery of glucose, the precursor of L-lactic acid;
2. a normally perfused liver yields glucose at a reasonable rate via metabolism of L-lactic acid.

In Type B L-lactic acidosis, the diagnosis may be less obvious. There is usually evidence of a liver problem or of the intake of drugs that interfere with hepatic metabolism. Alternatively, there may be signs of a very large tumor load. In addition, the other causes of metabolic acidosis associated with an increase in the plasma anion gap should be ruled out (Table 3·1, page 71).

The diagnosis of Types A and B L-lactic acidosis can be confirmed with an enzymatic determination of the plasma L-lactate concentration.

TREATMENT OF L-LACTIC ACIDOSIS

- In Type A L-lactic acidosis, the rate of production of H^+ must be decreased.

Type A L-Lactic Acidosis

The only effective treatment of Type A L-lactic acidosis is to stop the production of H^+ by increasing the delivery of O_2. Elevation of blood pressure is usually required if hypotension is present. Blood, plasma, and solutions containing Na^+ and/or albumin are required in many patients with an inadequate circulating volume, depending on the etiology. In those with cardiogenic shock, myocardial function and thereby tissue perfusion should be improved. Less commonly, hypoxemia may require correction with O_2. Other considerations could include the resection of an ischemic area if it is necrotic; if not, restore its blood supply. Sepsis can cause several circulatory disturbances that lead to tissue hypoxia (it decreases O_2 delivery and interferes with O_2 extraction). Treatment of the vasodilation caused by sepsis may involve vigorous expansion of the circulating volume.

$NaHCO_3$ therapy in Type A L-lactic acidosis: The use of $NaHCO_3$ in total anoxia is of little value because of the magnitude of the H^+ load (H^+ are produced at a rate of 72 mmol/min). Nevertheless, in cases in which hypoxia is marginal and potentially reversible, $NaHCO_3$ may "buy time" to improve myocardial function, though this issue remains controversial . The Na^+ load accompanying the HCO_3^- poses a major limit to this type of therapy, and the use of loop diuretics is rarely very beneficial in this regard. If time permits, dialysis against a HCO_3^- bath might be helpful. Hence, more imaginative adjuncts to therapy are required, and some of these are explored in the following questions.

Questions

(Discussions on pages 139–40)

3·12 If $NaHCO_3$ was given as treatment for L-lactic acidosis and there was no rise in the plasma [HCO_3^-], was the alkali of no help to that patient?

3·13 Why might the rate of L-lactic acid production rise with alkali therapy?

3·14 What metabolic adaptations prolong survival in hypoxic environments?

3·15 Why is the L-lactic acidosis of exercise better tolerated than the L-lactic acidosis of shock, even if the former is more severe in degree?

3·16 Is a rise in L-lactic acid production beneficial or harmful to a patient with L-lactic acidosis?

Type B L-lactic Acidosis

- Type B L-lactic acidosis does not have the same urgency as Type A L-lactic acidosis because it is not associated with a problem in generating ATP; in addition, the rate of H^+ accumulation is much lower.

Ethanol-induced L-lactic acidosis: One cause of L-lactic acidosis in an alcoholic is hepatic ethanol metabolism, which generates NADH and leads to the diversion of pyruvate to L-lactate (Figure 3·15). Ethanol metabolism must be ongoing for this form of L-lactic acidosis to occur; furthermore, because all other tissues can oxidize the L-lactate produced, the degree of L-lactic acidosis should be mild if ethanol metabolism is the sole cause. No specific treatment is required.

There are other ways that an alcoholic may develop L-lactic acidosis. For example, L-lactic acid can be produced at the time of a convulsion or extreme agitation, which occur in delirium tremens. L-Lactic acid is removed at an extremely slow rate because of the ongoing oxidation of fat-derived fuels. Hence, this cause for overproduction of L-lactic acid might be remote from the clinical presentation.

A third cause of L-lactic acidosis in an alcoholic is thiamine deficiency (discussed below).

L-Lactic acidosis from thiamine deficiency: This form of L-lactic acidosis is entirely preventable by giving *thiamine*. The clinical setting in which this form of L-lactic acidosis is most prevalent is in the alcoholic or in patients with an inadequate nutritional intake. The major danger is not the acidosis but rather the CNS lesion that is caused when glycolysis is accelerated to supply ATP that is no longer supplied by the oxidation of ketoacids. The rate of glycolysis accelerates when ketoacids disappear following a rise in insulin levels (after glucose and NaCl are given).

Thiamine (vitamin B_1):
A cofactor for pyruvate dehydrogenase (PDH). A deficiency of thiamine prevents glucose from being oxidized aerobically (Figure 3·14, page 98). Thiamine is also a component of 2-oxoglutarate dehydrogenase, a TCA cycle enzyme; this enzyme is less affected by minor deficits of thiamine.

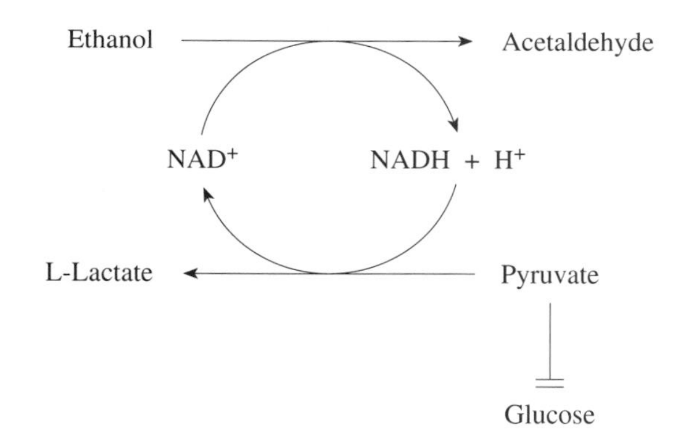

Figure 3·15
Ethanol induced L-lactic acidosis
The metabolism of ethanol is required to raise the NADH/NAD^+, which favors the conversion of pyruvate to L-lactate, not to glucose.

Drug-induced L-lactic acidosis: The drugs in question (Table 3·12, page 100) accumulate if intake is high or excretion is low (especially when prerenal failure is caused by the use of metformin or phenformin). Although the degree of L-lactic acidosis may be severe, chances of survival are good. The measures required are neutralization of the excess H^+ with $NaHCO_3$, slowing of the H^+ production with insulin in the case of phenformin-induced L-lactic acidosis, acceleration of L-lactate metabolism (via *dichloroacetate*), and elimination of the offending drug by renal excretion or dialysis.

Dichloroacetate:
A drug that activates PDH.

Section for the More Curious

An interesting example of drug-induced L-lactic acidosis was described in patients with AIDS who were treated with zidovudine (AZT). Several hypotheses have been offered to explain the pathophysiology of this L-lactic acidosis. Many of these patients have a mitochondrial lesion: a mitochondrial myopathy with ragged red fibers and coarse granular deposits on histology. Biochemical studies have shown that prolonged or high doses of AZT may induce a decrease in enzymes in the electron transport system (ETS) of mitochondria. This pathophysiology will be explored further in the discussions of Questions 3·17 and 3·18.

There are also a large number of conditions, many of them inborn errors of metabolism, that have in common a degree of L-lactic acidosis and the same characteristic findings on muscle biopsy: a mitochondrial myopathy. The general feeling is that there is a defect in mitochondrial ATP generation; hence, more L-lactic acid accumulates when the glycolytic rate is high because of an increased demand for turnover of ATP (Table 3·13).

Questions

(Discussions on pages 140–41)

3·17 A 34-year-old male patient with AIDS has extreme muscular weakness and chronic L-lactic acidosis related to AZT. He had a mitochondrial myopathy on biopsy. When he exercises, his L-lactic acidosis does not become more severe. Did the lesion in his muscle mitochondria cause his L-lactic acidosis?

3·18 The L-lactic acidosis of the patient in Question 3·17 was greatly aggravated by ethanol intake. What might this effect of ethanol imply?

Table 3·13
POSSIBLE LESIONS IN LACTIC ACIDOSIS WITH MITOCHONDRIAL MYOPATHY

Associated with more severe L-lactic acidosis during exercise

- Inadequate generation of NADH in mitochondria
  - Low activity of enzymes in the TCA cycle
  - Failure to supply enough acetyl-CoA in mitochondria (low oxidation of fatty acids and/or pyruvate)
- Problem with net generation of ATP for the cell
  - Low activity of the electron transport system
  - Problems with oxidative phosphorylation
  - Problems with the adenine nucleotide transporter
  - Uncouplers of oxidative phosphorylation

Not necessarily associated with more severe L-lactic acidosis during exercise

- Defect in redox transport such that the NADH/NAD^+ ratio in cytoplasm reflects that of mitochondria (Figure 3·16)

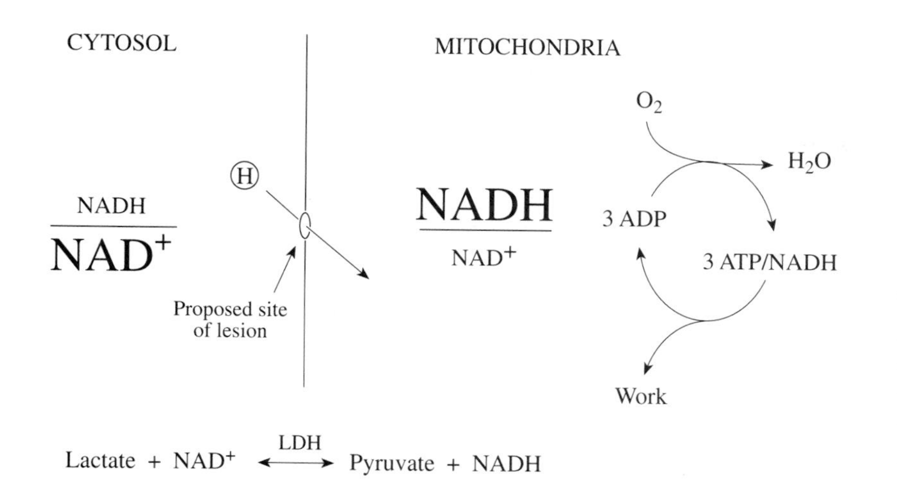

Figure 3·16
Possible defect in redox transport and L-lactic acidosis
This figure depicts the proposed site of the lesion induced by AZT. Normally, the $NADH/NAD^+$ ratio is much lower in the cytosol than in the mitochondria. This difference in ratio is maintained by an energy-dependent transport step in the inner mitochondrial membrane (glutamine/aspartate antiporter). Once the reducing power enters the mitochondria, it is transported down the electron transport system and leads to the conversion of ADP to ATP. Should the above barrier for reducing power be compromised, the high $NADH/NAD^+$ ratio in mitochondria might be "transmitted" to the cytosol and result in much higher L-lactate levels for a given concentration of pyruvate (the LDH reaction is close to thermodynamic equilibrium; LDH = lactate dehydrogenase). Reproduced with permission (*JASN* 3: 1212–1219, 1992).

Organic Acid Load from the GI Tract (D-Lactic Acidosis)

Certain bacteria in the GI tract may convert carbohydrate (cellulose) into organic acids. Three factors lead to the overgrowth of bacteria and give them ample time to generate organic acids. First, slow GI transit (from blind loops, obstruction, drugs decreasing GI motility) leads to bacterial growth; second, a change of the normal flora, which occurs with antibiotic therapy, can lead to a large population of bacteria that can form D-lactic acid and other organic acids. Since humans metabolize the D-lactic acid isomer more slowly than L-lactate and production rates can be very rapid, a severe degree of acidosis can result. Third, if these bacteria have an adequate supply of nutrients, they may produce D-lactic acid at rates that exceed its removal. Hence, feeding of carbohydrate-rich food can aggravate D-lactic acidosis in a patient with GI bacterial overgrowth.

There are three additional points that should be noted with respect to D-lactic acidosis:

1. The usual laboratory test for "lactate" is specific for the L-lactate isomer. Hence, with D-lactic acidosis, the laboratory report for "lactate" will not be elevated. To confirm this diagnosis, measure the plasma D-lactate concentration with a specific enzymatic assay.
2. GI bacteria produce amines and other compounds that may cause clinical symptoms related to central nervous system (CNS) dysfunction (personality changes, gait changes, confusion, etc.). Be wary of the diagnosis in the absence of CNS abnormalities.
3. Some of the D-lactate will be lost in the urine if the GFR is not too low. Hence, the degree of rise in the plasma anion gap may not be as high as expected for the fall in the plasma $[HCO_3^-]$.

Focus treatment on the gastrointestinal problem and ensure that the patients do not die of severe metabolic acidosis (give $NaHCO_3$ and possibly insulin or dichloroacetate, if necessary; see the margin).

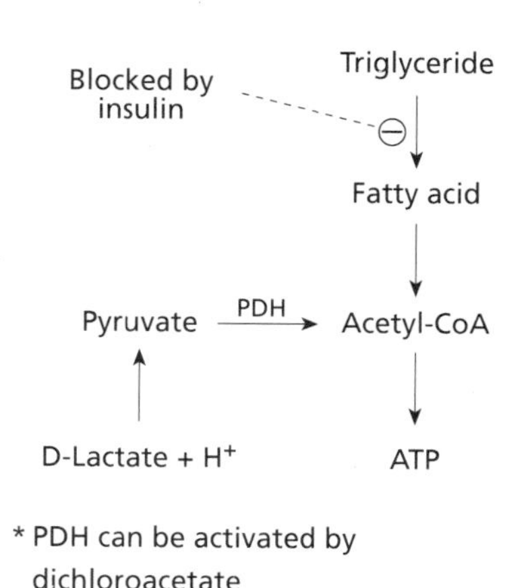

* PDH can be activated by dichloroacetate

Question

(Discussion on page 142)

3·19 If the same number of bacteria are growing in the small bowel rather than the colon, why might more D-lactic acid be formed?

Metabolic Acidosis Caused by Toxins

THOSE WITH AN INCREASED PLASMA OSMOLAL GAP

- Suspect the presence of toxins when an intoxicated patient has metabolic acidosis with an increased plasma anion gap.
- If the *plasma osmolal gap* is increased, treat the patient with ethanol.
- Confirm the diagnosis with a specific assay for the offending alcohols.

Plasma osmolal gap (measured osmolality – calculated osmolality):
Calculated osmolality = 2 × [Na^+] + [glucose] + [urea], in mmol/l.
If values are in mg/dl, convert [glucose] to mmol/l by dividing [glucose] by 18, and [urea] by 2.8 (see Table 2·3, page 52).

Methanol (wood alcohol):
An alcohol that is used as thinner for shellac and varnish, as windscreen and gasoline antifreeze, and as fuel for alcohol-burning devices.

Methanol ⟶ Formic Acid:
One oz (32 ml) of methanol is close to 32 g. Because the molecular weight of methanol is 32, 1 oz of methanol can yield 1000 mmol of formic acid. Hepatic alcohol dehydrogenase has a maximal activity of close to 7000 mmol/day, but constraints on the removal of NADH will cause a much lower rate of H^+ production (< 1 mmol/min).

Methyl Alcohol (Methanol) Intoxication

Methanol is an inexpensive and widely available alcohol. Overall, methanol should always be suspected in the differential diagnosis of the etiology of metabolic acidosis with an increased anion gap in plasma, especially if the ECF volume of the patient is not contracted and renal failure is not present. Methanol itself is nontoxic, but it is metabolized to a toxic product (formaldehyde) by alcohol dehydrogenase (Table 3·14). One ounce of methanol can yield 1000 mmol of formic acid; when methanol levels in blood exceed 50 mmol/l, 1 mmol of H^+ can theoretically be produced per minute (see the margin). The products of metabolism are particularly toxic to the CNS, especially the optic nerve.

Therapy: The affinity of ethanol for alcohol dehydrogenase is 100-fold greater than either methanol or ethylene glycol. Therefore, the principle of therapy is to diminish the metabolism of methanol with ethanol and thus prevent formation of the toxic products formaldehyde and formic acid. In addition, one must then promote the removal of methanol by dialysis.

1. **Arrest methanol metabolism with ethanol:**
   Establish therapeutic ethanol levels (100 mg/dl, or 22 mmol/l) by giving 0.6 g ethanol/kg IV or orally or give 4 oz whiskey orally. Maintain a therapeutic level of ethanol in the blood for the duration of the intoxication. Give chronic drinkers 0.15 g/kg/hr IV or 2 oz whiskey per hour orally. Give nondrinkers 0.07 g/kg/hr IV or 1 oz whiskey per hour orally. Monitor the ethanol levels during therapy.

2. **Correct the acidemia:**
   At high blood methanol levels, acidemia may be severe, although the rate of production of formic acid is not usually that rapid in most cases.

Nevertheless, if acidemia is severe, therapy with $NaHCO_3$ might be indicated.

3. **Remove methanol:**
If levels of methanol exceed 50 mg/dl (15 mmol/l), dialysis should be instituted. Hemodialysis is most efficient, but, if not available, peritoneal dialysis removes some methanol. Ethanol, which is also removed by dialysis, must be replaced. Adding ethanol to the dialysate is the easiest way to ensure adequate blood levels.

Clinical pearls:

- Suspect methanol intoxication if there is:
  1. a history of ingestion in an intoxicated patient;
  2. the sweet odor of methanol;
  3. an increased plasma osmolal gap in a patient with metabolic acidosis and an increased plasma anion gap.

  The diagnosis is confirmed by finding elevated levels of methanol in blood.
- Because chronic drinkers have higher levels of alcohol dehydrogenase, they may metabolize methanol more rapidly. They are therefore exposed to increased toxicity. On the other hand, if they have ingested ethanol as part of the intoxication, they will be protected as long as they have high ethanol levels in their blood.

Table 3·14
PRODUCTS OF METABOLISM OF ETHANOL, METHANOL, AND ETHYLENE GLYCOL

Because ethanol is the preferred substrate, competition prevents the metabolism of methanol and ethylene glycol. Although ethylene glycol is not really an alcohol it can be considered as a "dialcohol" because it has two adjacent carbons, each with a hydroxyl group. (AlcDH refers to hepatic alcohol dehydrogenase.)

Alcohol	AlcDH →	Aldehyde	→	Carboxylic acid
Ethanol	→	Acetaldehyde	→	Acetic acid
Methanol	→	Formaldehyde	→	Formic acid
Ethylene glycol	→	Glycoaldehyde	→	Glycolic acid
				↓
				Glyoxylic acid
				↓
				Oxalic acid

Ethylene Glycol (Antifreeze) Intoxication

- Suspect ethylene glycol intoxication if the patient is intoxicated, has metabolic acidosis with an increased plasma anion gap and osmolal gap, urine oxalate crystals, and acute tubular necrosis.

Ethylene glycol toxicity is a "chilling thought."
B. S.

Ethylene glycol, which is readily available, relatively inexpensive, and pleasant tasting, might be ingested as an intoxicant. It causes fulminant metabolic acidosis, severe CNS toxicity, and acute tubular necrosis. As with methanol, toxicity results from the products of ethylene glycol metabolism (Table 3·14). Following the initial toxicity associated with profound metabolic acidosis and CNS manifestations (confusion, coma, seizures), patients may develop congestive heart failure during therapy because of the large load of $NaHCO_3$ given coupled with acute renal failure. Those who survive usually have acute tubular necrosis that is generally of the oliguric form. Ethylene glycol intoxication should always be suspected in patients

Toxins from ethylene glycol:
One product of the metabolism of ethylene glycol is oxalic acid. This acid precipitates as its Ca^{2+} salt and might be a basis of the development of acute tubular necrosis.

with metabolic acidosis and an increased plasma anion gap, especially if the patient appears intoxicated and denies intake of ethanol or if the odor of ethanol is not evident. The index of suspicion should be increased greatly by finding oxalate crystals in the urine. As with methanol, finding an increased plasma osmolal gap is helpful in the absence of ethanol (see the discussion of Question 3·20, page 142). The diagnosis is confirmed by detecting ethylene glycol in the blood.

Therapy: The principles of initial therapy using ethanol administration for ethylene glycol intoxication are identical to those for methanol (see the margin). The additional complication of acute tubular necrosis may limit the quantity of $NaHCO_3$ that can be given, since the patient with this form of intoxication is generally oliguric. The use of nasogastric suction and pentagastrin stimulation of gastric acid secretion may lessen both the acidemia and the pulmonary edema while dialysis is being arranged (see page 82). Early dialysis is critical because of the acute renal failure. All patients who have ingested this toxin should probably be dialyzed because of the toxicity. As with methanol, hemodialysis is the preferred means of removing ethylene glycol; however, if it is not available, peritoneal dialysis should be implemented. Again, ethanol levels should be maintained during hemodialysis by infusion or addition to the bath.

Kinetics of alcohol metabolism:
If the concentration of alcohol is several times higher than the K_m, the enzyme proceeds at close to maximum velocity. As the concentration falls below the K_m, the rate of removal falls appreciably. Since the K_m for ethanol is 100-fold lower than for methanol or ethylene glycol, lower levels of ethanol prevent the metabolism of higher levels of methanol or ethylene glycol.

Methanol and ethylene glycol intoxications are characterized by higher rates of H^+ production at high blood levels; at blood levels below 40 mmol/l, rates decline appreciably. Therefore, at a low degree of intoxication, the agent can be removed by renal excretion without undergoing metabolism.

THOSE WITH A NORMAL PLASMA OSMOLAL GAP

Toluene Inhalation (Glue Sniffing)

- The patient who has sniffed glue has metabolic acidosis, but the plasma anion gap is usually not elevated appreciably.
- Hypokalemia is often present.
- There is an increased osmolal gap in the urine (NH_4^+ + hippurate).

Toluene intoxication has a variable presentation, depending on the clinical setting. While these patients do not have an increased plasma osmolal gap, they share certain clinical and metabolic features with the alcohol intoxications. They might even "enter the back door" of this classification by having an increased plasma osmolal gap from ingestion of ethanol as part of their intoxication.

It has now been recognized that the metabolic acidosis of toluene intoxication is indeed due to an acid load (hippuric acid); it had previously been incorrectly classified as distal renal tubular acidosis because of its frequent presentation as a normal plasma anion gap type of metabolic acidosis associated with hypokalemia and a urine pH close to 6.0.

Toluene is methylbenzene, a volatile compound that is inhaled during the sniffing of glue and that accumulates in fat. Its primary routes of removal are metabolism in the liver and exhalation via the lungs. For every toluene metabolized, one H^+ is added. Metabolism of toluene (Figure 3·17) leads to the formation of benzoic acid and the eventual excretion of hippurate in the urine. When hippurate is excreted in the urine, a cation—NH_4^+, K^+, or Na^+—must also be excreted. To the extent that hippurate is excreted with NH_4^+, the acid load is neutralized by renal production of HCO_3^-; excretion with K^+ will result in metabolic acidosis and hypokalemia. The excretion of hippurate with Na^+ will deplete the ECF volume and decrease the GFR. As a result, the rates of excretion of both NH_4^+ and hippurate will decrease; these low rates of excretion will prolong the acidemia and possibly increase the plasma anion gap (see the margin).

If Na^+ intake is maintained and ECF volume contraction is avoided, toluene intoxication should present as metabolic acidosis with a normal plasma anion gap. On the other hand, if ECF volume contraction is severe, hypokalemia might be significant, the plasma anion gap might increase, and the acidemia might be more severe (from reduced HCO_3^- generation that is due to decreased NH_4^+ excretion).

One must recognize that despite the lack of an increased plasma anion gap, toluene intoxication causes acidemia from an organic acid load. The specific clue to the diagnosis—excretion of NH_4^+ at an increased rate—may be identified in different ways, depending on the cation(s) accompanying the hippurate (see the discussion of Question 3·21, pages 142–43, and the margin).

Note:
If hippurate is excreted with K^+ that were derived from the ICF in conjunction with entry of Na^+ into cells, the ECF volume will be contracted further.

Note:
Some patients with toluene intoxication have been reported to have a low rate of NH_4^+ excretion. In these cases, there is some other process that reduces either NH_3 availability in the renal medullary interstitium or distal nephron H^+ secretion. In these cases, hippurate will be excreted with Na^+ and K^+, and the urine net charge will be very positive (the sum of Na^+ and K^+ greatly exceeds Cl^-).

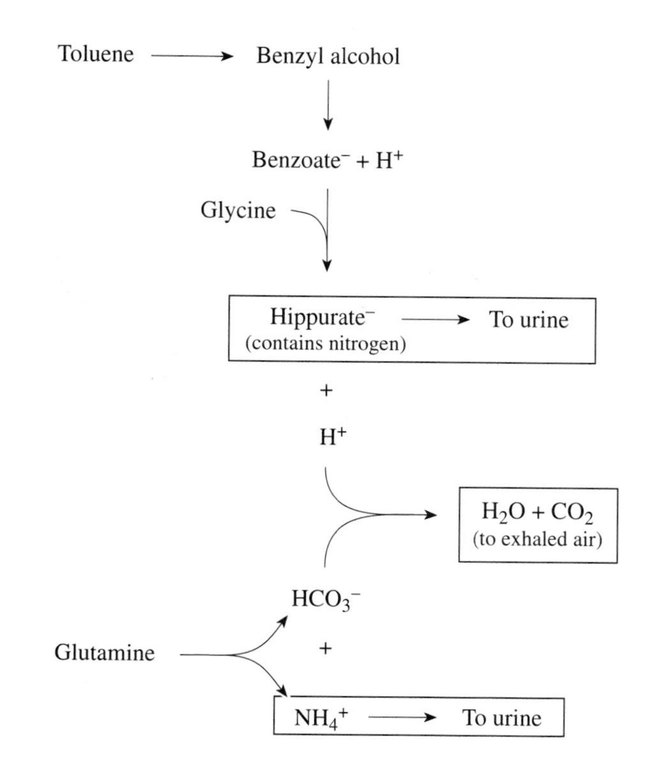

Figure 3·17
Metabolism of toluene
Metabolism of toluene occurs in the liver using a series of cytoplasmic enzymes, cytochrome P_{450}, alcohol and aldehyde dehydrogenases. Hippuric acid is formed in hepatic mitochondria. When hippurate is excreted with NH_4^+, there is no acid-base impact. Reproduced with permission from *Clinical Detective Stories.*

Hippurate excretion:
Hippurate is excreted at a rapid rate (it is both filtered and secreted), but peak excretion is often 24 hours after exposure to toluene; this delay suggests the need for enzyme induction in the liver (cytochrome P_{450}). Those who sniff glue frequently have had enzyme induction and therefore may have more severe acidemia as a result of rapid metabolism.

Therapy: Therapy should be implemented in several directions simultaneously.

1. Correct the K^+ deficits, which may be profound.
2. Correct the ECF volume depletion using a combination of NaCl and $NaHCO_3$. If hypokalemia is significant, delay aggressive $NaHCO_3$ therapy until the administration of K^+ because $NaHCO_3$ will aggravate the degree of hypokalemia.
3. In most cases, renal NH_4^+ excretion is enhanced and the acidemia will therefore correct itself over several days without $NaHCO_3$ therapy. If the acidemia is severe at presentation, use some $NaHCO_3$ with the above caveat concerning $[K^+]$. If the acidemia persists for several days, reassess the rate of excretion of NH_4^+.

Questions

(Discussions on pages 142–43)

3·20 A 26-year-old intoxicated man is brought to the emergency room. His friends state that he ingested several ounces of methanol. Other than intoxication, his physical exam was normal. His lab results were as follows:

Na^+	mmol/l	142	$[H^+]$	nmol/l	40
K^+	mmol/l	3.7	P_aco_2	mm Hg	40
Cl^-	mmol/l	105	Osmolality	mosm/kg	360
HCO_3^-	mmol/l	25	Glucose	mmol/l (mg/dl)	7 (126)
Urea	mmol/l (mg/dl)	5 (14)	Albumin	g/l	40

Should one take the allegation of methanol seriously? Explain your reasoning. What should the course of action be?

3·21 What is the specific clue suggesting glue sniffing as the diagnosis in a patient presenting with a normal plasma anion gap type of metabolic acidosis and hypokalemia?

3·22 What factors contribute to the hypokalemia and K^+ depletion in glue sniffers?

ASA = acetylsalicylic acid.
ASA^- = salicylate anions.
HASA = nonionized salicylic acid.

Salicylate Intoxication

- Respiratory alkalosis usually accompanies ASA intoxication, but metabolic acidosis may be prominent in children.
- The problem is ASA^- levels in tissue.
- The treatment is to promote excretion of ASA^- and avoid both acidemia and severe alkalemia.

Although ASA intoxication is very common, it rarely causes an appreciable degree of metabolic acidosis. When metabolic acidosis does occur, children are the most likely to be affected; the younger the child, the more likely the metabolic acidosis. The most common acid-base disturbance associated with ASA intoxication is respiratory alkalosis from central stimulation of respiration. Acid-base disturbances tend to accompany acute ASA intoxication but are less prominent in chronic ASA intoxication (see the margin).

Clinical note:
Because the toxic level of ASA^- is only 3–5 mmol/l, the plasma anion gap associated with a severe degree of metabolic acidosis is to a minor extent the result of ASA^-; ketoacid anions, L-lactate anions, and possibly other unidentified organic anions are the major causes for an elevated anion gap. Other findings could include adult respiratory distress syndrome (ARDS), a bleeding disorder (low vitamin K), and hypokalemia.

Diagnosis: The diagnosis of ASA intoxication might be suspected from a history of ingestion or symptoms of tinnitus and lightheadedness and the presence of a respiratory alkalosis complicating the metabolic acidosis. The suspicion is increased by finding unexplained ketosis (ASA^- activates hepatic lipase), hypouricemia (high-dose ASA^- is uricosuric) and an increased urine net charge from ASA^- excretion (Na^+ and K^+ in urine greatly exceed Cl^-). The diagnosis is confirmed by detecting ASA^- in the blood.

Treatment: Generally, metabolic acidosis is not a serious feature of ASA intoxication. Dialysis should be instituted for ASA^- levels above 90 mg/dl (6 mmol/l) and should be considered for levels greater than 60 mg/dl (4 mmol/l). In the absence of severe metabolic acidosis, the therapeutic efforts in ASA intoxication are to promote ASA^- excretion via the following maneuvers:

Methyl salicylate (oil of wintergreen):
Ingestion of oil of wintergreen provides a large load of readily absorbed ASA^- and leads to toxic ASA^- levels. Treatment must be aggressive because patients can worsen "before your eyes."

1. **Alkali therapy:**
   If the patient with ASA intoxication has metabolic acidosis, acidemia should be corrected because it increases the concentration of HASA in the blood. Because this uncharged form crosses cell membranes, its diffusion into brain cells is facilitated, and toxicity is promoted (Figure 3·18).
   In an analogous manner, an alkaline urine pH promotes ASA^- excretion by converting HASA to ASA^- and thereby retards HASA reabsorption by nonionic diffusion. The problem with aggressive HCO_3^- therapy is that the patient with respiratory alkalosis may become very alkalemic. In severe intoxications, hemodialysis or, in its absence, peritoneal dialysis should be used. Infusion of ½ isotonic saline containing 50 mmol of $NaHCO_3$ per liter at a rate of 300–500 ml/hr may produce an alkaline diuresis. If the patient cannot tolerate that Na^+ load, a loop diuretic may be used to promote Na^+ excretion. The blood pH should be monitored hourly, and if it exceeds 7.55, 250 mg acetazolamide should be given to promote HCO_3^- excretion.

2. **Use of acetazolamide:**
   Acetazolamide, a carbonic anhydrase inhibitor (CAI), has been advocated in the therapy for ASA^- intoxication, supposedly because of its ability to alkalinize the urine (alkalinization of the urine should enhance ASA^- excretion; Figure 3·18); however, its use has also been condemned. What are the issues? One detrimental effect relates to protein binding. Acetazolamide competes with ASA^- for binding to albumin and thus may enhance toxicity by increasing the free ASA^- concentration. Second, there is a possibility of inducing acidemia with acetazolamide by means of excess excretion of HCO_3^- in the urine. Acidemia will increase the amount of nonionized HASA in the blood, so that

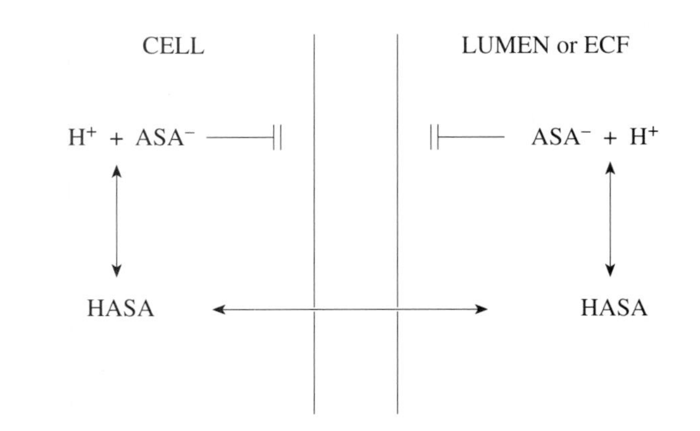

Figure 3·18
Role of alkali in ASA intoxication
Alkali therapy increases the concentration of ionized ASA^- in the lumen of the proximal convoluted tubule and in the ECF. This higher concentration should reduce diffusion of ASA^- + HASA into cells and thereby reduce the toxicity of ASA^- (the bulk of ASA^- is in the anionic rather than the HASA form).

more of this acid enters cells and toxicity increases. On the other hand, the benefit is that acetazolamide enhances ASA^- excretion, but the mechanism may not be by nonionic diffusion as a result of alkalinization of the urine (see the margin).

Increased ASA^- excretion with acetazolamide:
In the proximal tubule, the impact of acetazolamide is to increase, not lower, the $[H^+]$ of tubular fluid via inhibition of luminal carbonic anhydrase; both the $[H^+]$ and $[HCO_3^-]$ rise (see Figure 3·19).

Another important consideration is how much acetazolamide to use. The patient might develop acidemia from the HCO_3^- loss; acidemia will promote higher levels of ASA^- in brain cells and thereby increase toxicity. Measuring the levels of NH_4^+ in the urine indicates that 250 mg of acetazolamide has a tubular effect for 15–20 hours. Therefore, very little drug is necessary to achieve beneficial effects.

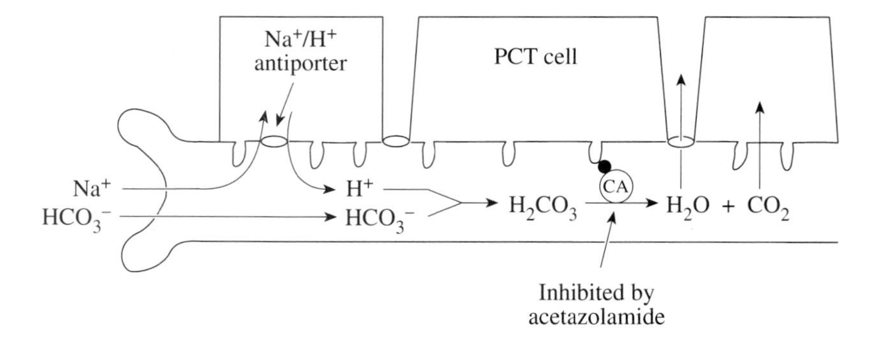

Figure 3·19
Acetazolamide and the decreased reabsorption of ASA^- in the proximal tubule
Acetazolamide inhibits the carbonic anhydrase (CA) on the luminal membrane of the PCT and thereby raises the $[H^+]$ in the lumen. Inhibition of ASA^- reabsorption is not due to tubular fluid alkalinization and is therefore more likely via another basis—perhaps inhibition of reabsorption on the organic acid transporter.

3. Dialysis:

If levels of ASA^- exceed 60 mg/dl (4 mmol/l), dialysis should be considered, particularly if further absorption is anticipated. In patients with decreased levels of consciousness, dialysis should also be considered because of the poor prognosis. Hemodialysis is more efficient in removal of ASA^-, but peritoneal dialysis can achieve substantial removal of this toxin.

Question

(Discussion on pages 143–44)

3·23 Patients A, B, and C listed below each have metabolic acidosis and an increased plasma anion gap. Which one has renal failure, which has methanol intoxication, and which has D-lactic acidosis?

Patient		**A**	**B**	**C**
Calculated Osmolality		290	290	320
$2 \times$ plasma [Na^+]	mmol/l	280	280	280
Urea	mmol/l	5	5	35
Glucose	mmol/l	5	5	5
Measured Osmolality		290	320	320

Metabolic Acidosis With a Normal Plasma Anion Gap

- Classify metabolic acidosis with a normal plasma anion gap by renal response:
  - associated with increased loss of HCO_3^-;
  - associated with low excretion of NH_4^+.

The normal renal response to acidemia is to reabsorb all of the filtered HCO_3^- and to increase new HCO_3^- generation by increasing the excretion of NH_4^+ in the urine. When the metabolic acidosis is due to the inability of the kidney to respond normally to acidemia, the clues are easily found by assessing renal HCO_3^- reabsorption and NH_4^+ excretion. When either of these functions is compromised, there is a renal cause of metabolic acidosis that is termed "renal tubular acidosis (RTA)" (Table 3·15 and Figure 3·20).

Table 3·15
METABOLIC ACIDOSIS WITH A NORMAL PLASMA ANION GAP

Excessive Excretion of HCO_3^-
- Proximal RTA
- Acetazolamide ingestion

Increased Excretion of NH_4^+
- Loss of HCO_3^- via the GI tract
- Ingestion of HCl or NH_4Cl
- Overproduction of acids with the rapid excretion of their conjugate base
  - Glue sniffing
  - Ketoacidosis with marked ketonuria
- After hypocapnia

Low Excretion of NH_4^+ (distal RTA)
- Reduced NH_3 available in the medullary interstitium
  - Decreased ammoniagenesis (from a low GFR or hyperkalemia)
  - Medullary interstitial disease
- Reduced collecting duct H^+ secretion
  - H^+ pump "failure"
  - H^+ back-leak (lumen to cell)
  - Failure of voltage augmentation of distal H^+ secretion

Difficulty with proximal RTA:
In proximal RTA, the excretion of NH_4^+ is much lower than expected for the degree of chronic acidemia. The acidosis is caused by:
- low reabsorption of HCO_3^-;
- relatively low excretion of NH_4^+.

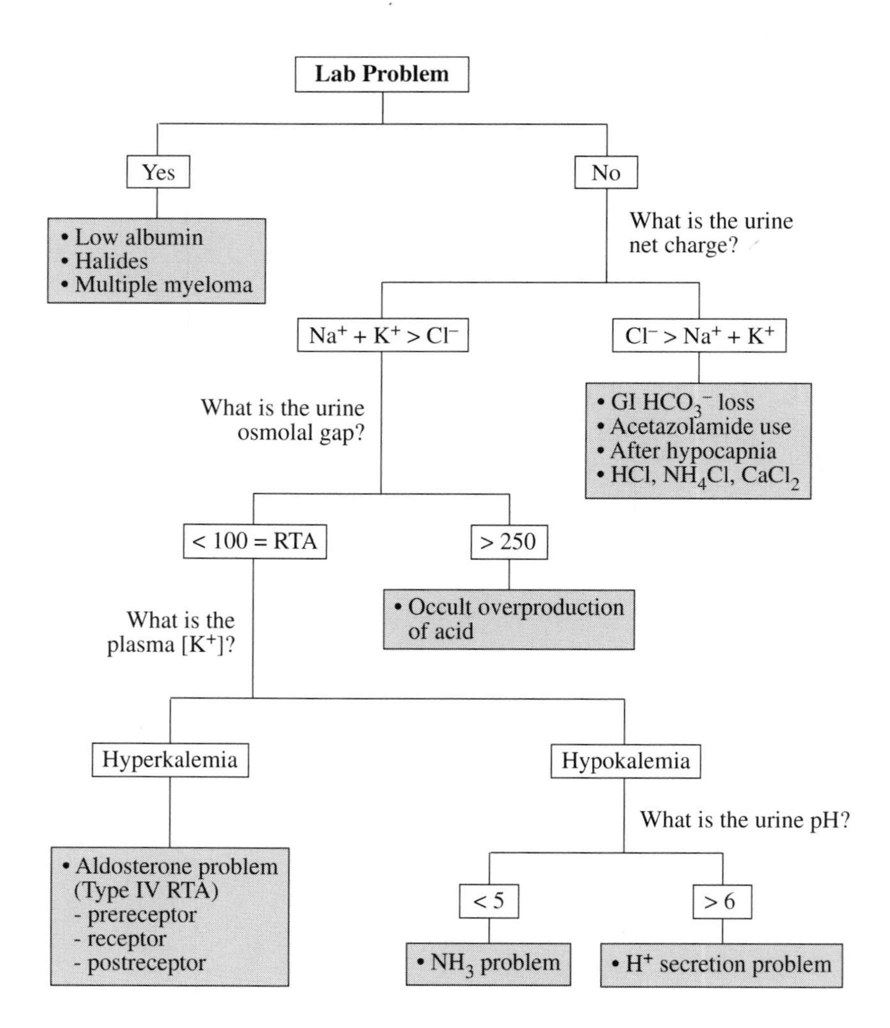

Figure 3·20
Approach to the patient with metabolic acidosis and a normal plasma anion gap

INCREASED RENAL NH_4^+ EXCRETION

An increased excretion of NH_4^+ is identified by either a negative urine net charge (urine $[Cl^-] > [Na^+] + [K^+]$) or a high urine osmolal gap (page 54). These patients have a "nonrenal" basis of their metabolic acidosis, since the enhanced renal NH_4^+ excretion should be sufficient to prevent metabolic acidosis from developing from the normal daily H^+ load derived from the diet.

Laxative abuse:
Patients who abuse laxatives may not admit to their habit and thus present a diagnostic challenge, a disorder that resembles distal RTA. The problem is resolved by the presence or absence of a low rate of excretion of NH_4^+.

Gastrointestinal HCO_3^- Loss

Diarrhea, unless greater than four liters per day, is not sufficient to cause a significant degree of metabolic acidosis because the normal kidney can generate 200 mmol of HCO_3^- per day as a result of enhanced excretion of NH_4^+ (stimulated by hypokalemia and/or a mild degree of chronic metabolic acidosis). Accordingly, the three factors associated with a severe degree of metabolic acidosis in the patient with diarrhea are copious diarrhea, reduced rates of NH_4^+ excretion, and excessive production of organic acids by bacteria in the GI tract. Diarrhea is easily diagnosed; pooling of HCO_3^--rich fluid in the bowel lumen (ileus), however, may not be so evident.

Distal RTA vs diarrhea:
Patients with diarrhea often have hypokalemia, ECF volume depletion, a mild degree of metabolic acidosis, and a urine pH that is close to 6.0; therefore, their condition could be confused with distal RTA. Again, the difference is the low rate of excretion of NH_4^+ in distal RTA.

Question

(Discussion on page 144)

3·24 An 80-year-old man with a history of "pyelonephritis" developed diarrhea after a course of antibiotics. On the basis of the following results, a diagnosis of distal RTA was made. Is it correct?

Plasma			Urine		
Na^+	mmol/l	134	Na^+	mmol/l	10
K^+	mmol/l	2.8	K^+	mmol/l	40
Cl^-	mmol/l	115	Cl^-	mmol/l	100
HCO_3^-	mmol/l	10	Osmolality	mosm/kg H_2O	800
H^+	nmol/l	62	Urea	mmol/l	300
pH		7.20	pH		5.9

Acid Ingestion

If the anion of an acid is Cl^-, its intake can cause metabolic acidosis with no increase in the plasma anion gap. HCl, NH_4Cl, lysine$^-$, and arginine-HCl may cause this disorder.

Other Causes

There are several situations in which a patient can have metabolic acidosis with a normal plasma anion gap, no overproduction of acids, and a high rate of excretion of NH_4^+. One example is a subject who ingests acetazolamide, which causes bicarbonaturia. A second example is recovery from chronic hypocapnia. With chronic hypocapnia, there is suppression of renal

NH_4^+ excretion, and the plasma [HCO_3^-] will fall. The patient is not acidemic initially because of the low P_aco_2. When the stimulus for hyperventilation is removed, the P_aco_2 will rise. The [HCO_3^-] will then be low, and metabolic acidosis with a normal anion gap will be present; NH_4^+ excretion should rise from the stimulus of acidemia. Within a few days, the enhanced renal new HCO_3^- generation will return the plasma [HCO_3^-] to normal. A third example is the so-called expansion acidosis. When the ECF volume is expanded with solutions lacking HCO_3^- (isotonic saline), the plasma [HCO_3^-] will fall. To understand this occurrence, one must distinguish between HCO_3^- content and HCO_3^- concentration in the ECF. In the table below, the patient described had a normal plasma [HCO_3^-] despite severe ECF volume contraction; thus, the HCO_3^- content in the ECF was reduced but this reduction might not have been appreciated (line 2). Reexpansion of the ECF volume with normal saline (a HCO_3^--free solution) made the reduced HCO_3^- content evident because the [HCO_3^-] was now lower (line 3). For the [HCO_3^-] to return to normal in this patient, he must either make new HCO_3^- (excrete NH_4^+) or receive $NaHCO_3$.

Condition	ECF volume (liters)	[HCO_3^-] (mmol/l)	HCO_3^- content (mmol)
Normal	15	24	360
Contracted ECF volume	10	24	240
Restored ECF volume	15	16	240

INADEQUATE INDIRECT REABSORPTION OF FILTERED HCO_3^- (PROXIMAL RTA)

NH_4^+ excretion in proximal RTA:
The excretion of NH_4^+ is on the low side in most patients with proximal RTA unless there is a second cause for metabolic acidosis (e.g., chronic diarrhea), in which case the rate of excretion of NH_4^+ may rise towards the expected rate.

Fanconi syndrome:
A defect in proximal tubular reabsorption that leads to glycosuria, aminoaciduria, increased excretion of uric acid and phosphate, and proximal renal tubular acidosis.

The diagnosis of proximal RTA hinges on the demonstration of impaired indirect reabsorption of filtered HCO_3^- (Tables 3·16 and 3·17). Indirect reabsorption of filtered HCO_3^- is achieved by proximal tubular H^+ secretion. A defect in this H^+ secretion results in a metabolic acidosis with no increase in the plasma anion gap and is due initially to the excretion of $NaHCO_3$ in the urine (called proximal RTA, or Type II RTA). Later on, the acidosis is maintained because the rate of excretion of NH_4^+ is low considering the fact that chronic metabolic acidosis is present and the urine pH is low (see the margin). The H^+ secretory defect may be isolated, but if it occurs in concert with other transport defects, the *Fanconi syndrome* may be the the appropriate diagnosis.

Distal H^+ secretion does not have the capacity to reabsorb all the HCO_3^- delivered as a result of the proximal deficit in H^+ secretion, and HCO_3^- excretion ensues if $NaHCO_3$ is given. Notwithstanding, patients with this disorder in steady state have a chronic metabolic acidosis, no bicarbonaturia, and a low excretion of NH_4^+ with a very low urine pH.

Possible Basis of the Findings in Proximal RTA

Possibly, the proximal cells of patients with proximal RTA have a defect that makes these cells uniquely more alkaline. This hypothesis could explain the cardinal features of low reabsorption of HCO_3^-: low synthesis of NH_4^+ despite chronic metabolic acidosis, and a high *excretion of citrate* (unique in metabolic acidosis with acidemia). Finally, it could help explain the absence of medullary nephrocalcinosis in untreated patients with isolated proximal RTA.

Excretion of citrate:
Citrate disappears from the urine in patients with metabolic acidosis. The sole exception to this rule is proximal RTA.

Table 3·16
DIAGNOSTIC FEATURES IN PROXIMAL RTA

The filtered load of HCO_3^- is 4500 mmol/day (see the margin). In all of the examples below, the GFR is presumed to remain normal at 180 liters/day. The major lesion in proximal RTA is reduced proximal H^+ secretion (lines 2–4). Since distal H^+ secretion is of low capacity, all the extra HCO_3^- delivered are not reabsorbed, and the urine pH is > 7.0. If an extra acid load is present, NH_4^+ excretion can rise and augment new HCO_3^- formation (line 4).

State	Filtered HCO_3^-	Proximal HCO_3^- Reabsorbed	Distal HCO_3^- Delivery*	HCO_3^- Excretion	NH_4^+ Excretion
	(mmol/day)				
Normal	4500	4000	500	0	30
Proximal RTA					
Onset	4500	3000	1500	>100	0
Established	3600	3000	600	0	20
+ Acid load	3000	2700*	300	0	60

* We assume that distal HCO_3^- delivery must be at least 300 mmol/day, so only 2700 mmol of HCO_3^- can be absorbed proximally when an extra acid load is given.

Calculation:
Normal filtered load of HCO_3^-
= GFR × plasma [HCO_3^-]
= 180 liters/day × 25 mmol/l
= 4500 mmol/day.

Table 3·17
CONDITIONS LEADING TO DECREASED INDIRECT BICARBONATE REABSORPTION IN THE PROXIMAL CONVOLUTED TUBULE

Proximal Renal Tubular Acidosis
- Fanconi syndrome
  - Genetic disorders, including cystinosis, galactosemia, hereditary fructose intolerance, Wilson's disease, Lowe's syndrome, tyrosinemia
  - Toxin-induced disorders from exogenous toxins (including heavy metals, outdated tetracycline, streptozocin) and endogenous toxins (dysproteinemias, including multiple myeloma, etc.)
  - Other disorders, including sporadic, transient proximal RTA of infants
  - Disorders secondary to other renal diseases, including amyloidosis, renal transplantation, autoimmune diseases such as chronic active hepatitis, Sjögren's syndrome, etc.
- Isolated proximal RTA

Carbonic Anhydrase Inhibitors

Other Conditions
- Hyperparathyroidism, hypocalcemia, vitamin D deficiency

Technique to Diagnose Proximal RTA

When the patient is acidemic, administer $NaHCO_3$ and monitor the plasma [HCO_3^-] and urine pH. If the urine pH becomes alkaline (> 7.0) while the plasma [HCO_3^-] remains low, impaired proximal H^+ secretion may be present. With continued HCO_3^- administration, the urine pH rises further, and distal H^+ secretion can be assessed by measuring the urine P_{CO_2} (see page 55 and Figure 3·21). Finally, in patients with proximal RTA, the [HCO_3^-]

Urine pH, [HCO_3^-], and [NH_4^+] in proximal RTA:
Because patients are in steady state, 24-hour net acid excretion in proximal RTA is the same as in normal individuals (i.e., HCO_3^- gain equals HCO_3^- loss but at a lower plasma [HCO_3^-]).

Fractional excretion of HCO_3^-:

$$\frac{[HCO_3^-]_{urine} / [HCO_3^-]_{plasma}}{[creatinine]_{urine} / [creatinine]_{plasma}}$$

in plasma falls promptly to subnormal levels once the infusion stops. One should measure the *fractional excretion of HCO_3^-* at this point; it will be greater than 15% in proximal RTA.

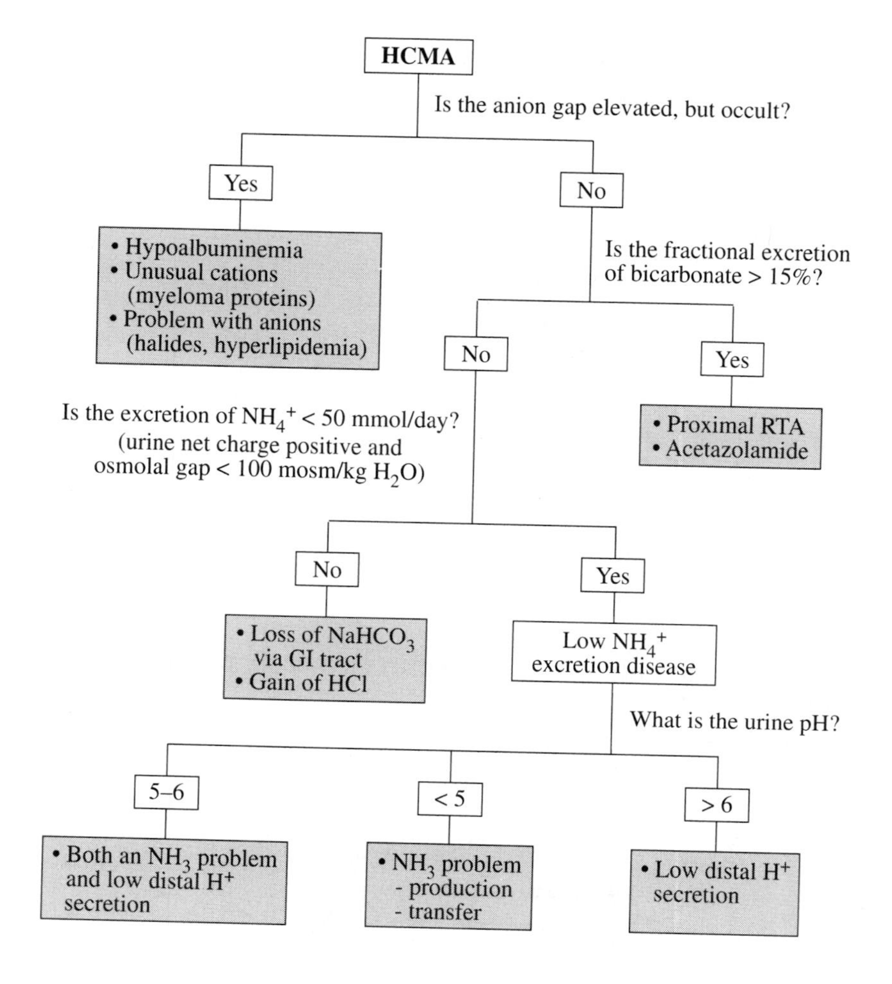

Figure 3·21
Approach to the diagnosis of renal tubular acidosis
The final diagnoses are shown in the shaded boxes. Reproduced with permission from *The ACID truth and BASIC facts with a Sweet Touch, an enLYTEnment.* (HCMA = hyperchloremic metabolic acidosis.)

Questions

(Discussions on page 144)

3·25 What are the diagnostic features of disorders with reduced indirect reabsorption of filtered HCO_3^-?

3·26 If a patient with metabolic acidosis is taking acetazolamide, how will the urine test results change once this diuretic is no longer acting (but metabolic acidosis persists)?

REDUCED RENAL NH_4^+ EXCRETION (DISTAL RTA)

- In distal RTA, there is metabolic acidosis and an NH_4^+ excretion defect from:
  - low NH_3 availability in the renal medullary interstitium;
  - low H^+ secretion in the collecting duct.

Distal RTA represents a heterogeneous group of disorders characterized by reduced excretion of NH_4^+ and failure to regenerate the needed HCO_3^- (consumed by the daily H^+ load). Although a number of titles have been applied to these defects (Type I RTA, Type IV RTA), we do not use them because they provide no etiologic insight and in fact may mislead the reader into grouping lesions with different pathophysiologies (Table 3·18).

A high urine pH has been proposed by some as the gold standard in the diagnosis of distal RTA; we discourage this approach because the urine pH may sometimes be misleading (Figure 3·22). The urine pH is certainly useful in identifying bicarbonaturia (pH > 7.0), which helps in the diagnosis of proximal RTA. Also, a very low urine pH (< 5.0) in a patient with metabolic acidosis and a low rate of excretion of NH_4^+ strongly suggests low availability of NH_3 as the basis of the defect. Our approach to the differential diagnosis of distal RTA is shown in Figure 3·21.

Clinical pearls:

- RTA should stand for **R**ecognize **T**he **A**mmonium excretion defect.
- Forget the urine pH for now!
  - The urine net charge and osmolal gap are much more useful than the urine pH in determining whether or not the rate of excretion of NH_4^+ is low.
  - The urine pH is useful in determining why the rate of excretion of NH_4^+ is low.

Note:
The subgroup of RTA that is due to hyperkalemia has been termed "Type IV RTA" by others, but we see no benefit and some disadvantages in this terminology.

Table 3·18
NOMENCLATURE USED IN THE CLASSIFICATION OF RTA

Readers are encouraged to think of the classification in terms of the pathophysiology rather than numerical labels.

Current	Preferred
Lesion: Reduced indirect reabsorption of filtered HCO_3^-	
Type II RTA or Proximal RTA	Proximal H^+ secretory defect
Lesion: Low excretion of NH_4^+	
Type I RTA, Distal RTA, or Classic RTA	Decreased NH_4^+ excretion from: - low medullary NH_3 concentration - low net distal H^+ secretion
Type IV RTA	Decreased NH_4^+ excretion related to hyperkalemia

Decreased NH_3 in the Medullary Interstitium

There are two ways to examine why the rate of excretion of NH_4^+ is low (Table 3·19 and the margin). Hyperkalemia and a very low GFR are the most common causes of a reduced ammoniagenesis; medullary diseases such as pyelonephritis and analgesic nephrotoxicity are common causes of medullary dysfunction. The distinctive features are metabolic acidosis and a normal plasma anion gap, low rates of NH_4^+ excretion, and a low urine pH (arbitrarily set at < 5.3). These patients can have a very low urine pH

Notes:

- There are two factors that lead to a low rate of excretion of NH_4^+: a low $[NH_3]$ in the renal medulla and a low transfer of NH_3 into the lumen of the collecting duct (i.e., low H^+ secretion in the collecting duct).
- There are two reasons for a low $[NH_3]$ in the medullary interstitium: a low production of NH_4^+ and/or a defect in medullary function.

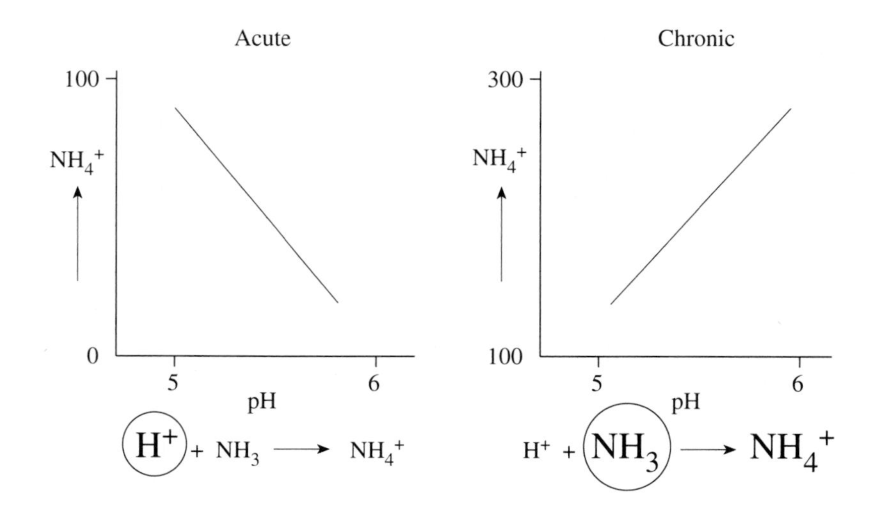

Figure 3·22
Urine pH and the excretion of NH_4^+
In acute acidosis, the rate of NH_4^+ excretion is higher when the urine pH is lower; the increased [H^+] traps more NH_3 in the lumen as NH_4^+. Thus, the rate of H^+ secretion exceeds that of NH_3. In chronic metabolic acidosis, both the H^+ secretory rate and the NH_3 availability are greatly increased. The increase in NH_3 is due to augmented ammoniagenesis and is relatively larger than the increment in H^+ secretion. Thus, the urine NH_4^+ excretion rate is increased in conjunction with a higher urine pH because of the increased trapping of H^+ by NH_3. Note the difference in scale for NH_4^+, as shown on in the y-axis of each panel.

(often < 5.0) because the H^+ secretory mechanism of their collecting ducts may be normal. Nevertheless, because they have little NH_3 available to bind the free H^+, they have a reduced overall capacity to excrete NH_4^+. Therefore, despite the low urine pH, there is very little renal H^+ excretion. This normal H^+ secretory mechanism may be revealed by finding a high urine P_{CO_2} with alkali loading (page 55).

Table 3·19
CAUSES FOR A LOW RATE OF EXCRETION OF NH_4^+ IN A PATIENT WITH METABOLIC ACIDOSIS

Patients in group 1 will be prone to have a low urine pH, and the urine pH will tend to be greater than 6.0 in patients in group 2.

1. **Low NH_3 in the renal medullary interstitium**
   - Low production of NH_4^+ in the proximal convoluted tubule
     - Low turnover of ATP (low GFR, i.e., renal failure)
     - Hyperkalemia
     - Low availability of glutamine (e.g., malnutrition)
     - High availability of fat-derived fuels (e.g., TPN)
     - Alkaline cell (possibly isolated proximal RTA)
   - Problems with functions in the loop of Henle
     - Interstitial diseases, hyperkalemia (see Figure 1·17, page 42)

2. **Low transfer of NH_3 into the lumen of the collecting duct**
   - H^+ATPase units with low activity
     - Medullary destruction
     - Inhibition by immune complexes (e.g., Sjögren's disease)
   - H^+ATPase units without a stimulator
     - Less lumen negative voltage (e.g., Cl^- shunt, possibly lithium)
     - Low aldosterone bioactivity
   - Back-leak of H^+
     - Amphotericin B

Decreased Transfer of NH_3 to the Lumen of the Collecting Duct

Less NH_3 is transferred to the lumen of the collecting duct as a result of defects in the H^+ secretory mechanism of the collecting duct. These defects can be considered in four subgroups; examples of each subgroup are shown in Table 3·20.

1. Problems with the H^+ATPase pump.
2. Problems with voltage augmentation of this pump.
3. Low NH_3 availability to neutralize luminal H^+ (a high luminal $[H^+]$ lessens H^+ pumping).
4. An abnormal back-leak of H^+ from the lumen into the cell.

Voltage augmentation of the H^+ pump:
H^+ secretion is "electrogenic" in that it renders the lumen voltage positive. It is enhanced by Na^+ reabsorption, which renders the transepithelial voltage negative. On the other hand, if Na^+ reabsorption is "electroneutral," i.e., accompanied by Cl^- reabsorption (Cl^- shunt), H^+ secretion will not be enhanced. This same principle applies to K^+ secretion, which takes advantage of the lumen negative potential difference to enhance its net secretion.

The above four subgroups have in common a low rate of NH_4^+ excretion. All but the third group tend to have a high urine pH (> 6). Groups 1 and 2 can also be distinguished because patients with these problems should have a low urine Pco_2 (< 50 mm Hg) in alkaline urine, yet the urine Pco_2 in the other groups should exceed 70 mm Hg. The low-voltage stimulation group tends to have reduced Na^+ reabsorption in the collecting duct, a problem that also interferes with K^+ excretion (a low transtubular $[K^+]$ gradient; see Figure 9·11, page 341); these patients therefore have hyperkalemia, which lowers NH_4^+ production and excretion.

Table 3·20
PATHOPHYSIOLOGIC APPROACH TO DISTAL RTA

Defect	Causes of Lesions	Diagnostic Features
• H^+ATPase	• Mineralocorticoid deficiency • Medullary damage or infiltration • Immunological basis (e.g., Sjögren's syndrome) • Congenital disorders	Low urine Pco_2, High urine pH
• Voltage augmentation of H^+ secretion	• Low distal Na^+ delivery (ECF volume contraction, congestive heart failure, cirrhosis) • Inhibitors of Na^+ reabsorption (e.g., amiloride, lithium, trimethoprim) • Lack of stimulators (e.g., aldosterone deficiency or blockade) • Excess Cl^- permeability	Hyperkalemia with a low transtubular $[K^+]$ gradient (TTKG)
• Raising luminal $[H^+]$ by low NH_3 availability	• Low NH_4^+ production (e.g., hyperkalemia, low GFR) • Low medullary NH_3 (tubulointerstitial diseases)	Low urine pH
• Back-leak of H^+	• Increased H^+ permeability (e.g., amphotericin B)	High urine pH, High urine Pco_2

Diseases involving the renal medulla can destroy collecting duct cells and thereby cause a low H^+ pump activity. Such a lesion also lowers NH_3 availability in the medullary interstitium. Hence, the excretion of NH_4^+ is low, but the urine pH can be less than 5.3 if the NH_3 defect predominates, greater than 6 if the H^+ secretory defect predominates, or between 5 and 6 if neither predominates. In fact, a urine pH of 5.5 would suggest a combined NH_3 and H^+ defect (Figure 3·21, page 118).

Hypokalemia and distal RTA:
In patients with impaired H^+ secretion and normal Na^+ reabsorption in the collecting duct, there will be a significant lumen negative voltage and a higher $[HCO_3^-]$ in the lumen, both of which facilitate K^+ secretion in this nephron segment.

Urine pH, H^+ secretion and NH_4^+ excretion:
In the absence of NH_3, the major H^+ acceptor in the collecting duct lumen, even reduced H^+ secretion may lower the urine pH. Nevertheless, with excess NH_3 present (e.g., in prolonged fasting), even normal H^+ secretion may be associated with a pH > 5.8. Hence, the urine pH may be less than useful in detecting the quantity of NH_4^+ being excreted (Figure 3·22, page 120).

Importance of the Plasma $[K^+]$ in the Diagnosis of Distal RTA

Hypokalemia: When the cause of distal RTA is low H^+ secretion in the collecting duct rather than low NH_3 availability, hypokalemia is often present. This hypokalemia is due to an unexpectedly high rate of K^+ excretion and is associated with a urine pH that is greater than 5.8 (see the margin).

Hyperkalemia: Hyperkalemia may be associated with metabolic acidosis and a low rate of NH_4^+ excretion (Table 3·21); this condition has also been referred to as "Type IV distal RTA." The basis of the reduced NH_4^+ excretion is inhibition of NH_4^+ production by hyperkalemia and low reabsorption of NH_4^+ in the thick ascending limb of the loop of Henle (see the discussion of Question 1·24, page 42).

Table 3·21
HYPERKALEMIA, METABOLIC ACIDOSIS, AND LOW URINE NH_4^+

Cause	Features	Basis
Decreased Na^+ reabsorption		
• Low distal delivery	• Urine $[Na^+]$ < 10 mmol/l	• ECF volume depletion • Severe hypoalbuminemia • Congestive heart failure
• Decreased mineralocorticoid bioactivity	• Urine pH < 5.3 as a result of hyperkalemia, which causes low NH_3 synthesis	• Hyporeninemia, • Converting enzyme inhibitor • Aldosterone antagonists • Adrenal gland problem
Renal failure	• Low GFR	• Low NH_4^+ production
Decreased distal H^+ secretion	• Urine pH > 6	• Interstitial disease • Drugs (e.g., amiloride, lithium, trimethoprim) • Immune basis
Chloride shunt	• ECF volume expansion, • Low urine $[K^+]$, • Urine pH < 5.3 as a result of hyperkalemia	• Dissipation of lumen negative transepithelial potential difference (possibly caused by cyclosporine)

Hyperkalemia and reduced H^+ secretion coexist for several reasons (Table 3·22). Each circumstance can be identified by specific diagnostic features. First, there may be insufficient distal Na^+ resorption because of low Na^+ delivery to the "cortical distal nephron." As a result, aldosterone has a smaller quantity of Na^+ on which to act, so that less secretion of K^+ and H^+ occurs here. The low rate of excretion of NH_4^+ should disappear when more Na^+ are delivered distally (following an infusion of NaCl or the administration of a loop diuretic). A second subgroup has low aldosterone bioactivity and subsequent hyperkalemia. The low rate of NH_4^+ excretion may disappear when hyperkalemia is treated (see the margin). In yet other examples, there is a failure to generate a lumen-negative transtubular voltage. The most common of these defects is the failure to reabsorb filtered Na^+ at the collecting duct because of a failure to respond to aldosterone or to the presence of drugs that block aldosterone action (e.g., amiloride). Less commonly, a Cl^- shunt type of lesion might be present. The third subgroup is characterized by a renal medullary lesion that impairs medullary function.

Hyperkalemia and metabolic acidosis can also be present in renal failure (discussed below).

Table 3·22
IMPAIRED VOLTAGE AUGMENTATION OF H^+ AND K^+ SECRETION IN THE COLLECTING DUCT

Low distal Na^+ delivery
- ECF volume contraction
- Hypoalbuminemia, cirrhosis, nephrotic syndrome

Low aldosterone bioactivity
- Prereceptor defects (e.g., ACE inhibitor, problems with the adrenal gland or low renin)
- Receptor blockage (e.g., spironolactone)
- Postreceptor defects (e.g., amiloride, Cl^- shunt)

Decreased end-organ function
- Interstitial nephritis
- Obstructive nephropathy

Mineralocorticoid deficiency and H^+ excretion:
Mineralocorticoid deficiencies may bring about decreased NH_4^+ excretion by two different means, hyperkalemia and impaired Na^+ reabsorption, either of which may predominate.
1. Hyperkalemia may decrease the amount of NH_3 in the renal medullary interstitium; as a result, the urine pH will be low. NH_4^+ excretion will increase with reduction of the hyperkalemia.
2. Impaired Na^+ reabsorption results in a failure of voltage augmentation of H^+ secretion. Consequently, the urine pH will be > 6, the urine Pco_2 will not be appropriately elevated, and NH_4^+ excretion will not increase with a fall in the plasma $[K^+]$ (NH_4^+ excretion will increase with hormone replacement, however, as will the transtubular $[K^+]$ gradient).

Metabolic Acidosis in Renal Failure

- Metabolic acidosis in renal failure progresses slowly, is associated with an increased plasma anion gap, and is usually associated with hyperkalemia.

Angiotensin-converting enzyme (ACE) inhibitor:
An inhibitor that prevents the formation of angiotensin II, the active stimulator of aldosterone release.

The acidemia associated with renal failure is due to the usual H^+ load from the diet (1 mmol/kg) coupled with a failure of the kidney to generate new HCO_3^- from a reduced rate of synthesis and excretion of NH_4^+ (Figure 3·2, page 124). Because the body accumulates approximately 70 mmol of H^+ per day, the $[HCO_3^-]$ can fall by approximately 2.3 mmol/l/day (see the margin). Nevertheless, this amount is a gross overestimate of the rate of H^+ accumulation in patients with chronic renal insufficiency because they still have residual net acid excretion.

Calculation:
A 70-kg person generates 70 mmol of H^+ daily. Approximately one-half is buffered in the ECF by the BBS. Therefore, 35 mmol are buffered in 15 liters. The expected decline in plasma $[HCO_3^-]$ is therefore 2.3 mmol/day.

The increase in the plasma anion gap is a result of the reduced GFR, with accumulation of anions (e.g., HPO_4^{-2}) (Figure 3·2). Usually the anion gap does not rise appreciably until the GFR has fallen to 20% of normal. If one compares the magnitude of the dietary H^+ load with the degree of reduction in NH_4^+ excretion and in the GFR, one can see discrepancies between the degree of increase in the plasma anion gap and the magnitude and rate of fall of the plasma $[HCO_3^-]$.

The acidemia of renal failure may be complicated by hyperkalemia. While both the acidemia and the hyperkalemia will improve after treatment with $NaHCO_3$, patients with renal failure often have an expanded ECF volume; therapy with $NaHCO_3$ may therefore be hazardous without an option for Na^+ removal.

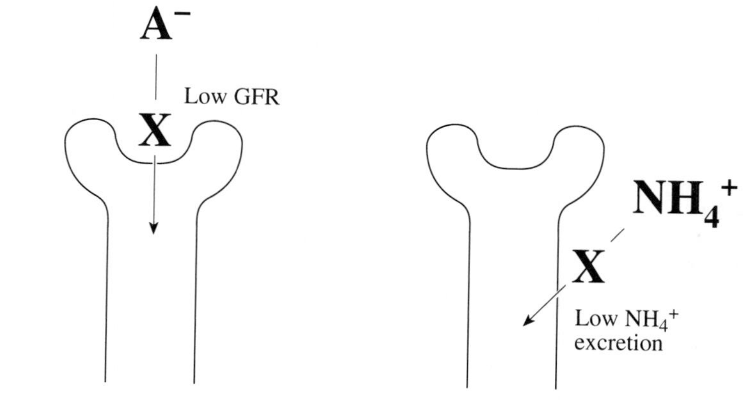

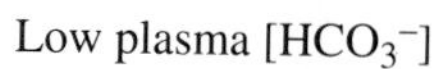

Figure 3-2
Basis of high plasma anion gap and acidosis in renal failure
The basis of the increased plasma anion gap (AG) is the low GFR with reduced excretion of anions such as phosphate or sulfate (left side). The acidosis is due to a low rate of excretion of NH_4^+ (right side).

PART D
REVIEW

Discussion of Introductory Case
To Make a Diagnosis, Step on the Gas

(Case presented on page 70)

What is (are) your acid-base diagnosis(es)?

Lee has a mixed acid-base disorder: metabolic acidosis ($[HCO_3^-]$ = 5 mmol/l with acidemia) and respiratory acidosis (during acidemia, a $[HCO_3^-]$ of 5 mmol/l should be accompanied by a Pco_2 of 20 mm Hg, not 30 mm Hg).

What would the plasma $[H^+]$ be if the respiratory response were appropriate?

If the P_aco_2 were the expected 20 mm Hg, the $[H^+]$ would be 96 nmol/l instead of 144 nmol/l (a blood pH of 7.02 instead of 6.84).

Henderson equation:

$$H^+ = \frac{24}{[HCO_3^-]} \times Pco_2$$

What are the likely causes for the metabolic acidosis in Lee?

Lee has a normal plasma anion gap and therefore has lost $NaHCO_3$. Suspect either reduced renal new HCO_3^- generation (low excretion of NH_4^+) or $NaHCO_3$ loss via the gastrointestinal tract or the urinary tract.

What is the significance of the hypokalemia?

Hypokalemia, unusual among the other causes of metabolic acidosis, is an important diagnostic clue. It suggests that the basis of the metabolic acidosis is most likely a reduced rate of renal NH_4^+ excretion or severe diarrhea. There are two other important aspects of hypokalemia: first, hypokalemia can lead to a cardiac arrhythmia and/or muscle weakness; second, as discussed below, it must be dealt with during treatment because it is likely to become more severe once the acidemia lessens in severity. With hypokalemia, $NaHCO_3$ therapy may be dangerous if it is not accompanied by aggressive K^+ replacement. In this case, the hypokalemia impairs respiratory muscle function and therefore contributes to the severe acidemia. If the hypokalemia is not treated as aggressively as the acidemia, respiratory arrest may occur.

As HCO_3^- appear in the ECF, H^+ exit from cells, and, to the extent that K^+ enter the ICF, the degree of hypokalemia could become much more severe. In patients with metabolic acidosis and a normal plasma anion gap, therapy with $NaHCO_3$ may be more essential than in patients with an increased plasma anion gap because metabolism of the unmeasured organic anions in the latter case leads to generation of HCO_3^-.

Must all patients with metabolic acidosis have a high plasma $[H^+]$?

No. Acidemia is not always present during metabolic acidosis; a coexistent metabolic or respiratory alkalosis may reduce the plasma $[H^+]$ to a normal or abnormally low value in a patient with metabolic acidosis. Metabolic acidosis and metabolic alkalosis are independent events, not mirror images. Consider the following example: a patient vomits and develops metabolic alkalosis with ECF volume contraction. As vomiting continues, he goes into shock and produces L-lactic acid. Although the blood pH is closer to normal, this patient is closer to death.

Can a patient have a low plasma [HCO_3^-] and not have metabolic acidosis?

Yes. If the Pco_2 is lowered by hyperventilation, the bicarbonate buffer equation is shifted to the right; the result is a lower [H^+] and [HCO_3^-] (i.e., respiratory alkalosis).

Is it possible to have a persistently alkaline urine (i.e., no renal HCO_3^- generation) and maintain acid-base balance?

Yes. Renal HCO_3^- generation is essential in maintaining acid-base balance only if there is ongoing net HCO_3^- consumption (i.e., a dietary H^+ load). If one consumes a diet that does not present an acid load (e.g., certain vegetarian diets), acid-base balance is maintained without renal HCO_3^- generation; in fact, maintaining this balance now requires HCO_3^- excretion in the urine.

Cases for Review

Case 3·1
Ken Has a Drinking "Problem"

(Case discussed on pages 129–30)

A 26-year-old man consumed an excessive quantity of alcohol during the past week; in the last two days, he has eaten little and has vomited on many occasions. He has no history of diabetes mellitus. Physical examination reveals marked ECF volume contraction. Alcohol is detected on his breath.

Blood			**Plasma**		
Glucose	mmol/l (mg/dl)	5 (90)	Na^+	mmol/l	140
BUN	mmol/l (mg/dl)	10 (28)	K^+	mmol/l	3.0
pH		7.30	Cl^-	mmol/l	93
H^+	nmol/l	50	HCO_3^-	mmol/l	15
P_aco_2	mm Hg	30			
			Ketones		Strongly positive

What is the total body Na^+ content?
Is there relative insulin deficiency?
Why is the patient hypokalemic, and to what degree is he K^+-depleted?
What are the acid-base diagnoses?
What are the primary considerations for therapy?
The plasma osmolality is 350 mosm/kg H_2O. Is the ICF volume high, low, or normal?
When ethanol is no longer present, what changes will be observed in the metabolic picture?

Case 3·2
An Unusual Type of Ketoacidosis

(Case discussed on pages 130–31)

A 21-year-old woman has had diabetes mellitus for two years and requires insulin. Six months ago, she presented with lethargy, malaise, headaches, and metabolic acidosis with a normal plasma anion gap. Her complaints and the acid-base disturbance have persisted for six months, so she was referred for further evaluation. She denies a past history of diarrhea, abdominal complaints, or ingestion of drugs (acetazolamide, halides, or HCl equivalents). She has no past history of renal disease.

While taking her usual 34 units of insulin per day, she frequently had glycosuria and ketonuria but no major increase in the plasma anion gap. Her physical examination was unremarkable, and her urinalysis showed no protein and a normal sediment. The results of her laboratory investigations are shown below.

Plasma					
Urea	mmol/l (mg/dl)	7 (20)	Na^+	mmol/l	136
Creatinine	μmol/l (mg/dl)	100 (0.9)	K^+	mmol/l	2.9
Glucose	mmol/l (mg/dl)	10.6 (190)	Cl^-	mmol/l	103
pH		7.35	HCO_3^-	mmol/l	19
$[H^+]$	nmol/l	45	Anion gap	mEq/l	14
$P_{a}CO_2$	mm Hg	35	β-HB	mmol/l	2.2
Urine					
Glucose	mmol/l	5	Na^+	mmol/l	47
Urea	mmol/l	50	K^+	mmol/l	60
pH		5.3	Cl^-	mmol/l	13
Osmolality	mosm/kg H_2O	680			

What is the differential diagnosis of her acid-base disorders?
What investigative plan would be appropriate?
Following the ingestion of $NaHCO_3$ and KCl, the plasma $[HCO_3^-]$ rose to 26 mmol/l and did not fall promptly after the intake of HCO_3^- was stopped; the urine PCO_2 was 90 mm Hg. What is the diagnosis now?
Why was this patient hypokalemic?

Case 3·3
Ketoacidosis, a Stroke of Bad Luck

(Case discussed on pages 131–32)

A 42-year-old man has two medical problems, hypertension and rare alcohol binges; he is not a chronic alcohol abuser. Last night he consumed half a bottle of whiskey. This morning he was found unconscious. Physical exam revealed coma and hemiparesis from an acute intracerebral hemorrhage. There was no ECF volume contraction. Laboratory results in plasma are summarized below; these values were essentially unchanged two hours later (before therapy was initiated). He produced only a small amount of urine in this two-hour period.

Plasma					
pH		6.96	Glucose	mmol/l (mg/dl)	9.0 (162)
P_aco_2	mm Hg	11	Urea	mmol/l (mg/dl)	5.0 (14)
HCO_3^-	mmol/l	3	Creatinine	μmol/l (mg/dl)	100 (0.8)
Anion gap	mEq/l	42	Osmolality	mosm/kg H_2O	305
Na^+	mmol/l	139	Ethanol	mmol/l	20
K^+	mmol/l	6.8	Ketone	(screen)	moderate

What is (are) the most likely cause(s) for his metabolic acidosis?
Are acids being produced rapidly?
What hormonal changes were involved?
Why did ketoacidosis develop so quickly?

Case 3·4
A Superstar of Severe Acidosis!

(Case discussed on pages 132–33)

A patient who had walked into the emergency room was found to have astounding results of blood tests. Physical examination revealed a near-normal ECF volume and hyperventilation. Clinical and laboratory tests for alcohol were negative. His [HCO_3^-] was 1 mmol/l, his pH was 6.69, and he had a markedly increased anion gap of 46 mEq/l. His GFR was not appreciably abnormal.

What are the most likely diagnoses?
What acid-base treatment would be most appropriate?

Case 3·5
Acute Popsicle Overdose

(Case discussed on pages 133–34)

A 56-year-old man developed diarrhea while traveling abroad for several months. He took antibiotics and a GI motility depressant. His clinical condition deteriorated and metabolic acidosis developed when he ate popsicles to quench his thirst. The only physical finding of note was confusion and poor coordination. Laboratory results revealed a mixed type of metabolic acidosis (pH was 7.20, [HCO_3^-] was 10 mmol/l, anion gap was increased by 7 mEq/l).

		Plasma	**Urine**
pH		7.20	5.2
P_aco_2	mm Hg	25	No data
HCO_3^-	mmol/l	10	0
Anion gap	mEq/l	19	101
Osmolal gap	mosm/kg H_2O	0	430
Albumin	g/l	38	No data
Ketoacids		Negative	Negative

Could chronic diarrhea have caused the acidosis?
If not, what other diagnosis is likely?
What role did the popsicles play?

Discussion of Cases

Discussion of Case 3·1 Ken Has a Drinking "Problem"

(Case presented on page 126)

What is the total body Na^+ content?

Since the ECF volume is markedly contracted, there is a large Na^+ deficit, which was probably partially due to renal Na^+ excretion ("dragged out by HCO_3^-") from vomiting.

Is there relative insulin deficiency?

Yes. Ketoacidosis signals relative insulin deficiency. If he does not have IDDM, the lack of insulin is probably due to the α-adrenergic response secondary to ECF volume contraction. Hyperglycemia is not present because ethanol has inhibited gluconeogenesis in the liver.

Why is the patient hypokalemic, and to what degree is he K^+-depleted?

The patient is severely K^+-depleted because hypokalemia is present during relative insulin deficiency (insulin deficiency is usually associated with hyperkalemia). Vomiting leads to large renal K^+ loss (see page 369).

What are the acid-base diagnoses?

The acidemia, low [HCO_3^-], and increased plasma anion gap all indicate that metabolic acidosis is present. The most likely cause is ketoacidosis plus some L-lactic acidosis from either ethanol or the low circulating volume. In addition, he has metabolic alkalosis from vomiting; the rise in plasma anion gap (32 – 12, or 20 mEq/l) is greater than the fall in the plasma [HCO_3^-] from normal (25 – 15 = 10 mmol/l).

What are the primary considerations for therapy?

Administer the following:
1. isotonic saline to reexpand the ECF volume,
2. KCl (40 mmol/l) to replace the K^+ deficit,
3. B vitamins to replace nutritional deficits.

Phosphate should be given if the patient is severely hypophosphatemic. Do not give insulin now, because the patient may suffer from acute hypokalemia (insulin will probably be released from β cells of the pancreas once the ECF volume is reexpanded).

There is controversy concerning glucose administration; in our opinion it should only be given to prevent hypoglycemia.

The plasma osmolality is 350 mosm/kg H_2O. Is the ICF volume high, low, or normal?

Although the plasma osmolality is 350 mosm/kg H_2O, the calculated value (2 $[Na^+]$ + mmol/l urea + mmol/l glucose) is 310 mosm/kg H_2O. The difference is mostly due to the presence of 40 mmol/l ethanol. Because the ethanol concentration (like that of urea) is equal in the ICF and ECF, it does not produce a water shift. The normal $[Na^+]$ and concentration of glucose suggest that the ICF volume is normal.

When ethanol is no longer present, what changes will be observed in the metabolic picture?

When alcohol is not present, the patient will still have ketoacidosis because of fatty acid mobilization. L-Lactate levels will not be elevated, and the concentration of glucose in plasma will be higher because of reduced impairment of glucogenesis.

Discussion of Case 3·2 An Unusual Type of Ketoacidosis

(Case presented on page 127)

What is the differential diagnosis of her acid-base disorders?

The patient's hypokalemia and metabolic acidosis suggest proximal RTA or reduced distal H^+ secretion. The urine net charge, which suggests little NH_4^+ excretion, supports a diagnosis of proximal or distal RTA. The differential diagnosis, at this point, would be the causes of low proximal or distal H^+ secretion associated with hypokalemia.

What investigative plan would be appropriate?

The plan would be to confirm the absence of proximal RTA with HCO_3^- administration and then confirm low distal H^+ secretion by measuring the rate of excretion of NH_4^+ and possibly the urine Pco_2.

Following the ingestion of $NaHCO_3$ and KCl, the plasma $[HCO_3^-]$ rose to 26 mmol/l and did not fall promptly after the intake of HCO_3^- was stopped; the urine Pco_2 was 90 mm Hg. What is the diagnosis now?

The HCO_3^- administration indeed ruled out proximal RTA. Nevertheless, the normal urine Pco_2 indicates that back-diffusion of H^+ from lumen to cell occurred when the urine pH was low (H^+ back-leak type of distal RTA) or that there was a normal distal H^+ secretory process and a more complex disease. Because the patient had not used amphotericin B, back-leak of H^+ in the collecting duct is an unlikely diagnosis. The other possibility is that we are being misled by the urine net charge. There may have been high NH_4^+ excretion that was in conjunction with an anion other than Cl^-. This possibility can be confirmed by measuring the urine osmolality and comparing it with the calculated urine osmolality.

The measured urine osmolality is 680 mosm/kg H_2O, but the calculated urine osmolality (2($[Na^+]$ + $[K^+]$) + glucose + urea) is 269 mosm/kg H_2O.

Therefore, there is indeed an "osmolal gap" in urine of 411 mosm/kg H_2O that indicates the presence of a large number of unmeasured osmoles (possibly NH_4^+ + β-HB^-). Urine is tested for NH_4^+ and β-HB^-; later, these values are confirmed to be 120 mmol/l and 234 mmol/l, respectively. Because the urine volume was greater than 1 liter/day, the patient had a normal renal response to acidemia and excreted close to 200 mmol of NH_4^+ per day. Ketoacidosis, which caused an ongoing acid load that resulted in acidemia, was not evident because of marked ketonuria (see Figure 3·3, page 73). With better diabetic control, the acidemia disappeared. Therefore, she had a renal tubular lesion in β-HB^- reabsorption in the presence of diabetic ketoacidosis.

Why was this patient hypokalemic?

In the face of hypokalemia, the patient had a urine [K^+] of 60 mmol/l. Back-correcting for medullary water abstraction, her [K^+] in the cortical collecting duct was close to 30 mmol/l, about 10-fold greater than her plasma [K^+]. Thus, aldosterone was released in response to ECF volume contraction, which was caused by renal Na^+ loss secondary to the ketonuria. Her hypokalemia was therefore secondary to renal K^+ loss. Furthermore, despite insulin deficiency, the plasma [K^+] was 2.9 mmol/l; this value suggests a very large K^+ deficit.

Note:
See pages 340–42 for a discussion of "back-correction" for medullary water abstraction to establish the [K^+] in the cortical collecting duct. This calculation allows the detection of mineralocorticoid actions, as reflected by the transtubular K^+ gradient (TTKG).

Discussion of Case 3·3
Ketoacidosis, a Stroke of Bad Luck

(Case presented on pages 127–28)

What is (are) the most likely cause(s) for his metabolic acidosis?

The first question to consider is whether or not acids are being overproduced. Because the patient's plasma anion gap is elevated markedly at 42 mEq/l (30 higher than expected) and renal failure is not present, he therefore has overproduction of acids. Given this story, AKA seems to be the most likely diagnosis, but his screening test for ketoacids in plasma is only moderately positive. This result suggests a mixture of L-lactic acidosis (most likely from ethanol) and ketoacidosis, although one cannot rule out the very unlikely diagnosis of D-lactic acidosis.

AKA = alcoholic ketoacidosis.

Follow-up data reveal very elevated levels of β-HB^- (12 mmol/l) and L-lactate (6 mmol/l), which confirm the clinical impression.

Are acids being produced rapidly?

There are two types of information needed to answer this question, the nature of the acid added and how many new anions are appearing.

1. **Nature of the acid:**
   L-Lactic acid, which is produced very quickly, requires poor delivery of O_2 to tissues. Given the clinical picture, the delivery of O_2 to all organs except part of the CNS is adequate, so this setting is probably not appropriate for rapid accumulation of acids.

2. **Net appearance of new anions:**
   For new anions to be present, there must be a rise in the plasma anion gap in the past hour or two (the anion gap did not rise) or the excretion

of more $Na^+ + K^+$ than Cl^-. Because the urine output is trivial, there is no major appearance of new anions in the urine.

In conclusion, it appears that the net rate of accumulation of acids is small at this point.

What hormonal changes were involved?

For ketoacidosis to have developed, there must have been a relative lack of insulin. Although the patient could have IDDM, the most likely basis of the lack of insulin is inhibited release of insulin from β cells by the α-adrenergic response to adrenaline. Adrenaline is released in large quantities secondary to the major CNS lesion (intracerebral hemorrhage), the so-called "cerebral diabetes."

Long-term follow-up reveals that he does not have diabetes mellitus.

Why did ketoacidosis develop so quickly?

1. **Increased production:**
   Usually there is a lag period before the rate of formation of ketoacids begins to rise. This patient may have "by-passed" the rate-limiting step (the supply of intramitochondrial fatty acyl-CoA in hepatocyte mitochondria) by making acetyl-CoA in hepatocytes from ethanol.

2. **Decreased utilization:**
   The major organ that utilizes ketoacids is the brain. Because the patient is comatose and has a major intracerebral lesion, the oxidation of ketoacids might be diminished appreciably.

Discussion of Case 3·4 A Superstar of Severe Acidosis

(Case presented on page 128)

What are the most likely diagnoses?

The patient has a very severe degree of metabolic acidosis. The production of acids should have the following characteristics:

1. **A relatively low rate of production of acids:**
   A rapid rate of production of acids would have killed the patient. This fact rules out (temporarily) production of L-lactic acid via anaerobic glycolysis.

2. **A near-normal ECF volume:**
   Ketoacidosis (DKA, AKA) would be an attractive diagnosis, but at this degree, it should be accompanied by an extremely contracted ECF volume. Given the absence of alcohols, both ketoacidosis and ingestion of alcohols are very unlikely diagnoses.

3. **Expected history for D-lactic acidosis:**
   There is no GI history to support the diagnosis of D-lactic acidosis, and the degree of acidosis is much more severe than one would expect with this diagnosis; hence, although D-lactic acidosis is a remote possibility, it is unlikely.

By exclusion, the most likely diagnosis is Type B L-lactic acidosis. We would suspect a small reduction in gluconeogenesis or a small deficit in vitamin B_1 (thiamine), since we need a diagnosis with a slow but steady net accumulation of acid. This suspicion was confirmed by finding that the L-lactate level in plasma exceeded 30 mmol/l and that he was taking metformin for treatment of his NIDDM (information provided later).

What acid-base treatment would be most appropriate?

The most important point to note is that although the numbers are alarming, the patient walked into the emergency room. Further, the rate of addition of new anions is small. Nevertheless, he would probably benefit from receiving some $NaHCO_3$. Doubling his current plasma $[HCO_3^-]$ (from 1 to 2 mmol/l) will raise his pH to 7.0 if his P_aco_2 stays constant. Tripling the $[HCO_3^-]$ will require 30 mmol of $NaHCO_3$ for his ECF (the ECF volume is 15 liters). In addition, he will need a large quantity of HCO_3^- for his ICF, but we cannot tell how much. Our guess is to give 100 mmol of $NaHCO_3$ and observe, but be prepared to give 200 mmol of $NaHCO_3$.

To increase the oxidation of L-lactate, administering dichloroacetate, an activator of PDH (Figure 12·1, page 430), would be a good choice in this setting.

Discussion of Case 3·5 Acute Popsicle Overdose

(Case presented on pages 128–29)

Could chronic diarrhea have caused the acidosis?

For diarrhea to be the sole cause of acidosis, the patient would have to lose 200 mmol of $NaHCO_3$ per day in the stool (see the margin). During chronic metabolic acidosis, the kidneys, if normal, will generate 200 mmol of HCO_3^- each day by excreting this quantity of NH_4^+. Hence, diarrhea per se is an unlikely cause. In addition, diarrhea-induced metabolic acidosis is not associated with an increase in the anion gap.

$[HCO_3^-]$ in diarrhea fluid:
This concentration is usually < 50 mmol/l. Therefore, to lose more than 200 mmol of HCO_3^- per day, the volume of diarrhea must exceed four liters per day.

If not, what other diagnosis is likely?

Two major causes come to mind in this setting, low excretion of NH_4^+ and overproduction of D-lactic acid by the intestinal bacteria.

1. **Low excretion of NH_4^+:**
   It is possible that the patient has a renal lesion that compromised the excretion of NH_4^+. Measuring the urine $[NH_4^+]$ did not reveal a low quantity (200 mmol excreted per day), so this less likely possibility was ruled out (see the margin).

Note:
Low excretion of NH_4^+ is not associated directly with a rise in the anion gap in plasma.

2. **Overproduction of D-lactic acid:**
   Several factors make this diagnosis very likely: the history of a GI problem; the ingestion of antibiotics to alter the GI flora; the ingestion of a motility suppressant to permit a longer time of incubation; the CNS disturbance (bacteria produce other toxins too); the unexpectedly small rise in the plasma anion gap (some D-lactate was excreted in the urine).

Confirm this diagnosis by measuring the D-lactate level in plasma; this metabolite was markedly elevated (10 mmol/l).

What role did the popsicles play?

The bacteria in his upper GI tract were starved. When fed sugar from the popsicles, they responded by producing D-lactic acid plus CNS toxins.

Summary of Main Points

ACID-BASE BALANCE

- Metabolism of proteins and vegetables yields a net of 1 mmol H^+/day/kg body weight. Excessive H^+ are produced during ischemia, if there is a lack of insulin, or during certain intoxications.
- Almost all H^+ are buffered by HCO_3^-, but, if the H^+ load is very large, an appreciable number of H^+ are buffered by intracellular proteins. A low Pco_2 signals the appropriate response to an acid load. Drop the 7 and the decimal point of the pH to see the expected value for the Pco_2; compare this value with the measured one.
- The expected response to a chronic acid load is excretion of more than 200 mmol of NH_4^+ per day and reabsorption of all filtered HCO_3^- by the kidney. Find NH_4^+ in urine by calculating the urine net charge and/or the urine osmolal gap.

DIAGNOSTIC PROCEDURES

- Use the anion gap in plasma to find anions produced with H^+.
- Quantitate new anions added to the body (multiply the rise in anion gap by the total body water) and to the urine (count the anions excreted without H^+ or NH_4^+ by multiplying ($Na^+ + K^+ - Cl^-$) in urine by urine volume).
- Use the osmolal gap in plasma to detect toxic precursors of acids (alcohols).
- Use the urine net charge and/or osmolal gap to detect NH_4^+ in the urine.
- The urine pH is good for detecting HCO_3^- in the urine but not good for detecting NH_4^+ in the urine. It is useful in determining why the [NH_4^+] is low in the urine.

OVERALL SUMMARY

Although metabolic acidosis may be the result of a large number of diverse disorders, these disorders can be sorted out easily with an organized approach, as outlined in Figure 3·4, page 76. Metabolic acidosis associated with increased H^+ production and with renal failure is identified by an increase in the plasma anion gap. The subgroup associated with methanol or ethylene glycol intoxication also has an increased osmolal gap.

The critical feature in discovering the basis of metabolic acidosis with a normal plasma anion gap is determining whether or not the excretion of NH_4^+ in the urine is appropriate. The urine [NH_4^+] is reflected by the apparent urine net charge ([Na^+] + [K^+] – [Cl^-]) when the urine pH is < 6.1. A negative apparent urine net charge (i.e., [Cl^-] greatly exceeds the sum of [Na^+] and [K^+]), indicates an appropriate urine [NH_4^+], and the diagnosis is $NaHCO_3$ loss, either via the GI tract or some other site (proximal RTA, acetazolamide, NH_4Cl administration). If the apparent net charge is positive (i.e., the sum of [Na^+] and [K^+] greatly exceeds [Cl^-]), the urine [NH_4^+] is low. A low level indicates one of three possibilities: the presence of RTA, diabetic ketoacidosis with marked β-hydroxybutyrate and NH_4^+ excretion, or toluene toxicity with hippurate and NH_4^+ excretion (revealed by the urine osmolal gap).

Discussion of Questions

3·1 Which patient has a primary respiratory acid-base disorder? How should each patient be managed from an acid-base point of view?

Patient	[H^+] nmol/l	pH	P_aco_2 (mm Hg)	[HCO_3^-] (mmol/l)
A	64	7.20	20	8
B	120	6.90	40	8
C	30	7.50	10	8

Case A represents the appropriate respiratory response to metabolic acidosis. Management involves determining the basis of the metabolic acidosis.

Case B has metabolic acidosis and respiratory acidosis. The coexistent respiratory acidosis has made the acidemia life-threatening, and intervention is indicated. If immediate correction of the basis of the respiratory acidosis is not possible, mechanical ventilation is advisable.

Case C has a significant degree of respiratory alkalosis; management depends on the basis of the hyperventilation.

3·2 Does the arterial or venous Pco_2 best reflect the degree of protonation of intracellular proteins during metabolic acidosis?

The venous Pco_2 best reflects the protonation of intracellular proteins. Because CO_2 must diffuse from tissues to plasma in capillaries, the

BBS = bicarbonate buffer system.

Pco_2 in tissues must be higher than that in venous blood. Therefore, the venous Pco_2 best reflects tissue Pco_2 and the degree of effectiveness of the BBS in the ICF. Since one does not measure the venous Pco_2 in vital organs, this value must be estimated. Remember that the mixed venous Pco_2 reflects the Pco_2 of the venous blood draining from organs with the largest blood supply. For example, the Pco_2 of venous blood draining from exercising muscles may be close to 100 mm Hg; at the same time, the Pco_2 of venous blood draining the from brain might be close to 46 mm Hg.

Arterial Pco_2 primarily reflects the effect of alveolar ventilation—removal of the CO_2 that is produced. It does not reflect the Pco_2 in the various organs.

3·3 How can reexpansion of the ECF volume affect the Pco_2 in vital organs?

The Pco_2 in venous blood is the result of the rate of production of CO_2 and its rate of removal via the blood. Consider the following example where the rate of production of CO_2 is 10 mmol/min and the cardiac output is 5 liters/min (normal) and 2.5 liters/min (reduced by 50% because of ECF volume contraction).

Cardiac output	CO_2 production	CO_2 carried	Venous Pco_2
5 liters/min	10 mmol/min	2 mmol/l	46 mm Hg
2.5 liters/min	10 mmol/min	4 mmol/l	60 mm Hg

Therefore, in this example, reexpansion of the ECF leads to a rise in cardiac output from 2.5 to 5 liters/min and thereby to a fall in venous Pco_2 from 60 to 46 mm Hg.

3·4 Some recommend that a 50:50 mixture of $NaHCO_3$ and Na_2CO_3 (carbicarb) be used as a source of alkali. If a patient has L-lactic acidosis and is producing 12 mmol of L-lactic acid per minute, how much less CO_2 (expressed in percentage form) will be produced by titrating the H^+ produced with carbicarb instead of $NaHCO_3$? (Assume 12 mmol of O_2 are consumed each minute by the body.)

Respiratory quotient = CO_2 produced/O_2 consumed.

$$0.8 = \frac{10 \text{ mmol of } CO_2}{12 \text{ mmol of } CO_2}$$

First, let us examine the rate of production of CO_2 before and after buffering. If the *respiratory quotient* is 0.8, approximately 10 mmol of CO_2 are produced each minute via aerobic metabolism. Titrating 12 mmol of H^+ with $NaHCO_3$ will yield 12 mmol of CO_2 and a total rate of CO_2 production of 22 mmol/min. Using carbicarb, for every 4 mmol of H^+ titrated, 3 mmol of CO_2 will be produced. Thus, to titrate 12 mmol of H^+, 9 mmol of CO_2 will be produced, and 19 mmol of CO_2 will be produced per minute if all the alkali is titrated. Hence, the difference in CO_2 production is small (22 vs 19), close to 15%, and will have a minor effect on the arterial Pco_2 unless alveolar ventilation is low and fixed.

3·5 Why is the rate of production of ketoacids so much lower than that of L-lactic acid if both are regulated by the rate of turnover of ATP?

The answer involves the restrictions set by the rate of consumption of oxygen and by stoichiometry. Ketogenesis is restricted to the liver, an organ with little variation in oxygen consumption. Stoichiometry is close to 6 ATP molecules generated per ketoacid formed. In contrast, L-lactic acid is produced by many organs. Consider skeletal muscle, which can increase its demand for O_2 20-fold and thus can have a much greater rate of consumption of O_2 than the liver. The rate of production of ATP and L-lactic acid, though still under ATP feedback control, can be enormous (H^+ production is perhaps 50-fold greater than in ketogenesis). In Type A L-lactic acidosis, some of the needs for ATP are not met by aerobic metabolism, hence the need for anaerobic metabolism. Therefore, one might conceptualize that L-lactic acidosis is being pushed by the need for ATP rather than being limited by it.

3·6 The rates of ketoacid production and removal are usually equal in a person who lacks insulin. Why is this equality beneficial?

Ketoacid production is designed to permit one to exist for long periods of time without the intake of carbohydrates. Ketoacids are formed in the liver so that the brain can oxidize a fat-derived fuel (it cannot oxidize fatty acids from the circulation at appreciable rates). This oxidation of ketoacids prevents major catabolism of lean body mass (the brain alone would oxidize the equivalent of 1–2 lb of muscle per day without ketoacids to oxidize, since gluconeogenesis (GNG) would have to provide the daily supply of glucose; see the margin). The danger of ketoacids is accumulation of H^+, but the ATP constraints on ketogenesis in the liver permit the rate of ketogenesis to be relatively small and to equal the rates of ketoacid removal in other organs (Figure 3·10, page 87). Hence, the degree of this ketoacidosis is constant and mild when fasting is the cause of low levels of insulin.

Facts:

- The brain consumes 120 g of glucose per day in the absence of ketoacids.
- 200 g of protein can yield 120 g of glucose.
- Lean body mass is 80% water (1 kg yields 200 g of protein).

3·7 What makes DKA severe in degree compared with the ketoacidosis of chronic fasting?

The answer is probably the lower rate of ketoacid removal by metabolism in DKA. Although the rate of ketogenesis is similar in DKA and fasting, coma and severe confusion appear with a severe degree of hyperglycemia. These disturbances, along with the low GFR (osmotic diuresis caused loss of Na^+), diminish O_2 consumption in DKA and thereby fuel oxidation in the two major organs that consume ketoacids.

3·8 In hepatic mitochondria, acetyl-CoA is formed from ethanol and from fatty acids at about the same rate. Why are these rates so similar?

The answer lies in the stoichiometry of ATP per acetyl-CoA formed.

C_{16} fatty acid $\longrightarrow$ 33 ATP + 8 Acetyl-CoA

Ethanol $\longrightarrow$ 4 ATP + Acetyl-CoA

Therefore, the yield is close to 4 mmol of ATP per acetyl-CoA in both cases (the numbers include the quantity of ATP needed to form CoA derivatives of fatty acids and acetic acid).

3·9 Must ethanol levels be elevated on admission for metabolism of ethanol to be an important cause of ketoacidosis?

No. Although ethanol levels were elevated when ketoacids were formed initially, the source of acetyl-CoA can now be fatty acids. Contraction of the ECF volume causes the release of adrenaline, which results in the stimulation of HSL and the subsequent release of more fatty acids.

3·10 An anoxic limb needs to regenerate 18 mmol of ATP per minute (25% of the ATP needed in the body) via anaerobic glycolysis. If the rest of the body was "persuaded" to oxidize L-lactate anions (+ H^+) to regenerate all needed ATP (54 mmol/min, equivalent to the utilization of 9 mmol of O_2), would L-lactic acid accumulate?

Note:
Athletes "cool down" by jogging. In doing so, they can oxidize more L-lactate anions because of the higher rate of turnover of ATP.

The normal rate of consumption of oxygen is 12 mmol/min, and it is reduced by 25% to 9 mmol/min. Given the stoichiometry of 3 mmol of O_2 per mmol of lactic acid oxidized, only 3 mmol of lactic acid could be oxidized by the rest of the body each minute. The generation of 18 mmol of ATP from glucose in the anoxic limb will result in the formation of 18 mmol of L-lactic acid. Because the maximum rate of glucogenesis is less than 15 mmol/min, L-lactic acidosis must get worse unless oxygen can be delivered to the anoxic limb.

Even if oxygen were delivered to the hypoxic leg and no more L-lactic acid formed, the rate of decline in the concentration of L-lactate anions in plasma would be very slow. In quantitative terms, a 70-kg patient with lactic acidosis (15 mmol/l) has a pool size of L-lactate of close to 450 mmol (assume a volume of distribution of lactate of 30 liters for simplicity). If all the ATP regenerated was derived from the oxidation of L-lactate anions, only 4 mmol of L-lactate anions could be oxidized per minute at rest. If we add a maximum rate of glucogenesis of 8 mmol/min, the overall clearance of L-lactate anions by metabolic routes would be 12 mmol/min. Hence, the degree of decline in L-lactic acid would be almost 4 mmol/l in 10 minutes, but this rate of metabolism of L-lactate anions is probably a gross overestimate because the brain consumes about 25% of O_2 at rest and it will burn glucose. Further, it is unlikely that 100% of the rest of the ATP generated will be from the oxidation of L-lactate anions.

3·11 If a patient has hypoxia but little glycogen in the liver, will L-lactic acidosis develop? If not, what changes would you expect to find in the concentration of metabolites in blood?

Development of L-lactic acidosis is not likely. To develop L-lactic acidosis, a supply of glucose is needed. Because glucose distributes in close to half the volume of L-lactate anion but yields two L-lactate anions on a molar basis, only a small rise in L-lactate anion is possible from this source in the absence of hyperglycemia. The usual source of L-lactic acid is muscle or liver glycogen. The former requires a specific stimulus (e.g., exercise) to be hydrolyzed, and there is little glycogen in the liver; this patient will therefore suffer from hypoglycemia and organ malfunction (from less regeneration of ATP) rather than L-lactic acidosis (Figure 3·23).

3·12 If $NaHCO_3$ was given as treatment for L-lactic acidosis and there was no rise in the plasma [HCO_3^-], was the alkali of no help to that patient?

To be a successful buffer, HCO_3^- must remove H^+ bound to proteins in the ICF or accelerate the rate of production of ATP. In terms of acid-base balance, if HCO_3^- were given and are no longer present, they were titrated or excreted (unlikely). To be titrated, a source of H^+ is needed. Two possible sources are the H^+ bound to proteins (buffered H^+) or new L-lactic acid formed. In both cases, this disappearance of H^+ and HCO_3^- was beneficial because ATP is formed whenever L-lactic acid is formed (Figure 3·23).

We focus on the plasma H^+ and [HCO_3^-] because they are easy to measure. It really does not matter how much the plasma [HCO_3^-] per se rises because it represents untitrated base; in most cases, the [H^+] in the ECF is unlikely to play an important direct role in toxicity.

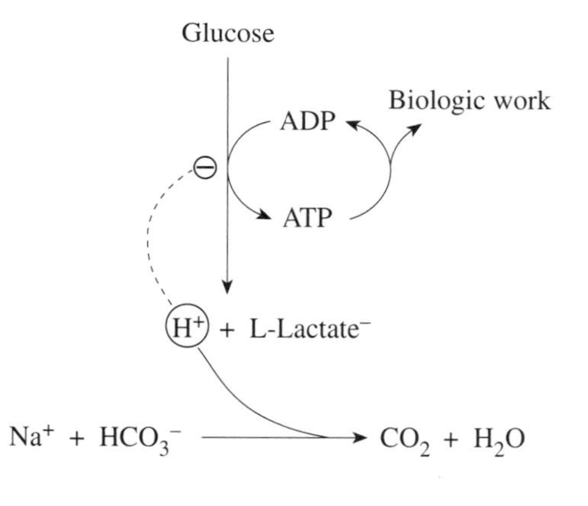

Figure 3·23
Production of L-lactic acid by anaerobic glycolysis
When L-lactic acid is produced, ATP is regenerated so that biologic work can be performed. This pathway is inhibited by a rise in the [H^+].

3·13 Why might the rate of L-lactic acid production rise with alkali therapy?

As background, two molecules of ATP are generated per molecule of glucose metabolized during anaerobic glycolysis. In contrast, 36 molecules of ATP are generated per molecule of glucose oxidized during aerobic metabolism. Hence, glycolytic flux must be 18-fold higher during anaerobic conditions to yield the same quantity of ATP. During anaerobic glycolysis, H^+ accumulate. Because H^+ inhibit the rate-limiting step in glycolysis (*PFK-1*), ATP regeneration will be compromised earlier in anaerobic conditions (faster flux is needed). Therefore, removal of H^+ with alkali can "de-inhibit" glycolysis and increase the rate of regeneration of ATP (Figure 3·23).

PFK-1 = phosphofructokinase-1.

3·14 What metabolic adaptations prolong survival in hypoxic environments?

To prolong survival in hypoxic environments, less anaerobic metabolism must occur, yet enough ATP must be available. Therefore, the demand for ATP (biologic work) must decline. Strategies in individual organs are summarized in Table 3·23. A general strategy is to lower the metabolic rate (hypothermia, hypothyroidism).

If you were a goldfish, you could make the end products of anaerobic metabolism uncharged (ethanol) rather than an anion such as L-lactate anion.

Table 3·23
DECREASING THE RATE OF METABOLISM IN SPECIFIC ORGANS

Organ	Strategy
Brain	Anaesthetic/sedative to reduce work Drug to decrease Na^+ permeability (adenosine in certain animals)
Kidney	Anything that lowers the GFR will lower Na^+ pumping
Muscle	Paralytic agent

3·15 Why is the L-lactic acidosis of exercise better tolerated than the L-lactic acidosis of shock, even if the former is more severe in degree?

In both the L-lactic acidosis of exercise and cardiogenic shock, L-lactic acid accumulates because the delivery of O_2 does not match the demand for it. A major difference is that in exercise, muscle undergoes hypoxia, but the brain does not. In addition, there is no reduction in the rate of cerebral blood flow.

Because tissue Pco_2 must be higher than that in the vein for diffusion of CO_2 to occur, the Pco_2 in the brain (and thereby the $[H^+]$ in the ICF) is much higher in cardiogenic shock than in exercise (see the margin). A higher $[H^+]$ can limit anaerobic glycolysis (Figure 3·23). Therefore, the supply of ATP is reduced and brain cells are more likely to die. This point is valid for all cells that are performing work.

Higher tissue Pco_2 in cardiogenic shock:
If the rate of production of CO_2 in the brain was similar in both settings, the lower cerebral blood flow rate in cardiogenic shock would require that each liter of blood flowing out of the brain carry more CO_2 and thus have a higher Pco_2 (see the discussion of Question 3·3, page 136).

3·16 Is a rise in L-lactic acid production beneficial or harmful to a patient with L-lactic acidosis?

During anaerobic glycolysis, glucose is consumed and both ATP and L-lactic acid are formed. Hence, dangers of glycolysis are H^+ accumulation and hypoglycemia; the advantage is ATP generation. If generation of more ATP occurs in the heart, contractility can increase and more O_2 can be pumped to other organs. Thus, more production of L-lactic acid can be beneficial at some times and detrimental at others (Figure 3·23).

Note:
This question is for experts.

3·17 A 34-year-old male patient with AIDS has extreme muscular weakness and chronic L-lactic acidosis related to AZT. He had a mitochondrial myopathy on biopsy. When he exercises, his L-lac-

tic acidosis does not become more severe. Did the lesion in his muscle mitochondria cause his L-lactic acidosis?

It has been stated that AZT causes a deletion of part of the mitochondrial system that regenerates ATP (the electron transport system). This disturbance is analogous to having hypoxia as the cause of L-lactic acidosis. In this setting, one would expect L-lactic acidosis to get worse if more regeneration of ATP is required. Because the severity does not increase, there must be another explanation for these results.

The problem is to devise a way in which more ATP can be generated in affected muscle fibers without the accumulation of more L-lactic acid. We offer the following speculations:

1. **More L-lactic acid was generated in affected fibers but was oxidized in adjacent normal fibers.**
   Although this theory is possible, we find it unattractive because of the stoichiometry described in the discussion of Question 3·10 (i.e., in terms of ATP turnover, so much more L-lactic acid is formed anaerobically than is oxidized aerobically). Further, one would not expect the L-lactic acidosis to be near steady state, as exhibited in the patient. Finally, the number of affected muscle fibers would have to be relatively small and perhaps not permit such predominant symptoms of weakness.

2. **Other organs removed the extra L-lactate anions.**
   While this theory is also possible, again we feel that it is unlikely because we do not know what would drive the higher flux rates. Surely at such high levels of L-lactate in plasma, metabolic processes for L-lactate anion removal are close to being saturated with their substrate, L-lactate.

3. **A lesion is present in mitochondria of affected myocytes that permits the rate of regeneration of ATP to increase but does not allow the accumulation of more L-lactic acid.**
   A rate-limiting lesion in the TCA cycle or the electron transport system would not have this effect. Instead, a lesion would have to permit a high enough level of pyruvate for oxidation and allow a very high level of L-lactate. A high $NADH/NAD^+$ ratio would therefore have to exist in the cytosol of myocytes. A possible example is shown in Figure 3·16, page 105.

3·18 The L-lactic acidosis of the patient in Question 3·17 was greatly aggravated by ethanol intake. What might this effect of ethanol imply?

When ethanol is metabolized, the $NADH/NAD^+$ in the cytosol of hepatocytes rises. This increase diverts pyruvate to L-lactate and may aggravate the degree of L-lactic acidosis if flux rates are particularly rapid (Figure 3·15, page 103). This patient might have had a very high rate of release of L-lactate anions from affected myocytes and now has a compromised removal of L-lactate anions via glucopaleogenesis in the liver.

Extra definitions:

We use three terms to describe the production of glucose in the liver.

1. Glucogenesis: Synthesis of glucose from all sources.
2. Gluconeogenesis: Synthesis of new glucose (e.g., from amino acids).
3. Glucopaleogenesis: Resynthesis of glucose using fuels derived from glucose molecules (e.g., L-lactate anions derived from circulating glucose).

3·19 If the same number of bacteria are growing in the small bowel rather than the colon, why might more D-lactic acid be formed?

The factors that determine the quantity of D-lactic acid formed are the number of bacteria, their nature, the amount of substrate (glucose, sucrose, etc.) available and the incubation conditions (e.g., pH and duration of incubation). In the small bowel, the supply of substrate and a higher pH (secretion of $NaHCO_3$) might favor the production of more organic acids. We cannot comment on the number and nature of the bacteria or the length of time that they dwell in the lumen of the bowel because these factors depend on specific details of the case.

3·20 A 26-year-old intoxicated man is brought to the emergency room. His friends state that he ingested several ounces of methanol. Other than intoxication, his physical exam was normal. His lab results were as follows:

Na^+	mmol/l	142	$[H^+]$	nmol/l	40
K^+	mmol/l	3.7	P_aco_2	mm Hg	40
Cl^-	mmol/l	105	Osmolality	mosm/kg	360
HCO_3^-	mmol/l	25	Glucose	mmol/l (mg/dl)	7 (126)
Urea	mmol/l (mg/dl)	5 (14)	Albumin	g/l	40

Should one take the allegation of methanol seriously? Explain your reasoning. What should the course of action be?

The absence of an increased plasma anion gap and no evidence of metabolic acidosis should diminish the suspicion of methanol intoxication; therefore, one should seek more evidence before embarking on dialysis, which has some risk (insertion of lines is associated with some morbidity). The osmolal gap is readily available. Because it is indeed elevated, some credence is added to the allegation of methanol poisoning; however, this elevation could result from the ingestion of any alcohol, including ethanol. With an increased plasma osmolal gap and the allegation of methanol ingestion, one must proceed with therapy until further data can be obtained. At the very least, one should give ethanol after sending blood and urine tests for toxin screen; then one should await confirmation by the laboratory. In the absence of either an increase in the anion gap or acidosis, it is perfectly safe to treat initially with just ethanol.

Laboratory results revealed that he indeed had toxic levels of methanol, and he also had elevated levels of isopropyl alcohol, but no ethanol was detected. Therefore, we assume that it was the ingestion of isopropyl alcohol that had delayed the metabolism of methanol by alcohol dehydrogenase and had protected him from both acidemia and the toxicity of formaldehyde and formic acid. He should undergo hemodialysis to remove methanol while maintaining high levels of ethanol in his plasma.

3·21 What is the specific clue suggesting glue sniffing as the diagnosis in a patient presenting with a normal plasma anion gap type of metabolic acidosis and hypokalemia?

The differential diagnosis in this problem includes three main possibilities: distal RTA, diarrhea, and toluene intoxication. A distant fourth would be diabetic ketoacidosis with excessive ketoacid anion excretion.

The important evidence needed to resolve the differential diagnosis is a critical assessment of the urine electrolytes and osmolality to establish the NH_4^+ excretion rate. One should also test for ketoacid anions in the urine.

With distal RTA, the $[Cl^-]$ does not exceed $[Na^+] + [K^+]$, and there is no osmolal gap in the urine (i.e., urine NH_4^+ excretion is not increased).

With diarrhea, the urine $[Cl^-]$ greatly exceeds $[Na^+] + [K^+]$ and indicates an increased NH_4^+ excretion.

With toluene toxicity, there may be significant Na^+ excretion despite ECF volume contraction because of the excretion of the anion hippurate (see pages 108–09 for more details). The $[Cl^-]$ is usually less than $[Na^+] + [K^+]$, but there will be a high urine osmolal gap because of the excretion of hippurate anions with NH_4^+.

With DKA, the qualitative test for ketoacid anions is positive and there is also an osmolal gap in the urine (due to $NH_4^+ + \beta\text{-HB}^-$).

3·22 What factors contribute to the hypokalemia and K^+ depletion in glue sniffers?

The formation and excretion of hippurate ultimately obliges the excretion of some Na^+. Because glue sniffers often consume a limited quantity of Na^+, they are frequently ECF-volume-depleted. Once ECF volume depletion is present, aldosterone is released. In response to aldosterone, Na^+ are reabsorbed in an electrogenic fashion because the luminal fluid contains primarily Na^+ and hippurate anions but only a small quantity of Cl^-. The excretion of K^+ ensues, resulting in K^+ depletion. As acidemia persists, NH_4^+ excretion increases. The bulk of hippurate is then excreted with NH_4^+; less is excreted with K^+ and Na^+.

3·23 Patients A, B, and C listed below each have metabolic acidosis and an increased plasma anion gap. Which one has renal failure, which has methanol intoxication, and which has D-lactic acidosis?

Patient	A	B	C
Calculated Osmolality	290	290	320
2 × plasma $[Na^+]$ (mmol/l)	280	280	280
Urea (mmol/l)	5	5	35
Glucose (mmol/l)	5	5	5
Measured Osmolality	290	320	320

The hallmark of renal failure is an elevated concentration of urea in plasma; only patient C can have renal failure because A and B have normal values for urea.

The hallmark of methanol intoxication is a large difference between measured and calculated osmolalities in plasma; only patient B has a large osmolal gap in plasma, so B has methanol intoxication.

In D-lactic acidosis, there is neither an elevated osmolal gap nor a high level of urea; hence, patient A has D-lactic acidosis.

3·24 An 80-year-old man with a history of "pyelonephritis" developed diarrhea after a course of antibiotics. On the basis of the following results, a diagnosis of distal RTA was made. Is it correct?

Plasma			Urine		
Na^+	mmol/l	134	Na^+	mmol/l	10
K^+	mmol/l	2.8	K^+	mmol/l	40
Cl^-	mmol/l	115	Cl^-	mmol/l	100
HCO_3^-	mmol/l	10	Osmolality	mosm/kg H_2O	800
H^+	nmol/l	62	Urea	mmol/l	300
pH		7.20	pH		5.9

Not likely. Both the urine net charge and the estimated osmolal gap in the urine imply that a high concentration of NH_4^+ is present in the urine. If he had distal RTA, the $[Na^+] + [K^+]$ would exceed the $[Cl^-]$ in the urine. His GI disease is the most likely basis of the metabolic acidosis with a normal renal NH_4^+ excretion.

3·25 What are the diagnostic features of disorders with reduced indirect reabsorption of filtered HCO_3^-?

Note:
The diagnostic features listed are those of isolated proximal RTA.

The major finding is metabolic acidosis with a normal plasma anion gap. The degree of acidosis is modest and is not really influenced by administration of $NaHCO_3$. The rate of excretion of NH_4^+ is much lower than expected for the chronic metabolic acidosis. Usually, in an isolated lesion, the urine pH is low, citrate excretion is not reduced, and there are no other renal findings. The patients studied have otherwise normal renal functions, are short in stature, and do not have nephrocalcinosis or hypokalemia if they have not received therapy.

3·26 If a patient with metabolic acidosis is taking acetazolamide, how will the urine test results change once this diuretic is no longer acting (but metabolic acidosis persists)?

The expected response in chronic metabolic acidosis is an increase the rate of excretion of NH_4^+. Therefore, expect to see 200 mmol of NH_4^+ excreted per day, a negative urine net charge ($Cl^- > (Na^+ + K^+)$), a high urine osmolal gap, and possibly a lower urine pH.

4

METABOLIC ALKALOSIS

PART C TREATMENT OF METABOLIC ALKALOSIS 166

PART D REVIEW 168

Objectives

- To provide the background so that the three components of the pathophysiology of metabolic alkalosis can be understood:
  1. generation of HCO_3^-;
  2. renal "permission" to maintain a high $[HCO_3^-]$ in plasma;
  3. intracellular K^+ deficit along with a gain of H^+ and Na^+ in the ICF.
- To identify the different clinical scenarios with metabolic alkalosis: those where a loss of Cl^- and gain of HCO_3^- play a prominent role (low urine Cl^-) and those where $NaHCO_3$ is retained in conjunction with the excretion of K^+ in the urine (urine Cl^- is not low).
- To emphasize that the treatment of metabolic alkalosis depends on the specific deficit involved; KCl is needed to replace a deficit of KCl, and NaCl will be needed if there is a deficit of NaCl.

Abbreviations:

A-a = alveolar minus arterial difference.
COPD = chronic obstructive pulmonary disease.
ECF = extracellular fluid.
GFR = glomerular filtration rate.
GI = gastrointestinal.
ICF = intracellular fluid.
JVP = jugular venous pressure.
P_aco_2 = partial pressure of CO_2 in arterial blood.
Pco_2 = partial pressure of CO_2.

Outline of Major Principles

1. The pathophysiology of metabolic alkalosis involves events in three major areas—the ECF, the ICF, and the urine. Each must be considered for a comprehensive understanding.
2. Metabolic alkalosis is not primarily an acid-base disorder; it is the net result of deficits in the ECF volume (NaCl) and in K^+ for the most part, with secondary acid-base changes.
3. Appropriate treatment of metabolic alkalosis requires a knowledge of the expected deficits of ions.

Introductory Case
Basically, Toby Is Not "OK"

(Case discussed on pages 168–69)

Toby, a 26-year-old dancer, complains of weakness. She denies vomiting and the intake of medications other than vitamins. Physical examination reveals a thin woman who has a contracted ECF volume. Laboratory results are listed below.

		Plasma	Random urine
Na^+	mmol/l	133	52
K^+	mmol/l	3.1	50
Cl^-	mmol/l	90	0
HCO_3^-	mmol/l	32	Not determined
pH		7.48	8.0

What acid-base disturbance is present?
Why is the [Na^+] in urine not lower, given the presence of ECF volume contraction?
Why is Toby hypokalemic?
What is the basis for the acid-base disturbance?

PART A
PATHOPHYSIOLOGY OF METABOLIC ALKALOSIS

- Metabolic alkalosis is present if the $[HCO_3^-]$ rises and the $[H^+]$ declines in the ECF.
- The $[HCO_3^-]$ = the ratio of the HCO_3^- content in the ECF to the volume of the ECF.
- The three components to assess are the ECF, the ICF, and the renal response.

Metabolic alkalosis, an acid-base disorder, has two hallmarks—an elevated $[HCO_3^-]$ and a lower $[H^+]$ in the ECF. Although this definition appears to be simple enough, it addresses only a part of the picture, the ECF. Two other aspects of this disorder also merit emphasis. First, implicit in this definition is the fact that renal mechanisms have been called into play to permit the ECF to have an elevated $[HCO_3^-]$. Second, though not so obvious, is the occurrence of important events in the ICF that may not be directly reflected by variables that are measured in the ECF or in the renal response. Accordingly, the events in the three areas of interest will be addressed in the following paragraphs.

Keys to the analysis of metabolic alkalosis:
- electroneutrality,
- stoichiometry,
- quantitative analysis.

Events in the ECF

- The plasma $[HCO_3^-]$ may be raised in two ways:
  1. by the addition of HCO_3^- to the ECF;
  2. by the loss of ECF volume.

ADDITION OF HCO_3^- TO THE ECF

With the constraints of electroneutrality, there are only two ways to add a specific anion (HCO_3^-) to a compartment: either loss of an anion such as Cl^- or retention of a cation such as Na^+.

Loss of Cl^-

Another anion must be lost from a compartment for HCO_3^- to "take its place," electrically speaking. In terms of the ECF, the only anion that is present in sufficient quantity to be lost is Cl^-. Hence, on mass balance, if a Cl^- is lost without a major cation (Na^+ or K^+), it will be lost with a H^+ or NH_4^+. Because a loss of H^+ or NH_4^+ is equivalent to a gain of HCO_3^- (Figure 4·1), the net effect is a loss of Cl^- along with a gain of HCO_3^-. Consider the swap of anions that takes place during vomiting or nasogastric suction. Imagine that you are viewing events at point X on the ECF side of

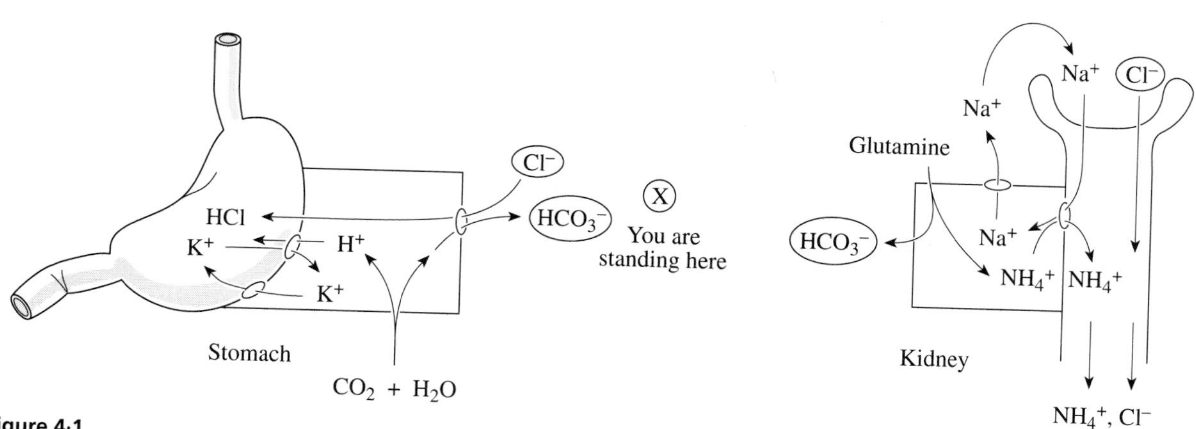

Figure 4·1
Loss of HCl or NH_4Cl is equivalent to a swap of Cl^- for HCO_3^-
The two organs capable of inducing a loss of Cl^- together with a gain of HCO_3^- are the stomach and the kidney; the stoichiometry is 1:1 for Cl^- loss and HCO_3^- gain.

the basolateral membrane of stomach cells (Figure 4·1). You will see Cl^- leaving the ECF and entering stomach cells; HCO_3^- will be moving in the opposite direction.

Two more steps are needed to complete the picture of the development of metabolic alkalosis associated with depletion of Cl^-.

1. **Bicarbonaturia:**
   As the [HCO_3^-] in plasma rises, more HCO_3^- are filtered by the kidney. Because not all this "extra" filtered HCO_3^- can be reabsorbed (reabsorption occurs mainly in the proximal convoluted tubule), HCO_3^- are excreted. To the extent that the HCO_3^- are lost with Na^+, a degree of contraction of the ECF volume occurs.

2. **Depletion of K^+:**
   When HCO_3^- are excreted in the urine, they may "drag out" K^+ (see Chapter 9, page 335). The resulting depletion of K^+ in the body indirectly contributes to an elevated [HCO_3^-] in the ECF.

Retention of $NaHCO_3$

The second way to add HCO_3^- to the ECF and maintain electroneutrality is to retain HCO_3^- along with Na^+. When Na^+ are retained in the ECF, its volume will be expanded. Hence, "permission of the kidneys" to retain extra Na^+ and HCO_3^- will be required (see the discussion of Question 4·1, page 176). It is also possible that some of the Na^+ retained will be located in the ICF (Na^+ enter the ICF while K^+ exit to the ECF with subsequent excretion of KCl). In this case, an underlying feature will be a deficit of K^+ (Figure 4·2).

It is possible that instead of ingesting $NaHCO_3$, a subject may consume the Na^+ (or K^+) salt of organic anions (e.g., potassium citrate, potassium malate, potassium acetate). Metabolism of these organic anions to neutral end products yields HCO_3^- (or removes H^+), as depicted in Figure 4·3.

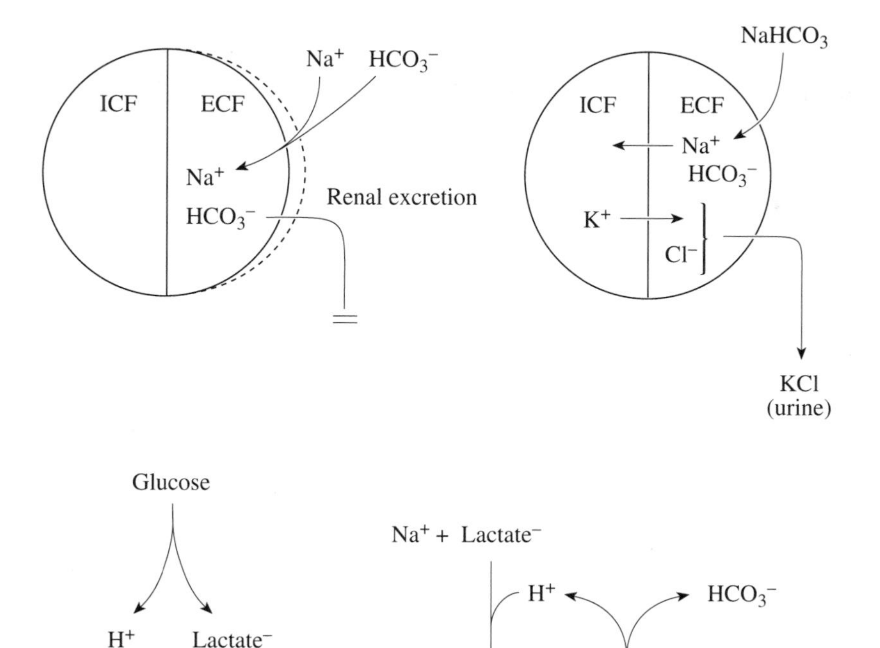

Figure 4·2
Retention of $NaHCO_3$ in the pathophysiology of metabolic alkalosis
For $NaHCO_3$ to be retained, either expansion of the ECF volume (left-hand figure) or K^+ depletion (right-hand figure) must develop.

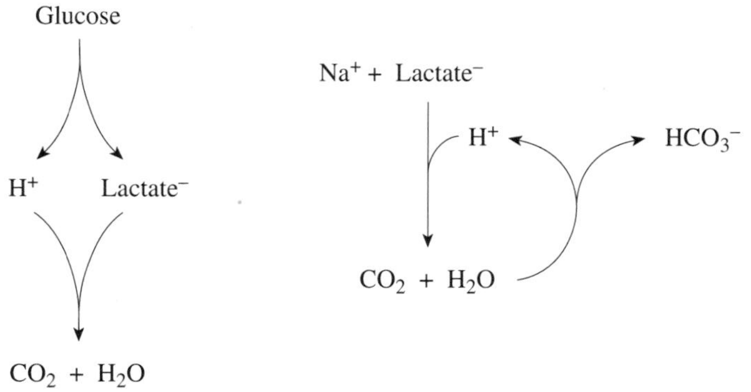

Glucose
H^+ Lactate$^-$
$CO_2 + H_2O$

Na^+ + Lactate$^-$
H^+ HCO_3^-
$CO_2 + H_2O$

Figure 4·3
Metabolism of organic anions and H^+ balance
Because organic anions of endogenous origin are synthesized from neutral compounds, metabolism of these anions just restores HCO_3^- balance (figure on left). In contrast, ingestion of the Na^+ or K^+ salts of organic anions yields new HCO_3^- (figure on right) and can produce a net gain of HCO_3^- for the body.

Questions

(Discussions on pages 176–78)

4·1 Why will the ingestion of $NaHCO_3$ not lead to the development of chronic metabolic alkalosis?

4·2 For net loss of HCl or NH_4Cl to be the sole cause of metabolic alkalosis, what must the mass balance for Na^+, K^+, and Cl^- be?

4·3 How many liters of emesis must be lost in order to raise the [HCO_3^-] in plasma by 10 mmol/l in a 70-kg adult?

4·4 If a patient has gastric drainage, how can the quantity of new HCO_3^- formation be minimized?

4·5 How can one explain the progressive rise in the [HCO_3^-] in plasma and K^+ depletion in a patient with metabolic alkalosis due to vomiting?

4·6 Why might the [HCO_3^-] in plasma rise in a patient with normovolemia who is undergoing plasmapheresis for rapidly progressive glomerulonephritis and renal failure?

CONTRACTION OF THE ECF VOLUME

If the ECF volume is contracted (from removal of NaCl and water), the $[HCO_3^-]$ in the ECF may rise without the addition of new HCO_3^-. In this situation, the $[HCO_3^-]$ rises because the HCO_3^- are distributed in a smaller volume—i.e., the content is unchanged, but the concentration is increased (see the discussion of Question 4·7).

Diuretic ingestion is commonly associated with metabolic alkalosis; diuretics cause NaCl loss in the urine and ECF volume contraction. They may also cause new HCO_3^- formation by increasing production and excretion of NH_4^+ consequent to hypokalemia. The ECF volume contraction results in enhanced proximal tubule H^+ secretion (angiotensin II), which enables the retention of the increased HCO_3^- (see the margin for details).

Metabolic alkalosis with diuretic use:

1. Loss of NaCl leads to ECF volume contraction.
2. ECF volume contraction leads to the release of aldosterone.
3. The delivery of NaCl to the CCD and the actions of aldosterone lead to excretion of K^+ and thereby to K^+ depletion.
4. K^+ depletion leads to intracellular acidosis.
5. Intracellular acidosis leads to new HCO_3^- generation (NH_4^+ excretion).
6. Metabolic alkalosis is thus due to:
   - ECF volume contraction;
   - K^+ depletion;
   - shift of H^+ into cells;
   - generation of new HCO_3^-;
   - enhanced renal reabsorption of HCO_3^- and a decreased GFR.

Questions

(Discussions on page 178)

4·7 How high might the $[HCO_3^-]$ in plasma rise if a patient has a modest degree of contraction of the ECF volume and no increase in the content of HCO_3^- in the body?

4·8 What impact does the loss of 200 mmol of Na^+ and Cl^- from diuretic action have on the content and concentration of HCO_3^- in the ECF? (For simplicity, assume no change in $[Na^+]$, which is 140 mmol/l.)

Mechanisms for Renal Retention of HCO_3^-

- The renal mechanisms for maintaining a high $[HCO_3^-]$ are mediated by a lower GFR and/or enhanced reabsorption of filtered HCO_3^-; both usually occur in a given patient.

If a normal person were to ingest substantial amounts of $NaHCO_3$, the $[HCO_3^-]$ in the ECF would increase temporarily before all this extra HCO_3^- would be excreted at a rapid rate (see discussion of Question 4·1, page 176). Because patients with metabolic alkalosis often have a plasma $[HCO_3^-]$ in excess of 35 mmol/l, they must have stimulated renal mechanisms to enable them to maintain this very elevated $[HCO_3^-]$ in the ECF. There are two mechanisms that might permit such a high plasma $[HCO_3^-]$ in the presence of normal kidney function—a decreased glomerular filtration rate (GFR) and enhanced reabsorption of HCO_3^-.

DECREASED GFR

One component of the renal mechanism for maintaining an elevated $[HCO_3^-]$ is a reduced filtered load of HCO_3^-. At times, there is a nearly proportionate reduction in the GFR; this reduction keeps the filtered load of HCO_3^- near normal (i.e., when the plasma [HCO_3^-] doubles, the GFR almost halves). This mechanism is an important component of the renal contribution to the maintenance of an elevated plasma [HCO_3^-] in patients with marked ECF volume contraction, especially if there is a large K^+ deficit. The basis of the decreased GFR in many cases is ECF volume contraction. In other cases, such as the metabolic alkalosis consequent to mineralocorticoid excess, however, there is no ECF volume contraction. Nevertheless, a decreased GFR has been demonstrated in animal models of this disorder and has been attributed to the associated hypokalemia.

GFR, decreased ECF volume, and K^+ deficit:
If the patient with ECF volume contraction, metabolic alkalosis, and a K^+ deficit did not have a decreased GFR, there would be increased $NaHCO_3$ delivery to the distal nephron and increased $KHCO_3$ excretion, which would result in a more marked K^+ deficit. Further K^+ depletion causes more ECF volume depletion because Na^+ enter the ICF when K^+ exit the ICF.

ENHANCED REABSORPTION OF HCO_3^-

A very important renal mechanism for maintaining an elevated [HCO_3^-] is increased reabsorption of filtered HCO_3^-. The major stimuli are low "effective" circulating volume and intracellular acidosis of cells of the proximal convoluted tubule. With respect to a low "effective" circulating volume, the major mediator identified is angiotensin II; this hormone leads to an activation of the Na^+/H^+ antiporter in the luminal membrane of proximal convoluted cells (Figure 4·4). Distal H^+ secretion may also be stimulated by mineralocorticoid secretion during ECF volume contraction, but the capacity for H^+ secretion in the distal tubule is small relative to proximal H^+ secretion. Regarding the fall in intracellular pH, both hypokalemia and a rise in Pco_2 raise the [H^+] in the ICF and could act in concert with the high angiotensin II levels to to promote the enhanced reabsorption of filtered HCO_3^-. It is not clear whether Cl^- deficiency per se plays any role in this regard. Some investigators who favor a more direct role of Cl^- deficiency propose that low luminal [Cl^-] in the collecting duct could promote Cl^- secretion via the Cl^-/HCO_3^- antiporter and thereby enhance the distal reabsorption of of HCO_3^-. For this hypothesis to be valid, the [Cl^-] would have to be very low in the lumen given the low K_m of this transporter for Cl^-, a requirement not strongly supported by experimental data.

Figure 4·4
Summary of renal events leading to an elevated [HCO_3^-]
The renal events in metabolic alkalosis are a lower filtered load of HCO_3^- (low GFR), more reabsorption of HCO_3^- in the proximal convoluted tubule (angiotensin II, hypokalemia, and an elevated Pco_2), and more excretion of NH_4^+ to generate metabolic alkalosis (hypokalemia).

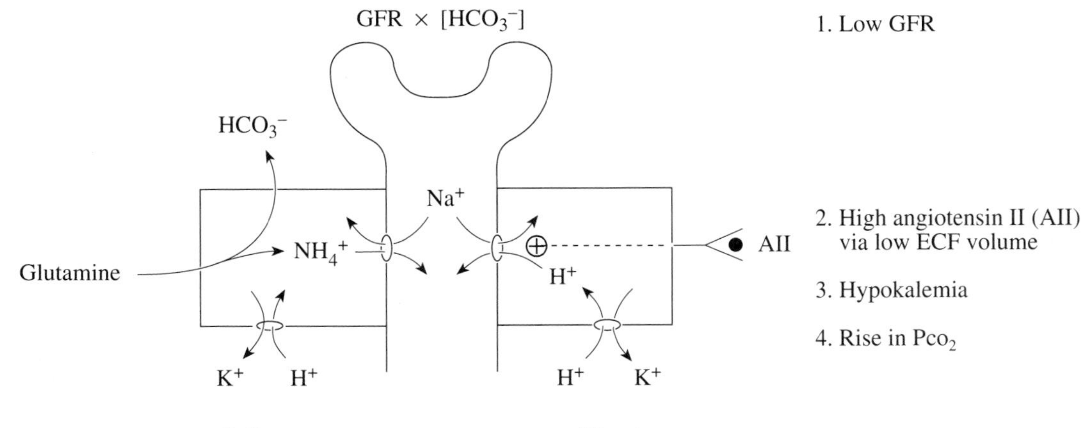

Decrease HCO_3^- excretion

1. Low GFR
2. High angiotensin II (AII) via low ECF volume
3. Hypokalemia
4. Rise in Pco_2

Events in the ICF

The simplest model of metabolic alkalosis in humans is that produced when HCl is removed from the stomach. In a study, normal human volunteers underwent this procedure for 4–5 days. All ion and water losses other than HCl were replaced. Examination of the mass balance data reveals that the ICF is an important site of involvement in metabolic alkalosis resulting from loss of HCl. The significance of the ICF is best illustrated by examining events in three phases.

LOSS OF HCl

As shown in Figure 4·1, loss of HCl results in a net gain of HCO_3^- in the body and metabolic alkalosis. Mass balance in man and in experimental animals reveals an initial deficit of Cl^- but not Na^+ or K^+, and the ECF volume changes to only a modest degree.

CHANGE OF MASS BALANCE

After losing HCl, subjects were permitted to stabilize in a post-drainage period of 5–7 days. Their deficit of Cl^- was not replaced. Over this period, all subjects developed a rather prominent degree of K^+ depletion. It is not clear from the study what anion was excreted with K^+. Nevertheless, the source of the K^+ was the ICF (the ECF did not contain enough K^+ to account for the negative balance). The likely events for the excretion of K^+ with an endogenous anion are depicted in Figure 4·5. The bottom line is that for every K^+ lost on mass balance without a Cl^-, a net gain of a H^+ (or loss of a HCO_3^-) occurred. The striking feature in the mass balance is the near-equimolar loss of K^+ and Cl^- (Table 4·1); there was little change in mass balance for Na^+. To provide mass balance and electroneutrality in the ICF and ECF, H^+ and Na^+ entered the ICF when K^+ exited.

Table 4·1
DEFICITS IN THE SELECTIVE DEPLETION OF HCL MODEL OF METABOLIC ALKALOSIS

Data were selected from studies involving selective loss of HCl (or a swap of HCO_3^- for Cl^-) in humans, dogs, and rats, followed by a post-drainage period. Note the near-equimolar losses of Cl^- and K^+ in each example. In each case, chronic metabolic alkalosis was present ($[HCO_3^-]$ was close to 35 mmol/l).

Variable	Species		
	Human (mmol)	Dog (mmol)	Rat (μmol)
Na^+	–22	9	–535
K^+	–213	–118	–2931
Cl^-	–199	–110	–2533
ECF volume	–0.5 liters	-	–10 ml
% change	–3	-	–12

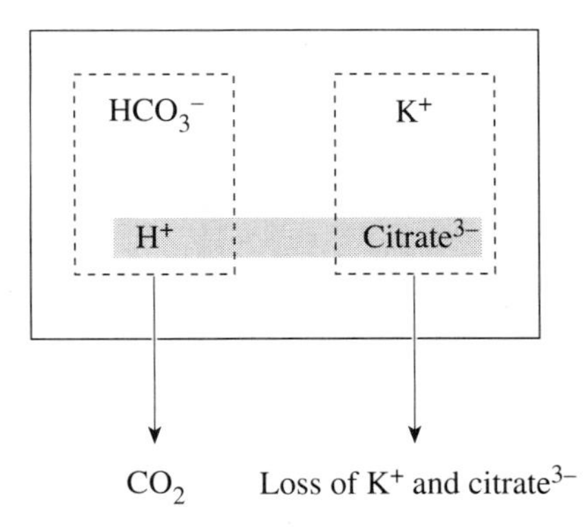

Figure 4·5
Loss of K^+ plus endogenous anion is equivalent to a gain of H^+
As indicated in the horizontal stippled bar, $citrate^{3-}$ are formed in the body along with protons. H^+ may titrate HCO_3^-, yielding CO_2, which is exhaled (dashed rectangle on the left). $Citrate^{3-}$ may be excreted with K^+ (dashed rectangle on the right side of the figure). Taken together, the excretion of K^+ and $citrate^{3-}$ represents the indirect loss of HCO_3^- from the body.

H^+ BALANCE DURING KCl DEPLETION

There is an equimolar loss of K^+ and Cl^-. Nevertheless, these two losses are not simultaneous. First, the loss of Cl^- without Na^+ or K^+ equates to a gain of HCO_3^- (Figure 4·1); second, the loss of K^+ without Cl^- equates to a gain of H^+ (Figure 4·5). Although there is acid-base balance, there is a high $[HCO_3^-]$ in the ECF that is largely due to an increase in content of HCO_3^- in this compartment (the ECF volume was not markedly contracted, Table 4·1). The explanation is that instead of being retained in the ECF (and lowering the $[HCO_3^-]$), H^+ enter cells concomitant with the exit of K^+ from the ICF. Ultimately, metabolic alkalosis associated with selective loss of HCl (the so-called Cl^-- depletion type of metabolic alkalosis) has the following features:

1. a higher content and concentration of HCO_3^- in the ECF;
2. K^+ depletion and intracellular acidosis;
3. enhanced reabsorption of HCO_3^- by the kidney.

Question

(Discussion on pages 178–79)

4·9 What pathophysiologic mechanisms must be involved to have Cl^--depletion metabolic alkalosis? What is the expected mass balance in each case?

PART B
CLINICAL CLUES

- Metabolic alkalosis occurs most commonly with the loss of gastric contents or diuretic use; mineralocorticoid excess is less commonly the sole cause of metabolic alkalosis.

If the basis for metabolic alkalosis can be established, the expected deficits can be approximated and appropriate therapy can be instituted. The most helpful aspects of the presentation are the history, an assessment of the ECF volume, and certain laboratory data—random urine electrolytes along with the plasma creatinine concentration and [K^+]; each will be discussed in turn.

History

Clinical pearls for Cl^- depletion and metabolic alkalosis:
- ECF volume is often contracted.
- Urine [Cl^-] < 20 mmol/l, unless diuretics are acting.

The most common causes of metabolic alkalosis are vomiting and diuretics; other causes are listed in Table 4·2. Among the items to explore in the history are issues such as previous investigations, eating habits, drug or unusual ingestions, body image, psychosocial aspects, a knowledge of previous kidney function, family history, GI complaints, hypertension, and the clinical setting (e.g., ventilation for chronic lung disease).

One might think that the diagnosis of metabolic alkalosis would be straightforward, and often it is. Nevertheless, a number of patients, because of personality disorders, are not willing to admit to the self-induction of vomiting or the abuse of diuretics. Therefore, the history cannot be relied upon entirely, and one must often resort to a series of laboratory tests and a degree of subterfuge to establish the diagnosis (Table 4·3).

Table 4·2
CAUSES OF METABOLIC ALKALOSIS

See Table 4·3 for more information concerning urine electrolytes in metabolic alkalosis.

Causes usually associated with a contracted ECF or "effective" circulating volume:

1. Low urine [Cl^-] (unless a diuretic is acting)
   - Loss of gastric secretions (e.g., from vomiting, nasogastric suction)
   - Remote use of diuretics
   - Delivery of nonreabsorbable anions plus a reason for Na^+ avidity
   - Post-hypercapnia (see also secondary hyperaldosteronism below)
   - Loss of NaCl via the GI tract (as in certain diarrheas, e.g., congenital Cl^- loss, some villous adenomas)
2. Persistent high urine [Cl^-]
   - Bartter's syndrome
   - Current diuretic use

Causes usually associated with a normal or expanded ECF or "effective" circulating volume:

1. Large reduction in GFR plus a source of HCO_3^-

Table 4·2 is continued on page 157.

- Examples include alkali ingestion, ingestion of ion-exchange resin plus nonreabsorbable alkali

2. Enhanced mineralocorticoid activity
   - Primary aldosteronism
   - Secondary hyperaldosteronism (examples include renal artery stenosis, malignant hypertension, renin-producing tumor, low effective arterial blood volume plus an alkali load)
   - Endogenous or exogenous mineralocorticoids, licorice ingestion, ACTH-driven mineralocorticoid secretion (see Chapter 10, page 370)

Causes that are difficult to classify with respect to ECF volume:

1. Hypomagnesemia
2. Excessive alkaline tide with Zollinger-Ellison syndrome

Table 4·3
CLUES TO IDENTIFY THE POSSIBLE CAUSES OF METABOLIC ALKALOSIS ASSOCIATED WITH A LOW "EFFECTIVE" CIRCULATING VOLUME

Cause	Diagnostic features
Recent vomiting	• Unmeasured anion in urine (HCO_3^-) • Urine pH > 7, [Cl^-] low, but [Na^+] high • Disturbed body image
Remote vomiting	• Urine [Na^+] and [Cl^-] both low
Recent diuretic	• Urine [Na^+] and [Cl^-] both high
Remote diuretic	• Urine [Na^+] and [Cl^-] both low
Post-hypercapnia	• Urine [Na^+] and [Cl^-] both low
Cl^- loss in diarrhea	• Urine pH < 6, [Na^+] and [Cl^-] both low; if hypokalemia induces NH_4^+ excretion, urine [Cl^-] will rise
Nonreabsorbable anion	• Urine [Cl^-] low, but [Na^+] high • Unmeasured anion in urine • Urine pH < 7

ECF Volume

ECF VOLUME CONTRACTION

The major causes of metabolic alkalosis—diuretic abuse or vomiting—usually present with a significant degree of ECF volume contraction (see Table 4·4 for cautions regarding clinical findings). ECF volume contraction contributes to both the generation and maintenance of an elevated plasma [HCO_3^-] either via a decreased denominator or increased numerator of the HCO_3^-/ECF volume ratio. As pointed out in the discussion of Question 4·7, page 178, a decrease in the denominator of this ratio will account for only a small portion of the elevation in the [HCO_3^-]. The primary influence is an increase in the content of HCO_3^- in the ECF.

In diuretic abuse, extra HCO_3^- are generated by the production and excretion of more NH_4^+. For this excretion to occur, hypokalemia must be present (hypokalemia augments renal ammoniagenesis and NH_4^+ excretion). Hypokalemia is also an expected sequela to the actions of aldosterone when the Na^+ salts of nonreabsorbable anions (e.g., carbenicillinates) are given to patients who have a contracted "effective" circulating volume.

In vomiting, the other common clinical cause of metabolic alkalosis in the setting of a low ECF volume, both the contracted ECF volume and HCO_3^- addition are the consequence of the loss of HCl. When HCl is lost, HCO_3^- replace Cl^- in the ECF. The resulting alkalemia depresses the reabsorption of HCO_3^- by the proximal tubule and leads to the excretion of some $NaHCO_3$ in the urine. The combined loss of HCl via gastric secretion and $NaHCO_3$ in the urine is equivalent to a net loss of NaCl and a tiny loss of $CO_2 + H_2O$ (equations below):

$$HCl \longrightarrow H^+ + Cl^- \longrightarrow \text{lost}$$
$$NaHCO_3 \longrightarrow HCO_3^- + Na^+ \longrightarrow \text{lost}$$
$$H^+ + HCO_3^- \longleftrightarrow CO_2 + H_2O$$

Result: Loss of $Na^+ + Cl^- + CO_2 + H_2O$

A second bout of vomiting will add more HCO_3^- in place of Cl^- (Figure 4·1), except now the patient will have a mild degree of ECF volume contraction. In this setting, there will be a release of renin, increased formation of angiotensin II, and a release of aldosterone. Although angiotensin II promotes the reabsorption of HCO_3^- in the proximal convoluted tubule, some $NaHCO_3$ will escape reabsorption and be delivered distally where it will lead to augmented excretion of K^+ (aldosterone and bicarbonaturia markedly augment kaliuresis).

Summary

Both vomiting and diuretic abuse result in the same findings of metabolic alkalosis, hypokalemia, and ECF volume contraction. Nevertheless, their routes to this setting differ. Notwithstanding, the therapy (replace the deficits) will be similar once the cause is no longer present (the reason for vomiting or the presence of diuretics).

When treating metabolic alkalosis associated with hypokalemia, one must also consider events in the ICF. In both examples cited above, the depletion of K^+ leads to a shift of cations Na^+ and H^+ into cells. As discussed previously, this shift not only exacerbates the degree of elevation of the $[HCO_3^-]$ in the ECF but also imposes a new demand on therapy: correction of the acidosis and K^+ depletion in the ICF via administration of KCl (see the margin).

Clinical pearls:

- The major deficits are Cl^- and K^+.
- A deficit of NaCl is commonly present as well.

ECF VOLUME EXPANSION

Patients with a normal or expanded ECF volume (Table 4·2) form another subgroup of those with metabolic alkalosis (see the margin). They often have hypertension and a modest degree of elevation of their $[HCO_3^-]$. The basis of their disturbance is hyperaldosteronism or enhanced mineralocorticoid actions. Although they are K^+-depleted, simple replacement of electrolyte deficits is not enough. The source of the mineralocorticoids must be addressed.

Note:

To maintain an elevated $[HCO_3^-]$ in the plasma, the GFR must be lower and/or reabsorption of HCO_3^- must be higher; both seem to be the result of the deficit of K^+.

Assessing the ECF volume in terms of circulating volume is a valuable means of diagnosing the basis of the metabolic alkalosis. The evaluation of the jugular venous pressure (JVP) is an important clue in this regard; however, be aware that JVP is a right-sided phenomenon that is used to gain insights into the pressure on the left side of the heart. Do not be misled by events that perturb the relationship between the right- and left-sided cardiac pressures (Table 4·4).

Table 4·4
CONDITIONS IN WHICH JUGULAR VENOUS PRESSURE (JVP) MAY NOT ACCURATELY REFLECT THE "EFFECTIVE" CIRCULATING VOLUME

Pulmonary hypertension
Chronic emphysema
Pulmonary emboli
Idiopathic pulmonary hypertension
Impaired right-ventricular emptying
Pulmonary-valve stenosis or insufficiency
Cardiomyopathy
Right-ventricular infarction
Tricuspid-valve insufficiency
Impaired right-ventricular filling
Tricuspid-valve stenosis
Superior-vena-cava syndrome
Tamponade
Poor left-ventricular output ("ineffective" circulating volume)
Aortic-valve stenosis or insufficiency
Myocardial infarction
Mitral-valve stenosis or insufficiency
Cardiomyopathy

Urine Electrolytes

A single random urine sample, or perhaps several random urine samples can usually unravel the pathophysiology of metabolic alkalosis (see the margin and Table 4·5).

Clinical pearls:

- There are no "normal" values for urine electrolytes, just expected values in a given clinical setting.
- In a patient with a contracted ECF volume, the expected values for urine [Na^+] and/or [Cl^-] are close to nil.

EXCRETION OF A SMALL QUANTITY OF Na^+ OR Cl^-

Metabolic alkalosis, ECF volume contraction, and similar low excretions of Na^+, K^+, and Cl^- could indicate the occurrence of remote vomiting, "yesterday's diuretics," or the prior intake of nonreabsorbable anions. Hence, to make a final diagnosis, one must interpret these urine results in conjunction with other information obtained via the history or repeated urine sampling. Identifying the basis of the ECF volume contraction is usually the key to making the diagnosis.

EXCRETION OF Na^+, BUT ONLY A SMALL QUANTITY OF Cl^-

Some patients with metabolic alkalosis have a low rate of excretion of Cl^-, yet the urine contains an abundant quantity of Na^+. The reason for the excretion of Na^+ is the presence of an anion that was not reabsorbed (this

Anion revealed by urine net charge: [Na^+] + [K^+] greatly exceeds [Cl^-].

anion is revealed by an increased net charge in the urine; see the margin). If the urine contains a large quantity of HCO_3^- (pH > 7), the patient must have a nonrenal source of HCO_3^- (the kidneys cannot excrete and generate HCO_3^- simultaneously). In this setting, vomiting or nasogastric suction is the most likely diagnosis if the patient is not ingesting HCO_3^- or organic anions. A low urine pH suggests the intake or generation of anions that are poorly reabsorbed by the kidneys.

Table 4·5
URINE ELECTROLYTES IN THE DIAGNOSIS OF METABOLIC ALKALOSIS

Values are taken from a random urine sample on presentation. "High" signifies > 20 mmol/l and "low" < 20 mmol/l. "Remote" means no recent vomiting or drug administration.

Clinical setting	Urine values				Comments
	$[Na^+]$	$[K^+]$	$[Cl^-]$	pH	
1. Low ECF volume (excluding edema)					
- Vomiting (recent)	High	High	Low	> 7	• If remote, Na^+, Cl^-, and pH will be low
- Diuretic (recent)	High	High	High	< 6	• If remote, same as for remote vomiting • Values may fluctuate as diuretic action wears off
- Nonreabsorbable anions (recent)	High	High	Low	< 6	• If remote, same as for remote vomiting
- Bartter's syndrome	High	High	High	5–7	• Values are high in every urine sample in the day
2. Normal or expanded ECF volume	High	High	High	5–8	• Hypertension
(see Table 4·2 for list of causes)					• History may change with different underlying diseases

EXCRETION OF Na^+ AND Cl^-

The expected pattern of excretion with a normal ECF volume is excretion of Na^+ and Cl^-. This same pattern may occur in patients with a contracted ECF volume if there is intrinsic renal disease or the lack of messenger to stimulate Na^+ reabsorption (see Chapter 6, pages 239–40). More specifically related to metabolic alkalosis, this pattern of electrolyte excretion suggests diuretic action or Bartter's syndrome.

Question

(Discussion on pages 179–80)

4·10 How can one distinguish between diuretic abuse and Bartter's syndrome using urine electrolytes?

Plasma Creatinine

Patients with ECF volume contraction will have elevated values for creatinine in plasma corrected for their body mass, especially if the degree of ECF volume contraction is marked. Previous values and the clinical setting make this interpretation relatively simple.

Patients with chronic renal failure usually have a normal or expanded ECF volume and a markedly elevated level of creatinine in plasma. Should such a patient ingest or be given alkali or organic anions, a small quantity of HCO_3^- will be filtered and most HCO_3^- will be retained in the body.

Questions

(Discussions on pages 180-81)

4·11 A very low GFR plays a prominent role in the metabolic alkalosis associated with the milk-alkali syndrome. What is the pathophysiology of this syndrome?

4·12 What is the pathophysiology of metabolic alkalosis associated with nonreabsorbable alkali and an ion-exchange resin?

4·13 What is the pathophysiology of metabolic alkalosis associated with a normal or expanded ECF volume?

Plasma [K^+]

Hypokalemia is usually present in most patients with metabolic alkalosis. It may play several roles in the pathophysiology. As mentioned earlier, by augmenting the production and excretion of NH_4^+, it may increase the content of HCO_3^- in the ECF. When intracellular shifts cause the cations Na^+ and H^+ to enter cells, the [HCO_3^-] in the ECF increases because of ECF volume contraction (Na^+ loss from the ECF) and loss of H^+ from the ECF. Hypokalemia also leads to a lower excretion of HCO_3^- because it may decrease the GFR and increase the reabsorption of HCO_3^- by the proximal convoluted tubule (intracellular acidosis).

Question

(Discussion on page 181)

4·14 In what circumstances might metabolic alkalosis be associated with hyperkalemia?

Clinical Approach

A clinical approach to a patient with metabolic alkalosis is outlined in Figure 4·6. The first step in determining the basis of metabolic alkalosis is

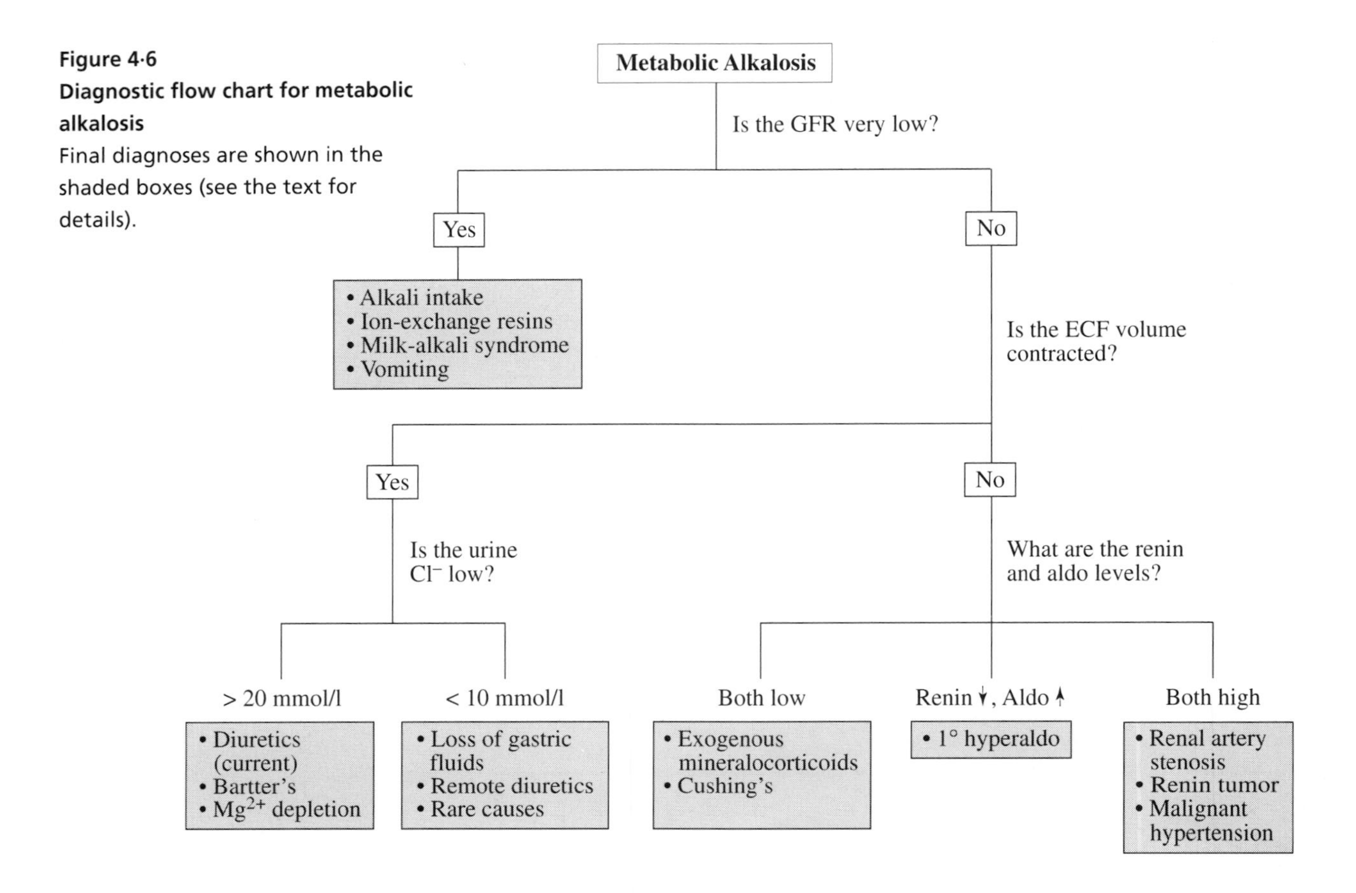

Figure 4·6
Diagnostic flow chart for metabolic alkalosis
Final diagnoses are shown in the shaded boxes (see the text for details).

to rule out chronic renal insufficiency (GFR < 25% of normal). If renal failure exists, the specific cause of the metabolic alkalosis should be evident from the history.

If a very low GFR plus alkali input are not the cause of metabolic alkalosis (and they usually are not), the ECF volume status is the next critical parameter to assess. The majority of patients have a contracted ECF volume and the [Cl^-] in their urine is very low (< 20 mmol/l). The differential diagnosis in this group with ECF volume contraction is indicated in Figure 4·6. If the history is not available, examine the urine electrolytes in a random urine sample to help in identifying the possible causes of the metabolic alkalosis (Table 4·5).

Examples of Less-Common Causes of Chronic Metabolic Alkalosis

CAUSES ASSOCIATED WITH ECF VOLUME CONTRACTION

The vast majority of patients with metabolic alkalosis will have either diuretic use or loss of gastric content as the basis. The following are the less-common causes of metabolic alkalosis.

Nonreabsorbable Anions

If a patient has a contracted ECF volume and takes a Na^+ salt with an anion that cannot be reabsorbed by the kidney (e.g., Na^+ carbenicillinate), the patient may develop metabolic alkalosis and hypokalemia. ECF volume contraction is usually present for an unrelated reason. Thus, there is a stimulus to Na^+ reabsorption, but all the Na^+ cannot be reabsorbed because of the presence of an anion that cannot be reabsorbed. In the cortical collecting duct, the actions of aldosterone cause Na^+ to be reabsorbed in conjunction with K^+ secretion. The result is hypokalemia, which, in turn, leads to increased production and excretion of NH_4^+ and an increased plasma [HCO_3^-] (a result of renal HCO_3^- generation and ECF volume contraction).

Drugs and substances causing metabolic alkalosis:
Drugs containing a poorly reabsorbable anion (e.g., carbenicillin), or a substance that can be metabolized to HCO_3^- (e.g., citrate^{3-}) can cause metabolic alkalosis in a subject with high renal avidity for Na^+.

The urine provides the clues to the diagnosis. The findings should be as follows:

- The [Cl^-] in urine should be low (< 20 mmol/l).
- The [Na^+] in urine will vary (if large, a recent load of nonreabsorbable anion was administered; if the intake of nonreabsorbable anions was discontinued, the [Na^+] in urine may be less than 10 mmol/l).
- The urine should contain a substantial concentration of unmeasured anion ([Na^+] + [K^+] > [Cl^-]). If the urine pH is alkaline, the anion is HCO_3^-; if it is acid, the patient has metabolic alkalosis associated with a nonreabsorbable anion.

Bartter's Syndrome

Patients with this unusual disorder are not hypertensive and have ECF volume depletion, renal NaCl wasting, metabolic alkalosis, hypokalemia, and usually renal Mg^{2+} wasting. The exact basis of the Na^+ and Cl^- wasting remains controversial; Bartter's syndrome is considered in more detail in Chapter 10, pages 368–69.

Loss of Cl^- in the Stool

The loss of Cl^--rich fluid in the stool (e.g., congenital chloridorrhea, a rare disorder) creates a situation identical to the diuretic-induced metabolic alkalosis, except that the urine always has a low [Na^+] and [Cl^-].

Post-hypercapnia

In the course of chronic hypercapnia, an increased [HCO_3^-] results in the loss of Cl^- in the urine (i.e., enhanced renal NH_4Cl excretion). If the patient has a contracted ECF volume when the hypercapnia resolves, there will be a stimulus for Na^+ reabsorption and H^+ secretion that results in the reabsorption of luminal HCO_3^-. The increase in the plasma [HCO_3^-] is thus maintained until the ECF volume is reexpanded with NaCl administration.

CAUSES THAT ARE COMMONLY ASSOCIATED WITH A NORMAL OR EXPANDED ECF VOLUME

Hyperaldosteronism

When patients with metabolic alkalosis fail to respond to the administration of KCl and NaCl, they are likely to have excessive mineralocorticoid activity of either exogenous or endogenous origin (Table 10·5, page 363). The associated hypokalemia is probably of major importance in both the generation and the maintenance of the metabolic alkalosis. Hypokalemia enhances ammoniagenesis, which enables renal new HCO_3^- formation, and it also causes an increased indirect reabsorption of HCO_3^- via the rise in proximal tubular intracellular [H^+]. In addition, it reduces the GFR and thereby maintains the elevated blood [HCO_3^-].

Alkali Loading

Under usual circumstances, $NaHCO_3$ loading leads to only a mild elevation in the plasma [HCO_3^-] because most of these HCO_3^- are excreted (see the discussion of Question 4·1 page 176). However, in the presence of Na^+ depletion or in renal failure, clinically important elevations of plasma [HCO_3^-] occur with $NaHCO_3$ administration (because of the lack of excretion of HCO_3^-).

Magnesium Depletion

Note:
Depending on the degree of Na^+ loss, some of the patients may have a contracted ECF volume.

Patients with Mg^{2+} depletion may have metabolic alkalosis and hypokalemia resembling a high mineralocorticoid state. The hypomagnesemia can be confirmed by plasma analysis. The usual clinical setting for this deficiency includes malabsorption, diarrhea, or the administration of drugs that act on the loop of Henle (e.g., cisplatin, loop diuretics, or aminoglycosides; see the margin). These patients must be distinguished from those with primary hyperaldosteronism who may also have Mg^{2+} deficiency.

Milk-Alkali Syndrome

Note:
If vomiting is excessive and there is still some renal function, the ECF volume may be contracted.

The milk-alkali syndrome (metabolic alkalosis, hypercalcemia, hypocalciuria, and renal insufficiency; refer to the discussion of Question 4·11, page 180) is due to the ingestion of large amounts of milk and absorbable antacids ($CaCO_3$). It has been mainly of historical interest; however, with the current emphasis on preventing osteoporosis by using $CaCO_3$ as a major source of Ca^+ supplementation, we may see a reappearance of this syndrome (see the margin).

Nonreabsorbable Alkali Ingestion With Ion-Exchange Resins

Ion-exchange resins are generally used in patients with renal insufficiency; when combined with nonreabsorbable alkali (aluminum or magnesium hydroxide) their use has resulted in metabolic alkalosis that resolves when either agent is discontinued. Refer to the discussion of Question 4·12, page 180.

Effect of Metabolic Alkalosis on Ventilation

Because the plasma $[H^+]$ is a major determinant of ventilation, one would expect metabolic alkalosis to depress ventilation. In fact, there is a linear relationship between the increasing plasma $[HCO_3^-]$ and the progressive increase in P_aco_2; the slope is approximately 0.7 (see the margin). Thus, when patients present with CO_2 retention and metabolic alkalosis, the metabolic alkalosis should be corrected before attributing the CO_2 retention to lung disease.

As hypoventilation develops, it is accompanied by hypoxia, which offsets the degree of respiratory suppression achieved (more severe respiratory suppression is observed in patients receiving O_2 supplementation when hypoxia is prevented). The reduced delivery of O_2 to tissues in metabolic alkalosis is further aggravated by the fact that alkalemia shifts the O_2-hemoglobin dissociation curve to the left; this shift increases the affinity of hemoglobin for O_2 (Figure 5·3, page 194).

Because patients with chronic lung diseases often take diuretics to cope with their Na^+ retention, they may develop metabolic alkalosis. The mixed acid-base disturbance may return their plasma $[H^+]$ to the normal range, but their clinical condition may worsen when they no longer have the acidemic drive to ventilate. It is important to note that these patients need not be alkalemic in order to experience the adverse effects of metabolic alkalosis in chronic respiratory acidosis (see the margin).

Clinical pearl:
For every mmol increase in plasma $[HCO_3^-]$ in patients with metabolic alkalosis, expect to see a 0.7 mm Hg increase in P_aco_2.

Note:
Data from eight patients with chronic respiratory acidosis prior to and following the partial correction of the metabolic alkalosis are provided in Table 4·6. The clinical condition (mental function and sense of well-being) was improved in association with the fall in P_aco_2 and the increase in P_ao_2.

Table 4·6
IMPACT OF METABOLIC ALKALOSIS ON PATIENTS WITH CO_2 RETENTION

Metabolic alkalosis	H^+ (nmol/l)	HCO_3^- (mmol/l)	P_aco_2 (mm Hg)	P_ao_2 (mm Hg)
Before correction	40	37	61	52
Partial correction	42	28	48	69

Legend for Table 4·6
There is little difference in the $[H^+]$ in plasma before and after partial correction of the coexisting metabolic alkalosis in patients with chronic respiratory acidosis. After correction of these disorders, the P_aco_2 is lower and the P_ao_2 is higher. The increase in the O_2 content of blood can be very large given the shape of the O_2 saturation: P_ao_2 curve (Figure 5·3, page 194).

PART C
TREATMENT OF METABOLIC ALKALOSIS

Low ECF or "Effective" Circulating Volume Group

- Identify the basis of the deficits.
- Replace the deficits of Na^+, K^+, and Cl^-.

Principles of therapy:
- NaCl administration is needed only if the ECF volume is contracted.
- K^+ administration is needed in virtually all cases to correct the ICF acidosis and K^+ deficit.
- Cl^- administration is needed to replace the deficit of Cl^-.

It goes without saying that patients with ECF volume contraction require Na^+ and Cl^- replacement. When their ECF volume is restored, the [HCO_3^-] will fall somewhat as a result of dilution; bicarbonaturia will occur if the ECF volume is overexpanded. Generally, the [HCO_3^-] in the ECF falls when NaCl is given, even if the K^+ deficits are not completely restored. Up to this point, the large K^+ deficit in cells and the intracellular acidosis have been ignored. Hence, NaCl administration is only partial therapy in these patients because it does not reverse the accompanying intracellular acidosis and K^+ deficit.

Some patients with metabolic alkalosis have little depletion of their ECF volume. Clearly, in these cases, NaCl should not be the linchpin of therapy. To treat the ICF acidosis and K^+ depletion, K^+ must be given with an anion that permits retention of this cation in the ICF. In most cases, KCl is administered because as K^+ enter the ICF, Na^+ and H^+ exit, for the most part. The H^+ titrate the excess HCO_3^- in the ECF and the extra NaCl is retained or excreted depending on the ECF volume status. Obviously, if some of the intracellular K^+ deficit represents loss of K^+ and phosphate, the entire deficit of K^+ cannot be replaced acutely. It must await the synthesis of intracellular phosphate esters, such as DNA, RNA, and phospholipids.

If the metabolic alkalosis is due to a reduction in "effective" circulating volume with ECF volume expansion (e.g., in a patient with congestive heart failure who is on diuretics), the patient will require therapy with KCl, and additional Na^+ should not be given (as K^+ enter the ICF, Na^+ leave, adding some Na^+ to the ECF). The patient must excrete the extra Na^+ in the body.

If the patient has been abusing diuretics or inducing vomiting, treatment of the underlying psychopathology will be the greatest challenge.

Question

(Discussion on page 181)

4·15 How accurate is the term "saline-sensitive metabolic alkalosis"?

High or Normal ECF Volume Group

- Identify the basis of the deficits.
- Replace K^+ (and possibly Mg^{2+}) deficits.
- Block aldosterone action.

Therapy is generally more difficult in patients with metabolic alkalosis and ECF volume expansion. If they have renal failure and are very alkalemic ($[H^+] < 20$ nmol/l, pH > 7.70) they should receive some H^+ in the form of HCl or NH_4Cl. If these patients are dialyzed, the bath should have the $[HCO_3^-]$ or concentration of acetate reduced to ameliorate the alkalemia.

If the patient has hyperadrenalism, agents that block Na^+ reabsorption in the collecting duct (e.g., amiloride) may play an important role in the therapy, as do mineralocorticoid antagonists (e.g., spirolactone). One must ensure that bowel sounds are present before using K^+ supplements and K^+-sparing diuretics (when the plasma $[K^+]$ rises, bowel motility may increase and a large quantity of K^+ may be absorbed quickly). Without the capability of renal excretion, a severe degree of hyperkalemia may develop. Even in the absence of an ileus, hyperkalemia may develop if aggressive K^+ replacement is given in conjunction with amiloride or spironolactone. In the acute setting, it is safest to order only one day of K^+ therapy at a time, noting the serum $[K^+]$ each day before ordering that day's therapy.

Patients with Mg^{2+} deficiency should have this deficiency corrected, but it is important to follow up on the effectiveness of the therapy because patients with hyperaldosteronism also have Mg^{2+} deficiency and thus have another basis for the metabolic alkalosis.

Acetazolamide is sometimes used to alleviate the alkalemia quickly (e.g., when weaning from a ventilator), but, because a large load of $NaHCO_3$-rich fluid will be delivered to the collecting duct, this agent often exaggerates K^+ loss. For this reason, we prefer an intravenous HCl preparation. An intravenous infusion of HCl also provides better control over the quantitative fall in $[HCO_3^-]$ than does acetazolamide.

If the patient is ventilated and has a fixed alveolar ventilation, the P_aco_2 may rise with H^+ administration; a small increase in P_aco_2 may not be detrimental if it is a transient phenomenon.

PART D REVIEW

Discussion of Introductory Case Basically, Toby Is Not "OK"

(Case presented on page 148)

What acid-base disturbance is present?

Toby has metabolic alkalosis because the [HCO_3^-] and the pH of plasma are both increased.

Why is the [Na^+] in the urine not lower, given the presence of ECF volume contraction?

For the [Na^+] in urine to be low, all the filtered Na^+ must be reabsorbed. There are two ways in which Na^+ can be reabsorbed: along with Cl^- and in conjunction with the secretion of the cations K^+ and/or H^+. In response to ECF volume contraction, the [Cl^-] in urine is close to 0; this extremely low value indicates appreciable NaCl reabsorption. Some Na^+ were not reabsorbed because they were excreted with an anion other than Cl^- (note the *urine net charge* of 102 mEq/l, as shown in the margin). The very high urine pH indicates that one of the "nonreabsorbed anions" is HCO_3^-. To the degree that the filtered load of HCO_3^- exceeds the tubular capacity to reabsorb it, HCO_3^- are excreted; this action obligates the coincident excretion of K^+ and/or Na^+. We cannot be sure if there is another "nonreabsorbable" anion other than HCO_3^- in the urine at this point. Therefore, bicarbonaturia is at least one cause of the unexpectedly high [Na^+].

Urine net charge:
Na^+ (52 mmol/l) + K^+ (50 mmol/l) - Cl^- (0 mmol/l)

Why is Toby hypokalemic?

Toby is hypokalemic because she has ECF volume contraction, which results in the release of renin. Renin, via angiotensin II formation, causes the release of aldosterone, which promotes K^+ secretion and Na^+ reabsorption in the cortical collecting duct. Adequate delivery of Na^+ to this segment is ensured by the presence of HCO_3^-. Another important effect of HCO_3^- in the cortical collecting duct is an augmented K^+ secretion, which raises the [K^+] in the lumen (see Chapter 9, page 335).

What is the basis for the acid-base disturbance?

Toby has metabolic alkalosis and coincident renal HCO_3^- excretion. If this condition represents a steady state (continued metabolic alkalosis despite bicarbonaturia), she must have an ongoing supply of HCO_3^- (one cannot lose HCO_3^- and maintain an elevated level in plasma without an ongoing source). If the source were exogenous ingestion of $NaHCO_3$, there would be ECF volume expansion, which is not the case. Hence, there is an endogenous generation of HCO_3^- that is not of renal origin (there is renal HCO_3^- loss, not generation). Therefore, Toby must be vomiting. The alkaline urine pH, the very low [Cl^-] in urine, and the metabolic alkalosis are pathognomonic of metabolic alkalosis secondary to loss of HCl via the GI tract (Figure 4·1, page 150).

Cases for Review

Case 4·1
Metabolic Alkalosis in the Beautiful People

(Case discussed on pages 171–72)

Farrah, a beautiful person, is concerned about her body image so she diets most of the time. Her food intake is erratic and consists mainly of vegetables and fruits; she consumes little meat or table salt. She jogs 60 km per week and is asymptomatic. When she volunteered for a clinical research project, she was surprised to find that she was hypokalemic (2.7 mmol/l) because she was normokalemic (4.0 mmol/l) six months before. She denied vomiting and the use of diuretics or laxatives. Her blood pressure was on the low side (90/55 mm Hg) and she had subtle signs that suggested a contracted ECF volume.

		Plasma	Urine
Na^+	mmol/l	138	63
K^+	mmol/l	2.7	34
Cl^-	mmol/l	96	0
HCO_3^-	mmol/l	30	0
Creatinine	μmol/l (mg/dl)	60 (0.7)	-
Osmolality	mosm/kg H_2O	287	563
pH		7.45	5.6

Do these urine values suggest that Farrah is a surreptitious vomiter or diuretic abuser?
What is the basis for the low urine [Cl^-]?
Why isn't her excretion of K^+ lower?
Do you think she has an adrenal tumor?
What advice would you give to Farrah?

Case 4·2
Solly Has Transient Metabolic Alkalosis

(Case discussed on pages 172–73)

Solly has had abdominal pain and profuse diarrhea for many months. More recently, he has vomited on occasion and has suffered from episodic tingling and weakness. He took antacids to relieve his abdominal pain, but their beneficial effect was transitory. He has been to the hospital on several occasions, each with a similar story—a set of "laboratory errors" because his condition reverts toward normal without therapy and because there are no physical findings of note. The laboratory values are as follows.

Plasma		Admission	4 hours later
pH		Not done	7.50
P_aco_2	mm Hg	Not done	48
HCO_3^-	mmol/l	62	40
K^+	mmol/l	3.1	3.6
Anion gap	mEq/l	15	13
Creatinine	μmol/l (mg/dl)	200 (2.3)	

In the four hours in the emergency room, the urine was alkaline but its volume was small. Why was metabolic alkalosis present?
How did it ameliorate spontaneously?
What is wrong with Solly and how should he be treated?

Case 4·3
Mr. Greene Looks Green

(Case discussed on pages 173–74)

Mr. Greene is 42 years old and is a chronic alcoholic. He was brought to the emergency room, obviously intoxicated. He had been lying in the park in a pool of vomitus. On physical examination, he was unkempt and incoherent. He had a markedly contracted ECF volume, was febrile (39° C), and had evidence of pneumonia. Laboratory data are summarized below.

Plasma			Plasma		
Na^+	mmol/l	130	H^+	nmol/l	30
K^+	mmol/l	2.9	pH		7.53
Cl^-	mmol/l	80	P_aco_2	mm Hg	25
HCO_3^-	mmol/l	20	P_ao_2	mm Hg	60
Creatinine	μmol/l (mg/dl)	120 (1.4)	Albumin	g/dl	38
Urea	mmol/l (mg/dl)	12 (34)			
Glucose	mmol/l (mg/dl)	15 (270)			
Osmolality	mosm/kg H_2O	320			
Serum ketones		weakly positive (undiluted specimen)			

What are the acid-base diagnoses?
What are the priorities for therapy?
What risks do you anticipate?

Discussion of Cases

Discussion of Case 4·1
Metabolic Alkalosis in the Beautiful People

(Case presented on page 169)

It is useful to begin with the diagnostic approach provided in Figure 4·6, page 162. The GFR is not very low, and Farrah has a contracted ECF volume. The [Cl^-] in urine now becomes a critical crossroad in the diagnostic flow chart; her [Cl^-] in urine is very low. Our differential diagnosis therefore includes: loss of gastric fluids, remote diuretic use, and rare causes of metabolic alkalosis.

Do these urine values suggest that Farrah is a surreptitious vomiter or diuretic abuser?

The history could be compatible with vomiting, even if it is denied. The key to the diagnosis is in the random urine sample because blood tests confirm metabolic alkalosis and hypokalemia. The urine should be "Cl^--free," and it is. As a result of recent vomiting, the urine could have abundant Na^+; if vomiting was remote, however, the urine would be Na^+-free. The urine suggests the possibility of recent vomiting, all except the urine pH, which says "no bicarbonaturia." Therefore, the urine electrolytes are not typical of vomiting but do suggest a degree of ECF volume contraction and organic anion excretion. The specific nature of the organic anion is unclear. It could be of dietary or metabolic origin.

Using the same type of analysis, because the urine has Na^+ but not Cl^-, recent or remote intake of diuretics is not the best simple answer.

What is the basis for the low urine [Cl^-]?

A low ECF volume is the cause of the low [Cl^-] in urine. Farrah consumes little salt, but loses NaCl via sweating.

Why isn't her excretion of K^+ lower?

She should have high levels of aldosterone as a result of the low ECF volume. Given the delivery of Na^+ with a poorly reabsorbed anion to the terminal cortical collecting duct, kaliuresis should be augmented.

Do you think she has an adrenal tumor?

No. The mineralocorticoid activity is explained by her ECF volume contraction. If she had an adrenal tumor, the Na^+ retention would be a primary event, and she would have ECF volume expansion and hypertension.

What advice would you give to Farrah?

Eating more NaCl and KCl should lead, ultimately, to a normal set of plasma electrolytes. Once Farrah follows these instructions, excessive NaCl retention will not occur and her results will normalize (see the margin).

Note:
When subjects with contracted ECF volumes are given large loads of NaCl, they often have excessive retention of salt, and edema may develop. This phenomenon should be transient (lasting only a few days).

Final diagnosis: Farrah has a negative NaCl balance because of poor dietary intake and nonrenal loss. She also has an unusual organic anion load that is probably from her diet. These findings mimic vomiting, except for the urine pH. Some physicians may believe that Farrah is a surreptitious vomiter, but Farrah still denies such a practice and the data available are not characteristic of vomiting (excretion of an unidentified organic anion rather than HCO_3^-).

Discussion of Case 4·2 Solly Has Transient Metabolic Alkalosis

(Case presented on pages 169–70)

The diagnostic quandary in this case is that the plasma [HCO_3^-] changes dramatically in a four-hour period with no obvious external source of HCO_3^- gain or loss.

Why was metabolic alkalosis present?

To have a high plasma [HCO_3^-], either HCO_3^- must be added or the ECF volume must be contracted. Solly's ECF volume had not been expanded earlier, nor is it very contracted now, so the problem is gain of HCO_3^-.

The HCO_3^- gain was most likely nonrenal because renal HCO_3^- loss, not gain, is present now. He denied any source of exogenous HCO_3^-. The most likely source of endogenous alkali is HCl loss (Figure 4·1, page 150). Perhaps Solly had trapped a large quantity of HCl in the lumen of his stomach, causing severe metabolic alkalosis.

How did metabolic alkalosis ameliorate spontaneously?

Although the urine was alkaline, its volume was small so there was little bicarbonaturia. If bicarbonaturia was the basis for the fall in plasma [HCO_3^-], so much Na^+ and K^+ would be lost that shock and an extreme degree of K^+ depletion would be anticipated. The [HCO_3^-] in the ECF fell by 22 mmol/l, and, if we assume an ECF volume of 15 liters, we must account for the loss of almost 330 mmol of HCO_3^-. Because there was no evidence of organic anion accumulation or excretion, the best answer we can give is that the HCl that was "hiding" in his stomach was partially reabsorbed four hours later; the result was a decrease in the [HCO_3^-] in the ECF with no evidence of external gain or loss of HCO_3^- (the [Cl^-] rose in plasma and the anion gap did not fall appreciably because Cl^- were reabsorbed from the GI tract).

What is wrong with Solly and how should he be treated?

Solly suffers from Zollinger-Ellison syndrome with excessive secretion of HCl. The diarrhea is due to HCl-induced damage to the duodenal mucosa plus denaturation of pancreatic enzymes and bile salt precipitation. Gastrin levels should be measured (see the margin).

Note:
Solly's gastrin levels, which were 10-fold higher than the upper limit of normal, confirm the presence of a gastrinoma; the offending tumor was resected and Solly is now fine.

Final diagnosis: Solly had excessive gastric HCl secretion; dissociation of the HCO_3^- generation and HCl reabsorption led to a transient severe metabolic alkalosis. This case has implications for the renal handling of HCO_3^-. Little excretion occurs if there is even a minor degree of ECF volume contraction; the low rate of excretion of HCO_3^- was due to a lower filtered load (low GFR, elevated plasma creatinine) and, more importantly, to more reabsorption of HCO_3^- by the kidney. Consider how acidemic, K^+-depleted, and ECF-volume-contracted he would have become if his kidneys had been "fooled" into excreting some of the "excess" HCO_3^- that accumulated in the ECF (see the margin). Finally, this case might help in explaining the wide variation in the normal values for HCO_3^- in plasma in normal individuals.

The alligator:
An alligator consumes 25% of its body weight in a meal. Food is retained in its stomach for days so that the high HCl load can dissolve the bones that it consumes. The plasma [HCO_3^-] may rise twofold to threefold, and the plasma [Cl^-] will decrease by a similar amount. There is little bicarbonaturia, and the plasma electrolytes return spontaneously to normal.

Discussion of Case 4·3 Mr. Greene Looks Green

(Case presented on page 170)

What are the acid-base diagnoses?

Mr. Greene has an increased plasma anion gap and a low [HCO_3^-]; therefore, he has metabolic acidosis. However, the increase in the plasma anion gap is 18 mEq/l (30–12, the normal value, given that his concentration of albumin is normal), but the fall in [HCO_3^-] is only 5 mmol/l. Therefore, a second process is increasing the [HCO_3^-]—metabolic alkalosis. The only other process that could increase the plasma [HCO_3^-] is respiratory acidosis, which was not present (low P_aco_2).

With a [HCO_3^-] of 20 mmol/l from metabolic acidosis, one would expect the P_aco_2 to be reduced to approximately 35 mm Hg; because it is 25 mm Hg in this case, Mr. Greene also has respiratory alkalosis. The final blood gas abnormality is a low Po_2 with an increased A–a difference (58 mm Hg) (see Chapter 5, page 193).

Mr. Greene has three acid-base diagnoses: metabolic acidosis with an increased anion gap, metabolic alkalosis (from vomiting), and respiratory alkalosis (from pneumonia). The basis of the metabolic acidosis, given the clinical setting, is most likely alcoholic ketoacidosis. Other diagnostic possibilities to consider are methyl alcohol or ethylene glycol intoxication and L-lactic acidosis resulting from low oxygen delivery, ethanol metabolism, or thiamine deficiency. If he has ketoacidosis, one might expect the plasma ketones to be strongly positive in diluted samples (the plasma level would be equal to the increase in anion gap, or 18 mmol/l); his plasma ketones, however, are weakly positive in an undiluted sample. In the setting of alcohol metabolism or hypoxia-induced L-lactic acidosis, one may have a false negative test for ketones because of the preponderance of β-hydroxybutyrate (see page 97). Also unexpected in ketoacidosis is his hypokalemia; one expects hyperkalemia with ketoacidosis (insulin deficiency).

What are the priorities for therapy?

1. Restore the ECF volume.
2. Replenish the K^+ deficit.
3. Avoid thiamine deficiency.

Mr. Greene has profound contraction of ECF volume because of vomiting and loss of $NaHCO_3$ and $KHCO_3$ in the urine. His deficits are NaCl and

KCl, and he requires both. Aggressive restoration of ECF volume using nothing but normal saline will run the risk of turning off the catecholamine suppression of insulin release and may aggravate the degree of hypokalemia because insulin secretion leads to the entry of K^+ into the ICF. A reasonable IV fluid would be 1 liter of isotonic saline plus 20 mmol of KCl given over the first 30 minutes. The next IV solution should be isotonic to Mr. Greene and could be 0.45% saline that contains 40–60 mEq/l KCl (depending on the plasma $[K^+]$ at that time).

Also send off blood samples to establish the specific diagnosis:

1. β-hydroxybutyrate;
2. methanol, ethanol, and ethylene glycol levels (the high value for the plasma osmolal gap is in part or entirely due to ethanol);
3. L-lactate.

What risks do you anticipate?

The first major risk of acute therapy is thiamine deficiency. Mr. Greene should receive thiamine intravenously. The second danger is overlooking the profound K^+ depletion that could cause life-threatening hypokalemia with vigorous ECF volume expansion. This risk would be magnified greatly with the administration or endogenous release of insulin. In this case, there is no immediate need for insulin; probably Mr. Green will secrete endogenous insulin once the ECF volume is restored.

The remaining issue to address is the cause of the wide A–a difference, which is most likely due to aspiration pneumonia; clarification and specific therapy will be necessary.

Summary of Main Points

- Metabolic alkalosis is present when there is a rise in the plasma [HCO_3^-] together with a fall in the [H^+]. Having an increased [HCO_3^-] involves either a source of new HCO_3^- and/or ECF volume contraction.
- Renal mechanisms are necessary to permit a sustained elevation in the [HCO_3^-] in plasma.
- Of the important events occurring in the ICF, the principal one is a loss of K^+ along with a gain of Na^+ and H^+. Hence, metabolic alkalosis in the ECF is usually accompanied by intracellular acidosis and K^+ depletion.
- Clinically, there are two major subgroups of patients with metabolic alkalosis: the most common group responds to administration of KCl and NaCl and exhibits ECF volume contraction. This form of metabolic alkalosis is caused most often by vomiting or diuretics and is characterized by a very low rates of excretion of Cl^- (when diuretics are not acting). Patients may not admit that they induce vomiting or abuse diuretics.

  The other group does not respond to administration of Cl^- salts and is often hypertensive. This type of metabolic alkalosis occurs most commonly with high aldosterone states. Other diagnoses of persistent metabolic alkalosis after administration of NaCl and KCl include Mg^{2+} depletion, Bartter's syndrome, and renal failure with a $NaHCO_3$ load. Therapy is usually more difficult with these types of metabolic alkalosis and requires that the specific etiology be identified and addressed.
- Patients with COPD and CO_2 retention may also have metabolic alkalosis. Because metabolic alkalosis might have a significant adverse effect on a patient's clinical state, it should be corrected.

Discussion of Questions

4·1 Why will the ingestion of $NaHCO_3$ not lead to the development of chronic metabolic alkalosis?

When the plasma [HCO_3^-] is elevated and the GFR is normal, more HCO_3^- are filtered. Reabsorption of HCO_3^- must be stimulated to retain these extra HCO_3^-. Consider the two components of $NaHCO_3$ separately:

1. Na^+: Intake of $NaHCO_3$ tends to expand the ECF volume, which should lead to a decline in renin and thereby angiotensin II levels. Because angiotensin II normally stimulates the reabsorption of HCO_3^-, fewer, not more, HCO_3^- should be reabsorbed in this setting.
2. HCO_3^-: The alkalemia will lead to a more alkaline proximal cell, which, in turn, will also depress the reabsorption of more HCO_3^- while alkalemia persists.

Hence, taken together, the intake of $NaHCO_3$ does not give the kidney "permission" to reabsorb more HCO_3^- than normal. To do so, one needs a lower GFR or stimulation of the proximal Na^+/H^+ antiporter (high angiotensin II or intracellular acidosis, related to hypokalemia or a high Pco_2).

4·2 For net loss of HCl or NH_4Cl to be the sole cause of metabolic alkalosis, what must the mass balance for Na^+, K^+, and Cl^- be?

The net loss of HCl or NH_4Cl requires that the negative balance for Cl^- exceed that for Na^+ plus K^+.

4·3 How many liters of emesis must be lost in order to raise the [HCO_3^-] in plasma by 10 mmol/l in a 70-kg adult?

Assume for simplicity that all the HCO_3^- retained (10 mmol/l) remained in the ECF (15 liters in a 70-kg person). Therefore, 150 mmol of HCO_3^- are needed. If the loss of HCl in the stomach was isosmotic (150 mmol of H^+ per liter and 150 mmol of Cl^- per liter), the net loss of emesis would be 1 liter. This amount is an underestimate because some HCO_3^- will distribute in the ICF. Hence, this degree of metabolic alkalosis is not due to a single episode of vomiting.

4·4 If a patient has gastric drainage, how can the quantity of new HCO_3^- formation be minimized?

Because a patient with gastric drainage can lose large amounts of HCl, metabolic alkalosis may result. The most effective way to reduce the H^+ loss is to administer an H_2-receptor blocker to inhibit secretion of HCl by the gastric mucosa. Although this procedure markedly reduces the HCl loss, some NaCl or $NaHCO_3$ can be lost and must be replaced to avoid ECF volume contraction and a resulting change in plasma [HCO_3^-].

Another way to reduce the acid-base impact of large HCl losses is to keep the patient's ECF volume well expanded with NaCl and replace any K^+ losses with KCl so that the $NaHCO_3$ is excreted (if renal function is adequate). If the renal function is poor, administer H^+ in the form of HCl or NH_4Cl to titrate the excess HCO_3^- and replace the deficit of Cl^-.

Note:
If the patient has severe liver disease, the use of NH_4Cl is not advisable.

4·5 How can one explain the progressive rise in the [HCO_3^-] in plasma and K^+ depletion in a patient with metabolic alkalosis due to vomiting?

With vomiting, there is a net loss of Cl^-, which are replaced by HCO_3^- (Figure 4·1). Given the same ECF volume and GFR, the filtered load of HCO_3^- rises, but proximal H^+ secretion does not rise to the same degree (no depletion of ECF volume or hypokalemia, but alkalemia is present). Hence, because HCO_3^- are less reabsorbable in the distal nephron, this increased filtered load of HCO_3^- results in $NaHCO_3$ excretion, which will lead to a small degree of ECF volume contraction. The urine contains Na^+ despite the ECF volume depletion because Na^+ accompany HCO_3^-, which cannot be reabsorbed. With the next episode of vomiting, the same sequence of events occurs—the [HCO_3^-] rises further in the ECF, and some bicarbonaturia occurs along with some Na^+ loss and more K^+ loss (the low ECF volume leads to aldosterone release and augments the excretion of K^+). This time the [HCO_3^-] in the ECF is slightly higher than previously. The degree of ECF volume depletion depends on the Na^+ intake between episodes of vomiting and the magnitude of the Na^+ losses (both GI and renal). With each episode of vomiting, there is transient bicarbonaturia as the "renal threshold" for HCO_3^- reabsorption is exceeded by the gastric HCO_3^- generation. The [HCO_3^-] in the ECF progressively rises, however, because the bicarbonaturia is not sufficient to excrete all the HCO_3^- generated. The progressive ECF volume depletion limits the amount of HCO_3^- filtered per unit of time (the filtered load), and the stimulation of H^+ secretion by angiotensin II causes an increased reabsorption of HCO_3^-. Therefore, the patient with recurrent vomiting will have chronic metabolic alkalosis, hypokalemia, and ECF volume contraction.

The urine composition varies, depending on when the patient is evaluated. After the patient vomits, the urine contains Na^+, K^+, and HCO_3^- but no Cl^- (as a result of ECF volume contraction and Cl^- depletion, all the Cl^- are reabsorbed); the urine pH is alkaline. Between episodes, the urine contains little Na^+ or Cl^- (from ECF volume contraction) and some K^+ (excreted with SO_4^{2-} and HPO_4^{2-}). The urine pH is low because of an enhanced secretion of H^+.

4·6 Why might the [HCO_3^-] in plasma rise in a patient with normovolemia who is undergoing plasmapheresis for rapidly progressive glomerulonephritis and renal failure?

That the patient's [HCO_3^-] in the ECF is increasing with no evidence of ECF volume contraction indicates an increased content of HCO_3^- in the ECF. In the absence of endogenous HCO_3^- generation (no vomiting and no renal HCO_3^- generation resulting from renal failure), the extra HCO_3^- must be due to exogenous bicarbonate—HCO_3^- mobilized from bone or from a resolving ileus, or the metabolic production of HCO_3^-. Assume that there is no ileus and that the patient is not receiving HCO_3^- per se. Daily plasmapheresis with plasma replacement (and several blood transfusions) will provide a large load of citrate^{3-} with Na^+ (the anticoagulant). The metabolism of citrate^{3-} to CO_2 consumes H^+, yielding HCO_3^-. The renal failure will prevent appreciable $NaHCO_3$ excretion.

4·7 How high might the [HCO_3^-] in plasma rise if a patient has a modest degree of contraction of the ECF volume and no increase in the content of HCO_3^- in the body?

Assume a 10% loss of ECF volume for easy math. With no change in the content of HCO_3^-, the [HCO_3^-] will rise 10% from 25 to 27.5 mmol/l.

4·8 What impact does the loss of 200 mmol of Na^+ and Cl^- from diuretic action have on the content and concentration of HCO_3^- in the ECF? (For simplicity, assume no change in [Na^+], which is 140 mmol/l.)

Note:
Loss of 200 mmol of Na^+ is equivalent to a loss of 1.5 liters of ECF, a 10% contraction. A 30% contraction could result in cardiovascular collapse.

If we assume the patient weighs 70 kg with a [Na^+] in the ECF that equals 140 mmol/l, the total amount of Na^+ in the body is 15 liters × 140 mmol/l, or 2100 mmol. If 200 mmol of Na^+ are lost, the total amount of Na^+ in the body will be 1900 mmol, and if the [Na^+] in the ECF remains at 140 mmol/l, the ECF volume will be 13.6 liters (1900/140). The HCO_3^- content, which was 25 mmol/l × 15 liters, or 375 mmol, is now contained in 13.6 liters; therefore, the new [HCO_3^-] is 28 mmol/l.

Therefore, it is easy to appreciate that with the loss of NaCl, the [HCO_3^-] in the ECF can rise with no change in the content of HCO_3^- in the ECF, because of a decrease in ECF volume. This process contributes to the metabolic alkalosis that results from diuretic use but is quantitatively small.

4·9 What pathophysiologic mechanisms must be involved to have Cl^--depletion metabolic alkalosis? What is the expected mass balance in each case?

Loss of HCl or NH_4Cl: To develop metabolic alkalosis, either the content of HCO_3^- in the ECF must rise or the ECF volume must decline. For the former to occur with a depletion of Cl^-, the pathophysiologic mechanism is loss of Cl^- along with H^+ (e.g., vomiting, nasogastric suction) or with NH_4^+ in the urine (hypokalemia consequent to diuretic actions). The expected mass balance is loss of more Cl^- than $Na^+ + K^+$.

Loss of NaCl: Loss of NaCl will be the pathophysiologic mechanism for alkalosis resulting from contraction of the ECF volume. The mass balance is equimolar loss of Na^+ and Cl^-.

Loss of KCl: Loss KCl can be the pathophysiologic mechanism for two aspects of Cl^--depletion metabolic alkalosis, depending on which cations entered the ICF in exchange for the K^+ that were lost. If Na^+ entered, there would be contraction of the ECF volume; if H^+ entered, there would be alkalosis of the ECF together with acidosis of the ICF. In either case, the mass balance would be equimolar loss of K^+ and Cl^-.

4·10 How can one distinguish between diuretic abuse and Bartter's syndrome using urine electrolytes?

There are two groups of patients with metabolic alkalosis, ECF volume contraction, and a $[Cl^-]$ in urine that exceeds 20 mmol/l—those who use diuretics and those who have Bartter's syndrome.

Diuretic use: Diuretic use (abuse) is the most likely cause of metabolic alkalosis with ECF volume contraction and a high $[Cl^-]$ in urine. Although a patient may deny using a diuretic, its presence can usually be identified by measuring the electrolytes on serial random urine samples. Some samples will be free of Na^+ and Cl^-, indicating the patient's ability to conserve Na^+ and Cl^- (some time has passed since the diuretic was consumed); other samples will have abundant Na^+ and Cl^- following more recent diuretic ingestion (see Tables 4·5 and 4·7). The patient who has ECF volume contraction and is wasting Na^+ should have excessive K^+ loss in the urine (from the mineralocorticoid secretion that is due to ECF volume contraction); if the $[K^+]$ in urine is not high, suspect the ingestion of a diuretic with "K^+-sparing" qualities. The other clues suggesting the diagnosis of diuretic abuse are: access to diuretics (paramedical personnel), patients with distorted body image, elevated uric acid in serum, and fluctuating body weight. The diagnosis can be confirmed with a direct assay of the urine for the diuretic when the $[Na^+]$ and $[Cl^-]$ in the urine are high.

Clinical pearls:
With hypokalemia, low ECF volume, and urine $[Cl^-] > 20$ mmol/l, suspect:
- Diuretics (urine $[Na^+]$ may be < 20 mmol/l at times);
- Bartter's syndrome (urine $[Na^+]$ is not < 20 mmol/l).

Table 4·7
SERIAL URINE ELECTROLYTES IN A PATIENT WITH ECF VOLUME CONTRACTION, INDICATING DIURETIC ABUSE

		Diuretics		
		Acting	Not acting	Acting
Na^+	(mmol/l)	30	10	42
K^+	(mmol/l)	43	13	59
Cl^-	(mmol/l)	85	3	108

Bartter's syndrome: Patients with Bartter's syndrome have metabolic alkalosis, ECF volume contraction, and inappropriately high Na^+ and Cl^- excretion via the urine. Serial observations reveal persistent Na^+ and Cl^- wasting and a relatively constant body weight, both of which contrast with a diagnosis of diuretic abuse. In patients with Bartter's syndrome, there is a much more profound degree of K^+ depletion and a much higher TTKG. Some patients also have hypomagnesemia and persistent renal Mg^{2+} wasting. In these patients, it is extremely difficult, if not impossible, to return the plasma $[K^+]$ to normal with K^+ supplements.

4·11 A very low GFR plays a prominent role in the metabolic alkalosis associated with the milk-alkali syndrome. What is the pathophysiology of this syndrome?

The basis of this disorder is the ingestion and absorption of large amounts of calcium and HCO_3^-. The patients then excrete large amounts of calcium and HCO_3^- in their urine. This excretion leads to nephrocalcinosis (calcium oxalate is less soluble in alkaline urine) and progressive renal function impairment. As renal function diminishes, the ability to excrete the large calcium load is reduced, and these patients actually develop hypercalcemia with soft tissue calcification. The hypercalcemia may cause vomiting, which leads to metabolic alkalosis and also stimulates indirect reabsorption of HCO_3^- in the proximal convoluted tubule if there is a contracted ECF volume. The decreased GFR also causes impaired renal excretion of HCO_3^-. With the continued ingestion of HCO_3^-, the plasma $[HCO_3^-]$ progressively rises. The metabolic alkalosis and hypercalcemia will resolve when ingestion of the calcium and alkali ceases.

4·12 What is the pathophysiology of metabolic alkalosis associated with nonreabsorbable alkali and an ion-exchange resin?

Ion-exchange resins are given to patients with renal failure to control hyperkalemia (they exchange Na^+ or Ca^{2+} for K^+ in the GI tract). Patients with renal failure also receive nonreabsorbable alkali as phosphate-binding agents (e.g., aluminum hydroxide).

The proposed mechanisms of the metabolic alkalosis are as follows:

- intake of Ca^{2+}, Mg^{2+}, or Al^{3+} salts of organic anions or their hydroxides;
- intake of an ion-exchange resin that is not in a H^+ form (i.e., with Na^+ or Ca^{2+});
- metabolism of organic anions to HCO_3^-;
- formation of a complex of Ca^{2+}, Mg^{2+}, or Al^{3+} with resin in the GI tract, so that there is no GI HCO_3^- loss via this route (absorption of Na^+ and OH^-). Under normal circumstances, Ca^{2+}, Mg^{2+} or Al^{3+} are excreted in the feces as carbonate salts; this process consumes HCO_3^-, so that acid-base balance results. In this case, however, if the patients have a very low GFR, they cannot excrete the HCO_3^- and could become progressively alkalemic.

Note:
The conversion of absorbed OH^- to HCO_3^- is catalyzed by carbonic anhydrase.

$$CO_2 + OH^- \longrightarrow HCO_3^-$$

4·13 What is the pathophysiology of metabolic alkalosis associated with a normal or expanded ECF volume?

In all cases of metabolic alkalosis with a normal or expanded ECF volume, Na^+ and Cl^- are delivered to the cortical collecting duct in the presence of mineralocorticoid activity. The result is the reabsorption of Na^+ along with the excretion of Cl^- and K^+. Because of the K^+ excretion, hypokalemia develops and, as a result, ammoniagenesis and new HCO_3^- generation are stimulated (see the margin). Thereafter, enhanced reabsorption of HCO_3^- and perhaps a lower filtered load (low GFR secondary to the hypokalemia) will lead to a progressive rise in the $[HCO_3^-]$ in plasma. The basis of the mineralocorticoid activity varies, depending on the diagnosis, as outlined below.

Note:
Hypokalemia induces a shift of K^+ out of cells together with a shift of H^+ (and Na^+) into cells; this shift of H^+ contributes to a higher plasma $[HCO_3^-]$ and intracellular acidosis.

Diagnosis	Basis of mineralocorticoid activity
1° aldosteronism	Autonomous secretion (adenoma, hyperplasia)
Glucocorticoid excess	Autonomous ACTH secretion, drugs mimicking aldosterone actions (see Table 10·5, page 363)
Renal-artery stenosis	Secondary to hyperreninemia
Magnesium-depletion	Mechanism unclear

4·14 In what circumstances might metabolic alkalosis be associated with hyperkalemia?

In almost all causes of metabolic alkalosis, hypokalemia from renal loss of K^+ is an expected (necessary) finding. If renal failure is present, hyperkalemia could be present; alternatively, look for a sudden shift of K^+ out of cells (e.g., necrosis or lack of insulin).

4·15 How accurate is the term "saline-sensitive metabolic alkalosis"?

The three components of metabolic alkalosis:

1. ECF alkalosis
2. ICF acidosis and K^+ depletion
3. Renal "permission" to have a high $[HCO_3^-]$ in plasma

Many nephrologists classify metabolic alkalosis as saline-sensitive and saline-resistant, but is this terminology entirely correct? The only times that the ECF, ICF, and renal alterations are "corrected" solely with NaCl are:

1. when the entire deficit is NaCl, a very rare event;
2. when the deficit is HCl (or NH_4Cl), not common clinically (Table 4·1, page 154).

In these cases, the Cl^- ingested replace the deficit of Cl^-, and the excess Na^+ and HCO_3^- are excreted.

If enough NaCl is given to overexpand the ECF volume (a usual occurrence), the $[HCO_3^-]$ will fall from dilution and there will be renal excretion of HCO_3^- consequent to the expanded ECF volume. Some view this decrease in the plasma $[HCO_3^-]$ as "correction" of metabolic alkalosis (the plasma $[HCO_3^-]$ is now normal), but the abnormal composition of the ICF persists (low K^+, high H^+, high Na^+). The composition of the ICF will be corrected only when the deficit of K^+ is replaced. It is for this reason that we prefer the term "chloride-plus-cation-responsive metabolic alkalosis"; the cation(s) lost along with Cl^- (Na^+ or K^+) must be determined clinically.

5

RESPIRATORY ACID-BASE DISTURBANCES

PART C
RESPIRATORY ALKALOSIS 201

PART D
P_aCO_2 IN METABOLIC ACID-BASE DISORDERS 203

PART E
REVIEW 204

Objectives

- To provide an understanding of the factors regulating the Pco_2 of the blood (arterial and venous) and the impact of a change in Pco_2 on the $[H^+]$ in the ECF and ICF.
- To provide an understanding of the differences in the acid-base status of acute vs chronic respiratory acid-base disturbances. In doing so, we present a diagnostic approach that will enable recognition of *respiratory acid-base disturbances*, diagnosis of the underlying disorder, and recognition of coexistent acid-base disorders.
- To provide an appreciation of the value and the pitfalls in assessing the difference between the Po_2 of alveolar air and that of the arterial blood.

Abbreviations:
A-a Po_2 = difference between alveolar and arterial Po_2.
BBS = bicarbonate buffer system.
DKA = diabetic ketoacidosis.
2,3-DPG = 2,3-diphosphoglycerate.
ECF = extracellular fluid.
F_iO_2 = percentage of inspired air that is O_2.
ICF = intracellular fluid.
P_aco_2 = partial pressure of CO_2 in arterial blood.
PDS = physiological dead space.
RQ = respiratory quotient.

Respiratory acid-base disturbances: Conditions in which an abnormality in the arterial Pco_2 is observed. Although the normal P_aco_2 is 40 mm Hg, the P_aco_2 varies with pathophysiological circumstances (e.g., in a patient with metabolic acidosis and a $[HCO_3^-]$ of 10 mmol/l, the expected P_aco_2 should be 25 mm Hg). Therefore, the P_aco_2 must be evaluated in conjunction with other clinical information.

Outline of Major Principles

1. The arterial Pco_2 (P_aco_2) reflects the concentration of CO_2 in alveolar air required for balance between CO_2 production (metabolism) and CO_2 removal (ventilation).
2. Normal ventilation is mediated by interaction among the central respiratory centers, peripheral chemoreceptors, respiratory muscles, and lung parenchyma.
3. The acid-base status of an acute respiratory acid-base disorder differs greatly from that of a chronic respiratory acid-base disorder because of variations in renal NH_4^+ excretion and/or HCO_3^- reabsorption; these variations influence the plasma $[HCO_3^-]$.
4. Rules:

   Acute respiratory acidosis: For every mm Hg increase in P_aco_2 from 40 mm Hg, expect a 0.8 nmol/l increase in the $[H^+]$ from 40 nmol/l. The plasma $[HCO_3^-]$ rises 2.5 mmol/l with a doubling of P_aco_2.

   Chronic respiratory acidosis: For every mm Hg increase in P_aco_2 from 40 mm Hg, expect a 0.3 nmol/l increase in the $[H^+]$ and a 0.3 mmol/l increase in the plasma $[HCO_3^-]$.

   Acute respiratory alkalosis: For every mm Hg decrease in P_aco_2 from 40 mm Hg, expect a 0.8 nmol/l decrease in the $[H^+]$ from 40 nmol/l.

   Chronic respiratory alkalosis: For every mm Hg decrease in P_aco_2 from 40 mm Hg, expect a 0.2 nmol/l decrease in the $[H^+]$ from 40 nmol/l and a 0.5 mmol/l decrease in the plasma $[HCO_3^-]$ from 25 mmol/l.

Introductory Case
Hack's Future Is Up in Smoke

(Case discussed on pages 204–05)

Anatomical dead space:
The part of the airway that never takes part in gas exchange (approximately 1 ml/lb body weight).

Alveolar dead space:
The part of the alveoli that does not take part in gas exchange because of disease.

Physiological dead space (PDS):
Anatomical + alveolar dead space.

Tidal volume:
The volume of inhaled air.

Alveolar ventilation:
Tidal volume – PDS × respiration rate.

Hack, a 64-year-old man with a long history of chronic obstructive lung disease, lives at home but requires O_2 therapy. In a 24-hour period, he developed a cough and shortness of breath, became confused, and was taken to the hospital. A diagnosis of pneumonia and acute respiratory failure was made, and he was intubated and ventilated. The results two days later are shown below. He was weaned from the ventilator on day four of treatment and remained on O_2 by mask.

	Hack's usual values	Admission values	Treatment	
			Day 2	Day 4
H^+ (nmol/l)	46	63	40	64
pH	7.34	7.20	7.40	7.19
P_aco_2 (mm Hg)	60	80	40	70
P_ao_2 (mm Hg)	60	35	180	55

What prompted the changes from the usual values to those seen on admission?
What happened to permit the changes in the blood gases while Hack was on the ventilator on day two of therapy?
Why are his post-extubation values on day four of therapy so different from his original values?

PART A
BACKGROUND

Overview of CO_2 Homeostasis

- CO_2 excretion = alveolar ventilation × $[CO_2]$ in alveolar air.

	CO_2 excretion	Alveolar ventilation	$[CO_2]$ in alveolar air
Normal	10 mmol/min	5 liters/min	2 mmol/l
Chronic respiratory acidosis	10 mmol/min	3 liters/min	3.3 mmol/l

The major end product of oxidative metabolism is CO_2. When carbohydrates or proteins are oxidized, 1 mmol of CO_2 is produced for every mmol of O_2 consumed (the *respiratory quotient*, or RQ, is 1.0). In contrast, less CO_2 is formed per O_2 consumed when fat is oxidized; in this case, the RQ is 0.7. On a typical Western diet, the usual RQ is close to 0.8. To place these numbers in a quantitative perspective, normal adults consume 12 mmol of O_2 per minute and produce 10 mmol of CO_2 per minute. The concentrations of O_2 and CO_2 in alveolar air are close to 6 and 2 mmol/l, respectively, and O_2 consumption and CO_2 production occur in close to a 1:1 ratio. Therefore, because the supply of O_2 at the level of the alveolus markedly exceeds demand, control of the rate of ventilation is via changes in the $[CO_2]$ rather than in the $[O_2]$. Hence, we shall focus on CO_2 in the remainder of this chapter.

Respiratory quotient (RQ): The quantity of CO_2 produced divided by the quantity of O_2 consumed. The RQ helps one deduce which type of fuel is being oxidized.

Normal metabolism results in the production of 10 mmol of CO_2 per minute at rest. This CO_2 leaves cells and enters venous capillary blood for transport to the lungs. Because cardiac output is 5 liters per minute at rest, venous blood must carry an extra 2 mmol of CO_2 per liter (10 mmol/min ÷ 5 liters/min) as compared with arterial blood. This 10 mmol of CO_2 is exhaled in 5 liters of alveolar ventilation per minute. If the alveolar ventilation is doubled to 10 liters per minute for any reason (e.g., metabolic acidosis or salicylate ingestion) with no change in production of CO_2, the Pco_2 of alveolar air and arterial blood will fall by 50%. Thus, with twice the alveolar ventilation rate, the same amount of CO_2 can be exhaled, but at half the concentration in each liter of alveolar air. Conversely, as alveolar ventilation falls, the concentration of CO_2 in alveolar air must rise (as will the P_aco_2) to remove the CO_2 produced daily (compare with serum creatinine and the GFR; when the GFR halves, the concentration of creatinine in plasma will double).

Questions

(Discussions on pages 209–10)

Room air:

21% O_2

$[O_2]$ = 8.7 mmol/l.

5·1 Assume that the consumption of O_2 is 12 mmol/min, that alveolar ventilation is 5 liters/min, and that 21% of air is O_2. Why is the Po_2 of alveolar air 100 mm Hg and not 150 mm Hg, as it is in room air? How many mmol of O_2 are in one liter of alveolar air? What percentage of the O_2 delivered to the alveoli is extracted?

5·2 Fish exchange gases through gills by taking up the O_2 dissolved in water and adding CO_2 to that water. CO_2 is 30-fold (for easy math) more soluble in water than O_2. What does this solubility imply for the P_aco_2 and the pH of blood in fish in steady state?

5·3 By what proportion must alveolar ventilatory capacity decline to have a P_aco_2 of 50 vs the expected 40 mm Hg?

In addition to alveolar ventilation, the other major factor determining the P_aco_2 is the rate of CO_2 production, which is determined by:

1. metabolic work (the need to regenerate ATP; see Tables 5·1 and 5·2);
2. mechanical work (patients who are cachectic may produce less CO_2 and have a lower P_aco_2 when hyperventilating);
3. fuels being utilized (oxidation of carbohydrates yields more CO_2 relative to ATP production than does the oxidation of fat-derived fuels; see Table 5·2).

Legend for Table 5·1

The values for the production of CO_2 are shown as mmol/min and are representative values for a 70-kg adult. The values for altered rates are rough estimates and are for illustrative purposes only. Reproduced with permission (*Nephron*. 64:514–17, 1993.)

Table 5·1
CLINICAL SETTINGS RESULTING IN ALTERED CO_2 PRODUCTION

State	Organ	Usual CO_2 production rate	Altered CO_2 production
Coma/anesthesia	Brain	3	1.5
Low GFR	Kidney	2	< 1
Cachexia/paralysis	Muscle	2.4	< 1
Vigorous exercise	Muscle	2.4	160
Ketogenesis	Liver	2.4	0

Legend for Table 5·2

The oxidation of carbohydrates produces more CO_2 than does the oxidation of fat-derived fuels when viewed in terms of the yield of ATP. (Table 5·2 is reproduced with permission from *Nephron*.)

Table 5·2
IMPORTANCE OF THE METABOLIC FUEL UTILIZED IN DETERMINING THE RATE OF CO_2 PRODUCTION

Fuel	mmol CO_2/100 mmol ATP	Products
Carbohydrate	16.7	$CO_2 + H_2O$
Fatty acids	12.2	$CO_2 + H_2O$
Fatty acids	0	Ketoacids
Ethanol	11.1	$CO_2 + H_2O$
Ethanol	0	Ketoacids
Ethanol	0	Acetic Acid

Physiology of O_2 and CO_2 Transport

OXYGEN TRANSPORT

Oxygen is transported bound to hemoglobin, with four molecules of O_2 bound per molecule of hemoglobin. Normally there are 2.25 mmol (140 g) of hemoglobin per liter of blood. The affinity of hemoglobin for O_2 is high but can be reduced by elevated concentrations of H^+, CO_2, and 2,3-diphosphoglycerate (2,3-DPG).

O_2 transport and concentration of hemoglobin:
Blood can carry close to 9 mmol of O_2 per liter. Even if the hemoglobin is decreased by a factor of 12, delivery of O_2 to organs can still be maintained because the extraction of O_2 from each liter of blood can increase threefold and the cardiac output can increase fourfold. Therefore, anemia per se will never be the sole cause of hypoxia-induced lactic acidosis at rest.

Effect of H^+ (Bohr Effect)

An increase in the [H^+] in plasma enhances the discharge of O_2 from hemoglobin because H^+ bind to hemoglobin and lessen hemoglobin's affinity for O_2. This effect is important during exercise because when L-lactic acid is produced, the [H^+] rises and facilitates the release of O_2 at the tissue level (Figure 5·1).

Effect of CO_2

An increase in the Pco_2 in tissue facilitates release of O_2 from hemoglobin by two mechanisms: a rise in [H^+] and the formation of *carbamates*.

Carbamates:
When CO_2 is exposed to an R-NH_2 group, but not an R-NH_3^+ group, (e.g., the terminal amino group on globin molecules), a spontaneous reaction occurs that yields an anionic carbamate (see Chapter 9, page 348).

Effect of 2,3-DPG

When 2,3-DPG binds to hemoglobin, the affinity of hemoglobin for O_2 is reduced. Increases in 2,3-DPG in red blood cells result in increased deliv-

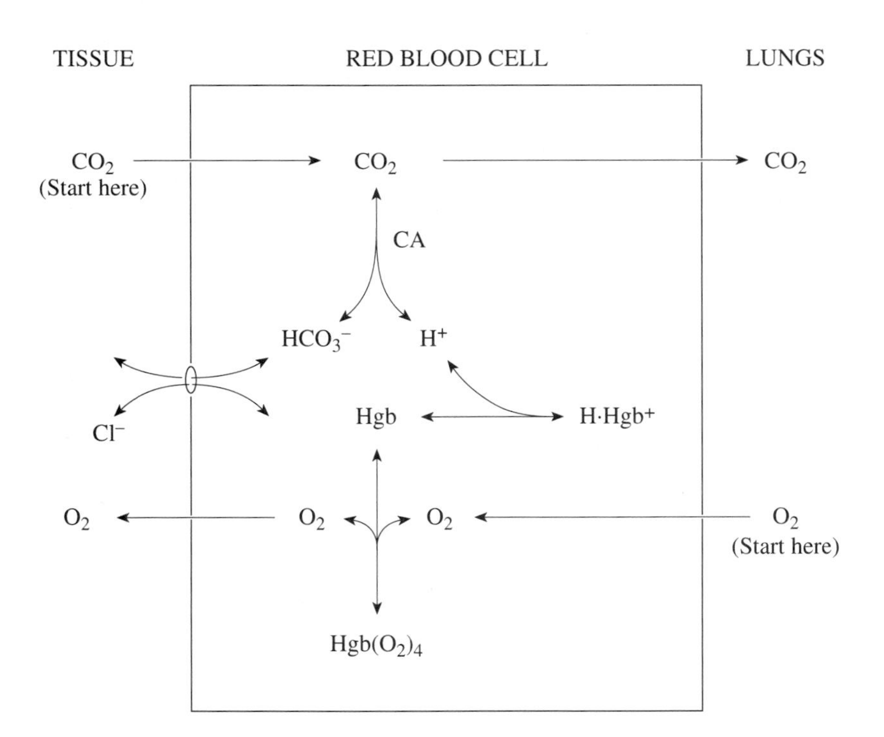

Figure 5·1
Interplay of CO_2 and O_2 transport
For CO_2 transport, begin at the upper left. CO_2 diffuses from the tissues into the red blood cells (RBC), where, under the influence of carbonic anhydrase (CA), HCO_3^- and H^+ are formed. The H^+ bind to hemoglobin (Hgb) and promote the unloading of O_2, which diffuses into tissues.

For O_2 transport, begin at the lower right. O_2 diffuses from the alveoli into the RBC. Binding of O_2 to hemoglobin promotes the exit of CO_2 from the blood. See the text for additional details.

ery of O_2 at any given Po_2. The signal for a rise in 2,3-DPG is the alkalemia that results from hyperventilation (e.g., high altitude, exercise, anemia). These effects require many hours to take place.

Effect of Temperature

When the temperature rises, the affinity of hemoglobin for O_2 falls. This lower affinity aids the down-loading of O_2 at a higher Po_2 in muscle capillaries during vigorous exercise because there is a large rise in the temperature of contracting muscles.

TRANSPORT OF CO_2

At the tissue level, 10 mmol of CO_2 per minute diffuse into red blood cells. The carbonic anhydrase in these cells converts the CO_2 to H^+ and HCO_3^- (Figure 5·1). Maintenance of a low Pco_2 in the red blood cell aids further diffusion of CO_2. The HCO_3^- formed (8.5 mmol/min) is transported into the plasma in exchange for Cl^- ("chloride shift"), and the H^+ bind to hemoglobin.

Note:
CO_2 diffuses between blood and the alveolar space much more rapidly than does O_2.

Once at the lung, the process is reversed (see the margin). The lower Pco_2 of alveolar air aids in the diffusion of CO_2 from blood to the alveolus. The high Po_2 of alveolar air promotes the binding of O_2 to hemoglobin, which leads to the dissociation of H^+ bound to hemoglobin. These H^+ combine with the HCO_3^- that were in plasma (reversal of the "chloride shift"), and the resultant CO_2 diffuses into the alveoli. The net result is oxygenation in concert with CO_2 unloading in the lung.

ARTERIAL OR VENOUS BLOOD FOR ANALYSIS?

Mixed venous Pco_2:
The Pco_2 of central venous blood.

- The *mixed venous Pco_2* reflects the tissue Pco_2 and hence the metabolic and/or circulatory status. The arterial Pco_2, on the other hand, reflects the alveolar Pco_2 and hence largely the ventilating function of the lungs.

Although it is "traditional" to use arterial blood for blood gas analysis, the sample used should depend on the information required. The venous Pco_2 reflects the tissue Pco_2; it is determined by the rate of production of CO_2 and the rate of blood flow. Thus, the Pco_2 of the venous blood from a specific limb can reflect local rather than systemic conditions. A high mixed venous Pco_2 can occur under three circumstances:

1. a high rate of aerobic metabolism (e.g., exercise);
2. a very high rate of H^+ formation (formation of CO_2 as a result of buffering of H^+ by HCO_3^-, e.g., during a sprint);
3. a reduced cardiac output, in which case, the same amount of CO_2 is formed by an organ but is added to a smaller volume of blood flowing through that organ.

Questions

(Discussions on pages 210–11)

5·4 Why is the venous Pco_2 much higher than the arterial Pco_2 in each of the following: exercise, a convulsion, and DKA?

DKA = diabetic ketoacidosis.

5·5 Why might the rate of production of CO_2 be lower than normal in a patient with DKA?

Pulmonary Physiology

To maintain a constant P_aco_2, the lungs must remove the 10 mmol of CO_2 produced per minute by normal metabolism. It is important to appreciate that the body produces a huge amount of CO_2—more than 14,000 mmol/day. Furthermore, with CO_2 accumulation, H^+ and HCO_3^- are produced in equimolar amounts (Figure 5·2) even though the $[HCO_3^-]$ normally exceeds the $[H^+]$ by 10^6-fold. The lungs therefore have a critical role in the maintenance of acid-base balance. Failure to remove this CO_2 leads to the generation of H^+ by displacement of the bicarbonate buffer system equilibrium to the left; acidemia (respiratory acidosis) then ensues. Excessive removal of CO_2 results in alkalemia (respiratory alkalosis) via displacement of the equilibrium to the right (Figure 5·2).

Point of emphasis:
H^+ and HCO_3^- are produced in a 1:1 ratio from CO_2, but their relative concentrations are close to $1{:}10^6$, respectively.

The three components of pulmonary function that regulate gas exchange are ventilation, diffusion, and perfusion. Abnormalities of CO_2 homeostasis are generally related to derangement of ventilation. We shall therefore focus on the physiology and control of ventilation (see the margin for comments concerning a patient with fixed volume ventilation).

P_aco_2 change resulting from change in CO_2 production:
In patients with fixed alveolar ventilation (on ventilator without patient triggering), the P_aco_2 will increase:

1. when a patient with metabolic acidosis produces more CO_2 in response to treatment with $NaHCO_3$;
2. when a fasted patient (metabolizing fatty acids) is fed and begins to oxidize carbohydrates.

CONTROL OF VENTILATION

Although the normal P_aco_2 is 38–42 mm Hg, in states of metabolic acidosis and hypoxia, one expects to see a lower $[CO_2]$ or Pco_2 in arterial blood. Similarly, in association with metabolic alkalosis, there should be a modest rise in the P_aco_2. Therefore, the P_aco_2 must be assessed in the context of the pathophysiologic state of the patient with consideration of the existing stimulators and suppressors of ventilation.

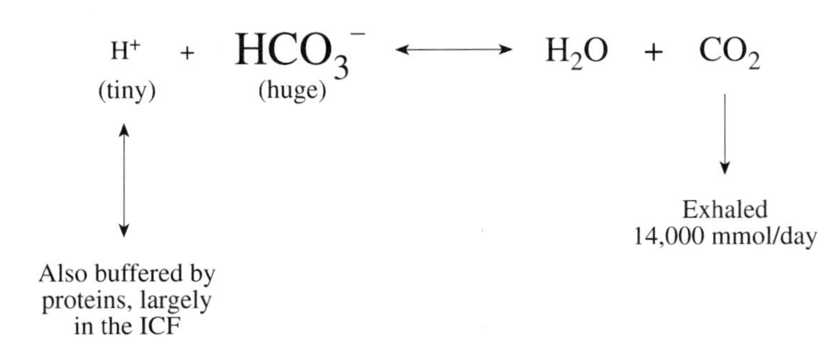

Figure 5·2
Bicarbonate buffer system and respiratory acid-base disorders
The key principle is that the $[H^+]$ is tiny, and the $[HCO_3^-]$ is 10^6-fold larger. Shifting this equilibrium requires a 1:1 stoichiometry that results in an enormous H^+ load relative to the basal level of H^+.

The three components of the control of ventilation are central control, receptors for CO_2 and O_2, and the respiratory muscles and their workload.

The Central Control of Ventilation

Several centers in the medulla and in the pons, which are often referred to as the "central respiratory centers," control the rhythmic nature of ventilation. In addition, voluntary control of breathing is mediated via the cerebral cortex and is capable of overriding the brain stem, within limits.

Receptors

- There are both central and peripheral chemoreceptors that play an important role in the control of ventilation.

Note:
There are also numerous receptors in the lung parenchyma (pulmonary capillary, or "J" receptors) that respond to a variety of stimuli, such as inflammation or edema, by leading to an increase in the central ventilatory drive. Thus, patients with lung disease characteristically hyperventilate and have a reduced P_aco_2, at least until the disease becomes quite severe.

The central chemoreceptors, located on the ventral surface of the medulla, stimulate ventilation primarily in response to increases in the $[H^+]$ of the cerebrospinal fluid and also in response to increases in Pco_2 (thought to be mediated by a rise in the $[H^+]$). Ventilation is inhibited when the $[H^+]$ falls.

The peripheral chemoreceptors, primarily found in the carotid bodies, are responsible for the ventilatory response to a low P_ao_2. These receptors also play a less important role in the ventilatory response to increased P_aco_2 and acidemia.

Question

(Discussion on page 211)

Note:
This question is for the more curious.

5·6 How might peripheral chemoreceptors detect a lower P_ao_2?

Respiratory Muscles and Their Workload

The final determinants of ventilation, given the appropriate signals, are the respiratory muscles, their innervation, and the magnitude of the resistance that they must overcome. A severe catabolic state, muscle disease, or metabolic disorders such as hypokalemia and hypophosphatemia may reduce the function of the respiratory muscles. Nerve damage (e.g., phrenic nerve) or changes in either the elastic load (lung and chest wall compliance, e.g., pulmonary fibrosis) or the resistive load (airway resistance, e.g., chronic bronchitis) may also increase the work of ventilation. Therefore, for any given level of central stimulation, the respiratory muscles may not be able to respond adequately. Ventilation will then decrease and lead to a rise in the P_aco_2.

The Alveolar-Arterial Po_2 Difference

- Using the *alveolar-arterial Po_2 difference* (Po_2 in alveolar air – Po_2 in arterial blood) to evaluate the exchange of O_2 between alveolar air and blood enables assessment of the causes of hypoxemia. Alveolar Po_2 is calculated as inspired Po_2 – 1.25 × arterial Pco_2. The normal value, which depends on age, is up to 15 mm Hg.
- The major difficulties with this approach are that the content of O_2 and Po_2 are related in a sigmoid fashion, not a linear fashion, and that the calculation is affected by several nonpulmonary parameters.

Alveolar-arterial Po_2 difference:

1. The difference in Po_2 between the alveolar air and arterial blood is referred to as the "A-a gradient." In truth this calculation should be a "difference" rather than a "gradient" because diffusion of nonelectrolytes is involved.
2. There is a problem with the A-a Po_2 difference because the relationship between Po_2 and O_2 content of blood is not linear (Figure 5·3).

Although respiratory acid-base disorders are defined by changes in the P_aco_2, important clinical information can also be derived by interpreting the P_ao_2. The arterial Po_2 is a function of both the Po_2 of alveolar air and the diffusion of O_2 across the alveolar capillary membrane.

Air is approximately 79% nitrogen and 21% O_2. In the lungs, CO_2 and water vapor are part of the "nonnitrogen" gases. Therefore, if one is breathing room air, as the Pco_2 of alveolar air rises, the Po_2 must fall. Similarly, for any given inspired O_2 tension, a reduction in alveolar Pco_2 (hyperventilation) leads to an increase in alveolar and, hence, arterial Po_2.

One can calculate the alveolar Po_2 using the abbreviated alveolar gas equation (equation below). Thereafter, examining the difference between the alveolar Po_2 and the arterial Po_2 (A-a Po_2 difference) becomes a useful diagnostic tool. However, this approach has several pitfalls; they are discussed below.

$$\text{Alveolar air } Po_2 = \text{Inspired air } Po_2 - (P_aco_2)/RQ$$
$$= \text{Inspired air } Po_2 - (P_aco_2)/0.8$$

Note:
Dividing by 0.8 is the same as multiplying by 1.25.

UTILITY OF THE A-a Po_2 DIFFERENCE

The calculation of the A-a Po_2 difference allows one to estimate how much of the derangement in arterial Po_2 is due to a change in alveolar Pco_2 (ventilation) and how much is due to reduced transfer of O_2 from alveolus to blood (intrinsic lung disease). One must have an accurate estimate of the Po_2 of the inspired air to calculate the A-a difference. If air is 21% O_2, barometric pressure is 760 mm Hg, and water vapor pressure is 47 mm Hg, the Po_2 of inspired air is 0.21 (760 – 47), or 150 mm Hg. The alveolar Po_2 can be estimated from the abbreviated alveolar gas equation.

There are two major types of pulmonary lesions that cause the P_ao_2 to be substantially lower than that of alveolar air:

1. Blood could pass from the pulmonary artery to the pulmonary vein without perfusing alveoli that have a high Po_2 (i.e., a shunt that prevents a good exchange of air). In reality, most lung diseases that cause hypoxemia have numerous small areas of shunting as well as areas of nonven-

tilated, nonperfused lung; together, these lesions lead to ventilation-perfusion mismatch.

2. There might be a barrier to diffusion of O_2 from alveolar air to the capillaries in lungs. The magnitude of the A-a difference is a parameter to be evaluated when trying to decide if a pulmonary condition is improving or worsening. The A-a difference can also clarify whether hypoxemia is due to lung disease or central suppression of ventilation. In the latter case, the A-a difference should be normal (corrected for the sigmoid vs linear relationship, Figure 5·3).

Impact of the content of O_2 in shunted blood on the A-a difference: Assume that arterial blood has 9 mmol of O_2 per liter and that 10% of the blood in the pulmonary artery bypasses aerated alveoli via a shunt into the pulmonary vein. The content of O_2 in the blood in the pulmonary artery is 6 mmol/l. After this 10% shunt, arterial blood would contain 8.7 mmol of O_2 per liter (0.9 liters with 9 mmol/l + 0.1 liter with 6 mmol/l). In a second example, assume that blood in the pulmonary artery contains 3 mmol of O_2 per liter. After the 10% shunt, the arterial blood would have 8.4 mmol of O_2 per liter instead of 9 mmol. Assume that the new P_aO_2 in the first instance would be 95 mm Hg and that it would be 65 mm Hg in the second example. The corresponding A-a differences would be 5 and 35 mm Hg.

PITFALLS TO RECOGNIZE IN THE USE OF THE A-a DIFFERENCE

The difference between the alveolar and the arterial Po_2 is normally less than 15 mm Hg, but this value increases with age. While the A-a difference is widely used clinically, there are several pitfalls that must be kept in mind:

1. The A-a difference utilizes the Po_2 instead of reflecting the content of O_2 (O_2 saturation). Thus, the same reduction in O_2 content will have a different impact on the Po_2 at different sites on the oxygen-hemoglobin dissociation curve because this function is sigmoid rather than linear (Figure 5·3).

2. In a fixed volume of shunt from pulmonary artery to pulmonary vein, the arterial Po_2 is strongly influenced by the content of O_2 in the blood in the pulmonary artery (see the margin).

3. The cardiac output is important in determining the extent to which a fixed shunt from pulmonary artery to pulmonary vein has affected the arterial Po_2 (see Question 5·7).

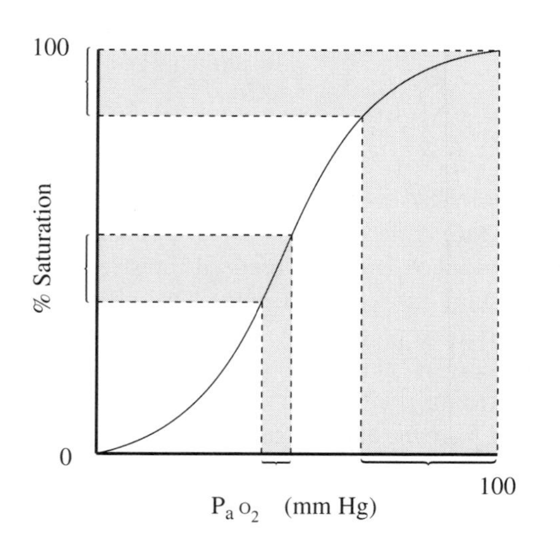

Figure 5·3
The oxygen-hemoglobin dissociation curve
Note that the shape of this curve is not linear, so a given absolute change in % saturation (shaded areas from the y-axis) leads to a very different change in P_aO_2 in the two examples illustrated in this figure.

Question

(Discussion on page 211)

5·7 A patient has a 0.1 liter/min shunt from pulmonary artery to pulmonary vein. Would the A-a difference be different if the cardiac output was 5 vs 2 liters per minute? (Assume no other abnormality.)

4. The Po_2 of inspired air must be known. When patients are receiving O_2 by mask or nasal prongs, the inspired Po_2 may not be known with sufficient accuracy. Therefore, the A-a difference is most useful when patients are breathing room air or are on ventilators with a measured content of inspired Po_2.
5. In the calculation of the alveolar Po_2, one must estimate the amount of O_2 removed and replaced by CO_2. To do so, one uses the arterial Pco_2 and assumes an RQ of 0.8. The RQ could be 1 if carbohydrate is the only type of fuel being metabolized (see the margin for an example).
6. The determination of the alveolar Po_2 assumes a steady state, so if an acute event has occurred (a sudden change in alveolar ventilation or an acute production of CO_2 from the administration of $NaHCO_3$ in L-lactic acidosis), an error will be introduced.
7. The normal A-a O_2 difference is increased with increasing F_iO_2 (percentage of inspired air that is O_2), even in patients on a ventilator.

Effect of the RQ on the A-a difference:

1. Assume an RQ of 0.8:
   Alveolar O_2 = 150 – (40/0.8)
   = 100 mm Hg
2. Assume an RQ of 1:
   Alveolar O_2 = 150 – (40/1)
   = 110 mm Hg

Therefore, the A-a differs by 10 mm Hg.

Questions

(Discussions on page 212)

Note:
These questions illustrate the clinical application and value of the A-a difference.

5·8 A patient, comatose from a drug overdose, has the following blood gases. The basis of the hypoxia is thought to be aspiration. Is this diagnosis correct?

pH		7.24	Pco_2	mm Hg	64
H^+	nmol/l	58	P_ao_2	mm Hg	66

5·9 Two days later the patient in Question 5·8 is awake but coughing. The following blood gases are obtained. Because his P_ao_2 has improved, he is declared ready for discharge. Is this decision appropriate?

pH		7.60	P_aco_2	mm Hg	24
H^+	nmol/l	25	P_ao_2	mm Hg	70

Respiratory Acid-Base Disorders

- Respiratory acidosis is an increased P_aco_2 and $[H^+]$ in plasma.
- Respiratory alkalosis is a decreased P_aco_2 and $[H^+]$ in plasma.
- Chronic respiratory disorders have a renal response that causes:
  - increased plasma $[HCO_3^-]$ in acidosis;
  - decreased plasma $[HCO_3^-]$ in alkalosis.

Clinical pearls:

1. It is too difficult to assess the P_aco_2 simply by clinical examination. Therefore, blood gas analysis is the key to the diagnosis of respiratory acid-base disorders.
2. In a patient who is breathing room air, adding the Po_2 and P_aco_2 and subtracting the total from 150 will give a quick estimate of the A-a difference.

Renal response to chronic respiratory acid-base change:
Acidemia from respiratory acidosis increases NH_4^+ excretion on a transient basis. This increase leads to a higher $[HCO_3^-]$ in plasma. As a result, the filtered load of HCO_3^- rises, as does the amount that is indirectly reabsorbed. Therefore, the plasma $[HCO_3^-]$ is higher than normal in chronic respiratory acidosis. The opposite occurs in chronic respiratory alkalosis.

Respiratory acid-base disorders arise from primary changes in P_aco_2. There is a very large flux of CO_2 relative to the $[CO_2]$ in the plasma (i.e., 10 mmol of CO_2 are produced per minute, yet the P_aco_2 and H_2CO_3 are only 1.2 mmol of CO_2 per liter of blood and usually vary by less than 10% on a moment-to-moment basis). If a discrepancy develops transiently between production and removal of CO_2, the resultant change in P_aco_2 will then displace the BBS equilibrium (Figure 5·2, page 191). Accumulation of CO_2 results in an increased $[H^+]$ (respiratory acidosis). A fall in P_aco_2 displaces the equilibrium and results in a fall in the $[H^+]$ (respiratory alkalosis).

In chronic respiratory acidosis, an increase in the indirect reabsorption of HCO_3^- by the proximal convoluted tubule permits an increase in the plasma $[HCO_3^-]$; in contrast, a fall in the plasma $[HCO_3^-]$ occurs during chronic respiratory alkalosis. Thus, chronic respiratory acid-base disturbances have a different steady-state plasma $[HCO_3^-]$, and hence $[H^+]$, than do the acute respiratory acid-base disorders. It is therefore important for the clinician to clarify, on clinical grounds, whether the acid-base disturbance is acute or chronic in origin.

Buffering of H^+ in Respiratory Acidosis

BUFFERING IN THE ICF

The BBS is the major buffer in the ECF, and its effectiveness is compromised with a defect in CO_2 removal (it cannot adjust the $[CO_2]$, as shown in Figure 5·2).

The ICF buffers consist of approximately equal proportions of bicarbonate and nonbicarbonate buffers; only the latter are effective buffers of H_2CO_3. A rise in P_aco_2 leads to binding of H^+ in the ICF and to formation of HCO_3^- (in the equations below, B^o represents the histidine buffers in intracellular proteins; see pages 17–22 for complete discussion of ECF and ICF buffers). When H_2CO_3 is buffered in cells, HCO_3^- are formed in the ICF. Because some HCO_3^- are exported into the ECF, a small rise in the $[HCO_3^-]$ occurs in the ECF.

$$CO_2 + H_2O \longleftrightarrow H^+ + HCO_3^-$$

$$H^+ + B^o \longleftrightarrow H \bullet B^+$$

Sum (equations above): $$CO_2 + H_2O + B^o \longleftrightarrow H \bullet B^+ + HCO_3^-$$

QUANTITATIVE ANALYSIS

Given that the ICF buffer capacity is large, the quantity of H^+ buffered in the ICF depends on the rise in the $[H^+]$ in cells, which is proportional to the rise in the tissue Pco_2. Therefore, the rise in the $[HCO_3^-]$ in the ICF (and hence in the ECF) that accompanies a rise in Pco_2 is proportional to the relative increment in Pco_2 rather than the absolute increment. In the example in the margin, with different original values for the Pco_2, a 20 mm Hg increase in Pco_2 has a very different impact on the $[H^+]$ in cells; therefore, the larger the percent change in Pco_2, the larger the generation of HCO_3^- from nonbicarbonate buffers in cells.

Example:
Consider a 20 mm Hg rise in Pco_2 under two circumstances:

1. If the Pco_2 rises from 20 to 40 mm Hg, there is a 100% change.
2. If the Pco_2 rises from 60 to 80 mm Hg, there is a 33% change.

EMPIRIC OBSERVATIONS

The patient with acute retention of CO_2 has acidemia, an elevated P_aco_2, and a slight rise in the plasma $[HCO_3^-]$; in contrast, the rise in plasma $[HCO_3^-]$ is greater and the rise in $[H^+]$ is smaller in chronic respiratory acid-base disorders (Figure 5·4).

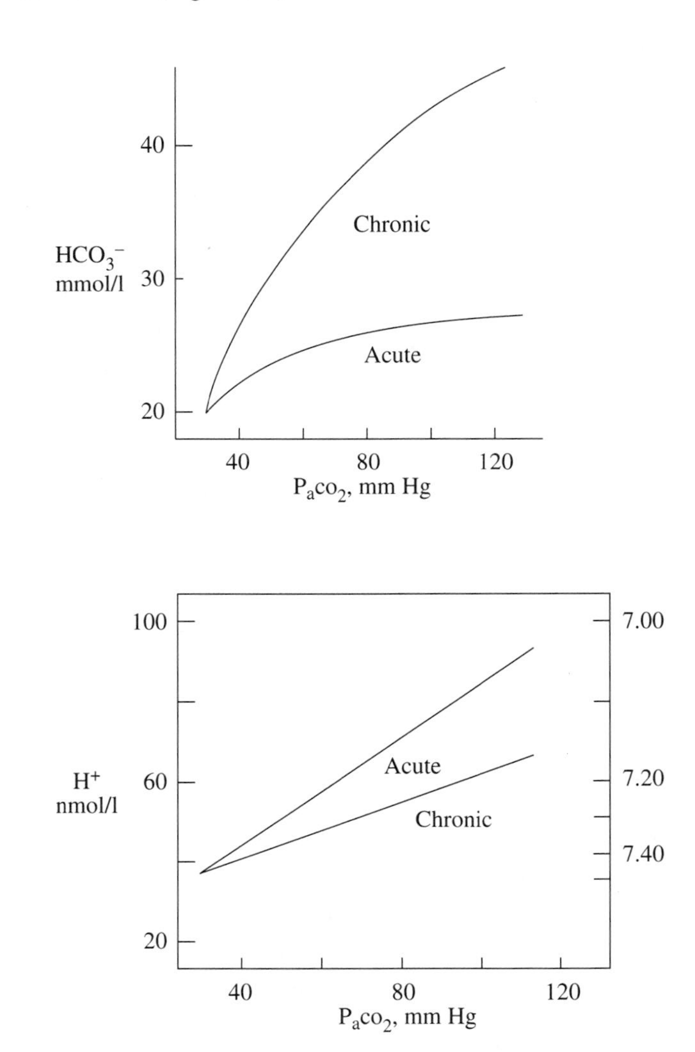

Figure 5·4
Empiric relationship between the P_aco_2 and the $[HCO_3^-]$ or $[H^+]$ in plasma
Note the different relationships with acute vs chronic changes in P_aco_2. The flatter slope of the H^+ vs Pco_2 regression line in chronic respiratory acidosis reflects the renal generation and retention of HCO_3^-.

PART B
RESPIRATORY ACIDOSIS

Clinical Approach

Respiratory acidosis is the result of alveolar hypoventilation. The clinician's problem is to establish the basis of the hypoventilation. Patients who hypoventilate can be divided into two groups: those who will not breathe (defective stimulus) and those who cannot breathe (defective equipment). In addition, patients with a fixed alveolar ventilation (i.e., those on ventilators) develop increased P_aco_2 if they have an increased rate of production of CO_2 or an increase in dead space (e.g., pulmonary embolus).

PATIENTS WHO WILL NOT BREATHE BECAUSE OF DEFECTS IN NEUROLOGICAL INPUT

Some patients may hypoventilate because they lack the normal respiratory neurologic input at one of three levels: first, cerebral nonrhythmic neurons; second, brain stem rhythmic function; third, upper airway reflexes (Table 5·3).

Table 5·3
PATIENTS WHO WILL NOT BREATHE

Category	Cause	Diagnostic test results
Cerebral nonrhythmic neurons	• Post-hypoxic brain damage • Cerebral trauma • Intracranial disease • Psychotropic drugs	• Inability to cough, talk, or hold breath
Brain stem rhythmic function	• Brain stem herniation • Encephalitis • Central sleep apnea • Metabolic alkalosis (severe) • Drugs (sedatives, narcotics)	• Abnormal ventilatory response to changes in inhaled CO_2 and O_2
Upper airway reflexes	• Bulbar palsy • Anterior horn cell lesions • Disruption of airway	• Inability to swallow • Absent nasal, tracheal, and pharyngeal reflexes

PATIENTS WHO CANNOT BREATHE BECAUSE OF RESPIRATORY OR MUSCULAR DISORDERS

The causes of inability to breathe fall into three categories (Table 5·4): first, primary muscle disorders; second, increased elastic work (restrictive disease); and third, increased resistance to flow (obstructive disease).

Table 5·4
PATIENTS WHO CANNOT BREATHE

Category	Cause	Diagnostic test results
Respiratory muscle disease	• Myasthenia • Relaxant drugs • Muscular dystrophy • Muscle fatigue or paralysis	• Reduced maximum inspiratory and expiratory pressures
Increased elastic work	• Interstitial lung disease • Pulmonary fibrosis	• Reduced lung volumes • Reduced lung compliance • Normal flow rates
Increased resistance to flow	• Asthma • Bronchitis • Emphysema • Upper airway obstruction	• Reduced flow rates • Increased airway resistance • Increased lung compliance • Abnormal flow-volume loop

Acid-Base Aspects

Respiratory acidosis occurs when ventilation transiently fails to remove the CO_2 produced by normal metabolism. As a result, the alveolar Pco_2 rises and increases the P_aco_2. At this new arterial and alveolar Pco_2, the CO_2 produced can now be removed despite the reduced ventilation.

Clinical pearl:
Make the diagnosis of acute or chronic disease on a clinical basis, not on laboratory values.

Diagnostic Approach

The diagnostic approach to respiratory acidosis is outlined in Figure 5·5. First, decide if the patient has chronic lung disease by the history, physical exam, and available past records. Then, compare the acid-base status with that expected for that acid-base disorder. If a discrepancy exists, a mixed disorder is present.

In acute and chronic respiratory acidosis in patients who were previously normal, an empirical linear relationship has been found between the $[H^+]$ and the P_aco_2 (see Figure 5·4 and the margin). In patients who do not begin with a normal acid-base state (e.g., they have coexisting metabolic acidosis or alkalosis), the plasma $[HCO_3^-]$ rises 2.5 mmol/l for a twofold change in P_aco_2.

Acute respiratory acidosis:
For every mm Hg increase in P_aco_2 from 40 mm Hg, expect a 0.8 nmol/l increase in $[H^+]$ from 40 nmol/l.

Chronic respiratory acidosis:
For every mm Hg increase in P_aco_2 from 40 mm Hg, expect a 0.3 nmol/l increase in $[H^+]$; for every mm Hg increase in P_aco_2, expect a 0.3 mmol/l increase in the plasma $[HCO_3^-]$.

Questions

(Discussions on pages 212–14)

5·10 When cold-blooded animals live at 37°C, their P_aco_2 is close to 40 mm Hg and their pH is close to 7.4. At colder temperatures, their pH rises, and their $[HCO_3^-]$ in plasma remains constant. What might these values imply for regulation of ventilation and pH?

Note:
Question 5·10 is for the more curious.

5·11 Three patients all experience an acute increase in P_aco_2 to 80 mm Hg. Each has a different $[H^+]$. What is the acid-base status in each case?

Patient	$[H^+]$	pH
A	96 nmol/l	7.02
B	70 nmol/l	7.25
C	50 nmol/l	7.30

5·12 A patient with chronic stable obstructive lung disease (P_aco_2 = 60 mm Hg, plasma $[HCO_3^-]$ = 31 mmol/l) developed shortness of breath and was admitted with a diagnosis of congestive heart failure. During treatment with diuretics and O_2, he vomited a few times, but his chest X-ray showed improvement. The following blood gases were obtained after therapy: $[H^+]$ = 38 nmol/l, P_aco_2 = 80 mm Hg, P_ao_2 = 70 mm Hg. What is the basis of the increased CO_2 retention? What therapy would be appropriate?

Figure 5·5
Diagnostic approach to respiratory acidosis
In a patient with an elevated P_aco_2, if there is no evidence of chronic lung disease or a chronic central reason for hypoventilation, assume that the patient has acute respiratory acidosis. The final diagnoses are shown in the shaded boxes.

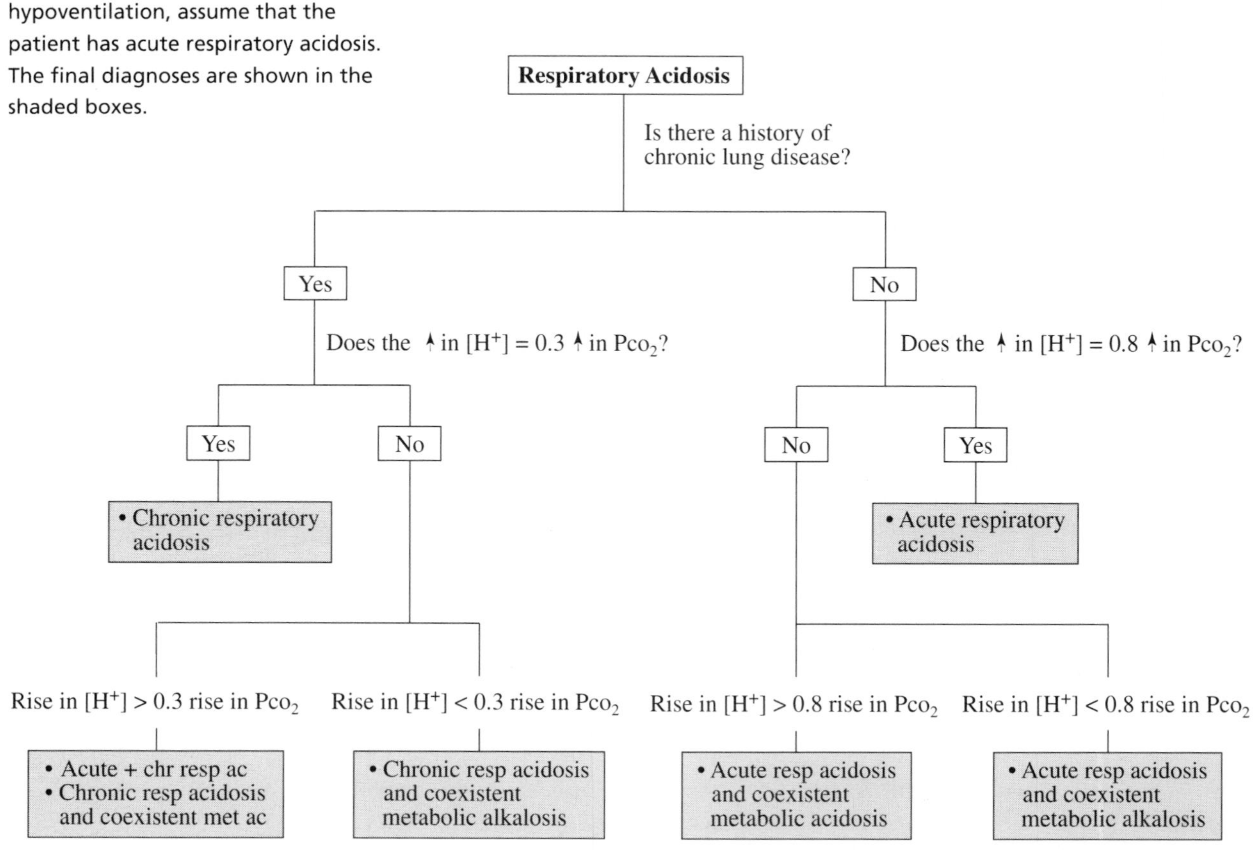

PART C
RESPIRATORY ALKALOSIS

Clinical Approach

Respiratory alkalosis is a common abnormality that is often ignored. Its mortality rate in the hospital, which may well be greater than that for respiratory acidosis, reflects the importance of the underlying disease process. Hypocapnia is often difficult to recognize clinically, and the diagnosis is often made only by determination of blood gases.

Respiratory alkalosis occurs when the ventilatory removal of CO_2 transiently exceeds its rate of production: thus, the alveolar and arterial Pco_2 fall. At this lower level of P_aco_2 , the daily production of CO_2 is then removed by the increased ventilation, which leads to a new steady state.

A fall in tissue Pco_2 has an important impact on the $[H^+]$ in the ICF in moderate metabolic acidosis. The associated decrease in $[H^+]$ results in back-titration of the protonated ICF proteins and shifts the majority of the ICF buffering to HCO_3^-. This action may preserve the structure and function of the ICF proteins. In the alkalemic patient, hyperventilation is likely to make these intracellular proteins less positively charged than normal, a change that could lead to altered function.

Respiratory alkalosis may result from stimulation of the peripheral chemoreceptors (hypoxia or hypotension), the afferent pulmonary reflexes (intrinsic pulmonary disease), or central stimulation by a host of stimuli (Table 5·5).

Table 5·5
CAUSES OF RESPIRATORY ALKALOSIS

Hypoxia
Intrinsic pulmonary disease, high altitude, congestive heart failure, congenital heart disease (cyanotic)
Pulmonary receptor stimulation
Pneumonia, pulmonary embolism, asthma, pulmonary fibrosis, pulmonary edema
Drugs
Salicylates (the most common), nikethamide, catecholamines, theophylline, progesterone
CNS disorders
Subarachnoid hemorrhage, Cheyne-Stokes respiration, primary hyperventilation syndrome
Miscellaneous
Psychogenic hyperventilation, cirrhosis, fever, gram-negative sepsis, recovery from metabolic acidosis, pregnancy

Acid-Base Aspects

The acid-base impact of respiratory alkalosis is analogous to respiratory acidosis in that the acute and chronic states differ because of the role of the kidneys kidneys in altering the concentration of HCO_3^- in the body.

ACUTE RESPIRATORY ALKALOSIS

Acute respiratory alkalosis:
For every mm Hg reduction in P_aco_2 from 40 mm Hg, expect a 0.8 nmol/l decrease in $[H^+]$ from 40 nmol/l.

Chronic respiratory alkalosis:
For every mm Hg reduction in P_aco_2 from 40 mm Hg, expect a 0.2 nmol/l fall in $[H^+]$ from 40 nmol/l, and a 0.5 mmol/l fall in $[HCO_3^-]$ from 25 mmol/l.

In *acute respiratory alkalosis*, displacement of the bicarbonate buffer system to the right leads to a reduction in both the $[H^+]$ and the $[HCO_3^-]$ (Figure 5·2, page 191). The impact on the $[H^+]$ in the ECF in acute respiratory alkalosis is virtually identical to that in acute respiratory acidosis (although opposite in direction).

CHRONIC RESPIRATORY ALKALOSIS

In *chronic respiratory alkalosis*, there is a temporary small suppression of renal NH_4^+ production and excretion, and the $[HCO_3^-]$ in the ECF falls (H^+ of dietary origin continue to consume HCO_3^- without equivalent renal formation of new HCO_3^-) until the plasma $[H^+]$ approaches normal. This acid-base disorder is the only one in which a normal plasma $[H^+]$ might be expected.

Diagnostic Approach

The diagnostic approach to respiratory alkalosis is detailed in Figure 5·6. One begins by deciding on clinical grounds whether or not there is a disease process associated with chronic respiratory alkalosis; if not, the patient is presumed to have acute respiratory alkalosis.

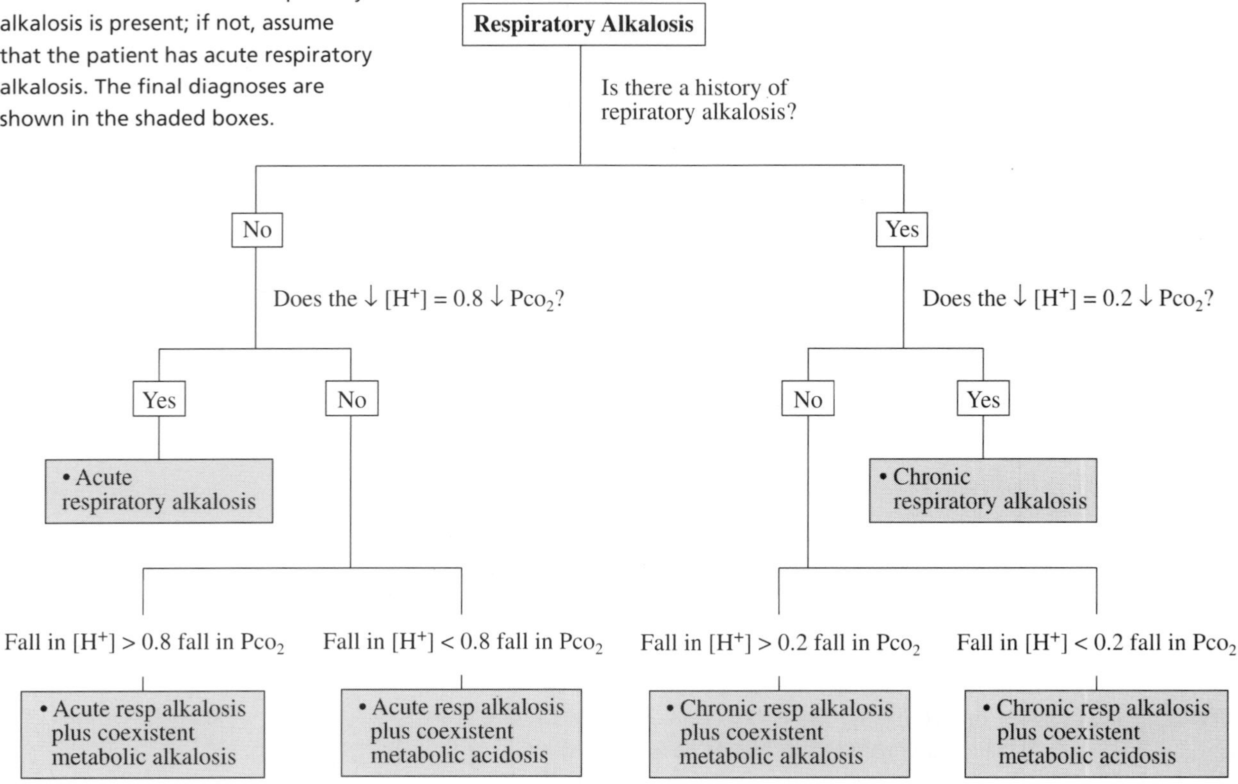

Figure 5·6
Diagnostic approach to respiratory alkalosis
In a patient with a reduced P_aco_2, decide whether a disease process that is associated with chronic respiratory alkalosis is present; if not, assume that the patient has acute respiratory alkalosis. The final diagnoses are shown in the shaded boxes.

PART D
P_aco_2 IN METABOLIC ACID-BASE DISORDERS

The normal P_aco_2 at a $[HCO_3^-]$ of 25 mmol/l is 40 mm Hg; in contrast, the expected P_aco_2 in metabolic acidosis at a plasma $[HCO_3^-]$ of 10 mmol/l is 25 rather than 40 mm Hg (see page 57). Therefore, at a $[HCO_3^-]$ of 10 mmol/l, a patient with a P_aco_2 of 40 mm Hg has respiratory acidosis in addition to metabolic acidosis.

A failure to hyperventilate appropriately in metabolic acidosis has a dramatic impact on the plasma $[H^+]$. In the example above, the patient has a $[H^+]$ of 60 nmol/l (pH 7.22) with a P_aco_2 of 25 (see the margin); with a P_aco_2 of 40 mm Hg, the $[H^+]$ is 96 nmol/l (pH 7.02). The impact on the ICF buffering, however, may be much more important both physiologically and clinically than the impact on the plasma $[H^+]$. Hyperventilation in metabolic acidosis transfers the H^+ burden from the ICF protein buffers to the BBS (see pages 19–21 for more details). Because ICF buffering on histidine may change the structure and function of ICF proteins, it is most desirable for the BBS to buffer as great a quantity of H^+ as possible.

Calculations:

$$[H^+] = \frac{24}{[HCO_3^-]} \times Pco_2$$

$$60 = \frac{24}{10} \times 25$$

$$96 = \frac{24}{10} \times 40$$

Patients with chronic obstructive pulmonary disease are often on diuretics and may have a coexistent metabolic alkalosis. Because H^+ stimulate ventilation, a lower $[H^+]$ can make the hypoventilation more severe. Interestingly, correction of the metabolic alkalosis in these patients does not result in a large change in their $[H^+]$; instead, there is a significant fall in both the $[HCO_3^-]$ and P_aco_2, coupled with an increase in P_ao_2. These changes are associated with a dramatic clinical improvement. It is tempting to speculate that the clinical improvement is due, in part, to the reduction in H^+ buffering on the ICF proteins.

Net charge on ICF proteins in metabolic alkalosis:
The $[H^+]$ in the ICF may be elevated as a result of hypokalemia, a higher P_aco_2, and the lower blood flow rate (higher venous Pco_2). If, in addition, respiratory acidosis is present, these proteins could bear an even more positive net charge.

Questions

(Discussions on pages 214–15)

5·13 A 30-year-old businessman had just returned from Europe when he suddenly developed a severe left-sided pleuritic chest pain and hemoptysis. He had no history of chest disease and exercised regularly. He was cyanotic, he had an elevated jugular venous pressure (8 cm above the sternal angle), and his blood pressure was 80/50 mm Hg. The blood gases were $[H^+]$ = 40 nmol/l (pH 7.40), P_aco_2 = 25 mm Hg, and P_ao_2 = 50 mm Hg. What is the most likely diagnosis?

5·14 A patient with cirrhosis of the liver was found in a confused state by his landlady. His physical examination was normal except for a low blood pressure and the stigmata of chronic liver disease. His laboratory results were as follows:

Na^+	mmol/l	133	H^+	nmol/l	36	pH 7.44
K^+	mmol/l	3.3	P_aco_2	mm Hg	20	
Cl^-	mmol/l	115	HCO_3^-	mmol/l	13	

The initial diagnosis was L-lactic acidosis secondary to severe hepatic insufficiency. Is this diagnosis appropriate? If not, why, and what is the most likely diagnosis?

PART E REVIEW

Discussion of Introductory Case Hack's Future Is Up in Smoke

(Case presented on page 186)

What prompted the changes from the usual values to those seen on admission?

Note:

$$[H^+] = \frac{24}{[HCO_3^-]} \times Pco_2$$

$$46 = \frac{24}{[HCO_3^-]} \times 60$$

Hack's usual plasma $[HCO_3^-]$ can be calculated as shown in the margin. His usual $[H^+]$ in simple chronic respiratory acidosis with a P_aco_2 of 60 mm Hg is 46 mmol/l because the rise in Pco_2 of 20 mm Hg × 0.3 results in the addition of 6 nmol/l to the normal value of H^+ (40 nmol/l).

When Hack was admitted, he had an acute increase in P_aco_2 to 80 mm Hg with profound hypoxia. The $[HCO_3^-]$, calculated as above, was still 31 mmol/l. Therefore, the severe acidemia was largely due to the presence of acute respiratory acidosis combined with chronic respiratory acidosis. There may have been an added element of metabolic acidosis that could be confirmed by evaluating the plasma anion gap. The acute respiratory acidosis was due to pneumonia.

What happened to permit the changes in the blood gases while Hack was on the ventilator on day two of therapy?

While Hack was ventilated on day two of therapy, his acid-base values returned to normal: $[H^+]$ = 40 nmol/l, $[HCO_3^-]$ = 24 mmol/l, P_aco_2 = 40 mm Hg. Although one can achieve a normal acid-base status with artificial ventilation, we know that Hack has severe lung disease and will not be able to maintain a P_aco_2 of 40 mm Hg once he is extubated. Therefore, it is inappropriate to ventilate him to this level; he should be maintained with a P_aco_2 of 60 mm Hg (his chronic steady-state value) to preserve his renal adaptation to chronic respiratory acidosis. In this case, his kidneys were "fooled" into thinking that his lungs were normal. He therefore excreted the extra HCO_3^-, and his plasma $[HCO_3^-]$ fell to 24 mmol/l instead of the 31 mmol/l that was present prior to the acute illness.

Why are his post-extubation values on day four of therapy so different from his original values?

With extubation on day four of therapy, Hack went from a normal acid-base state to acute respiratory acidosis with a P_aco_2 of 70 mm Hg. The $[H^+]$ was appropriate for acute respiratory acidosis: the rise in $[H^+]$ was 30 mm Hg × 0.8, or 24 nmol/l, which is a $[H^+]$ of 64 nmol/l. His plasma $[HCO_3^-]$ was 24 × 70/64, or 26 mmol/l, which rose appropriately for acute respiratory acidosis.

On day four of therapy, his P_aco_2 was slightly higher than his chronic steady state (70 vs 60 mm Hg). This increase may represent an incomplete resolution of the pneumonia or may indicate permanent destruction of lung tissue. Time will tell.

Cases for Review

Case 5·1
Annie-Abigale Has Chronic Lung Disease

(Case discussed on pages 206–07)

An 87-year-old lady with chronic obstructive lung disease and congestive heart failure with marked edema was treated with a diuretic and lost 5 liters of ECF. Her mental state deteriorated following treatment. On physical exam, she was obtunded; she had all her previous lung findings except that her edema disappeared. A summary of results is provided below. Note that she developed an extremely high P_aco_2 breathing room air. The diuresis caused little rise in net acid excretion (data not shown).

Parameter		Steady state	Post-diuretic
H^+	nmol/l	51	44
pH		7.29	7.34
P_aco_2	mm Hg	60	85
P_ao_2	mm Hg	50	39
A-a	mm Hg	25	5
HCO_3^-	mmol/l	28	44
K^+	mmol/l	4.1	3.1
Anion gap	mEq/l	11	16
Creatinine	μmol/l (mg/dl)	57 (0.7)	66 (0.8)

What are the acid-base disorders after diuresis?
Why did the $[HCO_3^-]$ and P_aco_2 rise so markedly?
Why did the A-a difference fall?
What should the therapy be?

Case 5·2
Is Doreen a "Blowhard"?

(Case discussed on pages 207–08)

Doreen, aged 84, was transferred from a nursing home for evaluation of chest pain. She has Alzheimer's disease and end-stage renal disease that requires chronic ambulatory peritoneal dialysis (CAPD) nd the oral intake of calcium carbonate. She does not take other drugs. All investigations, including her physical examination, have been negative so far. She is afebrile and there is no evidence of deep and/or extremely rapid respirations.

Laboratory results reveal an incidental finding of an extreme degree of alkalemia (pH 7.70). The remainder of the pertinent values are shown below.

$[H^+]$	nmol/l	20	pH		7.70
HCO_3^-	mmol/l	26	Anion gap	mEq/l	14
P_aCO_2	mm Hg	22	Albumin	g/l	32
P_aO_2	mm Hg	107			

What is (are) the acid-base disorder(s)?
What is the basis of the acid-base disorder(s)?
What other investigations are in order to clarify the basis of respiratory alkalosis?

Discussion of Cases

Discussion of Case 5·1
Annie-Abigale Has Chronic Lung Disease

(Case presented on page 205)

What are the acid-base disorders after diuresis?

The acid-base disturbance before therapy was chronic respiratory acidosis. Two major changes occurred with diuresis: there was a large rise in the $[HCO_3^-]$ in plasma (metabolic alkalosis) and in the P_aco_2 (coexistent acute and chronic respiratory acidosis). There could have been a minor component of metabolic acidosis as well because the plasma anion gap also rose.

Why did the $[HCO_3^-]$ and P_aco_2 rise so markedly?

The elevation in the $[HCO_3^-]$ was not due to intake of the HCO_3^- or excretion of net acid. Hence, it may have been due to an internal shift of H^+ when K^+ exited from cells (minor) or to a loss of ECF volume, (i.e., the same content of HCO_3^- in a smaller ECF volume, a major effect that causes a "contraction" metabolic alkalosis; see the margin).

The rise in P_aco_2 would have also lead to a modest rise in the $[HCO_3^-]$ (2.5 mmol/l per doubling of the P_aco_2). This rise in P_aco_2 was probably the result of suppressed respiration caused by reduction in the stimulus of acidemia and was possibly also due to weakness of respiratory muscles (hypokalemia).

Calculation:

- Assume her normal ECF volume is 12 liters.
- She had 5 liters of edema fluid that was lost, so her original ECF volume was 17 liters.
- Her initial content of HCO_3^- was 476 mmol (17 liters × 28 mmol of HCO_3^-).
- After the diuresis, the 476 mmol of HCO_3^- would have dissolved in 12 liters and her new $[HCO_3^-]$ would have been 40 mmol/l.

Why did the A-a difference fall?

The major problem here is the use of the Po_2 scale vs the oxygen content (or saturation of hemoglobin) scale. The difference between alveolar and arterial Po_2 was greater before therapy (75 to 50 mm Hg) than after diuresis (44 to 39 mm Hg)—a misleading A-a difference. If one recalculates the A-a difference values using O_2 content rather than Po_2, they are 8.4 to 7.2 mmol/l (1.2 mmol/l difference) prior to therapy and 6.7 to 5.8 mmol/l (0.9 mmol/l difference) after therapy. Therefore, because of the shape of the oxygen-hemoglobin dissociation curve, what appeared to be a large difference between the pre- and post-treatment values when the Po_2 was measured is actually very small when the O_2 content is used.

What should the therapy be?

The aim of therapy is to lower the P_aco_2 and $[HCO_3^-]$ in arterial blood. The first step is to replace the deficit of K^+ to see if ventilation will improve (it did not).

The next step is to administer 250 mg acetazolamide, a carbonic anhydrase inhibitor diuretic, to induce a degree of bicarbonaturia; losses of K^+ in the urine should be replaced. With bicarbonaturia, the Pco_2 and $[HCO_3^-]$ in arterial blood returned to her steady-state values within 24 hours. The patient then felt better.

Another option would have been to administer HCl (NH_4Cl) to titrate HCO_3^-, but, because CO_2 production will rise transiently, this mode of therapy is possibly less desirable.

Discussion of Case 5·2 Is Doreen a "Blowhard"?

(Case presented on page 206)

What is (are) the acid-base disorder(s)?

The pH of blood is very alkaline and the P_aco_2 is very low, so the most important acid-base diagnosis is respiratory alkalosis. Her plasma $[HCO_3^-]$ is higher than one might expect for acute or chronic respiratory alkalosis, so she has a coexistent metabolic alkalosis.

What is the basis of the acid-base disorder(s)?

1. Respiratory alkalosis may have resulted from increased ventilation and a low production of CO_2.

   Increased ventilation: A low P_aco_2 indicates an abnormal drive to respiration. Organic diseases to consider are pulmonary diseases (pulmonary embolism is a good bet, but any intrinsic pulmonary, CNS , or liver disease is possible). The A-a difference is slightly increased (15 mm Hg, see the margin) but is in keeping with her advanced age.

   The other considerations are her anxiousness about her health and the possibility of "occult" sepsis, brain disease, or lung disease. Her plasma levels of Ca^{2+}, Mg^{2+}, phosphate, Na^+, and glucose are all normal, so a specific lesion is not identified.

Calculation:
A-a Po_2 = 150 – (107 + (22/0.8))
= 15 mm Hg

Low production of CO_2: The quandary for the clinician is that she has a very low P_aco_2, yet shows little clinical evidence of an extreme degree of hyperventilation. The next step is to determine if she has a low rate of production of CO_2.

On a general basis, she is not hypothermic, so hypothyroidism should be considered. With respect to specific organs, her Alzheimer's disease may decrease CO_2 production in her brain. Her renal failure limits CO_2 production by her kidneys. She is very inactive (lies in bed), so her muscles produce little CO_2. Therefore, one basis for her low P_aco_2 is low production of CO_2.

Summary: Her respiratory alkalosis is very severe because she has a pathological drive to ventilation in the setting of very low production of CO_2.

2. Her metabolic alkalosis probably reflects the intake of alkali (calcium carbonate) in the presence of renal failure, as well as the metabolism of the D- and L-lactate anions from the CAPD fluid.

What additional investigations are in order to clarify the basis of the respiratory alkalosis?

First, one should seek a possible basis of the abnormal drive to respiration (e.g., if sepsis and pulmonary embolism are present). She should also be investigated for occult hypothyroidism. Confirm that her rate of production of CO_2 is really low by measuring her rates of O_2 consumption and CO_2 production.

Summary of Main Points

- Alterations in P_aco_2 have a major impact on the BBS; a P_aco_2 that is too high renders the BBS ineffective and increases buffering by intracellular proteins (the converse applies to a low P_aco_2). Both may have significant clinical impact.
- Examining the arterial blood Pco_2 permits one to assess the ventilation function of the lungs; the venous Pco_2 reflects the tissue Pco_2 and hence provides information about the metabolic rate, blood flow rate, and/or buffering of a H^+ load.
- In chronic respiratory acidosis and alkalosis, the kidneys return the $[H^+]$ towards normal by transiently increasing or decreasing the rate of NH_4^+ excretion and thereby increasing the plasma $[HCO_3^-]$ in chronic respiratory acidosis or decreasing the plasma $[HCO_3^-]$ in chronic respiratory alkalosis.
- In both acute and chronic respiratory acidosis and alkalosis, there are predictable physiological responses in the $[H^+]$ and $[HCO_3^-]$ that enable the clinical detection of mixed acid-base disorders.
- Metabolic alkalosis may coexist with chronic respiratory acidosis (because of the use of diuretics) and may have a significant deleterious impact on the clinical status of the patient.

Discussion of Questions

5·1 Assume that the consumption of O_2 is 12 mmol/min, that alveolar ventilation is 5 liters/min, and that 21% of air is O_2.

Why is the Po_2 of alveolar air 100 mm Hg and not 150 mm Hg, as it is in room air?

The usual Po_2 in humidified inspired air is (760 – 47) mm Hg × 0.21, or 150 mm Hg. In the alveoli, nitrogen gas is still 79% of total air, but CO_2 occupies some of the space left for O_2. If the RQ is 0.8 (see the margin), a Pco_2 of 40 mm Hg is derived from 50 mm Hg O_2. Therefore, alveolar air has two-thirds of the content of O_2 vs inspired air (100 vs 150 mm Hg).

Note:
RQ = CO_2 produced/O_2 consumed. Therefore, with an RQ of 0.8, 40 mm Hg of CO_2 are derived from 50 mm Hg of O_2.

How many mmol of O_2 are in one liter of alveolar air?

Calculate the mmol of O_2 provided by 100 mm Hg Po_2:

- One liter of air contains 210 ml of O_2 (21%), but alveolar air has two-thirds of this volume (or concentration) of O_2 (Po_2 is 100 vs 150 mm Hg). Therefore, alveolar air has 140 ml of O_2 (2/3 × 210 ml).

- For every 22.4 ml of gas at standard temperature and pressure, there is 1 mmol of that gas. Accordingly, one liter of air contains close to 9 mmol of O_2, and one liter of alveolar air contains close to 6 mmol of O_2.

What percentage of the O_2 delivered to the alveoli is extracted?

Since alveolar ventilation is 5 liters/min, close to 45 mmol of O_2 enter alveoli each minute, but only 12 mmol are "extracted." Hence, there is a large reserve, and 27% of O_2 ($100 \times 12/45$) is extracted.

5·2 Fish exchange gases through gills by taking up the O_2 dissolved in water and adding CO_2 to that water. CO_2 is 30-fold (for easy math) more soluble in water than O_2. What does this solubility imply for the P_aco_2 and the pH of blood in fish in steady state?

If the concentrations of O_2 and CO_2 are equal in the capillary blood of the gills, the solution bathing the gills (lake water) will wash away 30-fold more CO_2 than the amount of O_2 that it will supply. Accordingly, since the amounts of CO_2 produced and O_2 consumed are roughly equal (RQ = 1), the $[CO_2]$ must be 1/30 that of O_2. This relationship requires a very low $[CO_2]$ (Pco_2) in the arterial blood of fish (values close to several mm Hg).

Henderson equation:

$$[H^+] = \frac{24}{[HCO_3^-]} \times Pco_2$$

Given the P_aco_2, either the fish will have a very alkaline blood (Henderson equation), or a very low $[HCO_3^-]$ in plasma. The fish usually elects to have a very low $[HCO_3^-]$. In acid-base terms, it does not have an important bicarbonate buffer system.

5·3 By what proportion must alveolar ventilatory capacity decline to have a P_aco_2 of 50 vs the expected 40 mm Hg?

A simple answer, but one we believe to be less correct, is to assume that there is only a 20% reduction in alveolar ventilation (5 – 1) liters/min/5 liters/min. Consider the following analogy: normal individuals excrete 40 mmol of NH_4^+ per day but can increase this excretion to 200 mmol/day with chronic metabolic acidosis. If, during chronic metabolic acidosis, the kidneys excrete only 30 mmol of NH_4^+ per day, a loss of more than 80% of the capacity to excrete NH_4^+ has occurred.

Note:
- CO_2 excretion can be 160 mmol/min.
- Each liter of alveolar air contains 2 mmol of CO_2 (at a Pco_2 of 40 mm Hg).
- Therefore, alveolar ventilation was 80 liters/min.

Because a rise in P_aco_2 should stimulate alveolar ventilation, we suggest that in our analogy the entire capacity for enhanced alveolar ventilation has been inhibited. If the expected alveolar ventilation was as high as possible (80 liters/min; see the margin and Table 5·1, page 188), the percent decline in potential alveolar ventilation was (160 – 4) liters/min/160 liters/min, or close to 97.5%. Thus, the answer depends on what you think the denominator of the observed vs expected alveolar ventilation should be.

5·4 Why is the venous Pco_2 much higher than the arterial Pco_2 in each of the following: exercise, a convulsion, and DKA?

Exercise: In essence, because the athlete is hyperventilating, the arterial Pco_2 is not high. The muscle mass, which is actively burning glucose and glycogen, generates CO_2 to meet the need to regenerate ATP. Therefore, the Pco_2 in tissues is high; because this CO_2 must diffuse into venous blood, the Pco_2 remains very high in venous blood.

Convulsion: With a convulsion, there is vigorous muscle activity associated with insufficient O_2 to meet the body's needs. Anaerobic glycolysis ensues and generates L-lactic acid. Burning glucose to L-lactate anions generates ATP and H^+ without consuming O_2. Hence, the production of CO_2 is very high from both aerobic and anaerobic metabolism and results in a high venous Pco_2.

DKA = diabetic ketoacidosis

DKA: In DKA, the arterial Pco_2 is low because of the hyperventilation that results from the acidemia. At the tissue level, the flow of blood is very slow because of the contracted ECF volume, so each liter must carry more CO_2. The venous Pco_2 can therefore be very high.

Note:
Ventilatory alkalosis (a very low P_aco_2) and respiratory acidosis (high tissue Pco_2) can occur simultaneously in DKA..

5·5 Why might the rate of production of CO_2 be lower than normal in a patient with DKA?

Patients with ketoacidosis have a lower rate of CO_2 production. Their use of fatty acids as a fuel produces less CO_2/ATP than would the use of proteins or carbohydrates (Table 5·2, page 188). Also, ketogenesis from fatty acids is not a CO_2-producing reaction. In addition, renal and cerebral ATP needs may be markedly reduced (decreased GFR and coma), so the kidneys and the brain may produce much less CO_2 than normal.

5·6 How might peripheral chemoreceptors detect a lower P_ao_2?

Note:
Question 5·6 is for the more curious.

The detection of P_ao_2 is an important function of the carotid body. We find the following hypothesis attractive: to detect a lower P_ao_2, unique cells in the carotid body have a change in their mitochondria. The last step in the electron transport system (the conversion of O_2 to H_2O) is the rate-limiting step in the regeneration of ATP; it is dependent on the activity of cytochrome oxidase. This enzyme in the carotid body has a unique property, a low affinity for O_2 (in all other cells, this enzyme has a very high affinity for O_2 because it is saturated with O_2 at very low Po_2 values). When the P_ao_2 falls below 70 mm Hg or so, cytochrome oxidase in these unique cells is no longer saturated with O_2. As a result, less ATP is regenerated, and Ca^{2+} leak out of their mitochondria and signal a drive to ventilation (see the margin).

Another hypothesis:
Cells of the carotid body have a K^+ channel that is regulated by the Po_2. Hence, a low Po_2 can induce an electrical signal to indicate a low P_ao_2.

5·7 A patient has a 0.1 liter/min shunt from pulmonary artery to pulmonary vein. Would the A-a difference be different if the cardiac output was 5 vs 2 liters per minute? (Assume no other abnormality.)

Yes. Since the volume of the shunt is 0.1 liter/min, mixing it with 5 liters of cardiac output would create a shunt that is 2% by volume. In contrast, with a very low cardiac output of 2 liters/min, the same shunt would be 5% by volume. The net effect of the A-a difference would be much larger with the lower cardiac output.

5·8 A patient, comatose from a drug overdose, has the following blood gases. The basis of the hypoxia is thought to be aspiration. Is this diagnosis correct?

pH		7.24	P_{CO_2}	mm Hg	64
H^+	nmol/l	58	P_{aO_2}	mm Hg	66

Note:
Multiplying by 1.25 is equivalent to dividing by 0.8.

The P_{O_2} of inspired air is (760 – 47) mm Hg × 0.21, or 150 mm Hg, when 47 mm Hg is the water vapor pressure. The alveolar P_{O_2} is 150 – (1.25 × 66), or 68 mm Hg. The arterial P_{O_2} is 66 mm Hg; therefore, the A-a difference is 2 mm Hg (68 – 66)—a normal value (the A-a difference is low as a result of the sigmoid shape of the oxygen-hemoglobin dissociation curve).

Henderson equation:

$$[H^+] = \frac{24}{[HCO_3^-]} \times P_{CO_2}$$

$$58 = \frac{24}{[HCO_3^-]} \times 64$$

With respect to the acid-base disturbance, the plasma $[HCO_3^-]$ is 24 × 64/58, or 26 mmol/l, and the $[H^+]$ is 58 nmol/l; both are expected values for acute respiratory acidosis. Together, acute CO_2 retention and hypoxia with a normal A-a difference suggest that the patient has decreased ventilation because of central respiratory suppression from drug overdose. Because intrinsic lung disease is not the problem, it is incorrect to postulate aspiration pneumonitis based on these values.

5·9 Two days later the patient in Question 5·8 is awake but coughing. The following blood gases are obtained. Because his P_{aO_2} has improved, he is declared ready for discharge. Is this decision appropriate?

pH		7.60	P_{aCO_2}	mm Hg	24
H^+	nmol/l	25	P_{aO_2}	mm Hg	70

Alveolar P_{O_2} = 150 – (24/0.8)
= 150 – 30
= 120 mm Hg

Measured arterial P_{O_2} = 70 mm Hg

The alveolar P_{O_2} in the second set of blood gases is 150 – 30, or 120 mm Hg. Therefore the A-a difference is 120 – 70, or 50 mm Hg. On this occasion, the patient has a frankly increased A-a difference, which indicates the presence of a disease that impairs O_2 exchange between alveoli and blood. The hypoxia is less evident because of hyperventilation (if the patient had a normal P_{aCO_2} and an A-a difference of 50 mm Hg, the P_{aO_2} would have been 50 mm Hg). Therefore, significant lung disease is now present and its basis must be established. Discharge without a diagnosis is clearly inappropriate.

Note:
Question 5·10 is for the more curious.

5·10 When cold-blooded animals live at 37°C, their P_{aCO_2} is close to 40 mm Hg and their pH is close to 7.4. At colder temperatures, their pH rises, and their $[HCO_3^-]$ in plasma remains constant. What might these values imply for regulation of ventilation and pH?

It appears that warm-blooded species behave as if they choose to defend their pH. It is equally valid to say that they defend the net charge on their proteins (governed by pH at a constant pK of the major buffer, the imidazole group on histidines). When the temperature falls, there is a considerable rise in the pK of the imidazole group (with a lower temperature, these groups behave as if they were weaker acids). Hence, to keep the same net charge on imidazole groups at different temperatures, the pH must change (the pH must rise with decreasing strength of acid (higher pK) to keep the same net charge on imidazoles).

Cold-blooded animals do not defend a specific intracellular pH. Rather, they behave as if they choose to defend the same net charge on their proteins by adjusting ventilation (called the aminostat hypothesis).

5·11 Three patients all experience an acute increase in P_aco_2 to 80 mm Hg. Each has a different $[H^+]$. What is the acid-base status in each case?

Patient	$[H^+]$	pH
A	96 nmol/l	7.02
B	70 nmol/l	7.25
C	50 nmol/l	7.30

The $[HCO_3^-]$ can be calculated in each case from the Henderson equation (see the margin). The $[HCO_3^-]$ for patients A, B, and C are 20, 27, and 38 mmol/l, respectively. Because each patient has acute respiratory acidosis, we would expect the plasma $[HCO_3^-]$ to increase 2.5 mmol/l in association with the acute doubling of the P_aco_2.

In fact, patient A has a $[HCO_3^-]$ that is below normal rather than increased; therefore, patient A has metabolic acidosis in addition to the acute respiratory acidosis.

In patient B, the rise in plasma $[HCO_3^-]$ is close to 2.5 mmol/l, so this case could be simple acute respiratory acidosis.

In patient C, the plasma $[HCO_3^-]$ is 38 mmol/l. Thus, patient C has metabolic alkalosis in addition to the acute respiratory acidosis (see Chapter 4).

Alternatively, we can calculate the expected $[H^+]$ for a patient with an acute increase in blood Pco_2 from 40 to 80 mm Hg on the basis of the $[H^+]$ equalling the original Pco_2 plus ($0.8 \times$ the rise in P_aco_2). The expected $[H^+]$ is 40 + 32, or 72 nmol, and the $[HCO_3^-]$ is $24 \times 80/72$, or 27 mmol/l—virtually identical to the values in patient B. Thus, patient B indeed has blood gas values compatible with acute respiratory acidosis. Both approaches are consistent.

Henderson equation:

$$[H^+] = \frac{24}{[HCO_3^-]} \times Pco_2$$

nmol/l mmol/l mm Hg

5·12 A patient with chronic stable obstructive lung disease (P_aco_2 = 60 mm Hg, plasma $[HCO_3^-]$ = 31 mmol/l) developed shortness of breath and was admitted with a diagnosis of congestive heart failure. During treatment with diuretics and O_2, he vomited a few times, but his chest X-ray showed improvement. The following blood gases were obtained after therapy: $[H^+]$ = 38 nmol/l, P_aco_2 = 80 mm Hg, P_ao_2 = 70 mm Hg. What is the basis of the increased CO_2 retention? What therapy would be appropriate?

The patient's chronic steady-state $[H^+]$ is 24 × 60/30, or 46 nmol/l, which is appropriate for chronic respiratory acidosis (rise in $[H^+]$ = 6 nmol/l = 0.3 × 20). In response to therapy, his P_aco_2 has risen, but his $[H^+]$ has fallen to 38 nmol/l and the $[HCO_3^-]$ is thus 24 × 80/38, or 50 mmol/l. This $[HCO_3^-]$ is much higher than it should be for chronic respiratory acidosis. Thus, the patient has a mixed disorder: chronic respiratory acidosis and metabolic alkalosis. The patient has developed metabolic alkalosis subsequent to the diuretic therapy and vomiting. This alkalosis has resulted in some suppression of ventilation. If we knew his previous A-a difference we could ensure that it had not changed and we could therefore be more certain that all of his CO_2 retention was due to the alkalosis and none was subsequent to aspiration. The O_2 administration has allowed hypoventilation to occur without increasing hypoxia (hypoxia might have provided an additional stimulus to ventilate).

The goal of therapy is to correct the metabolic alkalosis. The patient will require KCl therapy and a careful assessment of his ECF volume. If the diuretic therapy has been excessive and he has developed ECF volume contraction, he will require NaCl therapy; one must be cautious in view of the recent congestive heart failure.

5·13 A 30-year-old businessman had just returned from Europe when he suddenly developed a severe left-sided pleuritic chest pain and hemoptysis. He had no history of chest disease and exercised regularly. He was cyanotic, he had an elevated jugular venous pressure (8 cm above the sternal angle), and his blood pressure was 80/50 mm Hg. The blood gases were $[H^+]$ = 40 nmol/l (pH 7.40), P_aco_2 = 25 mm Hg, and P_ao_2 = 50 mm Hg. What is the most likely diagnosis?

Hemoptysis:
Coughing up "blood."

From the patient's history, there was no obvious chronic stimulus for hyperventilation, so he probably had acute respiratory alkalosis. The plasma $[HCO_3^-]$ was 15 mmol/l (see the margin). Therefore, if the patient had acute respiratory alkalosis with a P_aco_2 of 25 mm Hg, his plasma $[H^+]$ should have been 28 nmol/l (40 – [0.8 × 15]), not 40 nmol/l . His higher plasma $[H^+]$ and lower plasma $[HCO_3^-]$ indicate that he had metabolic acidosis (L-lactic acidosis caused by low oxygen delivery to tissues). Thus, the combination of respiratory alkalosis and metabolic acidosis resulted in a normal plasma $[H^+]$. The L-lactic acidosis would be substantiated by finding an increased anion gap in plasma.

Henderson equation:

$$[H^+] = \frac{24}{[HCO_3^-]} \times Pco_2$$

$$40 = \frac{24}{[HCO_3^-]} \times 25$$

$$[HCO_3^-] = 15 \text{ mmol/l}$$

This patient's history and physical findings strongly suggest a pulmonary embolism, with severe hemodynamic compromise (right-sided heart failure). His blood gases demonstrate hyperventilation and hypoxia and indicate a wide A-a difference (50 mm Hg); these findings are also in keeping with a large shunt resulting from pulmonary embolism.

5·14 A patient with cirrhosis of the liver was found in a confused state by his landlady. His physical examination was normal except for

a low blood pressure and the stigmata of chronic liver disease. His laboratory results were as follows:

Na^+	mmol/l	133	H^+	nmol/l	36	pH 7.44
K^+	mmol/l	3.3	P_aco_2	mm Hg	20	
Cl^-	mmol/l	115	HCO_3^-	mmol/l	13	

The initial diagnosis was L-lactic acidosis secondary to severe hepatic insufficiency. Is this diagnosis appropriate? If not, why, and what is the most likely diagnosis?

While the low [HCO_3^-] might initially suggest the presence of L-lactic acidosis, the anion gap is only 5 mEq/l. If this were L-lactic acidosis, the anion gap would be increased by 10 mEq/l, so the diagnosis is not correct.

Chronic liver disease is associated with chronic respiratory alkalosis. In chronic respiratory alkalosis, for every mm Hg fall in P_aco_2, one should see a 0.2 nmol/l fall in plasma [H^+]. In this case, the plasma [H^+] would be 40 nmol/l – (0.2 × 20), or 36 nmol/l. Therefore, this patient's acid-base status is consistent with chronic respiratory alkalosis. The reason for the low plasma anion gap may be hypoalbuminemia.

SECTION TWO

Sodium and Water

6

SODIUM AND WATER PHYSIOLOGY

PART C SODIUM PHYSIOLOGY 235

PART D REVIEW 241

Concepts in Sodium and Water Physiology:

Objectives

- To explain the forces that regulate the movement of water across cell membranes:
  Water moves across cell membranes in response to a change in *"effective" osmolality*; it is only those particles that are restricted to the ECF or the ICF that determine these volumes. Because there are twice as many particles in the ICF (primarily K^+ salts of macromolecular anions) as in the ECF (primarily Na^+ salts), the *ICF volume* is twofold larger than the *ECF volume*.
- To explain the distribution of the ECF volume:
  The content of Na^+ determines the ECF volume, for the most part.
  The major factor promoting the movement of ultrafiltrate from the intravascular to the interstitial space is hydrostatic pressure. The colloid osmotic pressure (largely the result of *albumin*) and lymphatic flow cause the ultrafiltrate to reenter the intravascular volume.
  The most important principle concerning Na^+ and water in the body is that control of the "effective" ECF volume takes precedence over the control of water balance (i.e., the body "hates" shock).
- To explain the regulation of water balance:
  Water intake is stimulated primarily by a high effective plasma osmolality and to a lesser degree by a fall in ECF volume. Notwithstanding, a positive stimulus of a low ECF volume will lead to thirst and the release of antidiuretic hormone (ADH) even if the "effective" osmolality is low.
  Water output is controlled in a reciprocal fashion by ADH—low values cause a water diuresis and vice versa.
- To explain the relationship between Na^+ balance and control of the ECF volume:
  Na^+ intake is stimulated by a low effective ECF volume, but this stimulus is weak.
  Control over the excretion of Na^+ is the primary means of defending the ECF volume. The signal is the "effective" circulating volume, and control is exerted by a number of factors (hormones, neuronal activity) that influence the quantity of Na^+ reabsorbed by the nephron.

Abbreviations used commonly in this chapter:

ADH = antidiuretic hormone.
ANF = atrial natriuretic factor.
CCD = cortical collecting duct.
COP = colloid osmotic pressure.
ECF = extracellular fluid.
ICF = intracellular fluid.

"Effective" osmolality (tonicity):
Osmolality refers to the number of particles dissolved in water. Water moves across cell membranes to ensure equal osmolalities in the ECF and ICF. Some particles determine the volume in a single compartment because they are restricted to that compartment (e.g., Na^+ in the ECF); these particles are called "effective osmoles." Other particles exist at equal concentrations in both the ECF and the ICF and do not influence water movement (e.g., urea); these particles are called "ineffective osmoles." A term used to describe the concentration of effective osmoles is "tonicity."

ICF volume:
The volume of water inside cells.

ECF volume:
The volume of water held outside cells by the osmotic force created by Na^+ and its attendant anions Cl^- and HCO_3^-.

Albumin:
The protein in plasma that is the principal determinant of colloid osmotic pressure.

Outline of Major Principles

Clinical pearl:
Note the different implications of the content and the concentration of Na^+.

1. Consider Na^+ and water separately because they are regulated independently.
2. The principal particle retained in the ECF is Na^+. The content of Na^+ in the ECF determines the ECF volume. The concentration of albumin and, to a lesser extent, its valence determine the movement of fluid from the interstitial to the intravascular space. Many hormones, renal nerves, and hemodynamic factors act in concert to regulate Na^+ balance and thereby the ECF volume.
3. The [Na^+] in the ECF reflects the ICF volume and thus relative water balance; *hyponatremia* results in swollen cells and *hypernatremia* results in shrunken cells. The [Na^+] in plasma does not indicate whether the ECF volume is normal, high, or low.
4. Thirst and the release of ADH are stimulated by shrunken cells and by ECF volume contraction. ADH is the major hormone controlling the excretion of water.

Hyponatremia:
A decreased [Na^+] in plasma and the ECF (< 135 mmol/l) that can be due to Na^+ loss or water gain.

Hypernatremia:
An elevated [Na^+] in plasma (> 144 mmol/l) that can be due to Na^+ gain or water loss.

Introductory Case
Lee Is Sweet and Not Salty

(Case discussed on page 241)

Lee, the patient with diabetes mellitus in poor control, returns (see pages 4, 45, 70). She complains of excessive urination and thirst. The points to focus on with respect to Na^+ and water pathophysiology are as follows:

1. her ECF volume is contracted;
2. she has hyponatremia (126 mmol/l) and hyperglycemia (50 mmol/l, 900 mg/d!);
3. her urine has a high [Na^+] (46 mmol/l), given her ECF volume contraction (see the margin); the urine osmolality is 525 mosm/kg H_2O and the flow rate is 3 ml/min.

Note:
When the ECF volume is contracted, the "expected" renal response is the excretion of as small a quantity of Na^+ as possible.

Why is her ECF volume contracted?
How did her physician know?
Why is she hyponatremic, given the composition of her urine?
What stimulated her thirst?
Why is her urine volume so high?

PART A
COMPOSITION OF BODY FLUIDS

- Water is 60% of body mass; two-thirds of body water is located in the ICF and one-third in the ECF.
- As the proportion of muscle to body weight declines, the percentage of water declines as well (muscle usually contains 50% of body water).

The most abundant constituent of the body is water; it accounts for approximately 60% of the body mass (Tables 6·1 and 6·2). This water is divided into two main compartments, ECF and ICF (Figure 6·1). In relating total body water to body weight, one presumes that the relative proportion of fat is constant (obviously, this assumption is not true as judged from a simple inspection of people in a shopping center, and a correction must be made for body composition because neutral fat does not dissolve in water). Females tend to have a lower water content per body mass (50% body weight vs 60% for males). Older people also tend to have a relatively smaller water content because of their relatively small proportion of muscle mass. Infants, however, store less adipose tissue and therefore have a higher proportion of water (70%).

The ECF consists of plasma fluid (4% body weight) and interstitial fluid (i.e., water in tissues between the cells, 16% body weight). Water in the abdominal cavity (*ascites*) or thoracic cavity (*pleural effusion*) is a component of the interstitial space. In certain disease states, fluid accumulates in the interstitial space of the ECF to an appreciable degree and is called *edema*, ascites, or pleural effusion.

Ascites:
Accumulation of an ultrafiltrate of plasma plus some albumin in the abdominal cavity.

Pleural effusion:
Accumulation of ultrafiltrate of plasma with or without albumin in the pleural cavity.

Edema:
Accumulation of an ultrafiltrate of plasma in the interstitial space (usually the dependent area via gravity).

Table 6·1
COMPOSITION OF THE BODY

Values are approximations for a 70-kg person. The amount of water present depends on the mass of adipose tissue. There is a larger percentage of water in infants (70%) and less in the obese, in females, and in the elderly, largely because of a variable proportion of muscle vs adipose tissue.

	Water (liters)	Protein (kg)	Na^+ (mmol)	K^+ (mmol)
Total body	45	6	2550	4560
ECF	15	0.3	2250	60
ICF	30	5.7	300	4500

Notes for the expert:
- The volume of the ICF is usually considered to be twice that of the ECF, but the data to support this claim are not clear-cut. It is equally possible that just over half of body water (55%) is intracellular and 45% is extracellular.
- During pregnancy, plasma osmolality is reduced by 10 mosm/kg H_2O and [Na^+] by 5 mmol/l, on average. This reduction seems to be associated with a resetting of the "osmostat."

Concept:
1. Water moves to osmotic equilibrium across cell membranes. Because Na^+ are restricted primarily to the ECF, the content of Na^+ deterines the ECF volume.

Table 6·2
COMPOSITION OF THE ECF
Values are approximations in a 70-kg adult.

Compartment	Water	Other constituents (liters)
Total ECF	15	• Contains 230 g of albumin and 2250 mmol of Na^+
- Interstitial volume	12	• Contains one-fourth of the concentration of albumin in plasma, close to 50% of total albumin
- Plasma volume	3	• Contains 120 g of albumin; exists with 2 liters of red blood cells in blood volume

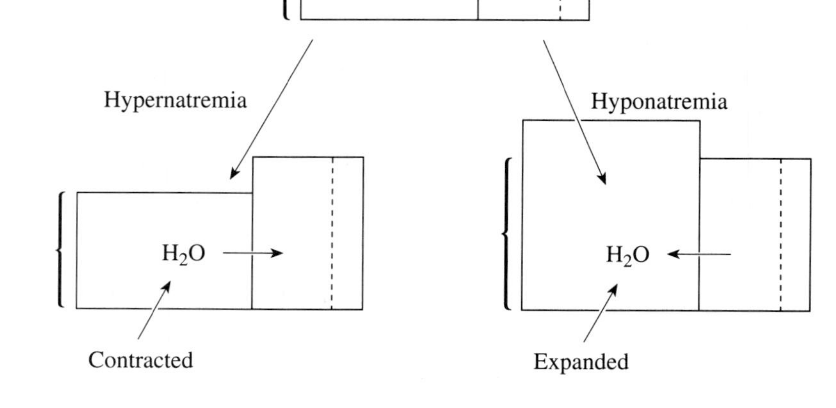

Figure 6·1
Body fluid compartments
The normal distribution of the ECF and ICF is shown in the top rectangle. Note that an ultrafiltrate of plasma moves across the capillary membrane, and water moves across cell membranes. During hypernatremia, the ICF volume is contracted; the converse occurs during hyponatremia. The bracket reflects the normal height of the ICF volume.

Note:
The ECF volume cannot be predicted a priori from the change in $[Na^+]$ in plasma (see text).

Influence of Electrolytes on Compartment Volumes

- Particles restricted to a compartment determine its volume. Na^+ (and Cl^- plus HCO_3^-) determine ECF volume. Large *macromolecular* anions plus K^+ largely determine the ICF volume.

Macromolecular:
Referring to a compound that has many charged groups (phosphate in DNA, RNA, etc.), but only a few particles.

Water crosses cell membranes rapidly (through pores or channels) to achieve osmotic equilibrium; however, not all materials dissolved in water disperse equally in the ICF and ECF because there are differences in permeability, transporters, and active pumps that regulate their distribution (Table 6·3).

Table 6·3

A) COMPOSITION OF THE ECF AND ICF

The data are expressed as mmol/kg of cellular water (approximate values for skeletal muscle).

	ECF	ICF
Na^+	141	10
K^+	4.1	120–150
Cl^-	113	3
HCO_3^-	26	10
Phosphate	2.0	140

B) DISTRIBUTION AND COMPOSITION OF THE ICF

Values are approximations for a 70-kg adult.

	Water liters	K^+ mmol	Na^+ mmol
Muscle	22	3300	220
Brain, liver, and kidneys	2.5	375	25
Other	5.5	825	55
Total	30	4500	300

C) SIZE OF VARIOUS BODY FLUID COMPARTMENTS

Values are reported for a 70-kg normal male. We have arbitrarily selected ECF and ICF volumes of 15 and 30 liters, respectively.

Compartment	% Body weight	Volume (liters)
Body	60–65	45
ICF	40–45	30
ECF	20–23	15
- interstitial	16	12
- plasma	4	3
Blood	7	5

Note:
The concentrations in the ECF are in mmol/kg H_2O; it is difficult to be certain about the concentrations in the ICF for technical reasons.

Distribution of Water Across Cell Membranes

- Water (without Na^+) crosses cell membranes until the osmolality (particle/H_2O ratio) is equal on both sides of that membrane.
- *Tonicity* ("effective" osmolality) = total osmolality – (urea + alcohol, both in mmol/l).
- The total number of particles in the ICF in most cells rarely changes, but changes do occur in brain cells during chronic shrinking or swelling.
- Bottom line: the content of Na^+ determines the ECF volume, and the concentration of Na^+ ($[Na^+]$) in the ECF reflects the ICF volume.

Tonicity ("effective" osmolality):
The osmolality resulting from the restriction of particles to a compartment. This restriction determines the volume of that compartment.

$[Na^+]$:
The ratio of Na^+ to water reflects the ICF volume; from a conceptual viewpoint, it is the only time that the denominator of a ratio is the most important of the components of that ratio.

Clinical points:

- Particles like urea or alcohol cross cell membranes rapidly so that their concentrations are equal in the ICF and ECF; thus, they do not change the "effective" osmolality or tonicity and do not induce water movement between the ECF and ICF (Figure 6·2).
- Brain cells are virtually the only ones that regulate their volume by changing the number of intracellular particles.

Water distribution depends on the number of particles restricted to the ICF or to the ECF (Figure 6·2). These particles account for the "effective" osmolality, or tonicity, in these compartments. The particles restricted to the ECF (which thereby control the ECF volume) are Na^+ and its attendant anions (Cl^- plus HCO_3^-). It is not that Na^+ cannot cross the cell membrane (although permeability of cell membranes to Na^+ is relatively low compared with permeability to K^+); rather, any additional Na^+ that enter the ICF are actively transported out of the cell by the membrane enzyme Na^+K^+ATPase. Under normal conditions, it is believed that there are roughly twice as many particles in the ICF as in the ECF; it therefore follows that the ICF volume is twice as large as that of the ECF.

The particles that attract water into cells differ from one cell type to another. The major factor responsible for water movement into the cell is the retention of large macromolecular anions (organic phosphates) inside the cell. Although the macromolecules do not exert a large osmotic pressure (there are not a large number of particles), they bear a large anionic net charge and, as a result, have a large number of cations associated with them (primarily K^+) that not only provide electroneutrality but also account for the majority of the osmoles in the ICF (Table 6·3). Because the ICF macromolecules are largely organic phosphate esters (ATP, creatine phosphate, RNA, DNA, phospholipids, etc.) and are essential for cell function, only small net changes in their content occur. Thus, it follows that the total number of ICF particles is relatively "fixed" in number and charge; hence, changes in the particle/water ratio in the ICF usually come about only by a change in the water content of the ICF.

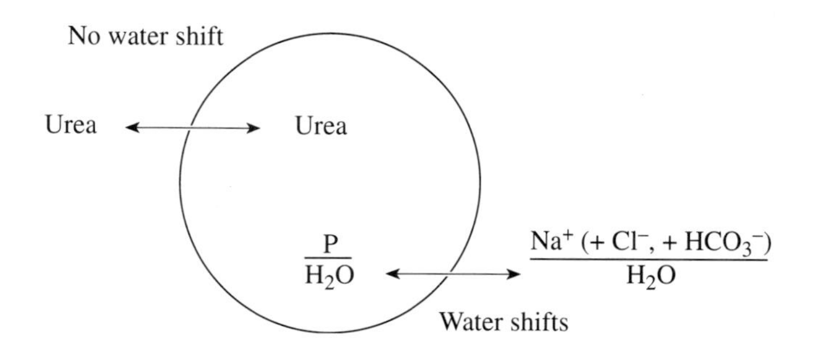

Figure 6·2
Review of factors regulating water distribution across cell membranes
The circle represents the cell membrane. Water crosses this membrane rapidly to achieve osmotic equilibrium. Particles such as urea (and alcohol) also cross this membrane rapidly, and their concentrations are equal in the ICF and the ECF; hence, they play no role in water distribution. The major particles restricted largely to the ECF are Na^+ and the anions Cl^- plus HCO_3^-; the particles (P) restricted primarily to the ICF are predominantly the K^+ (salts of organic phosphates).

Question

(Discussion on page 244)

6·1 How might one reconcile an osmolality of 285 mosm/kg H_2O in the ICF with a [K^+] in the ICF that is only 150 mmol/l (i.e., what constitutes the osmolality of 285 mosm/kg H_2O in the ICF)?

DEFENSE OF CELL VOLUME

The number of particles in the ICF is relatively constant for the majority of cells; however, brain cells can specifically defend against a large water shift by varying the number of their intracellular particles. This defense is advantageous because the brain is contained in a rigid box (the skull); if brain cells were to increase markedly in size, they might occupy too large a space in the skull. The increased pressure would cause herniation and diminished cerebral blood supply. In contrast, diminishing the ICF volume in the brain would shrink the brain, stretch vascular connections to the calvarium, and ultimately lead to an intracranial hemorrhage. Recall that the large intracellular macromolecular anions are essential compounds and are not expendable; hence, defense of a cell volume occurs only if there are a sufficient number of ions or non-electrolytes that can be translocated across cell membranes. A "loose" term used for these particles that change in the ICF of the brain is *idiogenic osmoles*.

Chronic hypernatremia and hyponatremia:
Disorders that are associated with regulation of cerebral cell volume. Rapid corrections (or overcorrection) may be associated with irreversible cell damage (osmotic demyelination; see Chapter 7, pages 266 and 270–71).

Idiogenic osmoles:
Some clinicians are fond of using this term to describe how certain cells of the brain may defend their ICF volume. We do not find this term "attractive" because it is deliberately ambiguous, the particles are not completely identified, the measurements of osmolality are somewhat inexact, and the normal particles that constitute the ICF osmoles are not defined (see the discussion of Question 6-1, page 244).

Regulatory Decrease in Brain Cell Volume

One mechanism used in the controlled return of swollen cells toward their original volume is the extrusion of electrolytes. Typically, it involves a decrease in the content of K^+ in the ICF. Cells vary in their amount of intracellular Cl^-, the major anion to be lost with K^+ if this loss is to shrink the ICF volume (i.e., the $[Cl^-]$ in the ICF is close to 3 mmol/l in muscle and 70 mmol/l in red blood cells). In addition, the time course for this effect varies greatly among species and cell types. Amino acids or small peptides, if present in large enough concentrations, may be extruded from cells as part of the regulatory decrease in volume. Alternatively, it is possible that a decrease in ICF volume could occur without ion extrusion if ions were bound and thus made "osmotically inactive" (see the margin note regarding idiogenic osmoles).

Note:
It is not certain whether the concentration of glucose rises in all types of brain cells during hyperglycemia (see Chapter 12 for more discussion).

Regulatory Increase in Brain Cell Volume

The mechanism for a gain in ICF volume during hypernatremia also varies among species and cell types. Typically, controlled return of shrunken cells toward their original volume involves an influx of Na^+. In certain cases, the furosemide-sensitive Na^+, K^+, 2 Cl^- cotransporter is involved, but, in others, the amiloride-sensitive Na^+/H^+ antiporter works with the Cl^-/HCO_3^- antiporter. It is also possible to change the number of organic compounds (e.g., amino acids or taurine, among others).

Distribution of an Ultrafiltrate Across Capillary Membranes

- Movement of an ultrafiltrate of plasma across capillary membranes does not cause water to shift between the ECF and ICF.
- Hydrostatic pressure is the major force moving fluid out of the capillary lumen, and the colloid osmotic pressure (largely the result of albumin) is the major force moving fluid into the capillary lumen.

Concept:
2. An ultrafiltrate moves between the vascular and interstitial spaces. This movement is controlled by the hydrostatic pressure and colloid osmotic pressure (largely due to albumin).

Colloid osmotic pressure (COP):
This force is due principally to the concentration of albumin in plasma; however, the quantitative relationships are interesting. The total COP is 25 mm Hg, 19 mm of which are due to dissolved protein and 6 mm of which are due to attracted Na^+ (the Donnan effect). The other plasma proteins account for 50% of the weight of proteins, but only 25% of the COP.

The Donnan equilibrium is based on differences in the concentrations of impermeant ions (largely anionic albumin) located in the vascular and interstitial compartments. Albumin attracts cations (largely Na^+) into the vascular compartment and repels anions (largely Cl^-) out of it. Since the $[Na^+]$ exceeds the $[Cl^-]$, the vascular space has a larger number of ionic species.

The factors controlling ultrafiltrate movement across the capillary membrane are shown in Figure 6·3; the major outward driving force is the hydrostatic pressure gradient. Hence, more ultrafiltrate exits when this hydrostatic pressure gradient increases. The hydrostatic pressure at the venous end of the capillary increases with venous hypertension (e.g., with venous obstruction, congestive heart failure). The major inward (interstitial to vascular) flow of fluid is driven by the colloid osmotic pressure (COP) difference, which is ultimately due to the higher concentration of albumin in the vascular fluid as compared with that in interstitial fluid. Interstitial fluid is also returned to the venous system via the lymphatics.

Because not all capillary pores are smaller than the diameter of plasma proteins, albumin leaks into the interstitial space. The concentration of albumin in the interstitial space is 10 g/l, and it exerts a COP of close to 5 mm Hg (the content of albumin in the interstitial space (10 g/l × 12 liters) is equal to that in vascular volume (40 g/l × 3 liters)).

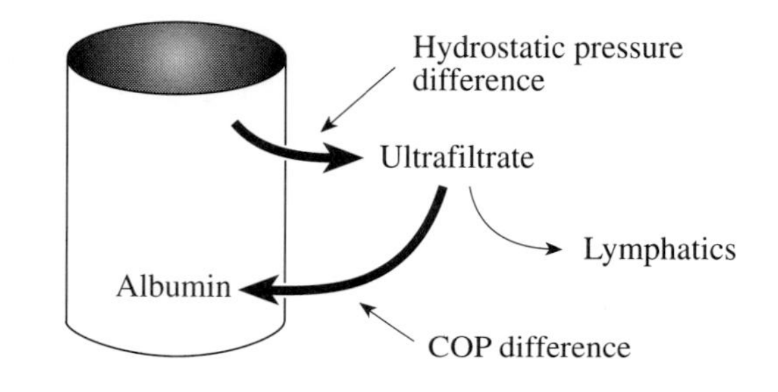

Figure 6·3
Review of factors controlling distribution of ultrafiltrate across the capillary membrane
There are two major forces to consider: a higher hydrostatic pressure causes fluid to leave the vascular space, and the colloid osmotic pressure (COP) causes fluid to enter the vascular space. Since the interstitial fluid volume is so much larger than the vascular volume, any time you detect expansion (edema) of the interstitial space, the patient will always have ECF volume expansion, even if the vascular volume is reduced (i.e., chronic hypoalbuminemia).

Questions

(Discussions on pages 244–46)

6·2 What IV solution would you give if you wanted all of it to stay in the ECF? What IV solution would you give to expand the ECF but contract the ICF volume?

6·3 What proportion of a liter of dextrose in water (D_5W) ends up in the ICF once glucose is metabolized?

6·4 A 70-kg person perspires and loses 4 liters of a NaCl solution; the $[Na^+]$ in this fluid is one-half that in the ECF. What changes should occur in ICF and ECF volumes and what will the resultant $[Na^+]$ be? (For this calculation, assume that the starting $[Na^+]$ is 150 mmol/kg H_2O and that the ECF and ICF volumes are 15 and 30 liters, respectively.)

6·5 What factors influence the volume of infused D_5W that will enter cells? How can more of this infused water enter cells? (These questions should illustrate principles governing fluid compartmentation.)

6·6 Will exchanging a K^+ in the ICF for a Na^+ in the ECF have a direct effect on the ICF and ECF volumes?

D_5W:
A commonly infused solution that contains 5 g of glucose per 100 ml of water, or 50 g/liter. The concentration of glucose is close to isosmolal, or 276 mmol/l.

6·7 What is the effect of hyperglycemia on a water shift between the ICF and ECF? For simplicity, assume that no insulin is present and that no excretions occur.

6·8 What would happen acutely to the ICF volume if the permeability of capillary membranes to albumin were to increase?

Clinical pearl:
In edematous states resulting from a low COP, one may increase the tissue hydrostatic pressure with support stockings and diminish the degree of edema.

PART B
WATER PHYSIOLOGY

- Defense of tonicity involves thirst and excretion or conservation of *electrolyte-free water*.
- The controls of tonicity are remarkably sensitive, responding to 1–2% changes.
- Change of tonicity is virtually synonymous with change in the [Na^+] in plasma. An increased tonicity results in thirst and a reduction in water excretion. Reduction of tonicity diminishes thirst and increases excretion of *osmole-free water*.

Electrolyte-free water:
A term used to describe water that is gained or lost from total body water instead of either the ECF or ICF. The solution can be electrolyte-free but can have a high osmolality if it contains a considerable amount of urea.

Osmole-free water:
A term used to describe water that is excreted without solutes. It is not the best term for describing the influence of urinary excretion on body water distribution (calculate electrolyte-free water for this purpose).

Overview

Assume a normal subject consumes one liter of pure water and that this liter mixes with all body fluids. To excrete this liter of water, it must be sensed, a message must be sent to the kidney, and then this liter of water must be segregated from the electrolytes (Na^+ salts) dissolved in it so that this excess pure water can be excreted.

Pure water per se cannot be excreted by the kidney because there must always be some dissolved particles such as electrolytes and/or urea (the minimum urine osmolality is close to 20–50 mosm/kg H_2O).

Clinical pearls:
- The main indicator of intracellular water volume is the [Na^+] in plasma.
- Changes in the [Na^+] in plasma primarily reflect water balance.
- If hypernatremia develops, there is a problem with thirst or access to water.

MECHANISMS OF EXCRETION

Sensor

The addition of pure water dilutes body constituents. In the ECF, this dilution is recognized clinically as hyponatremia. In contrast, in the ICF, the important compartment, this dilution is recognized as swelling of cells. The sensor (a cell that is sensitive to its volume) is located in the CNS and is an "osmoreceptor" linked to both the thirst center and the *antidiuretic hormone* (ADH) release center.

Messages

Swelling of osmoreceptor cells sends a message to the thirst center to diminish water intake and to the kidneys to excrete as much osmole-free water as possible. The messenger to the kidneys is ADH. With a surplus of water, there is little if any release of ADH, and its level in plasma declines markedly and rapidly.

Antidiuretic hormone (ADH):
A hormone that the brain synthesizes, stores, and releases. ADH acts on the collecting duct, making it per-

Concept:
3. The body regulates the content of Na^+ and H_2O in an independent fashion.

Renal Events

In the kidney, saline is filtered and some Na^+ and Cl^- are reabsorbed without water in the thick ascending limb of the loop of Henle and the distal nephron. The water remaining is excreted because the luminal membrane of the distal nephron has a low permeability to water when there are no actions of ADH.

Concept:
4. Water balance is the result of the interplay of thirst and ADH.

Control of Water Intake

Thirst is stimulated by an increased tonicity. Particles like urea are not involved in tonicity and are not sensed by the hypothalamic centers because they have similar concentrations in the ICF and ECF.

Contraction of the ECF volume is also a weak stimulus to thirst. In this regard, elevated levels of *angiotensin II* might operate as a signal. Other factors unrelated to a need for water may also stimulate water intake (e.g., dryness of the mouth, habit, culture, psyche, etc.). The major inhibitors of thirst are hypotonicity and ECF volume expansion.

Angiotensin II:
The vasoconstrictor formed when renin is produced (in response to ECF volume contraction). It also causes aldosterone to be released, more $NaHCO_3$ to be reabsorbed by the proximal tubule, and thirst to be stimulated.

Renin acts on angiotensinogen from the liver, forming angiotensin I, which is cleaved to angiotensin II by angiotensin-converting enzyme.

Clinical pearls:
1. When assessing a conscious patient with hypernatremia, first ask if the patient is thirsty to evaluate the intake component of water regulation.
2. If the ECF volume is contracted, thirst will be stimulated even if the tonicity of body fluids is low.

Control of Water Excretion

EXCRETION OF A DILUTE URINE

- Excretion of a dilute urine requires three steps:
  1. delivery of Na^+, Cl^-, and water to the diluting sites (delivery);
  2. reabsorption of solutes without water (separation or desalination);
  3. excretion of this electrolyte-free water (maintenance of separation).

When water is ingested without solutes, for balance, water without solutes must be excreted in the urine. There are several nephron sites where Na^+ and Cl^- are reabsorbed, but water is not. The most important of these sites is the thick ascending limb of the loop of Henle. Other segments are the early distal convoluted tubule (not sensitive to ADH), the late distal convoluted tubule, and the collecting duct (the latter requires absence of ADH to maintain impermeability to water). To excrete a dilute urine, three processes must occur. They are considered in quantitative terms in Figure 6-4.

1. **Delivery of saline to the thick ascending limb of the loop of Henle:** One-third of the GFR (60 liters/day) is delivered to diluting sites of the nephron (i.e., 40 times more than is needed to excrete the usual daily water load of 1.5 liters). Therefore, only a very major reduction in GFR can reduce delivery sufficiently to be the sole cause of a limited excretion of electrolyte-free water.

2. **Separation of salt and water (reabsorption of NaCl without water):** This separation begins in the thick ascending limb of the loop of Henle; there is a net reabsorption of Na^+ and Cl^- without water. This net transport of Na^+ and Cl^- is inhibited by *furosemide*, a diuretic that

Furosemide:
A loop diuretic that inhibits the reabsorption of Na^+ and Cl^- in the thick ascending limb of the loop of Henle. Since an electrical gradient develops (lumen positive), furosemide also inhibits the reabsorption of Mg^{2+} and Ca^{2+} here. For the luminal concentration of furosemide to be high, it must be secreted in the proximal convoluted tubule; in renal insufficiency or with drugs inhibiting proximal secretion, this diuretic is less potent. By inhibiting Na^+ and Cl^- reabsorption in the loop, loop diuretics can compromise the ability to dilute and to concentrate the urine.

binds to the luminal Na^+, K^+, 2 Cl^- cotransporter. In the collecting duct, the urine can be diluted further when NaCl is reabsorbed to a greater degree than water; ADH must be absent for this dilution to occur.

3. **Maintenance of separation:**
The hypotonic fluid that exits from the loop of Henle must not equilibrate with the isosmolar and hyperosmolar fluid surrounding the remainder of the nephron. These membranes remain relatively water-impermeable when ADH is absent; therefore, ADH secretion must cease when osmole-free water is to be excreted (Table 6·4). ADH can be released for a variety of reasons unrelated to tonicity (see Chapter 7, pages 263–64 for more discussion). Thus, a hyperosmolar urine can be excreted during a hypoosmolar state.

Bottom line:
- To assess medullary hyperosmolality, measure urine osmolality after ADH acts. The expected value is 1200 mosm/kg H_2O.
- To assess ADH action, you must know the medullary osmolality.
- To assess whether a given urine will lead to a rise or fall in the plasma [Na^+], examine the [Na^+] + [K^+] in the urine, not the urine osmolality. Compare this sum of electrolyte concentrations with that in plasma (see the margin).

Note:
If the K^+ in urine came from the ICF and Na^+ entered the ICF as K^+ exited, excretion of these K^+ in the urine would represent loss of Na^+ from the ECF.

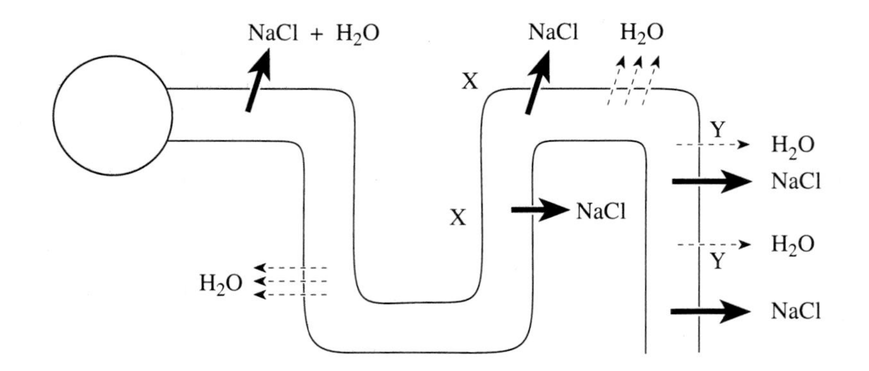

Figure 6·4
The excretion of a dilute urine
The structure is a stylized nephron. Isosmolal fluid is reabsorbed in the proximal tubule; electrolytes are reabsorbed without much water in the loop of Henle. Even in the absence of ADH, a considerable amount of water is reabsorbed in the distal nephron. Because more Na^+ than water may be reabsorbed in the CCD, the final urine may even be more hypoosmolal than the delivered fluid. An "X" designates nephron sites where NaCl is reabsorbed without water (diluting sites where ADH does not influence water permeability). The sites designated with a "Y" are nephron sites where NaCl is reabsorbed, but water may be reabsorbed if ADH is present. The details in segments involved are provided in Figure 6·5.

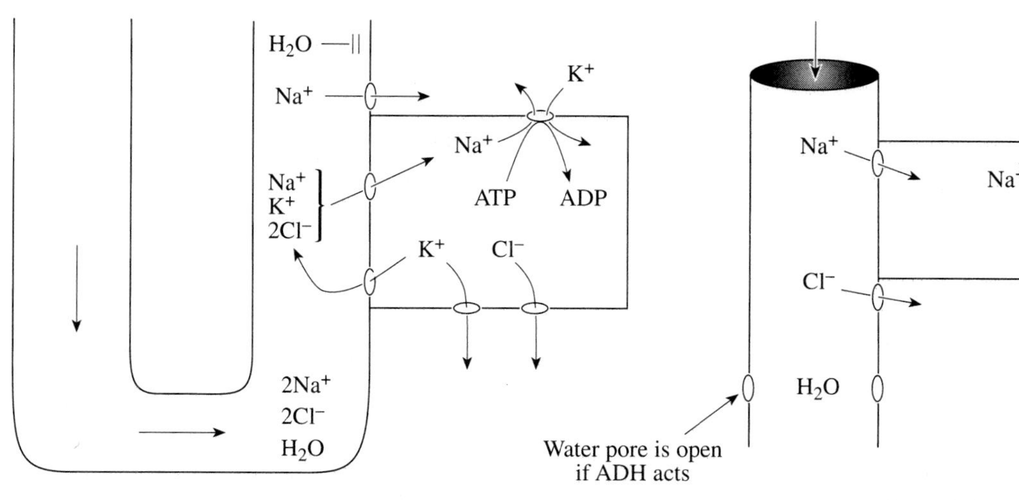

Figure 6·5
Formation of a dilute urine
The events in the thick ascending limb of the loop of Henle are shown on the left and events in the CCD are on the right.

Table 6·4
EFFECT OF WATER REABSORPTION ON VOLUME AND OSMOLALITY

These representative values do not include the reabsorption of isotonic saline. ADH makes the cortical and medullary collecting ducts permeable to water (Figures 6·4 and 6·5). The osmolality of the urine rises threefold in the cortex and a further fourfold in the medulla; however, the bulk (67%) of the water is reabsorbed in the cortex, so that solutes are not "washed out" in the hypertonic medulla.

Nephron site	Volume exiting (liters)	Volume reabsorbed (liters)	Osmolality (mosm/kg H_2O)	Rise in osmolality
End – loop of Henle	12	-	100	-
End – cortical collecting duct	4	8	300	Threefold
End – medullary collecting duct	1	3	1200	Fourfold

Questions

(Discussions on pages 246–47)

6·9 What mechanisms are responsible for the excretion of more water when furosemide is given?

6·10 What can limit the renal excretion of "osmole-free water" in a subject who drinks a large amount of water?

6·11 The plasma $[Na^+]$ was 160 mmol/l in a patient who was otherwise unaware of his problem. What aspect of his history should catch your attention?

6·12 Several days after a car accident, a patient has hypernatremia and a marked degree of ECF volume contraction. Three to four liters of urine are passed each day (osmolality = 420 mosm/kg H_2O). What is the most likely cause of polyuria?

6·13 How would you know that the failure to excrete "osmole-free water" was not due to an inadequate delivery of filtrate to the loop of Henle?

6·14 Two normal 70-kg persons excrete the following urine over the same time period:
Subject A, 1 liter of urine (1200 mosm/kg H_2O);
Subject B, 5 liters of slightly hypertonic urine (450 mosm/kg H_2O).

Both subjects have a total of 11 970 mosmoles (42 liters of body water and an osmolality of 285 mosm/kg H_2O) before the urine losses. If neither has any fluid intake, who will have the higher plasma osmolality?

EXCRETION OF CONCENTRATED URINE

- ADH is required for excretion of concentrated urine.

In the absence of water intake, the tonicity of body fluids rises because there is ongoing water loss. This rise in tonicity leads to cell shrinkage and thereby stimulates the release of ADH from the posterior pituitary gland. Binding of ADH to its receptors on the basolateral membrane of late distal convoluted tubular cells, as well as on cells of the cortical and medullary collecting duct, results in the formation of cyclic AMP, which increases the permeability of the collecting duct to water (Figure 6·6). Therefore, water is reabsorbed passively down an osmotic difference from the collecting duct lumen to the hyperosmolar renal medullary interstitium; as a result, a hyperosmolal urine is excreted and free water is conserved.

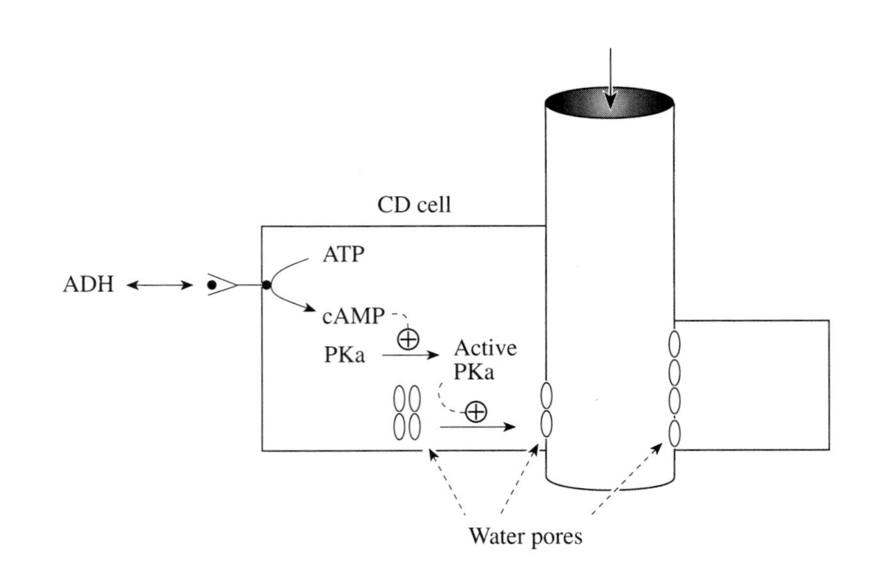

Figure 6·6
ADH actions on the distal nephron
The binding of ADH to its V_2 receptor on the basolateral membrane leads to activation of adenylate cyclase in these cells and thereby to a rise in cyclic AMP. The next step in the cascade is to activate a cAMP-dependent protein kinase (PKa = protein kinase a) and phosphorylate intracellular elements favoring the insertion of open water pores (channels) into the luminal membrane (open circles). In the medullary collecting duct, ADH also leads to the insertion of a pore that permits urea to diffuse across the luminal membrane (not shown).

V_2 receptor:
There are two ADH receptors: V_1 receptors are vascular and lead to an increase in blood pressure in response to ADH (vasopressin); V_2 receptors are in the distal nephron and lead to increased water permeability.

Components Required for a Concentrated Urine

- Urine osmolality = medullary function plus ADH action.
- Urine $[Na^+] + [K^+]$ = *renal impact on tonicity*.

Renal impact on tonicity: To analyze the impact of a given urine excretion on body tonicity, ignore the urine osmolality and consider only the osmoles critical for water distribution across cell membranes—Na^+ and possibly K^+ (plus accompanying anions)—and the water itself.

The following processes generate a hyperosmolal medullary interstitium.
Single effect: To excrete a concentrated urine, there must be a hyperosmolal medullary interstitium (1200 mosm/kg H_2O in humans). This hyperosmolality is the result of a single effect—the active transport of NaCl (via the Na^+, K^+, 2 Cl^- cotransporter, Figure 6·5, page 231) out of the lumen of the thick ascending limb of the loop of Henle. However, this process seems to achieve only 200 mosm/kg H_2O gradient in any given horizontal plane. Longitudinal flow permits a progressive rise in absolute osmolality.

Selective permeabilities: Differential permeabilities of the descending and ascending limbs of the loop of Henle help in generating a very hyperosmolal medullary interstitium. The descending limb is water-permeable (and relatively Na^+-impermeable). Hence, luminal osmolality rises from loss of water as fluid descends in the loop. In contrast, the ascending limb is more permeable to Na^+. Since the luminal $[Na^+]$ rose in the descending limb because of water abstraction, there is a favorable $[Na^+]$ difference for Na^+ to exit in the thin ascending limb of the loop of Henle.

Role of urea: In the cortex (late distal convoluted tubule and cortical collecting duct), water is reabsorbed under the influence of ADH. However, ADH does not induce urea permeability here. Thus, the concentration of urea in the luminal fluid rises progressively (same amount of urea but less water). In the medullary collecting duct lumen, there may now be a higher concentration of urea than in the medullary interstitium. Uniquely in this nephron segment, ADH induces urea permeability. Hence, a small amount of urea diffuses into the medullary interstitium, adding solute to this area. Urea accounts for about half of the osmolality of the medullary interstitium.

Anatomy of blood vessels: The vasa recta descend straight to the medullary tip, then turn around and ascend in a parallel fashion; they do not remove much material. They are called countercurrent exchange vessels.

In summary, the first three components generate medullary hyperosmolality and are called the countercurrent multiplier. The fourth component permits this hyperosmolality to remain. Low urea levels (low-protein diets), high vasa recta flow rates (renal vasodilation), lack of ADH, and inhibition of the Na^+, K^+, 2 Cl^- cotransporter all compromise the ability to generate a hyperosmolal medullary interstitium. In addition, diseases involving this area also limit the ability to excrete a concentrated urine.

Questions

(Discussions on pages 247–48)

6·15 In a patient with edema who has retained an excessive quantity of water, what is the total body Na^+ content and the plasma $[Na^+]$?

6·16 A patient has a contracted ECF volume and the following values in plasma and urine:

		Plasma	Urine
$[Na^+]$	mmol/l	130	60
$[K^+]$	mmol/l	5.0	20
Osmolality	mosm/kg H_2O	270	520

Why is hyponatremia present? Is the plasma ADH high, low, or normal in this patient?

CALCULATION OF A WATER DEFICIT OR SURPLUS

Note:
The $[Na^+]$ in plasma reveals the likely change in the ICF, but not the ECF volume.

In calculating a loss or gain of water, the simplest approach is to obtain quantitative estimates of ECF and ICF volume deficits. Consider this example. The plasma $[Na^+]$ in a 70-kg patient rose to 160 mmol/l from the normal value of 140 mmol/l because of a loss of water; his ECF volume, however, was normal on physical examination. His normal values were an ICF volume of 30 liters and an ECF volume of 15 liters.

Calculation of a Change in Water Balance

Notes:

1. Total effective osmoles in the ICF: 2 (140 mmol/l) × 30 liters = 8400 mosmoles.
2. Assumption: After the water loss, there is no change in number of particles in the ICF.
3. New ICF volume: Total osmoles in the ICF divided by osmolality equals new ICF volume.

$$\frac{8400 \text{ mosmoles}}{320 \text{ mosm/kg } H_2O} = 26.25 \text{ liters}$$

Change in ICF volume: The total number of particles in the ICF that affect water movement is 2 (plasma $[Na^+]$) × ICF water (30 liters), or 8400 mosmoles. The new "effective" osmolality is 320 mosm/kg H_2O (2 × new plasma $[Na^+]$ of 160 mmol/l). Dividing 8400 by 320 yields the new ICF volume of 26.25 liters, a fall of 3.75 liters (see the margin).

Change in ECF volume: This change is a clinical impression and cannot be calculated; it is not directly influenced by the plasma $[Na^+]$.

Change in total body water: Add the new volumes of the ICF and ECF; this patient needs a positive balance of 3.75 liters of water.

CALCULATION OF A CHANGE IN Na^+ BALANCE

Change in content of Na^+ in the ECF: This calculation is simply the product of the $[Na^+]$ in plasma and the estimated ECF volume on clinical grounds. One can ignore the Donnan distribution. In a normal 70-kg adult, the ECF volume is 15 liters and the plasma $[Na^+l$ is 140 mmol/l; hence, the ECF contains 2100 mmol of Na^+. In the example above, the ECF now contains 2400 mmol of Na^+ (15 liters × 160 mmol/l). Hence, this patient needs to lose a net of 300 mmol of Na^+ to achieve a balance for Na^+.

PART C
SODIUM PHYSIOLOGY

- The content of Na^+ defends the ECF volume because Na^+ are restricted to the ECF; Na^+ and the accompanying anions account for more than 90% of the ECF osmoles.
- Control of renal excretion of Na^+ is the major way to regulate the content of Na^+ in the body.

Control of Na^+ intake:
There is some evidence of stimulation of Na^+ intake when the ECF volume is low (craving salt).

Overview

People on a typical Western diet consume close to 150 mmol of NaCl each day. To remain in NaCl balance, 150 mmol of NaCl must be excreted daily. To do so, the extra Na^+ must be sensed, and a message must be sent to the kidney so that this extra NaCl can be excreted with or without water (depending on the independent controls on water balance). Three components are involved: a sensor of the "effective" arterial volume, the messengers to the kidney, and the intrarenal events.

Concept:
5. The kidney controls the content of Na^+ in the body. Changes in "effective" circulating volume signal the retention or excretion of Na^+.

SENSOR

When NaCl is retained, the ECF volume expands; the most important component of the ECF is the effective arterial volume. Hence it is not surprising that the sensors to detect important changes in the ECF volume are located in the arterial and central venous vessels. Once stimulated by hypervolemia, these sensors send messages to the kidney via renal nerves, circulating messengers, and by direct means (physical factors) to promote the excretion of NaCl, largely by decreasing its reabsorption.

Normal ECF Volume in More Detail

- The ECF volume is not a constant value in normal physiology.

It is obvious that the ECF volume should not be a simple static value as described for a 70-kg adult (15 liters total with 12 liters of interstitial volume and 3 liters of plasma volume). For example, people who eat a high-salt diet have a larger ECF volume and maintain this volume (remain in Na^+ balance) as long as NaCl intake is high. The converse applies to a person consuming a low-salt diet.

The ECF Volume in an Elite Athlete

A more dynamic way to think of the ECF volume is to consider an elite athlete who is in training. This athlete retains extra NaCl and water so that the ECF volume will be larger than normal. This larger ECF volume is retained, day in and day out, as long as training persists, even though exercise is performed over less than 10% of the day. A part of this ECF volume is retained in the venous system (in the vascular bed). A lower venous tone permits this blood to remain and not influence hemodynamics at rest. When exercise is anticipated, venoconstriction causes this blood to enter the "effective" vascular volume. Physicians may recognize this condition as *sports anemia*; the runner might recognize it by the weight loss that occurs along with a diuresis several days after exercise stops.

Sports anemia:
A lower hematocrit resulting from a normal RBC pool size within an increased plasma volume.

MESSAGES

Expansion of the "effective" arterial volume sends messages to the kidneys to excrete the extra NaCl (150 mmol in our example) but no more. Part of the message is delivered by renal nerves, part by hemodynamic or physical factors, and part via hormones. Because delivery of this message is such an important factor to regulate, it is not surprising to find many forms of regulation and, indeed, many regulators acting in concert to achieve this task. To be succinct, we shall only mention hormones in this section. The major hormones and their sites of action are summarized in Table 6·5. Their more detailed renal actions will be considered after the normal renal physiology of Na^+ handling has been discussed.

Table 6·5 Legend:
Only the major hormones are presented. Other hormones, such as an ouabain-like factor from the CNS, are not included because their physiologic role has yet to be established.

Table 6·5
HORMONES AND RENAL REABSORPTION OF NaCl

Hormone	Major stimulus	Major nephron site	Major effect
• Angiotensin II	• Low ECF volume, β-adrenergics via renin release	• Proximal convoluted tubule	• Enhanced reabsorption of $NaHCO_3$ and thereby NaCl
• Aldosterone	• Angiotensin II • Hyperkalemia	• Cortical distal nephron	• Reabsorption of NaCl • Secretion of K^+
• Atrial natriuretic factor	• Vascular volume expansion	• GFR • Medullary collecting duct	• Increased GFR • Reduced reabsorption of NaCl

Control of Na^+ Excretion

In the kidney, NaCl and water are filtered, but their reabsorptions are regulated independently. The regulation of Na^+ excretion is the most important factor in maintaining Na^+ balance. In a normal adult, close to 27 000 mmol of Na^+ are filtered each day. In our example, only 150 mmol need be excreted; therefore, more than 99% of these filtered Na^+ must be reabsorbed. Control of reabsorption is the key to regulation of balance for Na^+. It is important to recognize that filtration and reabsorption of Na^+ are linked so that the right amount is excreted no matter what the GFR is (within reason); the phenomenon whereby changes in GFR are accompanied by parallel changes in tubular Na^+ reabsorption is known as *glomeru-*

lar tubular balance. In the paragraphs to follow, we shall first consider the nephron sites where Na^+ are reabsorbed and then the role of hormones to modulate this segmental Na^+ reabsorption.

Glomerular-tubular balance:
A protective mechanism whereby tubular events feed back to regulate the GFR and protect the host from major NaCl losses. The macula densa, which senses increased NaCl delivery to the distal nephron, reduces the GFR by feeding back to the afferent arteriole via an interplay between angiotensin II and nitric oxide.

NEPHRON SITES INVOLVED IN THE TRANSPORT OF Na^+

The renal handling of Na^+ is best considered by examining the nephron segments individually (see also Table 6·6).

Na^+ Handling in the Proximal Tubule

Close to two-thirds of the Na^+ that are filtered are reabsorbed in the proximal tubule, and electroneutrality is maintained by either Cl^- reabsorption (larger) or by H^+ secretion (smaller). The epithelium of the proximal tubule is permeable to water, and the osmolality of the fluid leaving this tubule is the same as that of the ECF. The proximal tubular luminal membrane cannot generate a large transtubular $[Na^+]$ gradient; in fact, the maximum $[Na^+]$ gradient here is 33% and is generated during an *osmotic diuresis*. Na^+ reabsorption down the $[Na^+]$ gradient between the lumen and the ICF is produced by the Na^+K^+ATPase on the basolateral membrane of these cells. This lower intracellular $[Na^+]$ provides the driving force for nutrient and HCO_3^- reabsorption by Na^+-dependent transporters (Table 6·7).

Osmotic diuresis:
The excretion of Na^+, K^+, Cl^- and water in an osmotic diuresis is driven in part by an enhanced delivery of NaCl to the cortical distal nephron. Details are considered in the discussion of Question 6·17.

Na^+ Handling in the Loop of Henle

NaCl reabsorption in the loop of Henle enables the excretion of a concentrated or dilute urine (Figures 6·4 and 6·5). NaCl is actively transported in the thick ascending limb via the luminal electroneutral Na^+, K^+, 2 Cl^- cotransporter. Again, the driving force is provided by the basolateral Na^+K^+ATPase, but only half of the Na^+ use the transcellular route. Absolute Na^+ reabsorption in the loop of Henle varies directly with delivery. Net NaCl reabsorption is stimulated by ADH in the medullary thick ascending limb and is inhibited by loop diuretics. Water is not reabsorbed to an appreciable degree in this nephron segment.

Thiazide class of diuretics:
Diuretics that inhibit the reabsorption of Na^+ via the Na^+, Cl^- cotransporter in the distal convoluted tubule.

Na^+ Handling in the Distal Convoluted Tubule

The major function of this segment of the nephron is Na^+ balance and formation of electrolyte-free water. Na^+ are actively reabsorbed in this nephron segment along with Cl^-; high transepithelial $[Na^+]$ gradients can be maintained. Because this nephron site is sparingly permeable to water, the *thiazide class of diuretics*, which act on this site, lead to a compromised ability to excrete a load of free water.

Quantitative aspects of Na^+ handling in the cortical collecting duct:
In a 24-hour period, the cortical collecting duct:
- reabsorbs 1000 mmol of NaCl,
- permits the excretion of 60 mmol of K^+;
- secretes 20 mmol of NH_4^+, but this amount can rise to 100 mmol in chronic metabolic acidosis.

Na^+ Handling in the Collecting Ducts

The major functions of the cortical collecting duct are NaCl reabsorption and secretion of K^+ and H^+. Quantitative aspects are provided in the margin. The major step is Na^+ transport through a Na^+ specific channel. This channel is inhibited by the diuretic amiloride, a K^+-sparing type of diuretic.

Open probability:
A term used to imply that more channels and/or channels with greater conductivity are open.

Atrial natriuretic factor (ANF):
A hormone released from the right atrium of the heart when the central venous volume is high. It causes a fall in vascular resistance and promotes the excretion of NaCl.

Mineralocorticoids play an important role here by leading to a greater "*open probability*" of this Na^+ channel.

In the medulla, the major functions are concentration of the urine, NH_4^+ excretion, and Na^+ balance. The medullary collecting duct is capable of maintaining a large [Na^+] gradient across its epithelium. The medullary collecting duct is the site where final decisions are made about the excretion of NaCl. During ECF volume contraction, virtually all the Na^+ delivered are reabsorbed—a process that requires the absence of *atrial natriuretic factor (ANF)*. In contrast, with ECF volume expansion, a small quantity of the Na^+ delivered are reabsorbed because ANF is acting.

Question

(Discussion on page 248)

6·17 Why does glucosuria cause enhanced excretion of Na^+ and Cl^-?

Table 6·6 Legend:
Assuming a normal GFR (180 liters/day), 27 000 mmol of Na^+ are filtered per day. The following table depicts the events involved in the excretion of 150 mmol of Na^+ in one liter of urine.

***Note:**
The concentration of Na^+ in the luminal fluid depends upon whether water was reabsorbed as well. When ADH acts, the osmolality in the cortical collecting duct is equal to that in plasma. Given the concentrations of urea and K^+ in the luminal fluid, these values are reasonable guesses. The [Na^+] in the urine requires a urine volume of 1 liter.

Note:
CAI = Carbonic anhydrase inhibitors. These drugs do not block the Na^+/H^+ antiporter but inhibit indirect reabsorption of HCO_3^- by blocking the luminal carbonic anhydrase (see page 24 for details).

Table 6·6
WATER AND Na^+ REABSORPTION IN THE KIDNEY

Nephron site	Na^+ reabsorbed (mmol)	Na^+ remaining (mmol)	[Na^+] (mmol/l)
Proximal convoluted tubule	18 000	9000	150
Loop of Henle	6000	3000	45
Distal convoluted tubule	1600	1400	100 *
Cortical collecting duct	800	600	100 *
Medullary collecting duct	450	150	150 *

Table 6·7
SUMMARY OF TRANSPORTERS IN THE NEPHRON AND THE DIURETICS THAT INHIBIT THEM

Nephron site	Luminal transporter	Function	Diuretic
Proximal convoluted tubule	• Na^+/H^+ • Na^+-nutrient • NaCl	• Reabsorption of $NaHCO_3$ • Reabsorption of nutrients • ECF volume regulation	• CAI for $NaHCO_3$ reabsorption
Loop of Henle	• Na^+, K^+, 2 Cl^-	• NaCl reabsorption • Concentration and dilution	• Furosemide
Distal convoluted tubule	• NaCl	• NaCl reabsorption • Dilution	• Thiazide class
Cortical collecting duct	• Na^+ channel • NaCl reabsorption	• K^+ secretion	• Amiloride • Triamtererene • Spironolactone
	• H^+ATPase	• NH_4^+ excretion	• None in use
Medullary collecting duct	• NaCl • Na^+ channel • H^+ATPase • K^+/H^+ATPase	• Na^+ balance • Na^+ balance • NH_4^+ excretion • K^+ reabsorption	• ANF • Amiloride • None in use • None in use

REGULATION OF Na^+ EXCRETION

ECF Volume

When examining Na^+ excretion within the context of the ECF volume, expect to see urine that is virtually free of Na^+ (and Cl^-) when the ECF volume is contracted (see the margin and Table 6·8). Also, expect to see Na^+ excretion when the ECF volume is expanded. During euvolemia, the kidney excretes the dietary NaCl load. Hence, there are no normal values for urine Na^+ and Cl^-; these values must be interpreted relative to the physiologic state and dietary intake of the patient.

Note:
In a patient with ECF volume contraction, the urine can have a high $[Na^+]$ if Na^+ are excreted with a nonreabsorbed anion such as HCO_3^-, ketoacid anions, or drug-derived anions. In these examples, the $[Cl^-]$ in the urine will be low.

Anions in the Urine

Na^+ (and K^+) excretion can be influenced by the anion composition of the filtrate. For example, a patient who vomits delivers more HCO_3^- to the collecting duct than can be reabsorbed; the urine should contain Na^+ (and K^+) in conjunction with the HCO_3^- that are not reabsorbed even though the ECF volume may be contracted. ECF volume contraction is evident from the fact that the urine contains at most a small quantity of Cl^-.

Need to Excrete NH_4^+

During chronic metabolic acidosis resulting from laxative abuse, the urine may contain 200 mmol of NH_4^+. If the patient has a contracted ECF volume, the urine will not contain Na^+. Nevertheless, electroneutrality in the urine must be maintained and usually requires that Cl^- accompany the NH_4^+ in the urine. Hence, the urine is not Cl^--poor despite ECF volume contraction in this setting.

Table 6·8
URINE ELECTROLYTES IN A PATIENT WITH ECF VOLUME CONTRACTION
All values are in mmol/l.

		Electrolyte in the urine	
Condition	**Na^+**	**K^+**	**Cl^-**
Nonrenal or previous renal loss of NaCl	0–15	Variable	0–15
Vomiting			
- Recent	> 20	> 50	0–15
- Remote	0–15	Variable	0–15
Diuretics			
- Recent	> 20	> 20	> 20
- Remote	0–15	Variable	0–15
Renal disorders involving wasting of salt	> 20	Variable	> 20

ROLE OF HORMONES

The major hormones acting in the kidney to preserve Na^+ homeostasis are listed in Table 6·5 (page 236).

Angiotensin II

Angiotensin II is produced when renin is released from the juxtaglomerular apparatus. A low renal perfusion pressure leads to renin release, as does a rise in β_1-adrenergic agonists; both reflect a low ECF volume. The major renal effect of angiotensin II is to stimulate the reabsorption of $NaHCO_3$ in the proximal convoluted tubule. Because fluid must remain isosmotic here, water is reabsorbed, and the $[Cl^-]$ in the luminal fluid rises. This increase then creates a concentration difference for Cl^- and drives the passive reabsorption of Cl^- (dragging Na^+ for electroneutrality and water for isosmolality). The effects of this hormone are also discussed in Chapters 4 and 9 in the context of metabolic alkalosis and K^+ physiology. Angiotensin II is also a potent vasoconstrictor and selectively constricts efferent arterioles. Thus, angiotensin II leads to an increase in the filtration fraction and alters proximal reabsorption secondary to changes in physical factors.

Aldosterone

Aldosterone:
A hormone released by the adrenal cortex that leads to reabsorption of NaCl and secretion of K^+ in the cortical distal nephron.

Aldosterone has an important influence on renal Na^+ reabsorption in the distal nephron; it is responsible for 5% of the total Na^+ reabsorption. The secretion of aldosterone is stimulated by contraction of the "effective" circulating volume via angiotensin II and by hyperkalemia; the converse is also true. Quantitatively, the most important action of aldosterone is to lead to the reabsorption of NaCl. It also promotes the net secretion of K^+ and stimulates H^+ secretion.

Atrial Natriuretic Factor (ANF)

ANF leads to a natriuresis when the ECF volume is expanded. ANF acts by increasing the GFR and decreasing Na^+ reabsorption in the medullary collecting duct.

In addition to GFR, ANF, and aldosterone, other factors can influence renal Na^+ reabsorption. Peritubular capillary colloid osmotic pressure and renal perfusion pressure both modulate Na^+ reabsorption.

Questions

(Discussions on pages 248–51)

6·18 What is the volume of distribution of the common intravenous solutions?

6·19 What are the purposes for administration of the various IV solutions?

6·20 When should hypertonic NaCl be infused? How much should be given?

6·21 The ECF volume is expanded chronically in trained athletes. Why might this expansion be advantageous?

6·22 A person was in Na^+ balance on two occasions, eating and excreting 150 mmol of Na^+. On the first occasion, he had two kidneys (the GFR was 200 liters per day), and on the second, he had one kidney (the GFR was 100 liters per day). What was his fractional excretion of Na^+ on each occasion?

PART D
REVIEW

Discussion of Introductory Case
Lee Is Sweet and Not Salty

(Case presented on page 220)

Why is her ECF volume contracted?

Because of hyperglycemia, Lee has a glucose-induced osmotic diuresis, which results in an increased delivery of more volume and Na^+ out of the loop of Henle. If all of these Na^+ are not reabsorbed in the collecting duct, some will be excreted (note that her urine $[Na^+]$ is 46 mmol/l despite ECF volume depletion; see the discussion of Question 6·17, page 248).

How did her physician know?

She had the following physical signs of ECF volume depletion: low blood pressure, a postural fall in blood pressure, tachycardia, and a low jugular venous pressure along with the absence of edema or ascites.

Why is she hyponatremic, given the composition of her urine?

Her urine contains only 46 mmol/l Na^+; therefore, this urine is removing more water than Na^+ and would tend to make her hypernatremic if there was no entry of water into her ECF. There are two reasons why she has hyponatremia. First, she has an impermeant osmole (glucose) in her ECF that has raised the tonicity of the ECF. Movement of water from the ICF of those cells (muscle) requiring insulin for the entry of glucose has caused hyponatremia. The $[Na^+]$ will fall 1.4 to 1.5 mmol/l for every 5.5 mmol/l (100 mg/dl) above the normal concentration of glucose. Therefore, her $[Na^+]$ has fallen 11 mmol/l as a result of the hyperglycemia, close to the observed value of 126 mmol/l. The second reason for hyponatremia is that she is thirsty and is drinking a large quantity of water. The low ECF volume has led to the release of ADH, which prevents excretion of the ingested water.

What stimulated thirst?

Thirst was stimulated primarily by the contracted ECF volume. It is not clear whether hyperosmolality resulting from hyperglycemia stimulates thirst on a chronic basis.

Why is her urine volume so high?

The glucose excreted in her urine led to the high urine volume.

Summary of Main Points

BODY COMPARTMENTS

In a 70-kg person, total body water is 45 liters, with 30 liters in the ICF and 15 liters in the ECF.

Particles to be ignored:
Since urea and ethanol cross cell membranes readily, they have no impact on water distribution between the ECF and ICF.

- Approximately 60% of body weight is water; two-thirds is ICF and one-third is ECF.
- Water crosses cell membranes rapidly; therefore, the osmolality in the ECF equals that in the ICF.
- Particles that readily cross the cell membrane and achieve an equal concentration in the ECF and the ICF can be ignored with respect to the distribution of water during steady state.
- A major function of Na^+ (and Cl^- and HCO_3^-) is to keep water out of cells, thereby maintaining ECF volume (glucose acts like Na^+ for some cells only).
- The particles that determine ICF volume are large macromolecular anions and their attendant cations (K^+); these anions rarely leave the ICF.
- The $[Na^+]$ in the ECF (Na^+/H_2O) reflects the ICF volume of most cells; hypernatremia signals ICF contraction, and hyponatremia signals ICF expansion, except during hyperglycemia. The total Na^+ content (not the Na^+/H_2O) determines the ECF volume.
- An ultrafiltrate (not just water) crosses the capillary wall. A hydrostatic force (blood pressure) is the major outward-driving force, and vascular colloid osmotic pressure (albumin:saline ratio) is the major inward-driving force.

WATER PHYSIOLOGY

- There is an obligatory loss of 0.8 liters of water each day; therefore, mechanisms are needed to ensure this minimum water intake and to excrete any excess water.
- Adequate water intake is ensured by the CNS thirst center. A rise in tonicity of only 1–2% provides a powerful urge to drink.
- ADH is the hormone that limits water excretion. A rise in tonicity or a fall in the "effective" blood volume causes ADH release from the posterior pituitary.
  - For dilute urine to be excreted, ADH must be absent. The presence of ADH results in the excretion of a concentrated urine.
  - Assess ADH action on the kidney by two parameters—osmolality and volume. ADH produces a urine with maximum osmolality and usually a minimum volume. Conversely, in the absence of ADH, the urine has a minimum osmolality and maximum volume.

Na^+ PHYSIOLOGY

- Na^+ content determines the ECF volume. With ECF volume expansion, there is an excess of total body Na^+; with ECF volume contraction, there is a Na^+ deficit.
- The renal response to ECF volume contraction is the excretion of a urine that is free of Na^+ and/or Cl^-; failure to do so points toward a renal problem.
- When the ECF volume is low, the major renal mechanisms preventing Na^+ excretion are reduced GFR, enhanced proximal Na^+ reabsorption, aldosterone-stimulated Na^+ reabsorption (and K^+ excretion), and increased Na^+ reabsorption in the medullary collecting duct.
- When a Na^+ load is ingested, renal mechanisms are called into play to cause a natriuresis; these mechanisms include an increased GFR, a relatively diminished proximal Na^+ reabsorption (low angiotensin II), suppression of aldosterone release, and the release of ANF.
- Heart failure and hypoalbuminemia are major conditions in which excessive Na^+ are retained.
- Water and Na^+ are regulated independently.

CLINICAL APPLICATIONS

- One can recognize four distinct primary clinical abnormalities, two regarding Na^+ and two regarding water.

  Excess Na^+: The cardinal feature will be ECF volume expansion. Two prerequisites are required: an intake of Na^+ and "renal permission" to have an expanded ECF volume, which often means a low "effective" circulating volume (e.g., heart failure, hypoalbuminemia, or renal insufficiency).

  Deficit of Na^+: The cardinal feature will be a contracted ECF volume. Two major aspects should be identified: low intake of NaCl and/or excessive loss of Na^+ by renal or nonrenal routes. An examination of the urine electrolytes, plasma [K^+], and acid-base status will help in identifying the basis of this deficit.

  Excess water: The cardinal features are water intake and "renal permission," usually via ADH, to retain this extra water. Excess water implies swelling of cells and is recognized clinically by finding hyponatremia and hypoosmolality.

  Deficit of water: The cardinal features are a problem with water intake (lack of thirst, communication, and/or access

Note:
Clinical applications summary is continued on page 244.

to water) and loss of water by renal or nonrenal routes; urine osmolality, volume, and osmole excretion rates help in the differential diagnosis. A deficit of water implies cell shrinking and is recognized clinically by hypernatremia and hyperosmolality (not necessarily hyperosmolality alone if urea or alcohols have led to the high plasma osmolality).

Discussion of Questions

6·1 How might one reconcile an osmolality of 285 mosm/kg H_2O in the ICF with a [K^+] in the ICF that is only 150 mmol/l (i.e., what constitutes the osmolality of 285 mosm/kg H_2O in the ICF)?

Because water moves to osmotic equilibrium, the osmolality of the ICF is also 285 mosm/kg H_2O. The basis of the problem is that the anions in the ICF are largely macromolecular and will not make a large direct contribution to the osmolality of the ICF. Although K^+ may account for 150 of these mosmoles in a liter of ICF, we are unable to identify all 135 mosmoles of other intracellular particles. Other uncharged compounds might exist, or much of the water in cells may be *structural* instead of solvent water and the osmotic contribution of K^+ could be much higher.

Structural water:
Water that is bound to ionic groups or compounds in cells and is not able to dissolve materials in cells. For example, when magnesium sulfate ($MgSO_4$) is isolated in dry form, each crystal rapidly picks up water so that its natural form has bound water ($MgSO_4 \cdot 7\ H_2O$).

6·2 What IV solution would you give if you wanted all of it to stay in the ECF?

Isotonic saline is the solution to give to keep the entire volume in the ECF. If a patient has severe hyponatremia (e.g., 105 mmol/l), give a solution containing 105 mmol/l NaCl to expand the ECF quickly; this solution will stay in the ECF and not change the ICF volume.

What IV solution would you give to expand the ECF but contract the ICF volume?
To contract the ICF volume but expand the ECF volume, give saline that is hypertonic to the patient.

6·3 What proportion of a liter of dextrose in water (D_5W) ends up in the ICF once glucose is metabolized?

Water moves to achieve osmotic equilibrium between the ICF and ECF. Once the glucose is metabolized, it no longer requires consideration. Water will distribute in proportion to the existing ICF and ECF volumes. Therefore, two-thirds of the volume enters the ICF.

6·4 A 70-kg person perspires and loses 4 liters of a NaCl solution; the [Na^+] in this fluid is one-half that in the ECF. What changes should occur in ICF and ECF volumes and what will the resultant [Na^+] be? (For this calculation, assume that the starting [Na^+] is 150 mmol/kg H_2O and that the ECF and ICF volumes are 15 and 30 liters, respectively.)

For simplicity, divide the fluid loss into two components: isotonic saline and pure water. Because the [Na^+] is one-half of normal in this example, the 4 liters could be thought of as 2 liters of water and 2 liters of isotonic saline. The water loss occurs from the ICF and ECF in direct proportion to their existing volumes; thus two-thirds of the 2 liters of water is lost from the ICF (1.33 liters) and one-third from the ECF (0.67 liters). In contrast, the 2 liters of isotonic saline is lost exclusively from the ECF (no change in the Na^+/H_2O, the driving force for a water shift). Thus, the ECF is 15 – 0.67 liter – 2 liters, or 12.33 liters. Although we ignored the fact that the ECF volume was no longer half that of the ICF volume (a small error), we prefer this form of deductive calculation to the application of a formula. The final [Na^+] is the Na^+/water ratio in the ECF. We know the ECF volume (12.33 liters from above). The original [Na^+] is 150 mmol/l × 15 liters, or 2250 mmol, from which the patient lost 300 mmol of Na^+; therefore, the total Na^+ content is now 2250 – 300, or 1950 mmol. The resulting [Na^+] is 1950 mmol/12.33 liters, or almost 159 mmol/l. Hypernatremia signals a loss of ICF volume in virtually every case.

6·5 What factors influence the volume of infused D_5W that will enter cells?

Water moves to osmotic equilibrium. Once all the glucose is metabolized to triglycerides or to glycogen plus CO_2 and H_2O, we need only consider the water infused. Two-thirds of a liter of D_5W goes into the ICF.

How can more of this infused water enter cells?
If something is done in addition, such as decreasing the number of ECF particles (Na^+, Cl^-) or increasing the number of ICF particles, a larger proportion of infused water can enter cells.

6·6 Will exchanging a K^+ in the ICF for a Na^+ in the ECF have a direct effect on the ICF and ECF volumes?

To change compartment volumes, one must either change the ratio of water to "particles that count" or lose particles and water in their existing ratios in body compartments. Because no K^+ are excreted in this example, there is simply an exchange of particles, both of which are not bound in the ICF so no change in ECF or ICF volume should occur. If K^+ are excreted (with Cl^-), the ECF will lose particles and become contracted. In contrast, if K^+ exit and H^+ enter the ICF, the osmolality of the ICF will fall and water will move from the ICF to the ECF because H^+ are largely bound to proteins in the ICF.

6·7 What is the effect of hyperglycemia on a water shift between the ICF and ECF? For simplicity, assume that no insulin is present and that no excretions occur.

With hyperglycemia, the gain of particles is restricted to the ECF, for the most part. Therefore, hyperglycemia causes a shift of water from the ICF (cells that have a low concentration of glucose) to the ECF and into cells where the concentration of glucose is the same as that in the ECF. As discussed in Chapter 12, page 437, only those cells that depend on insulin for the transport of glucose and always have a very low concentration of glucose (e.g., muscle cells) are important in this regard.

6·8 What would happen acutely to the ICF volume if the permeability of capillary membranes to albumin were to increase?

Nothing would happen because the Na^+/H_2O ratio would not change. The distribution of ECF would change—more interstitial volume and less vascular volume.

6·9 What mechanisms are responsible for the excretion of more water when furosemide is given?

The effects are indirect. The most important action of furosemide is to block the reabsorption of NaCl in the thick ascending limb of the loop of Henle, a segment with low permeability to water. Because Na^+ are normally reabsorbed, and water is not, hypoosmolal fluid is created in the loop of Henle. This fluid normally enters the cortex. Under the influence of ADH, water is reabsorbed in the late distal convoluted tubule and collecting ducts, providing that the luminal fluid is hypoosmolal. Furosemide prevents the creation of desalinated, hypoosmolal fluid in the lumen and, as a result, prevents subsequent water reabsorption by this mechanism (Table 6·4, page 231).

6·10 What can limit the renal excretion of "osmole-free water" in a subject who drinks a large amount of water?

Assuming normal kidneys and full suppression of ADH, one excretes close to 30 mosmoles of urea and electrolytes in 1 liter of water. Hence, a salt-poor diet can limit the excretion of "free water" because loss of NaCl in the urine (see the margin) results in ECF volume depletion, which then leads to ADH release and diminished ability to excrete a dilute urine.

Note:
A water load expands the ECF and ICF volumes temporarily. In response to an expanded ECF volume, a natriuresis occurs. If NaCl is not ingested, the natriuresis and subsequent ECF volume contraction will prevent the complete excretion of the water load (because of the release of ADH).

6·11 The plasma [Na^+] was 160 mmol/l in a patient who was otherwise unaware of his problem. What aspect of his history should catch your attention?

No normal patient will permit hypernatremia to develop because thirst is such a powerful stimulus. Therefore, one would suspect a deficit in the thirst mechanism (local or general CNS problem). In cases where the thirst mechanism is intact, one may see hypernatremia if there is either a communication problem or restricted access to water.

6·12 Several days after a car accident, a patient has hypernatremia and a marked degree of ECF volume contraction. Three to four liters of urine are passed each day (osmolality = 420 mosm/kg H_2O). What is the most likely cause of polyuria?

The patient is undergoing an osmotic diuresis (excreting 1680 mosmoles per day, whereas the expected value is less than half this amount). Therefore, one would suspect an osmotic diuresis induced by glucose, urea, or mannitol. Look at the glucose and urea in blood and urine; check to see if mannitol was given.

6·13 How would you know that the failure to excrete "osmole-free water" was not due to an inadequate delivery of filtrate to the loop of Henle?

Look at the GFR. Normally less than one-third of the filtrate is delivered to the thick ascending limb of the loop of Henle. If the GFR is not markedly reduced, the lesion is due to something else or is in addition to a lower delivery of filtrate to the loop of Henle.

6·14 Two normal 70-kg persons excrete the following urine over the same time period:
Subject A, 1 liter of urine (1200 mosm/kg H_2O);
Subject B, 5 liters of slightly hypertonic urine (450 mosm/kg H_2O).

Both subjects have a total of 11 970 mosmoles (42 liters of body water and an osmolality of 285 mosm/kg H_2O) before the urine losses. If neither has any fluid intake, who will have the higher plasma osmolality?

Subject A has lost 1 liter and 1200 mosmoles; the new osmolality is 263 mosm/kg H_2O ((11 970 – 1200)/(42 – 1) liters). Subject B has lost 5 liters and 2250 mosmoles (450 mosm/kg $H_2O \times 5$ liters); the new osmolality is also 263 mosm/kg H_2O ((11 970 – 2250) mosmoles/(42 – 5) liters). Therefore, both subjects have the same osmolality. This calculation points out that one needs to know both the osmolality and the volume of urine in order to understand the contribution of renal excretion to body osmolality.

6·15 In a patient with edema who has retained an excessive quantity of water, what is the total body Na^+ content and the plasma $[Na^+]$?

The total body Na^+ content is increased in any patient with edema. In this patient, retention of excessive water produces hyponatremia.

6·16 A patient has a contracted ECF volume and the following values in plasma and urine:

		Plasma	Urine
$[Na^+]$	mmol/l	130	60
$[K^+]$	mmol/l	5	20
Osmolality	mosm/kg H_2O	270	520

Why is hyponatremia present?

Hyponatremia can be present if there is a deficit of Na^+ and/or a gain of water (see the margin).

Note:
In a patient who has a deficit of Na^+, there may be some water loss; if this loss is relatively small, hyponatremia can still be present.

Deficit of Na^+: The contracted ECF volume implies a deficit of Na^+. With ECF volume contraction, the urine should be free of Na^+, but it is not. Therefore, there is a renal cause for the deficit of Na^+; the clue to its basis is revealed by examining the $[K^+]$ in plasma and urine. A contracted ECF volume should cause aldosterone release and a high urine $[K^+]$. The high plasma $[K^+]$ and the relatively low urine $[K^+]$ suggest that aldosterone deficiency is a cause for the renal Na^+ loss. One underlying diagnosis in this patient is low aldosterone bioactivity.

Gain of water: The ECF volume contraction also causes ADH release, which accounts for the failure to excrete free water.

Is the plasma ADH high, low, or normal in this patient?

Because the urine osmolality is tenfold greater than minimum values, ADH is acting. Therefore, plasma ADH is high.

6·17 Why does glucosuria cause enhanced excretion of Na^+ and Cl^-?

There are two answers to this question, a short one and a longer one. The short answer is that glucose prevents the reabsorption of water in the earlier nephron segments because fluid here must remain isosmolal to plasma and more glucose is filtered than can be reabsorbed in the proximal convoluted tubule (PCT). This extra volume drags Na^+ and Cl^- to distal sites at a rate that exceeds their reabsorptive capacity.

The longer answer requires us to be more specific. In the PCT, if the degree of hyperglycemia is less than 50 mmol/l (900 mg/dl), there will be no change in proximal Na^+ reabsorption. When the fluid leaves the PCT, however, the composition differs; the $[Na^+]$ is lower (100 vs the usual 150 mmol/l), there is more glucose (100 mmol/l), and there may be a larger volume to be delivered to the loop of Henle. Fewer Na^+ can be reabsorbed in the loop of Henle because the $[Na^+]$ in each liter of fluid delivered to the loop is now smaller (100 vs the usual 150 mmol/l) and reabsorption in the loop proceeds to a limiting $[Na^+]$ of close to 30–40 mmol/l. Hence, both the larger volume and the increased number of osmoles (glucose) that cannot be reabsorbed in the loop of Henle lead to a larger volume of fluid with a higher osmolality being delivered to the distal nephron. Even when ADH acts, less water will be reabsorbed distally because the osmolality of fluid entering the distal convoluted tubule is higher than normal (Table 6·4). As a result, a much larger volume is delivered to the collecting duct and might compromise complete reabsorption of Na^+ and Cl^- in these final nephron segments.

6·18 What is the volume of distribution of the common intravenous solutions?

The major premise is that isotonic saline remains entirely in the ECF, and electrolyte-free water distributes in proportion to the ICF and ECF volumes (two-thirds in the ICF and one-third in the ECF). Accordingly, the distribution of infusates can be calculated by dividing each solution into equivalent volumes of isotonic NaCl and electrolyte-free water (i.e., 1 liter of 75 mmol/l saline is 0.5 liters of isotonic saline and 0.5 liters of electrolyte-free water; see Table 6·9).

Table 6·9
CHANGES FOLLOWING THE INFUSION OF ONE LITER OF COMMONLY USED INTRAVENOUS SOLUTIONS

Solution	Composition	Volume change (liters)	
		ECF	ICF
Saline			
0.9%	150 mmol Na^+, Cl^-	1	0
0.45%	75 mmol Na^+, Cl^-	0.67	0.33
3.0%	512 mmol Na^+, Cl^-	2.6	− 1.6
D_5W	276 mmol glucose	0.33	0.67
⅔ D_5W, ⅓ Saline	183 mmol glucose 50 mmol Na^+, Cl^-	0.55	0.45

Table 6·9 Legend:
Assume that all the glucose infused in D_5W was metabolized. A sample calculation is provided in the margin.

Sample calculation:

1. Calculate the total osmoles in the body and in the ECF (osmolality × total body water, one-third in the ECF).
2. Add osmoles added (Na^+ and Cl^-) to total body osmoles, then divide by new total body water (original and infused volumes).
3. Divide new total osmoles by new total volume to obtain new osmolality.
4. Divide total ECF osmoles (original osmoles plus added Na^+ and Cl^-) by new osmolality to obtain new ECF volume.
5. Subtract new ECF volume from new total body water to deduce new ICF volume.

6·19 What are the purposes for the administration of the various IV solutions?

There are four different impacts that intravenous infusions have on body fluid compartments.

1. **Expand the ECF only.**
   Isotonic saline is the solution to infuse in patients with ECF volume contraction; the tonicity should be isotonic to the patient, not the doctor. If the vascular volume is especially low, infuse albumin, plasma, or red blood cells.

2. **Expand the ICF with as little fluid to the ECF as possible.**
   Infuse water with a particle that prevents osmotic lysis of blood cells but disappears promptly. The water disperses, with two-thirds going to the ICF. Glucose is the most commonly used particle, but beware—not all patients can metabolize large quantities of glucose rapidly (see the margin). This technique is used in patients with ICF volume contraction (hypernatremia).

3. **Expand both the ICF and ECF volumes.**
   With salt and water deficits, a dilute saline is infused. The first option is half-normal saline (75 mmol/l, 0.45%), two-thirds of which is distributed in the ECF; the second option, a solution of two-thirds D_5W and one-third NaCl, distributes close to 50% in each compartment (Table 6·9). This latter solution requires removal of glucose.

4. **Remove water from cells.**
   Hypertonic saline (3%, 5%) causes a water shift from cells. However, a necessary concomitant event is expansion of the ECF volume. This therapy is used in patients with symptomatic hyponatremia (see the margin).

Rate of metabolism of glucose:
Glucose can be metabolized at a rate of only 0.2 g/kg body weight per hour in an ill or stressed patient, equivalent to 14 g in a 70-kg patient. Because 14 g is close to 30% of 50 g, do not infuse more than 300 ml of D_5W per hour in this setting; less should be infused if the patient is already hyperglycemic.

6·20 When should hypertonic NaCl be infused?

Hypertonic saline is used to draw water out of cells (especially brain cells) in patients with symptomatic hyponatremia (e.g., convulsions).

Use of diuretics to minimize ECF volume expansion:
To avoid excessive expansion of the ECF volume, give a potent diuretic (e.g., furosemide) and replace the water and K^+ that are lost.

The driving force for this water shift is a rise of tonicity in the ECF (a rise in the Na^+/H_2O ratio). This therapy has the major disadvantage of expanding the ECF volume (see page 281, Chapter 7).

How much should be given?
The answer is simple—enough to stop the symptoms (i.e., the convulsion). Usually such an amount will be sufficient to raise the [Na^+] in plasma to the level that it was before the convulsion (3–5 mmol/l). A little caution is needed here because the convulsion increases the number of particles in the ICF (glycogen $\longrightarrow$ H^+ + lactate anions, and this increase raises the plasma [Na^+].

6·21 The ECF volume is expanded chronically in trained athletes. Why might this expansion be advantageous?

During exercise, the heart needs a large stroke volume; blood must fill the expanded capillary beds of exercising muscle to deliver O_2 and to dissipate heat via the skin. Further, there must be sufficient volume in the ECF to permit a shift of water into cells because vigorous exercise promotes a breakdown of macromolecules (e.g., glycogen) to many small particles (e.g., lactate anions) in cells. In addition, because exercise generates excessive heat, the body loses water and some Na^+ via sweat. To accommodate all of these demands, elite athletes maintain much of their extra ECF volume in the venous capacitance vessels. The bottom line is that the ECF volume varies in normal physiology.

6·22 A person was in Na^+ balance on two occasions, eating and excreting 150 mmol of Na^+. On the first occasion, he had two kidneys (the GFR was 200 liters per day), and on the second, he had one kidney (the GFR was 100 liters per day). What was his fractional excretion of Na^+ on each occasion?

The fractional excretion of an electrolyte is used to analyze how the kidneys handle a specific ion. It is calculated by dividing the quantity excreted by the quantity filtered. Because it is usually expressed as a percent, the value is multiplied by 100.

1. **Fractional excretion when GFR was 200 liters per day:**
   GFR (200 liters per day) × [Na^+] in plasma (140 mmol/l) = 28 000 mmol.
   Excretion of Na^+ = 150 mmol.
   Therefore, fractional excretion was 0.50% (i.e., he was excreting .5% of the filtered load).

2. **Fractional excretion when GFR was 100 liters per day.**
   In this setting, the excretion of Na^+ was identical, but the filtered load was halved. Therefore, the fractional excretion was double, or 1%.

Interpretation of the fractional excretion of Na^+ in chronic renal insufficiency: Assume that the fall in the GFR is now 10-fold, but the intake and excretion of Na^+ are the same. For balance, the fractional excretion will have to be 10-fold higher than normal, or 5%.

This increase in fractional excretion means that a much smaller quantity of Na^+ can be reabsorbed compared with the filtered load and that changes in reabsorption are induced in the kidney without necessarily changing the ECF volume or external stimuli to the kidney. These adaptations, in turn, say something about flow rates per nephron and tubuloglomerular signals.

7

HYPONATREMIA

PART C
TREATMENT OF HYPONATREMIA 268

PART D
REVIEW 272

Objectives

- To provide an appreciation that *hyponatremia* indicates an increase in water relative to Na^+ and implies expansion of the ICF volume (unless the hyponatremia is due to hyperglycemia).
- To indicate the difference between tonicity and osmolality and the clinical relevance of this difference.
- To provide a classification of patients with hyponatremia based on plasma osmolality and to indicate the physiological, pharmacological, and pathological causes of ADH release.
- To provide a therapeutic approach to hyponatremia and to emphasize the potential risks associated with aggressive therapy.
- To emphasize the role of available osmoles in facilitating water excretion and to highlight the importance of the nature of the urine osmoles in determining the impact of urine output on the $[Na^+]$ in plasma.

Abbreviations:
ADH = antidiuretic hormone.
CCD = cortical collecting duct.
DI = diabetes insipidus.
ECF = extracellular fluid.
ICF = intracellular fluid.
SIADH = syndrome of inappropriate ADH secretion.
TURP = transurethral resection of the prostate.

Hyponatremia:
- Definition: $[Na^+]$ less than 136 mmol/l.
- Causes: Na^+ loss and/or water gain.
- Hyponatremia without cell swelling indicates a laboratory error caused by hyperlipidemia, hyperproteinemia, or the presence of hyperglycemia.
- When hyponatremia is present during hyperglycemia, some cells swell and others shrink.

Outline of Major Principles

1. With the exceptions of *pseudohyponatremia* and hyperglycemia, hyponatremia indicates a net water shift into cells and swelling of cells. This shift is especially important in the CNS because the brain is enclosed in a fixed space, and swelling causes symptoms.
2. In virtually every patient with chronic hyponatremia, one can find evidence of Na^+ depletion and water gain. A clinical aim is to find out which of these processes is the primary one.
3. Most commonly, hyponatremia is due to water gain. Water gain is almost always due to the failure to excrete the quantity of water ingested. The kidneys of a 70-kg adult can make 12 liters of "electrolyte-free water" per day; thus, excessive water intake alone is virtually never the primary cause of hyponatremia. Reduced "free water" excretion is usually due to the actions of ADH. The release of ADH from the posterior pituitary occurs when there is hypertonicity, a reduced "effective" circulatory volume, neural stimuli, or drug ingestion. A very low rate of excretion of osmoles may also limit water excretion.

   An uncommon cause of hyponatremia is Na^+ loss in the absence of water gain; this loss leads to ECF volume contraction. Examine the urine electrolytes when determining the cause of Na^+ loss. The urine should have a minimal $[Na^+]$ and/or $[Cl^-]$ if the kidneys are responding appropriately. Treatment consists of Na^+ replacement and attention to underlying problems. Beware of a rapid excretion of water once the ADH level falls.

Pseudohyponatremia:
A condition in which the Na^+ to water ratio in plasma is normal, but the lab reports hyponatremia because the nonaqueous phase of plasma (protein or lipids) is increased and Na^+ are distributed only in the aqueous phase (discussed in detail later).

Notes:
- Remember that the $[Na^+]$ is the Na^+ to H_2O ratio!
- Hyponatremia indicates swelling of cells. This is virtually the only time that the denominator of the ratio is more important than the numerator in providing critical information.
- A Na^+ deficit means ECF volume contraction, and a Na^+ excess means ECF volume expansion.

4. There can be no specific therapy for the hyponatremia because it can be associated with different diseases and clinical settings. The options for therapy are net water loss or Na^+ gain, depending on the ECF volume and whether or not the patient is symptomatic. The underlying disorder will require specific treatment.

5. The major aim of therapy—to shrink the ICF volume towards normal—involves raising the plasma [Na^+]. The major issue in chronic hyponatremia is to do it slowly—raise the plasma [Na^+] 0.5–1.0 mmol/l/hr, or 12 mmol/l/day. If the hyponatremia is acute or the patient is seriously symptomatic (i.e., is having a convulsion), raise the [Na^+] rapidly to a level where convulsions cease (a 3–5 mmol/l rise in plasma [Na^+] is usually required).

Introductory Case
Water, Water, Everywhere, So Not a Drop to Drink

(Case discussed on pages 272–73)

Five days ago, a previously healthy young man developed aseptic meningitis accompanied by frequent vomiting. He had taken no medications. Physical examination revealed ECF volume contraction on day three, but this volume was normal yesterday (day five). Laboratory results are summarized below.

		Day 3	Day 5
Plasma			
Na^+	mmol/l	130	117
K^+	mmol/l	3.1	4.0
Cl^-	mmol/l	87	83
HCO_3^-	mmol/l	33	24
Osmolality	mosm/kg H_2O	270	245
Glucose	mmol/l (mg/dl)	5.0 (90)	3.0 (54)
Creatinine	mmol/l (mg/dl)	200 (2.3)	90 (1.0)
Urine			
Na^+	mmol/l	60	50
Cl^-	mmol/l	6	48
Osmolality	mosm/kg H_2O	450	407

Could Na^+ loss have caused the hyponatremia on day three?
Why was the ADH level elevated on day three?
Why was the plasma [Na^+] lower on day five?
Was the ADH level higher on day five?
What are the best urine electrolytes to indicate whether the ECF volume is contracted?
Would your answer change if the patient had vomiting or diarrhea?

PART A CLINICAL APPROACH

Background

$[Na^+]$ = ICF VOLUME

The ICF volume, which can be deduced from the plasma $[Na^+]$, is contracted in hypernatremia and expanded in hyponatremia. The only exceptions to this rule are pseudohyponatremia and the addition of a "Na^+-like" particle to the ECF, such as glucose or mannitol (see the margin).

Note:
A "Na^+-like" particle is one that is largely restricted to the ECF.

Na^+ CONTENT = ECF VOLUME

Sodium, an osmole that is largely restricted to the ECF, keeps water out of cells; thus, the Na^+ content determines the ECF volume because Na^+ (and attending anions) are the principal solutes of the ECF.

DANGER OF HYPONATREMIA

In patients with hyponatremia and hypotonicity, the ICF volume is expanded, and swollen brain cells may compromise the blood supply to the brain. On a chronic basis, brain cells tend to shrink their volume back to almost normal size largely by ion extrusion, a process that has major implications for the speed of correction of hyponatremia (correction is slow if hyponatremia is chronic and the patient is asymptomatic).

Note:
Correction of hyponatremia should be slow in an asymptomatic patient with chronic hyponatremia.

RENAL RESPONSE TO Na^+ LOSS

The appropriate renal response to a Na^+ deficit (ECF volume contraction) is to avoid additional excretion of Na^+, Cl^-, or water in the urine. Check for these responses: the urine should have a very low $[Na^+]$ and $[Cl^-]$, and the urine osmolality should be high (a reduced "effective" vascular volume leads to ADH release).

Note:
Compare urine osmolality to the "expected" value, not to plasma osmolality.

RENAL RESPONSE TO WATER EXCESS

The appropriate renal response to water excess is to excrete the maximum volume of maximally dilute urine (20–80 mosm/kg H_2O). If this response is not observed, suspect that ADH is acting (seek the cause!) or that the kidneys are abnormal.

"Effective" osmolality of plasma:
The aspect of osmolality that causes a movement of water across cell membranes. When determining the "effective" osmolality of plasma, ignore the contribution of urea and ethanol because they do not influence water shifts across cell membranes.

Hyponatremia and a Normal "Effective" Plasma Osmolality

The approach to hyponatremia is summarized in Figure 7·1. It is based on whether or not the "effective" plasma osmolality (tonicity) is low. The first step in the assessment of the patient with hyponatremia is to rule out

Figure 7·1
Approach to hyponatremia
The final diagnoses are shown in the shaded boxes (see the text for details).

Note:
By raising the plasma osmolality, a high concentration of urea might lead one to the mistaken belief that pseudohyponatremia or hyperglycemia is present.

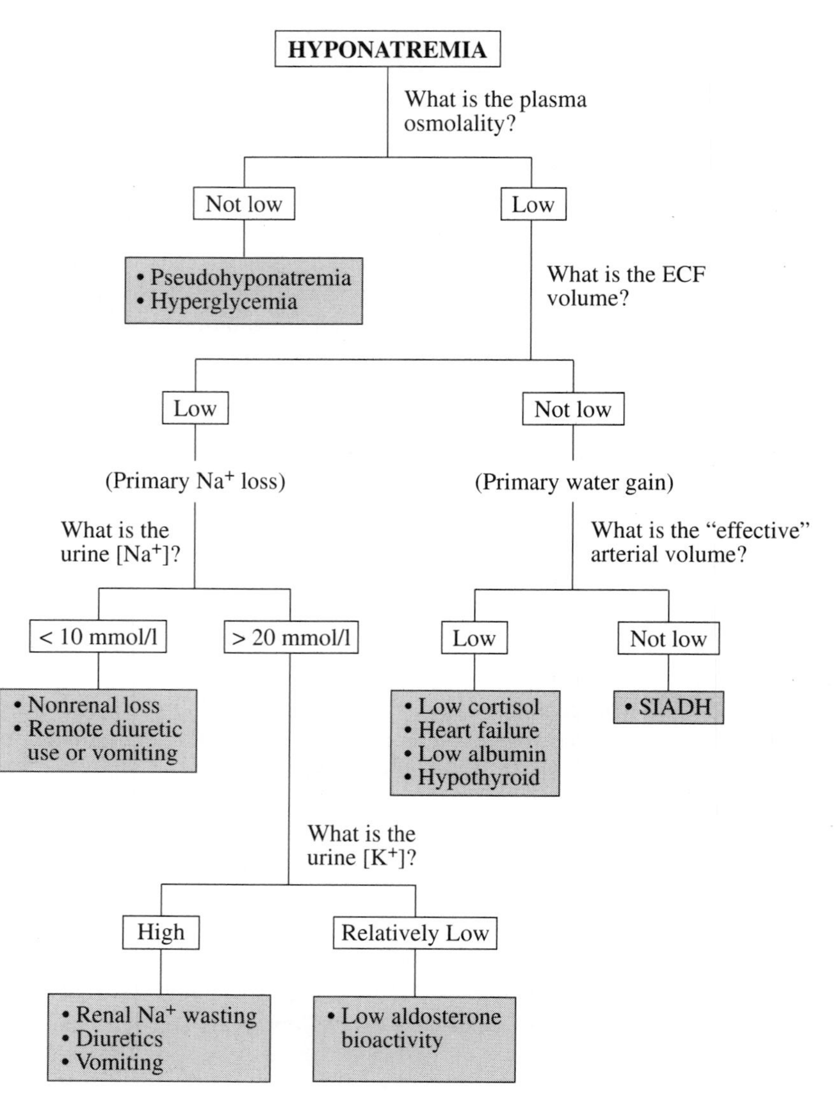

Note:
To use the urine electrolytes in a more sophisticated way to determine the basis for a contracted ECF volume, see Table 4·5, page 160.

pseudohyponatremia because it is not associated with the danger of cell swelling and requires no therapy for hyponatremia. In the presence of pseudohyponatremia, the "effective" plasma osmolality (Na^+: plasma water ratio) is normal. These patients are labelled as hyponatremic by the laboratory because the lab uses plasma volume, not plasma water, as the denominator of this ratio (see the discussion of Question 7·1, pages 281–82).

Question

(Discussion on page 282)

7·1 What laboratory methods to measure the $[Na^+]$ in plasma will provide a lower value if hyperlipidemia is present?

Hyponatremia and an Elevated "Effective" Plasma Osmolality

Hyperglycemia causes a shift of water from muscle cells; this shift results in hyponatremia with a relatively high plasma osmolality. It is important to recognize that water shifts caused by hyperglycemia do not affect all cells equally (see Figure 12·5, page 437). If mannitol was administered and retained, it could cause a similar type of hyponatremia. In contrast, a high osmolality from urea or ethanol does not cause hyponatremia from water shifts because these compounds are not "effective" osmoles at the cell membrane level (they rapidly achieve an equal concentration in the ICF and ECF).

Hyponatremia resulting from hyperglycemia:
In quantitative terms, expect a 1.3 mmol/l fall in the [Na^+] for every 100 mg/dl (5.5 mmol/l) rise in the glucose concentration; the [Na^+] falls 1 mmol/l with a 4 mmol/l (72 mg/dl) rise in blood glucose concentration. This calculation incorporates the fact that the glucose concentration in the ICF of the liver (and other noninsulin-sensitive cells with respect to glucose transport) is the same as that in the ECF. In this setting, the osmolality of plasma is not low.

Hyponatremia and a Low "Effective" Plasma Osmolality

The majority of patients with hyponatremia will have a low plasma osmolality and may suffer from swelling of brain cells. All causes of hyponatremia with a low "effective" osmolality have an expanded ICF volume. The brain is most affected because it is in a "closed box." Therapy is designed to reverse brain swelling. The price to pay for this therapy is expansion of the ECF volume.

To understand the cause of hyponatremia with a low plasma osmolality, the input and output of Na^+ and water must be examined independently. Typical amounts consumed by an adult on a Western diet are approximately 150 mmol of Na^+ per day and close to 1.5 liters of water per day; wide variations about this average should be expected. On the output side, to stay in balance, a normal subject would have to excrete 150 mmol of Na^+ in 1.5 liters of urine; thus, the 24-hour urine [Na^+] should be, on average, close to 100 mmol/l. Therefore, with a typical diet, hyponatremia can result if the [Na^+] in urine exceeds 100 mmol/l or if water is ingested without being excreted for a period of time. Note that the [Na^+] in the patient's diet and urine—not the osmolality—are critical factors in understanding the renal contribution to hyponatremia (see the discussion of Question 7·5 on page 283).

The causes of hypoosmolal hyponatremia can be divided into two major groups based on the ECF volume. If hyponatremia is associated with ECF volume contraction, Na^+ loss had occurred. If hyponatremia is associated with a normal or high ECF volume, look at the "effective" circulating volume. Low cardiac output suggests a circulatory problem, and normal values suggest autonomous ADH release (Table 7·1).

Hyponatremia with hypoosmolality:
The physiologic (or appropriate) stimuli for ADH release are hypertonicity and decreased "effective" circulating volume.

Notes:
- Measuring the urine osmolality helps one assess ADH action and the degree of medullary hyperosmolality.
- Measuring the [Na^+] + [K^+] in urine helps one determine if the urine excreted will raise or lower the plasma [Na^+].

Table 7·1
CAUSES OF HYPONATREMIA

Pseudohyponatremia (normal plasma osmolality)
- Hyperlipidemia, hyperproteinemia

Addition of particles largely restricted to the ECF
- Hyperglycemia, mannitol, sorbitol

Table 7·1 is continued on page 260.

Decreased "effective" circulating volume

- ECF volume contraction (loss of Na^+)
- ECF volume not low (edema states)
  - Low vascular volume
    - Hypoalbuminemia (liver, kidney, nutritional, or GI problem)
    - Leakage of albumin from capillaries (e.g., sepsis)
  - Low arterial volume but expanded venous volume, e.g., heart problems

Primary water gain (and secondary Na^+ loss)

- ADH release from the posterior pituitary without the stimulus of tonicity or "effective" circulatory volume
  - Drugs (see Table 7·3), excessive pain, nausea, vagal nerve stimulation, CNS disease, metabolic disorders (e.g., porphyria)
  - ADH from other sources (neoplasms, exogenous administration)
- Drugs potentiating ADH action (see Table 7·4)
- Drugs interfering with formation of free water (e.g., loop diuretics, thiazides)
- Decreased available osmoles for excretion ("tea and toast" diet; see Case 10·1, pages 377 and 379–80)

Questions

(Discussions on pages 282–84)

7·2 Match the four separate causes of hyponatremia with the critical parameter for diagnosis.

Etiology	Critical parameter measured
(1) Chronic diuretic intake	(a) Normal plasma osmolality
(2) Hyperlipidemia	(b) Normal ECF volume and low urine osmolality
(3) Excess ADH	(c) ECF volume contraction and hypokalemia
(4) Compulsive water consumption	(d) Low plasma and high urine osmolality

7·3 A diabetic patient has renal failure. His blood sugar is 1000 mg/dl (55 mmol/l), $[Na^+]$ is 127 mmol/l, and blood urea nitrogen is 70 mg/dl (urea is 25 mmol/l). What is his calculated osmolality? Has hyperglycemia changed his ECF and ICF volumes?

7·4 What is the average urine osmolality in 24-hour urine in a normal subject?

7·5 What conclusions can be drawn when the urine osmolality is high?

7·6 Three patients have hyponatremia (120 mmol/l). In case A it is due to water gain; in case B, Na^+ loss; in case C, hyperglycemia. Which patient(s) will have swelling of brain cells? Give reasons for your answer.

HYPONATREMIA AND A CONTRACTED ECF VOLUME

- Na^+ loss causes a low blood volume and ADH release.
- Water ingested is retained.
- Hence, the causes of hyponatremia are Na^+ loss (primary) and water retention (secondary).

Hyponatremia of clinical significance is almost always due to reduced water excretion in the face of a hypotonic intake. Loss of Na^+ without water could contribute to hyponatremia, but, without water retention, hyponatremia would be mild because loss of Na^+ would lead to a significant degree of contraction of the ECF volume. Hence, cardiovascular collapse limits the ability to develop a very severe degree of hyponatremia.

The mechanism for the Na^+-loss type of hyponatremia usually involves three steps. First, the loss of Na^+ lowers the numerator of the $Na^+:H_2O$ ratio. Second, and more importantly, ADH is released from the posterior pituitary in response to the lower blood volume. ADH prevents the excretion of dilute urine. Third, there must be water ingestion; thirst is stimulated by a low ECF volume. Thus, it is misleading to call this hyponatremia "depletional" because there is usually a "dilutional" component as well.

Causes of Na^+ loss:
Loss of Na^+ can occur via nonrenal and renal mechanisms. The nonrenal sources are usually obvious (gastrointestinal tract, skin) and should be suspected when the urine is virtually Na^+- and Cl^--free all the time. The urine $[K^+]$ (reflecting aldosterone actions) and osmolality (reflecting ADH actions) may be high despite hyponatremia.

The most common renal cause for Na^+ wasting is the ingestion of a diuretic; less often, renal salt wasting and/or an osmotic diuretic (glucose, urea) may cause Na^+ loss. The most useful index to discriminate among the number of kidney diseases associated with renal Na^+ wasting is an examination of K^+ excretion. A low $[K^+]$ in urine in the face of renal Na^+ loss and ECF volume contraction should suggest low aldosterone bioactivity (see pages 401–03 for details). In contrast, a high $[K^+]$ in urine with renal Na^+ wasting suggests that the abnormal loss of Na^+ occurs in the proximal convoluted tubule, loop of Henle, or early distal convoluted tubule.

HYPONATREMIA AND A NORMAL OR HIGH ECF VOLUME

- A patient need not have a reduced ECF volume to release ADH on a "volume basis."

Even if the total ECF volume is normal or high, the vascular component may not be "pumped in an optimum fashion" by the heart. The term "effective" circulating volume is used to refer to the critical component of the ECF required for adequate tissue perfusion (notice the vague terms in this description). In a more practical clinical sense, this "effective" volume is decreased either when the overall ECF volume is reduced or when the ECF volume is distributed such that there is insufficient volume in the vascular space. This latter maldistribution occurs in some edema states (e.g., hypoalbuminemia). A second type of maldistribution occurs when the arterial volume is low and the venous volume is high (when there is a primary decrease in cardiac function; see Table 7·2).

Table 7·2
CAUSES OF "EFFECTIVE" CIRCULATORY VOLUME CONTRACTION

1. **ECF volume depletion (overall Na^+ loss)**
   Nonrenal Na^+ loss
   - Gastrointestinal tract: e.g., vomiting, drainage, ileus, diarrhea
   - Skin: excessive sweating, burns
   - Hemorrhage

   Renal Na^+ loss
   - Diuretics, both pharmacologic or osmotic
   - Low aldosterone bioactivity
   - Tubular disorders
     - Proximal (Fanconi's syndrome, etc.)
     - Loop (Bartter's syndrome?)
     - Distal (interstitial disease, low aldosterone bioactivity)
2. **ECF volume normal but maldistributed**
   High interstitial volume and low vascular volume
   - Low plasma albumin (e.g., liver disease, nephrotic syndrome)
   - Leakage of albumin from capillaries

   Low arterial volume and high venous volume
   - Primary myocardial, valvular, or pericardial disease

Questions

(Discussions on pages 284–85)

7·7 Could Na^+ loss have occurred via the renal route if the urine is now Na^+-free?

7·8 What role might K^+ depletion play in determining the severity of hyponatremia?

7·9 How much water must a normal person drink to produce hyponatremia? Would it matter if this person is on a low-salt, low-protein diet?

7·10 Why do patients with adrenal insufficiency commonly present with hyponatremia?

Hyponatremia Associated with a Low "Effective" Circulating Volume

Pathophysiology:
- Na^+ maintain the ECF volume.
- Loss of Na^+ (ECF volume) stimulates ADH release.
- ADH prevents the excretion of free water.

ADH can be released and thirst stimulated if there is an important reduction in the volume of blood delivered to vital organs (water intake must be present to develop hyponatremia). This reduced volume is sensed by receptors in the arterial or venous limbs of the circulation. Hence, one need not have overall ECF volume contraction to have a lower total or arterial circulating volume.

Hyponatremia Associated with Autonomous Release of ADH

- ADH release in *SIADH* may be from the posterior pituitary or from a nonphysiologic source (e.g., a neoplasm).

SIADH:
A syndrome in which ADH is present and acting, but the two primary stimuli for its release (hypernatremia and low "effective" circulating volume) are absent.

In virtually all patients with hyponatremia and hypoosmolality, water gain is present and is due to water intake plus ADH action. If a low "effective" circulating volume is not present, ADH release is not under the usual physiological control and the diagnosis is called the "syndrome of inappropriate ADH secretion" (SIADH). The causes for this high ADH level can be divided into two major categories: ADH release from the normal gland (posterior pituitary) and ADH or ADH-like material from other sources. A list of causes of high ADH levels is presented in Table 7·3.

Table 7·3
ORIGIN OF THE HIGH ADH IN HYPONATREMIA

ADH released from the posterior pituitary gland in response to physiologic stimuli
- Cardiovascular (low "effective" circulating volume)
- Other diseases or organ damage leading to the release of ADH
  - Pulmonary or CNS lesions, endocrine disorders (e.g., hypothyroidism, hypoadrenalism), metabolic disorders (e.g., acute intermittent porphyria), excessive pain (e.g., during post-operative period), nausea, vomiting, or anxiety (psychotics).
- Drug-induced (see Table 7·4)

ADH released from other sources
- Solid neoplasms (especially oat-cell carcinoma of the lung), possibly granulomas such as tuberculosis
- Exogenous administration
  - ADH (e.g., treatment for diabetes insipidus)
  - Oxytocin for labor induction

Drugs that potentiate the actions of endogenous ADH
- Adenylate cyclase activation, phosphodiesterase inhibition, adenosine receptor antagonists, prostaglandin inhibitors (Table 7·4)

Table 7·4
DRUGS THAT MAY CAUSE HIGH ADH ACTIVITY

This table is only a partial listing because the list continues to grow.

Central stimulation of ADH release
- Drugs include: nicotine, morphine, clofibrate, tricyclic antidepressants, antineoplastic agents such as vincristine, cyclophosphamide (probably acting via nausea and emesis)

"ADH-like" agents such as oxytocin

Drugs that promote the actions of ADH on the kidney by increasing cyclic AMP levels or bioactivity
- Oral hypoglycemics (e.g., chlorpropamide), methylxanthines (e.g., caffeine, aminophylline), analgesics that inhibit prostaglandin synthesis (e.g., aspirin, indomethacin)

Note:
Prostaglandins antagonize ADH action by inhibiting the formation of cyclic AMP. The synthesis of prostaglandin, stimulated by ADH, diminishes the ADH-induced cyclic AMP rise. In addition, the vasodilator action of prostaglandins increases medullary flow, which lowers medullary hyperosmolality. These actions of prostaglandins promote the excretion of a less concentrated urine.

What are the consequences of a persistently elevated ADH level? In general, the clinical effect is to limit the excretion of osmole-free water. Thus, if water is not ingested or administered, the patient will suffer virtually no consequences from a high ADH level. When water is ingested, most is retained; the result is hyponatremia with initial ECF volume expansion. In response to this ECF volume expansion, Na^+ are excreted in the urine. Hence, the characteristic findings during the developmental stage are hypotonicity, hyponatremia, and the absence of ECF volume contraction. The urine has an osmolality that exceeds 20–80 mosm/kg H_2O and contains an appreciable quantity of Na^+. After attaining a steady state, the rate of excretion of Na^+ will vary depending on dietary intake (and excretion) of Na^+ and water.

In summary, before confirming the diagnosis of SIADH, be sure that there is no reduction of the "effective" circulating volume (check the urine electrolytes, see Table 6·8, page 239). Patients with SIADH can be classified into four subcategories depending on why ADH is present (Figure 7·2).

A patient with SIADH may also have ECF volume contraction for other reasons. In this case, the urine [Na^+] should be very low unless there is a renal basis for the Na^+ loss. In this setting, however, hyponatremia will not be corrected by a large water diuresis after the Na^+ deficit is replaced.

Question

(Discussion on page 285)

7·11 A patient with hyponatremia (127 mmol/l) excretes urine with an osmolality of 176 mosm/kg H_2O. How would you know if this excretion was due to a reset osmostat type of SIADH?

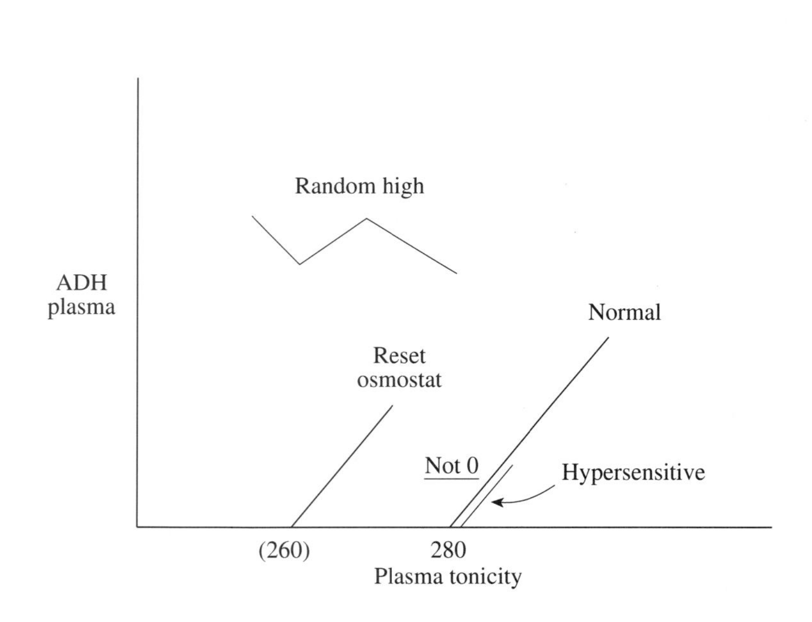

Figure 7·2
Subclassification of patients with SIADH (modified from the original by G. L. Robinson)
There are four subgroups of SIADH:

1. Random high autonomous ADH release (e.g., carcinoma of the lung).
   These patients represent 35–40% of of those with SIADH.
2. Reset osmostat.
   These patients have normal regulation of ADH release but it is focused around a hypotonic stimulus. About 33% of patients with SIADH fall into this subcategory.
3. Failure to suppress ADH totally with hypotonicity.
   About 15% of patients have this disorder, in which ADH release is normal at high tonicity but is not 0 at low tonicity.
4. About 15% of patients with SIADH have no problem with ADH secretion but their kidneys are either overly sensitive to it or there is an ADH-like material present.

Auxiliary tests in patients with SIADH: There are four laboratory values in plasma that help in the differential diagnosis of SIADH.

1. **Plasma [K^+]:**
   In SIADH, the plasma [K^+] is usually normal; therefore, abnormalities in the plasma [K^+] should alert one to seek other causes for hyponatremia.
   When the ECF volume is low, release of aldosterone results in renal K^+ loss. This loss is especially high when volume delivery to the terminal cortical collecting duct (CCD) is high. Therefore, diuretic-induced or vomiting-induced hyponatremia should be suspected if hypokalemia is present.
   In contrast to the above, suspect states with low levels of aldosterone if hyponatremia is accompanied by hyperkalemia.

Plasma [K^+]:
- Usually normal in SIADH.

2. **Plasma [HCO_3^-]:**
   In SIADH, the plasma [HCO_3^-] is usually normal, but it may be high in vomiting or in diuretic-induced hyponatremia (metabolic alkalosis).
   In contrast, hyponatremia of hypoaldosteronism is generally accompanied by a mild fall in the plasma [HCO_3^-] (to close to 20 mmol/l) because of a low excretion of NH_4^+ consequent to renal actions of hyperkalemia (see the discussion of Question 1·24, page 42).

Plasma [HCO_3^-]:
- Usually normal in SIADH.

3. **Concentration of urea in plasma:**
   In patients with SIADH, urea clearance increases, probably as a result of ECF volume expansion. This occurrence, together with dilution due to water retention, results in a fall in the concentration of urea in plasma (BUN).
   In contrast, hyponatremia from a low "effective" circulating volume is associated with a higher concentration of urea in plasma because the stimulus to ADH release (ECF volume contraction) also leads to a fall in the GFR. Because the changes in the concentration of creatinine in plasma are less pronounced, the ratio of urea to creatinine in plasma is likely to be low in SIADH and high with ECF volume contraction.

Plasma urea:
- Usually low in SIADH.

4. **Concentration of urate in plasma:**
   The urate level may also be quite low in patient with hyponatremia caused by SIADH.

Plasma urate:
- Usually low in SIADH.

PART B
CLINICAL SETTINGS

Hyponatremia is the most common electrolyte abnormality observed in a hospital setting; its incidence in a typical North American hospital is close to 1%. The most common scenario is the patient who releases ADH because of ECF volume contraction, pain, or emesis and then develops hyponatremia when a water load is consumed. In most cases, it is simply an incidental finding. In the paragraphs to follow, we have selected clinical settings where hyponatremia may become the dominant part of the picture.

Hyponatremia in the Infant

Apart from unique disorders such as inborn errors, hyponatremia in this setting is most commonly due to a loss of Na^+ (e.g., in diarrhea). ADH is released in response to the contracted ECF volume and leads to the retention of the water that is ingested. If a hyponatremic infant is fed sugar water to "rest the GI tract" and avoid "*dehydration*," this water will be retained. Thus, hyponatremia has two components—Na^+ loss and water gain—and its degree may be very severe. The primary aims during therapy are to reexpand the ECF rapidly and to return the ICF to normal slowly. The problem to be aware of is that ADH levels might fall precipitously once the ECF volume is reexpanded and may therefore yield a very rapid water diuresis. If the plasma [Na^+] rises too rapidly, *central pontine myelinolysis* could occur; to avoid this problem, one should have a fast acting ADH preparation for use.

Dehydration:
We dislike this term because it is ambiguous. To some, it means a lack of water, but to others, it means a deficit of ECF volume.

Central pontine myelinolysis (CPM):
A symmetrical demyelination, primarily in the base of the pons, that was first described in 1959. Clinically, there is a wide spectrum of presentations ranging from the absence of symptoms to a devastating disorder characterized by confusion, agitation, and, eventually, flaccid or spastic quadriparesis. This disorder may be accompanied by bulbar involvement. The diagnosis can be confirmed by MRI scanning, including coronal views. There is no known treatment, and the mortality and morbidity rates are high. CPM may occur in patients with disorders other than hyponatremia; it is observed after therapy for hyponatremia, and the incidence rises sharply if the hyponatremia is overcorrected. Hyponatremic patients in poor nutritional condition seem more likely to develop CPM.

Hyponatremia in Young Women Undergoing a Minor Surgical Procedure

Every year, cases appear in court for litigation with the following scenario. An otherwise healthy, often small woman becomes severely and acutely hyponatremic over the 24–48 hours following minor surgery. Some may experience brain swelling with herniation and ultimately develop central diabetes insipidus aggravated by hyperglycemia.

The issues seem to be as follows: ADH is released in response to the surgery, the low "effective" circulating volume associated with this setting, the pain or drugs administered during surgery, and the nausea and/or vomiting in the postoperative period. The problem occurs when Na^+-free water is administered and it cannot be excreted because of the actions of ADH. The typical IV solution is *D_5W*, and several liters are administered. This volume is relatively large given the small size of these women and/or the percentage of water in their body composition.

The natural history is easy to explain from a pathophysiologic point of view. First, the degree of hyponatremia is entirely predictable from the positive water balance. Second, herniation occurs with brain swelling; damage to the posterior pituitary system leads to central diabetes insipidus. Third,

D_5W:
A solution of glucose in water with the same osmolality as body fluids but not the same tonicity. It contains 5 g of glucose per 100 ml, or 50 g of glucose per liter.

the hyperglycemia reflects the very large infusion of glucose into a patient in whom glucose oxidation is very slow because of the catecholamine-induced metabolic responses (high release of fatty acids and inhibition of insulin release; see Chapter 12, page 433, for more discussion).

This syndrome is entirely preventable by selecting the appropriate intravenous fluids. Isotonic saline should be given initially in the operative and perioperative periods. Hypotonic saline or D_5W should be given only if the plasma [Na^+] rises above 140 mmol/l. More should be given if there is an unusual polyuria with a [Na^+] in urine that is less than 100 mmol/l.

The final issue is whether there is a unique cerebral susceptibility in this population to the ravages of sudden, massive brain swelling. While there is some evidence to support this view, it is not compelling at present.

Rationale for IV therapy in patients undergoing surgery.
There are two components to this therapy. First, one should ensure a normal "effective" circulating volume by infusing 1–2 liters of isotonic saline (the amount given depends on the nature of the operation and the size of the patient). Second, electrolyte-free water should be given only if the plasma [Na^+] exceeds 140 mmol/l because these patients have a restricted ability to excrete water (see page 269 regarding insensible loss of water).

Questions

(Discussions on pages 285–86)

7·12 If a person on a low-salt, low-protein diet had surgery, what features might be present in the postoperative period to permit more of the administered electrolyte-free water to be retained while ADH is acting?

7·13 A patient had surgery. Although the plasma [Na^+] was always in the range of 139–141 mmol/l, the preoperative plasma [Na^+] was 136 mmol/l. Why? The plasma [Na^+] was 128 mmol/l 36 hours after surgery. Why?

Hyponatremia After Prostate Surgery

During a transurethral resection of the prostate (TURP), the urologist may need to use electric cautery and lavage the prostatic bed. Hence, the lavage solution needs osmoles to avoid hemolysis (a variable quantity of fluid is absorbed systemically), but the fluid cannot contain electrolytes and conduct electricity. The "solution" for this problem is the use of organic osmolytes such as mannitol, sorbitol, glycerol, or glycine in half-osmolal to isosmolal strength. When these solutions are absorbed systemically, hyponatremia can result, but it is usually transient in nature. Although a severe degree of hyponatremia may develop, its net effect on swelling of brain cells may be mild. The shrinking caused by the added organic osmoles offsets the swelling from hypoosmolal hyponatremia. The features of this syndrome are discussed more fully in Case 7·4, pages 275, 279–80.

Question

(Discussion on page 286)

7·14 What properties of glycine make this compound unique in the pathogenesis of the post-TURP hyponatremia syndrome?

PART C
TREATMENT OF HYPONATREMIA

Note:
Because of limited space, we shall not deal with the diagnosis and treatment of the underlying disease state that has caused the hyponatremia.

- Hyponatremia is not a specific disease. It is a fluid and electrolyte "symptom."
- Two specific lines of attack should be considered:
  - Deal with the underlying disease (e.g., improve the arterial circulation if necessary).
  - Withhold water and promote water loss.
- Raise the $[Na^+]$ towards normal. Speed depends on the presence of symptoms and on the acuteness of hyponatremia; the actual plasma $[Na^+]$ is less important.

Status of the ECF Volume

For treatment of a patient with hyponatremia, the status of the ECF volume is important. The aims of therapy are to give Na^+ if the ECF volume is contracted and to remove water if the patient has a surplus of water (Table 7·5). The following examples should illustrate the principles involved.

Use a loop diuretic to cause a negative balance for water:

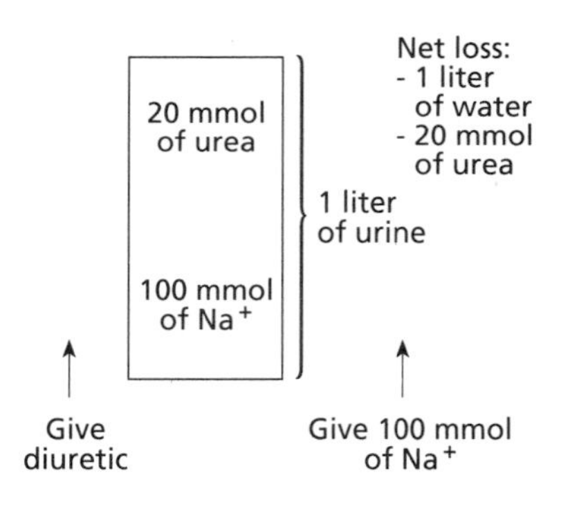

Table 7·5
TREATMENT OF A WATER SURPLUS

Restrict water intake.

Promote water loss independent of ADH.
- Give diuretics and replace electrolytes but not water (see the margin).
- Give osmoles (urea) to increase the volume of urine.

Reduce ADH levels.
- Correct the low "effective" circulating volume.
  - Give NaCl if the total ECF volume is low.
  - Give albumin if its concentration is very low.
  - Improve myocardial function (inotropic agents, afterload reduction, chronotropic agents).
  - Replace hormone deficiencies (glucocorticoids, thyroid hormone).
  - Remove drugs that promote ADH release or potentiate ADH action.

Administer antagonists to ADH.
- Give receptor blockers (new), lithium, demeclocycline.

HYPONATREMIA IN A PATIENT WITH A NORMAL ECF VOLUME

During oliguria:
Look at the $[Na^+]$ and $[K^+]$ in urine, not the osmolality, to determine if the urinary excretion is leading to a rise or fall in the $[Na^+]$ in plasma.

The major problem in this setting is a surplus of water in the ICF along with a deficit of Na^+ in the ECF. Therefore, the main aim of therapy is to lose water but not Na^+; in fact, some Na^+ must be retained to prevent contraction of the ECF volume. Treatment will involve limited intake of water, as well as the excretion of water without an overall loss of Na^+. Since the urine contains water plus Na^+, one should replace all urinary losses of Na^+;

do not permit the plasma [Na^+] to rise more than 12 mmol/l/day in chronic hyponatremia (some water administration may be required). One can rely on the diet to replace Na^+ unless the degree of hyponatremia is severe. In this case, Na^+ (or K^+) must be given (see the discussion of Case 10·1, pages 379–80).

If the urine output is not large enough, an option is to increase it with a loop diuretic and replace the losses of electrolytes and extra water. Another option is to give osmoles (urea) to facilitate the excretion of more water.

HYPONATREMIA IN A PATIENT WITH AN EXPANDED ECF VOLUME

The major problems in this setting are a surplus of Na^+ and a surplus of water. Therefore, the aim of therapy is to lose both Na^+ and water. The speed with which Na^+ should be lost depends on clinical features; it should be very rapid if there is acute pulmonary edema.

The [Na^+] in the urine should be lower than that of the plasma to cause a rise in the plasma [Na^+]. Adjust the negative balances for Na^+ and/or water to produce the desired slow rise in the plasma [Na^+].

Three other points merit emphasis. First, the intake of Na^+ and water should be nil. Second, urinary losses of Na^+ and water may have to be augmented with a loop diuretic, and part of the loss of Na^+ and/or water may have to be replaced to have the desired rise in the [Na^+] in plasma (0.5 mmol/l/hour). Third, the loss of K^+ will have to be replaced as well. In a patient who is hypokalemic, replacement of a deficit of K^+ will raise the plasma [Na^+] because as K^+ enter cells, Na^+ will exit.

HYPONATREMIA AND A CONTRACTED ECF VOLUME

The major problems in this setting are a deficit of Na^+ and water in the ECF, as well as a surplus of water in the ICF. Therefore, the aim of therapy is to give Na^+ to reexpand the ECF volume and to shift water out of cells. Since water moves rapidly across cell membranes, give 1 mmol Na^+ per liter of total body water to raise the plasma [Na^+] by 1 mmol/l. This treatment will expand the ECF volume by close to 2% (see the margin).

To expand the ECF volume much more rapidly (e.g., if the patient is in shock), give several liters of fluid with the same [Na^+] plus [K^+] as that of the plasma of the patient (see the margin).

A potential danger of this therapy is that ADH levels will fall promptly when the ECF volume is reexpanded, and too large a water diuresis could ensue. If a water diuresis begins, it can be curtailed by giving ADH.

Water Produced by Metabolism

The oxidation of carbohydrates and fat yields water as a product. Nevertheless, one need not worry about the volume produced because it is excreted along with CO_2 via the respiratory tract in a volume equal to that produced (see the margin).

Calculation:
In a 70-kg person with 45 liters of body water, 45 mmol of Na^+ will be given to raise the plasma [Na^+] by 1 mmol/l. Because water crosses cell membranes rapidly, the apparent volume of distribution of Na^+ is total body water even though Na^+ remain in the ECF. The patient normally has close to 2000 mmol of Na^+ in the ECF (15 liters × 140 mmol/l). Therefore, raising the Na^+ content 45 mmol is close to a 2% rise in Na^+ content (and ECF volume).

Clinical pearls:
1. Expand the ECF by infusing a solution that is isotonic to the patient rather than to the doctor.
2. The main reasons for a rapid rise in the [Na^+] in plasma are the infusion of hypertonic saline and/or a rapid excretion of free water; the latter is usually more important.
3. When restoring the ECF volume in a chronically hyponatremic patient, have a form of ADH with a short half-life available (e.g., aqueous vasopressin) to minimize a water diuresis and a rapid rise in plasma [Na^+] when endogenous ADH is turned off. An examination of the urine specific gravity and/or volume at the bedside will indicate when to give the ADH.

Production of H_2O and CO_2:
The rates of production of H_2O and CO_2 are 1:1 during the oxidation of fats or carbohydrates. The P_aCO_2 and PH_2O in alveolar air are similar (40 vs 47 mm Hg). The PCO_2 of air is trivial and the PH_2O may be close to 10–20 mm Hg, depending on atmospheric conditions. Hence, metabolically generated CO_2 and H_2O are both produced and lost in 1:1 proportions.

Questions

(Discussions on pages 286–87)

7·15 Which is more important in causing a rapid "correction" of hyponatremia when hypertonic saline is infused, Na^+ gain or water loss? (Assume that when a 70-kg patient with hyponatremia (100 mmol/l) and a near-normal ECF volume received 150 mmol of NaCl, the plasma [Na^+] rose to 115 mmol/l.)

7·16 What volume of an IV solution containing 152 mmol/l Na^+ (isotonic saline) or 856 mmol/l Na^+ (5% saline) must be infused to raise the plasma [Na^+] by 1 mmol/l in a 70-kg patient? (Ignore the water infused for simplicity.)

Specific Principles of Therapy

RAPIDITY OF CORRECTION OF HYPONATREMIA

- Do not overcorrect hyponatremia because CNS damage might occur.
- Raise the plasma [Na^+] 0.5–1 mmol/l/hr if the patient is not symptomatic or is mildly symptomatic.
- Raise the plasma [Na^+] 3–6 mmol/l by giving NaCl if the patient is convulsing.

Because overly rapid correction of severe hyponatremia may be associated with central pontine myelinolysis (CPM), a midbrain disorder, do not correct the hyponatremia too rapidly. This CNS lesion should be suspected when the patient does not recover sufficiently upon correction of hyponatremia or if there is a deterioration in the mental state in the week following therapy. There is a controversy concerning the rapidity of correction of hyponatremia; several review articles on this subject are provided in the suggested reading list at the end of the book.

Symptomatic hyponatremia depends on the plasma [Na^+] and, more importantly, on the rate at which hyponatremia occurred. Symptoms may be present at a [Na^+] of 120–125 mmol/l during very acute hyponatremia; when hyponatremia has developed slowly, symptoms may be absent at a [Na^+] below 110 mmol/l. Although there is no argument that severe symptomatic hyponatremia, especially if acute, must be treated promptly, asymptomatic hyponatremia should be treated much less vigorously.

The therapeutic approach to hyponatremia is a critical area since the potential exists to do more harm with the treatment than was caused by the disorder. An axiom that is relevant to hyponatremia is "beware of the acute recognition of a chronic disease." To determine how aggressively one should treat the hyponatremia, evaluate the severity of symptoms attributable to the hyponatremia.

There are three specific risks associated with the treatment of hyponatremia:

1. If the $[Na^+]$ rises quickly, the patient may develop CPM.
2. If the release of ADH is due to ECF volume contraction and the ECF volume is reexpanded, a water diuresis may ensue (ADH is turned off), and the $[Na^+]$ may rise quickly and cause CPM.
3. The NaCl administration may cause congestive heart failure.

THERAPEUTIC APPROACH IF THE PATIENT IS SYMPTOMATIC

The goal of therapy in the symptomatic patient is to shrink the swollen brain cells by increasing the plasma $[Na^+]$. First, stop all water intake and then raise the plasma $[Na^+]$ rapidly until the symptoms are improved or until you have increased the plasma $[Na^+]$ by 6 mmol/l, whichever comes first (see the sample calculation in the margin). After the plasma $[Na^+]$ has risen by 6 mmol/l, reduce the rate of rise to 0.5 mmol/l/hr and limit the daily rise in plasma $[Na^+]$ to 12 mmol/l/day.

If the ADH release was on the basis of ECF volume contraction, be prepared to give an ADH analogue when the patient has too large a water diuresis (see note #1 in the margin).

If the patient is at risk of developing congestive heart failure from the Na^+ load, give a loop diuretic to promote the loss of Na^+ and water.

If the patient's blood pressure is low, administer saline isotonic to the patient to increase the ECF volume acutely. This treatment will expand the ECF volume with no fluid exit from the ICF (see note #2 in the margin for an example).

Sample calculation:

- A 70-kg patient has 45 liters of total body water.
- To raise the $[Na^+]$ by 6 mmol/l, give 270 mmol of Na^+ (6×45).

Notes:

1. If a 70-kg patient with a $[Na^+]$ of 115 mmol/l has a 1 liter water diuresis, the $[Na^+]$ may rise to 118 mmol/l.
2. If the patient's plasma $[Na^+]$ is 115 mmol/l, equal volumes of isotonic saline (155 mmol/l) and 0.45% saline (75 mmol/l) would be ideal therapy with which to expand the ECF volume acutely.

THERAPEUTIC APPROACH IF THE PATIENT IS ASYMPTOMATIC

Look at the urine osmolality and the osmole excretion rate. Restriction of water intake is modestly effective and may be the only therapy necessary (assume 0.75 liters per day will be ingested). Water restriction will be sufficient in example 1 in the margin but not in example 2. If the plasma $[Na^+]$ does not rise, give saline, but limit the rate of increase to 0.5 mmol/l/hr (see the discussion of Question 7·17).

Example 1:
Urine osmolality is 600 mosm/kg H_2O, and 600 mosmoles are excreted today. Hence, the urine volume is one liter today.

Example 2:
Urine osmolality is 600 mosm/kg H_2O, and 300 mosmoles are excreted today. Hence, the urine volume is 0.5 liters today. To increase net water loss, give a diuretic and replace electrolyte losses (see page 268) or give urea (or protein) to increase the rate of excretion of osmoles.

Question

(Discussion on page 287)

7·17 If a 60-kg woman has a plasma $[Na^+]$ of 110 mmol/l, how quickly will her $[Na^+]$ rise with a restriction of water intake to 1 liter/day in fluids and food? (Assume that the urine osmolality is 600 mosm/kg H_2O and that she has 600 mosmoles to excrete.)

PART D REVIEW

Discussion of Introductory Case Water, Water, Everywhere, So Not a Drop to Drink

(Case presented on page 256)

Could Na^+ loss have caused the hyponatremia on day three?

Yes. Not only does Na^+ loss directly cause hyponatremia, it also leads to ECF volume contraction. In response to ECF volume contraction, ADH is released; the result—water retention—contributes to a greater degree of hyponatremia.

The physical examination revealed ECF volume contraction. Laboratory evidence to support this clinical impression was the elevated plasma creatinine, low plasma [K^+], and elevated plasma [HCO_3^-]. The urine [Na^+] of 60 mmol/l is not usual for ECF volume contraction. Perhaps the Na^+ excretion is obligated by the presence of a nonreabsorbable anion (HCO_3^-) that appears as a result of recent vomiting.

Why was the ADH level elevated on day three?

We know that the ADH level was elevated because the urine osmolality was 450 mosm/kg H_2O. A low ECF volume causes ADH release and thereby prevents free water loss in the urine.

Patients with meningitis can have ADH release independent of their tonicity and ECF volume status. Hence, there are at least two causes of a high ADH level, ECF volume contraction and a CNS lesion. Other causes could include drugs, vomiting, or the stress of the illness. Water intake or infusion in this situation could lead to more water retention and a more severe degree of hyponatremia.

Why was the plasma [Na^+] lower on day five?

The lower plasma [Na^+] on day five occurred because the patient lost more Na^+ and/or had a net water gain in the face of continuing actions of ADH. The improved ECF volume suggests that water gain is the basis—most likely the result of SIADH (meningitis).

Was the ADH level higher on day five?

The absolute ADH level cannot be deduced from the plasma [Na^+], but it was high enough to have prevented sufficient renal excretion of free water.

What are the best urine electrolytes to indicate whether the ECF volume is contracted?

Following NaCl loss, the urine should contain minimal quantities of Na^+ and Cl^- (usually less than 10 mmol/l if the kidney is functioning is normally).

Would your answer change if the patient had vomiting or diarrhea?

Vomiting: Vomiting results in the loss of Cl^- and a commensurate rise in the [HCO_3^-] in the ECF. The filtered load of HCO_3^- should rise, and some of this excess may be excreted; for electroneutrality, Na^+ and K^+ will also be lost in the urine. The resultant Na^+ loss will lead to ECF volume contraction, renin release, angiotensin II formation, and, thereby, aldosterone release. Aldosterone causes the K^+ loss in the urine. Hence, a very low [Cl^-] in urine is expected at all times, but the [Na^+] may not be low if bicarbonaturia persists.

Note:
The very low urine [Cl^-] is the best indicator of ECF volume contraction in a patient who vomits.

Diarrhea: A different urine electrolyte pattern occurs with diarrhea and ECF volume contraction. $NaHCO_3$ loss with diarrhea can also produce ECF volume contraction, but in this case there is a low plasma [HCO_3^-]. The kidneys respond to acidemia by excreting a large quantity of NH_4^+ (with Cl^-). Thus, the urine should have a very low [Na^+], which indicates the ECF volume contraction, but not a very low [Cl^-] because of the electroneutrality required for NH_4^+ excretion.

Note:
A very low urine [Na^+] is the best indicator of ECF volume contraction in a patient with diarrhea.

Cases for Review

Case 7·1
Jerry, the Executive

(Case discussed on pages 275–77)

Jerry is an entrepreneur. Because his mother, Emily of Case 10·1, was recently diagnosed as having hypertension, Jerry had his blood pressure checked and he had a similar degree of hypertension (160/110 mm Hg). Like his mother, he was started on a thiazide diuretic and asked to eat a diet without added salt. While on this regimen, his blood pressure was normalized, but routine electrolytes revealed two new abnormalities, a mild degree of hyponatremia (133 mmol/l) and hypokalemia (3.3 mmol/l). There were no obvious clinical findings of a contracted ECF volume. Other laboratory data are provided below.

Plasma		Usual values	Present values
Na^+	mmol/l	140	133
K^+	mmol/l	4.0	3.3
Cl^-	mmol/l	103	90
HCO_3^-	mmol/l	25	28
Creatinine	μmol/l (mg/dl)	88 (1.0)	110 (1.2)
Osmolality	mosm/kg H_2O	288	276

Is Jerry's ECF volume contracted?
Does Jerry now maintain mass balance for Na^+? If so, how?
Is the hyponatremia due to Na^+ loss, water gain, or both?
Why is Jerry's hyponatremia less severe than Emily's (Case 10·1)?
Why is Jerry's hypokalemia much less severe than Emily's (Case 10·1)?

Case 7·2
Hyponatremia and Hypoosmolal Urine

(Case discussed on page 277)

Cindy had surgery to remove a tumor of her pituitary many years ago. She did reasonably well over the ensuing years by taking the appropriate hormonal replacements for her anterior pituitary deficits. Although she drank and eliminated large volumes daily, this aspect of her problem was never investigated further or treated.

More recently, she developed inflammatory bowel disease, had a colectomy, and lost many liters of fluid daily via her ileostomy. On many occasions, she was admitted for episodes that are best characterized by profound contraction of her ECF volume. The most recent one was typical. She had marked ECF volume contraction, hyponatremia (134 mmol/l), a large urine volume (0.2 liters per hour), and a low urine osmolality (160 mosm/kg H_2O). She was given many liters of isotonic saline IV, and the following changes were observed in these parameters: the ECF volume was normal, the plasma [Na^+] was 138 mmol/l, the urine flow rate doubled, and the urine osmolality fell to 80 mosm/kg H_2O.

Does Cindy have central diabetes insipidus (DI), nephrogenic DI, or primary polydipsia?

Case 7·3
Hyponatremia and the Guru

(Case discussed on pages 278–79)

In order to cleanse her soul, Tracy's guru recommended that she wash away evil spirits by consuming nothing but water. Tracy followed this advice enthusiastically. She drank approximately one liter of water per hour, hour after hour while awake. Urine output came close to matching input initially but then fell off considerably. This procedure was repeated daily for many days. Because of other aspects of a behavior problem, Tracy was brought to the emergency room. Physical examination revealed no abnormalities apart from confusion and paranoid delusions. Laboratory results are summarized below.

		Plasma	**Urine**
Na^+	mmol/l	107	46
K^+	mmol/l	3.6	10
Cl^-	mmol/l	75	30
HCO_3^-	mmol/l	24	0
Urea	mmol/l (mg/dl)	1.0 (2.8)	34
Creatinine	μmol/l (mg/dl)	60 (0.7)	3000 (35)
Glucose	mmol/l (mg/dl)	3.3 (60)	0
Osmolality	mosm/kg H_2O	220	160

Is ADH acting now?
What limits free water excretion?
What role did the fasting play and how did it affect the urine osmolality?
What is the differential diagnosis of hyponatremia and a urine osmolality of 160 mosm/kg H_2O?

SUPPLEMENTAL INFORMATION

Another random urine sample revealed a urine with an osmolality of 50 mosm/kg H_2O; the plasma [Na^+] was 110 mmol/l.

How does this information help in the differential diagnosis?
What treatment should Tracy receive?

Case 7·4
The Post-TURP Syndrome

(Case discussed on pages 279–80)

Barry underwent a TURP for a very large prostate. The large prostatic bed was exposed to the irrigating solution; blood loss was higher than usual. The irrigating solution was 6 liters of an equal mixture of sorbitol and mannitol; final osmolality was 150 mosm/kg H_2O. Barry's plasma [Na^+] was 138 before surgery and 93 mmol/l six hours after surgery. His cerebral function did not change markedly, and his ECF volume was not greatly changed.

What is the basis of the hyponatremia?
What simple lab test will help in confirming the diagnosis?
What treatment should be ordered?

Discussion of Cases

Discussion of Case 7·1
Jerry, the Executive

(Case presented on page 273)

Is Jerry's ECF volume contracted?

Yes. Given the history of a low-salt diet and the thiazides, he was probably in negative Na^+ balance initially, and his total body Na^+ content is lower than when he began this regimen. Each liter of ECF has lost 7 mmol of Na^+, so he has lost at least 100 mmol of Na^+ from his ECF plus all the Na^+ lost in each liter of ECF lost (see the margin). Because one cannot detect a mild degree of ECF volume contraction on physical examination, other data are needed to support this impression. In the lab data, the hyponatremia (ADH action), hypokalemia (aldosterone action), and the higher values for creatinine and HCO_3^- in plasma support this impression.

Calculation:

- The fall in [Na^+] is 7 mmol/l (140 – 133 mmol/l).
- His ECF volume is normally 15 liters. Therefore, Na^+ loss is 7 liters × 15 mmol/l, or 105 mmol.
- If his ECF volume is now 14 liters, he lost an additional 133 mmol of Na^+.

Does Jerry now maintain mass balance for Na^+? If so, how?

Yes, in a chronic situation, one achieves a new steady state or mass balance. Two opposing forces are in operation. When the diuretic is not working, Jerry's kidneys, stimulated by the contracted ECF volume, actively reabsorb all filtered Na^+, so he retains all dietary Na^+ at this time. In contrast, he is in negative balance for Na^+ when the diuretic acts.

Is the hyponatremia due to Na^+ loss, water gain, or both?

Both. The diuretic and diet modification (see above) induced the Na^+ loss. Water gain is present because the contracted ECF volume has stimulated thirst and release of ADH.

Why is Jerry's hyponatremia less severe than Emily's (Case 10·1, pages 377 and 379–80)?

To answer this question, we must consider the factors that influence Na^+ and water.

Na^+: Both Jerry and Emily have a mild degree of ECF volume contraction, but Emily has a much greater deficit of Na^+ in her ECF. This deficit reflects the much greater negative mass balance for K^+ because she has shifted some Na^+ into her ICF (see below).

H_2O: Emily may or may not have a larger water intake (tea) relative to body mass, but this intake cannot be assessed. She probably has a much lower excretion of water. Two factors must be assessed, urine osmolality and the osmole excretion rate.

1. **Urine osmolality:**
   Emily should excrete more water because her urine osmolality is lower than Jerry's. In more detail, water excretion depends on the urine osmolality and the osmole excretion rate. With age, one cannot excrete as concentrated a urine as do younger people (urine osmolality was 402 mosm/kg H_2O in Emily and 850 mosm/kg H_2O in Jerry). For the same rate of osmole excretion, Emily excretes more water, but, as described below, she has many fewer osmoles to excrete.

$$\text{Urine volume} = \frac{\text{osmole excretion rate}}{\text{urine osmolality}}$$

2. **Osmole excretion rate:**
   Emily eats tea and toast (little urea produced) so she has far fewer osmoles for excretion and thus excretes a decreased amount of free water. Quantitatively, if Jerry excretes 850 mosmoles per day, he will excrete 1 liter of urine. Emily, on the other hand, excretes only 300 mosmoles, so she excretes 0.75 liters of water daily at her usual urine osmolality.

Why is Jerry's hypokalemia much less severe than Emily's (Case 10·1)?

Intake of K^+: Again, we must look to mass balance. Jerry eats more K^+ than Emily does.

Excretion of K^+: With respect to excretion, aldosterone may lead to similar concentrations of K^+ in their cortical collecting ducts (CCD), but Emily should have a lower volume delivered to the CCD (excretion = $[K^+] \times$ volume). The basis of this statement is as follows: volume delivery to the CCD depends on the number of osmoles excreted when ADH acts because the osmolality of luminal fluid approaches that of plasma. Although Emily has a lower plasma osmolality, she has a much, much lower rate of osmole excretion because of her dietary intake (see above).

Overall: Up to this point, it appears that Jerry should have a greater K^+ deficit. Nevertheless, when the diuretic acts, Emily and Jerry both deliver high volumes of fluid to the CCD. Thus, Emily will excrete almost as much K^+, but the low dietary K^+ may account for her greater negative K^+ balance. Hence, the much lower dietary K^+ in Emily is probably the most important reason for her larger deficit of K^+.

Discussion of Case 7·2 Hyponatremia and Hypoosmolal Urine

(Case presented on page 274)

Does Cindy have central DI, nephrogenic DI, or primary polydipsia?

The issue here is to explain the presence of hypoosmolal urine (160 mosm/kg H_2O) when Cindy was hyponatremic. We shall consider the following three diagnoses:

Central DI: This diagnosis can be established by finding a urine with a very low osmolality and large volume when ADH should be acting. The two major stimuli to ADH release are hypernatremia (not present) and marked ECF volume contraction (present on admission). Thus, it appears that ADH release was abnormally low, given her ECF volume status. This view is supported by two pieces of evidence: first, she had a fall in urine osmolality when the ECF volume was reexpanded (80 mosm/kg H_2O); this decline in urine osmolality was accompanied by a doubling of her urine output. Second, by history, she had a pituitary tumor resected. Hence, we believe she had partial central DI. This view was confirmed later when ADH was given and the urine osmolality rose acutely to 456 mosm/kg H_2O (this value is not her maximum attainable urine osmolality because several days of protein intake plus avoidance of polyuria are needed to reestablish medullary hyperosmolality).

Nephrogenic DI: The diagnosis of nephrogenic DI is possible, but, given the above renal response to ADH, it is no longer tenable.

Primary polydipsia: Patients with central DI usually have polyuria caused by a water diuresis. If they can gain access to water, they usually have a slightly elevated plasma [Na^+]; without adequate water intake, frank hypernatremia is expected. Cindy was accustomed to drinking large quantities of water, independent of thirst. When water output was curtailed somewhat (50%) by the release of ADH (which causes a rise in urine osmolality) and by the lower GFR (which lowers the delivery of hypotonic fluid to the distal nephron), she went into positive water balance and became hyponatremic. Therefore, in this setting, she also has polydipsia that is due in part to ECF volume contraction.

Discussion of Case 7·3 Hyponatremia and the Guru

(Case presented on pages 274–75)

Is ADH acting now?

Yes, the urine osmolality exceeds the expected value of 20–80 mosm/kg H_2O.

What limits free water excretion?

Free water excretion can be as high as 10% of the GFR. Since the minimum urine osmolality is 20–30 mosm/kg H_2O, each of the following might limit free water excretion:

1. low GFR;
2. number of osmoles available for excretion;
3. ADH or ADH-like drugs.

What role did the fasting play and how did it affect the urine osmolality?

The number of osmoles available for excretion is very important. If someone drank 20 liters of water per day, that person would need 400–600 mosmoles to excrete this water load (20 liters × 30 mosm/kg H_2O). If a low-protein and low-salt diet were ingested, there would be insufficient osmoles provided from the diet to excrete the water load, so endogenous osmoles would be required. When Na^+ are excreted without intake, the ECF volume tends to fall. As a result, ADH is released and less water can now be excreted. As water is retained, the ECF volume reexpands, ADH is suppressed, and the urine osmolality falls again for a while until the cycle repeats itself. Hence, the urine osmolality may be very low or surprisingly high in the various stages of this clinical story.

What is the differential diagnosis of hyponatremia and a urine osmolality of 160 mosm/kg H_2O?

With this osmolality, ADH is present and acting. The differential diagnosis is summarized in Table 7·6 and includes polydipsia and a mild degree of contraction of her ECF volume or some other stimulus for ADH release (e.g., drugs, vomiting). Alternatively, Tracy could have SIADH of the reset osmostat type. Finally, Tracy could conceivably have central DI and be treated with an ADH preparation. She may drink in response to nonosmotic stimuli.

Table 7·6
THE DIFFERENTIAL DIAGNOSIS OF HYPONATREMIA WITH A URINE OSMOLALITY OF 160 mosm/kg H_2O

Polydipsia in the face of mild ADH actions
• A mild degree of ECF volume contraction (usually psychogenic polydipsia)
• Psychosis, which causes ADH release that wanes in degree
• Intermittent vomiting and nausea
• Drugs that are taken intermittently or are weak stimulators of ADH release
Exogenous ADH given to a patient who either:
- is accustomed to drinking large quantities of water
- has a fall in the GFR, but did not decrease the intake of water
Reset osmostat type of SIADH

SUPPLEMENTAL INFORMATION

Another random urine sample revealed a urine with an osmolality of 50 mosm/kg H_2O; the plasma [Na^+] was 110 mmol/l.

How does this information help in the differential diagnosis?

Since the urine osmolality fell when the plasma [Na^+] rose, the data suggest that ADH was suppressed at this time. The data are most compatible with the input of NaCl in a patient with psychogenic polydipsia. The data are not consistent with reset osmostat, but it still could be possible if another stimulus for the release of ADH is being removed (psychosis) or if the effects of the exogenous ADH preparation wore off.

What treatment should Tracy receive?

Tracy needs negative water and positive Na^+ balance. It goes without saying that water intake must be curtailed. It is best to give judicious amounts of NaCl. The danger to anticipate is excessive water loss when endogenous ADH is suppressed by ECF volume reexpansion (see the margin). Slow down this rapid water loss with small amounts of ADH. The rate of rise of the plasma [Na^+] should be 12 mmol/l/day. Do not give large amounts of hypertonic saline without observing Tracy for a rapid water diuresis.

Note:
In a patient with psychogenic water drinking and ADH release in response to ECF volume contraction (caused by Na^+ loss because of no osmole intake), administration of NaCl will turn off ADH and result in dilution of the urine. A very large urine output might be dangerous if the plasma [Na^+] rises too quickly.

Discussion of Case 7·4 The Post-TURP Syndrome

(Case presented on page 275)

What is the basis of the hyponatremia?

Hyponatremia implies a loss of Na^+ or a gain of water in the ECF.

Loss of Na^+: If Barry simply had a loss of Na^+, he would have died of shock. Because his ECF volume is not greatly changed and he is very hyponatremic, there was a large loss of Na^+ (675 mmol, see the margin). His ECF volume is being maintained by the presence of a particle that has a distribution similar to Na^+ and keeps water from shifting from his ECF to his ICF.

Calculation:
With the data provided, it is easy to calculate the quantity of Na^+ lost from his ECF:
- The ECF volume remained the same.
- Each of the 15 liters of ECF lost 45 mmol of Na^+ ([Na^+] fell from 138 mmol/l to 93 mmol/l).
- Therefore, 15 liters × 45 mmol/l = 675 mmol of Na^+.

Gain of water in the ECF: Although a gain of water is the primary reason for hyponatremia, several points need to be explained:

1. Acute hyponatremia to this degree should have caused his brain to occupy “four-thirds” of his skull (impossible). Therefore, the basis of his hyponatremia is not just water gain. Barry's normal mentation despite such a severe and acute degree of hyponatremia is a clue of critical importance.

   The volume of water needed to lower his plasma [Na^+] from 138 to 93 mmol/l is close to 15 liters (see the margin). Because he was given only 6 liters of dilute solution, there is more to the answer.

Calculation:
- 2 [Na^+] × ICF volume (30 liters) = number of particles in his ICF.
- Assume that the number of particles in his ICF does not change.
- Therefore, 2 (138) × 30 liters = 2 (93) × new ICF volume. The new ICF volume is close to 45 liters, a rise of about 15 liters.

Notes:

- Barry had two reasons for hyponatremia: water retention from the hypoosmolal solution and the retention of mannitol plus sorbitol in his ECF.
- We doubt that the clinical impression of a normal ECF volume was accurate. The ECF volume was surely expanded.

2. Given the above, Barry must have retained (formed and/or was given) an osmole that remained in his ECF (like glucose) and drew water out of cells. This particle turned out to be the mannitol and sorbitol used in the irrigating solution for his prostate. It was given as hypoosmolal lavage solution (150 mosm/kg H_2O). Much of this fluid was probably absorbed via the exposed veins in his prostatic bed (see the margin).

What simple lab test will help in confirming the diagnosis?

The plasma osmolality should be much greater than twice the plasma [Na^+] + urea + glucose, all in mmol/l. It was!

What treatment should be ordered?

Barry is asymptomatic; he should be allowed to correct his hyponatremia himself by excreting the offending osmoles and the extra water load. The deficit of Na^+ that is present plus the additional loss of Na^+ induced by the ongoing osmotic diuresis must be replaced; to do so, simply maintain his ECF volume. This instance is probably the only one in which rapid correction of the plasma [Na^+] can occur and not be harmful. (To be sure that we made a correct analysis, the plasma [Na^+], osmolality, urinary excretions, and Barry's clinical state were followed closely. Barry's numbers moved in the appropriate direction and at the expected rate; Barry did not suffer a change in cerebral function.)

Summary of Main Points

- Hyponatremia is a very common fluid and electrolyte disorder. Expect it to develop in certain clinical situations associated with water administration and a decreased "effective" circulating volume. Lab results confirm its presence. The clinical approach should answer the following questions.
  1. Is there a decreased "effective" circulating volume?
  2. Is something promoting the release or the actions of ADH (e.g., drugs, nausea, vomiting)?
  3. Is there a lung or CNS lesion that is responsible for the release of ADH?
  4. Does the patient respond appropriately to a small water load?

If the answers to questions 1 and 2 are "no," then the patient has SIADH.

SUMMARY OF THERAPY

Aims in Therapy with a Water Overload

- Raise the plasma [Na^+] 0.5–1 mmol/l/hr to shrink the ICF volume.
- Bring the plasma [Na^+] to close to 125 mmol/l.

- Treat vigorously if the patient is symptomatic, but raise the plasma [Na^+] by less than 6 mmol/l acutely.
- Treatment should be more aggressive if the hyponatremia is acute.
- The severity of hyponatremia itself is not a critical determinant of aggressive therapy if the patient is asymptomatic.

Dangers in Therapy

- Too rapid correction or overcorrection of the water deficit may lead to central pontine myelinolysis.
- If ADH release is due solely to ECF volume contraction, rapid reexpansion of the ECF volume suppresses the release of ADH and may cause a very large water diuresis and possibly central pontine myelinolysis. The consequences of a suppressed release of ADH can be avoided, if anticipated, by giving ADH once there is a water diuresis of 1–2 liters.
- Rapid and/or severe overexpansion of the ECF volume endangers the patient

Therapeutic Options

- Stop water intake (a must).
- Promote water loss (without Na^+ loss). (Although this treatment may be possible, it is not practical.)
- Shift water from the ICF to the ECF with hypertonic saline if the patient needs or can withstand ECF volume expansion.
- Promote water and Na^+ output but replace ongoing electrolyte losses. Usually a loop diuretic is used to increase these losses; the urine usually contains 40–100 mmol/l Na^+ and 5–10 mmol/l K^+. These electrolyte losses should be replaced with a hypertonic solution to yield net water loss.

Overall

- Using hypertonic saline to raise the plasma [Na^+] from 110 to 125 mmol/l requires close to 600 mmol of positive balance for NaCl. This amount expands the ECF volume by 3–4 liters if there is no excretion of the infused Na^+ and water.

Discussion of Questions

7·1 What laboratory methods to measure the [Na^+] in plasma will provide a lower value if hyperlipidemia is present?

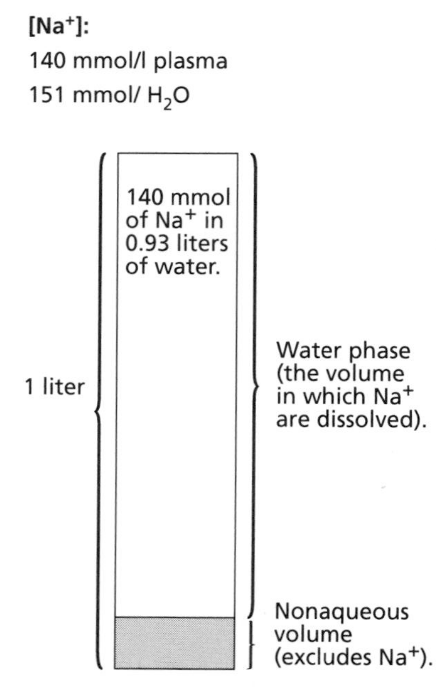

In plasma, Na^+ are dissolved only in the water phase (see the margin). The presence of excessive quantities of nonaqueous volume in the plasma causes a fall in the ratio of Na^+ to total volume. In hyperlipidemia or hyperproteinemia, there is a marked increase in the nonaqueous phase of plasma. Therefore, although the [Na^+] in the aqueous phase is normal and the plasma osmolality is normal, the quantity of Na^+ per plasma volume is low. Hence, the term "pseudohyponatremia" applies because the problem is apparent, not real. The diagnosis can be confirmed by inspecting plasma for lipids or by measuring the plasma osmolality; a normal value is found in pseudohyponatremia.

From the laboratory perspective, a number of different methods are used to detect the [Na^+] in plasma. Some of them require a major dilution of plasma (e.g., the flame photometer, certain types of ion-selective electrodes). Because the plasma is diluted, the original Na^+:plasma volume is reflected, and pseudohyponatremia will be evident. In contrast, if a Na^+-selective electrode or a conductance method is used on undiluted plasma, the [Na^+] (Na^+:H_2O ratio) will approach 152 mmol/kg H_2O. In this case, the machine back-calculates the value to 140 mmol/l in order not to confuse the clinician, but this procedure confuses the physiologist.

7·2 Match the four separate causes of hyponatremia with the critical parameter for diagnosis.

Etiology	Critical parameter measured
(1) Chronic diuretic intake	(a) Normal plasma osmolality
(2) Hyperlipidemia	(b) Normal ECF volume and low urine osmolality
(3) Excess ADH	(c) ECF volume contraction and hypokalemia
(4) Compulsive water consumption	(d) Low plasma and high urine osmolality

The answers are as follows:

1. Chronic diuretic intake may lead to a degree of ECF volume contraction and hypokalemia consequent to the actions of aldosterone (c).
2. In hyperlipidemia, pseudohyponatremia may occur and will be evident if the plasma osmolality is normal (a).
3. With hyponatremia from excessive ADH action, the plasma osmolality will be low and the urine osmolality will be high (d).
4. In the compulsive water drinker, the ECF volume will be normal and the urine osmolality may be very low (b).

7·3 A diabetic patient has renal failure. His blood sugar is 1000 mg/dl (55 mmol/l), [Na^+] is 127 mmol/l, and blood urea nitrogen is 70

mg/dl (urea is 25 mmol/l). What is his calculated osmolality? Has hyperglycemia changed his ECF and ICF volumes?

Hyperglycemia causes water to shift out of muscle cells (the bulk of the ICF volume and close to half of body water). Quantitatively, a 900 mg/dl rise in the concentration of glucose causes the $[Na^+]$ to fall by about 13 mmol/l (9×1.4) to 127 mmol/l (see Chapter 12, page 442 for the details of this calculation). The calculated osmolality in this case is $2 \times [Na^+] + [glucose]$ + urea in mmol/l, or 334 mosm/kg H_2O.

The ECF volume is expanded because the ICF volume of muscle has contracted. Because renal failure is present, no osmotic diuresis has occurred; hence, the usual clinical finding of ECF volume contraction with severe hyperglycemia is not present.

The liver ICF volume is expanded, the brain volume is probably close to normal, and the ICF volume of muscle is contracted (see Chapter 12, Figure 12·5, page 437).

7·4 What is the average urine osmolality in 24-hour urine in a normal subject?

A normal person produces 500 mmol of urea and consumes approximately 200 mmol of NaCl plus KCl (400 mosmoles); thus, the urine contains 900 mosmoles. If a typical urine volume is 1.5 liters, the osmolality of a 24-hour urine is close to 600 mosm/kg H_2O.

7·5 What conclusions can be drawn when the urine osmolality is high?

A high urine osmolality indicates that ADH is acting and that the renal medulla has a high osmolality. One cannot deduce the impact on the plasma $[Na^+]$ from just the urine osmolality; a high urine osmolality may lead to hypernatremia or hyponatremia. Consider the following two examples in which nothing was ingested. First, when urea is the sole osmole excreted in 1 liter of hyperosmolar urine (600 mosm/kg H_2O), a rise in the ECF $[Na^+]$ occurs (the Na^+:H_2O ratio rises) because water without Na^+ was excreted. In contrast, the excretion of a urine with the same osmolality and a $[Na^+]$ of 200 mmol/l (higher than the plasma Na^+:H_2O) favors the development of hyponatremia.

7·6 Three patients have hyponatremia (120 mmol/l). In case A it is due to water gain; in case B, Na^+ loss; in case C, hyperglycemia. Which patient(s) will have swelling of brain cells? Give reasons for your answer.

The $[Na^+]$ indicates the ICF volume if no additional particles (like glucose in Case C) are acting to shift water out of cells and if defense of the ICF volume (brain cells) has not occurred.

Cases A & B: The cells of all organs other than the brain are swollen to the same degree in both of these patients because the plasma $[Na^+]$ is 120 mmol/l in each case. With respect to brain cells, if the hyponatremia is chronic, brain cell size in both of these patients will be close to normal since particles should have been removed from their brain cells. If hyponatremia is acute, brain cells will swell in both of these patients.

Case C: Hyponatremia that is due to hyperglycemia will cause muscle cells to shrink, liver cells to swell, and brain cells to remain close to normal in size (see Chapter 12, Figure 12·5, page 437).

7·7 Could Na^+ loss have occurred via the renal route if the urine is now Na^+-free?

Yes. If a patient took a diuretic yesterday and had the natriuresis then (and did not ingest Na^+ today), that patient would have ECF volume contraction and the appropriate renal response of minimal Na^+ and Cl^- in the urine today. This example is presented to emphasize that one must suspect the antecedent intake of diuretics in patients with a contracted ECF volume (intake of diuretics might even be denied by the patient).

7·8 What role might K^+ depletion play in determining the severity of hyponatremia?

K^+ depletion is expected when there is hyponatremia and decreased ECF volume. A contracted ECF volume leads to the release of aldosterone, which augments K^+ excretion if delivery of Na^+ and volume to the CCD are sufficient. Where do these urinary K^+ come from? As shown in Chapter 9, Figure 9·12, page 347, many of these K^+ are derived from muscle ICF and require Na^+ to shift from the ECF to the ICF. This shift depletes Na^+ in the ECF and further accentuates the degree of hyponatremia (this portion of hyponatremia is corrected by KCl therapy, as discussed in response to Case 7·1, page 275–77). Therefore, in patients with hypokalemia and chronic hyponatremia, there is a risk to aggressive KCl replacement because it may lead to a rapid rise in plasma $[Na^+]$ and ECF volume as Na^+ exit the ICF.

7·9 How much water must a normal person drink to produce hyponatremia? Would it matter if this person is on a low-salt, low-protein diet?

A person with normal kidneys can excrete at least 12 liters of electrolyte-free water per day. Hence, water intake that exceeds this amount can be the sole cause of hyponatremia; this condition is called "psychogenic polydipsia." Recall that each liter of maximally dilute urine contains 20–80 mosmoles. To excrete 15 liters with a urine osmolality of 30 mosm/kg H_2O, a person needs 15 liters × 30 mosmoles per liter, or 450 mosmoles (close to the urea produced by a normal diet). Failure to provide urea osmoles as a result of oxidation of protein means that Na^+ must be excreted or water will be retained. If the diet does not contain sufficient NaCl, endogenous Na^+ will be excreted. As a result, the ECF volume will contract and ADH will be released. These actions obviously compromise further electrolyte-free water excretion and can lead to a severe degree of hyponatremia. This pattern can be seen in patients with an excessive beer intake.

Bartenders as physiologists: Bartenders who serve excretable osmoles (salty nuts and pickled eggs, which contain protein that can be converted to urea) are "practicing physiologists" because their patrons can now drink large volumes of beer without suffering from confusion, a danger of severe hyponatremia (beer contains only a few mmol of Na^+, K^+, and Cl^- but close to 1000 mmol of ethanol, an "osmole that does not count" with respect to water excretion; see Case 8·3, pages 311–312 and 314–15, for further discussion).

7·10 Why do patients with adrenal insufficiency commonly present with hyponatremia?

Hyponatremia is commonly seen in patients with adrenal insufficiency. The major mechanisms are renal Na^+ wasting and reduced water excretion secondary to ADH release, which is caused by ECF volume contraction (a low aldosterone consequence) and reduced cardiac performance (a low glucocorticoid consequence). It is no longer believed that glucocorticoids influence water permeability in the collecting duct.

Caution: Treatment with replacement doses of glucocorticoids and mineralocorticoids and replacement of the Na^+ deficit will correct the hyponatremia. Do not let this correction occur too rapidly; administration of ADH may be necessary to slow the rate of the water diuresis.

7·11 A patient with hyponatremia (127 mmol/l) excretes urine with an osmolality of 176 mosm/kg H_2O. How would you know if this excretion was due to a reset osmostat type of SIADH?

The presence of hyponatremia and a urine osmolality that is low, but not the expected minimal value of 20–80 mosm/kg H_2O, suggests that some ADH is present. There are three possible reasons for the low urine osmolality.

1. **A reset osmostat:**
   If the clinical setting suggests a reset osmostat type of SIADH, the patient, if given a water load, will be able to excrete it promptly. The patient will also have a much higher urine osmolality after water is restricted. The plasma [Na^+] should be similar before and after treatment. In response to exogenous ADH, the urine osmolality will rise.

2. **Partial therapy of central DI:**
   A low urine osmolality might occur if the ADH that was administered for treatment of central DI has almost "worn off" and if the patient drinks a large quantity of water out of habit. There should be an obvious history of central DI and a fall in urine osmolality with time. Rapid (too rapid) correction of hyponatremia should be anticipated and controlled with ADH administration.

3. **Psychogenic polydipsia:**
   The history will usually reveal a psychiatric problem and polydipsia. As discussed in Case 7·3, pages 278–79, a recurring cycle develops. With water ingestion, a maximally dilute urine will be excreted. If there is an inadequate osmole load for excretion, Na^+ will be lost. A fall in the ECF volume will ensue, ADH will be released, and there will be an excretion of urine with a somewhat higher osmolality. After the patient ingests water the next day, the cycle will repeat itself. The diagnosis is established by demonstrating that a normal plasma [Na^+] will occur once the ECF volume is restored (NaCl is given, but the rate of correction of hyponatremia must be slow because the hyponatremia is often chronic).

7·12 If a person on a low-salt, low-protein diet had surgery, what features might be present in the postoperative period to permit more of the administered electrolyte-free water to be retained while ADH is acting?

$$\text{Urine volume} = \frac{\text{osmoles excreted}}{\text{urine osmolality}}$$

Because ADH is fully active and the osmolality of fluid in the renal medullary interstitium does not change, the urine osmolality of a patient on a low-salt, low-protein diet could be equal to that of an average subject. Let us assume that both the patient and the average subject have a urine osmolality of 1000 mosm/kg H_2O. Therefore, the number of osmoles excreted determines the urine volume at a constant urine osmolality. If the patient on the low-salt, low-protein diet excretes close to 250 mosmoles, 0.25 liters would be excreted at a urine osmolality of 1000 mosm/kg H_2O. In contrast, if the average subject excretes the usual 900 mosmoles, the urine volume would be 0.9 liters. Hence, the low osmole excretion rate compromises the ability to excrete water.

7·13 A patient had surgery. Although the plasma [Na^+] was always in the range of 139–141 mmol/l, the preoperative plasma [Na^+] was 136 mmol/l. Why? The plasma [Na^+] was 128 mmol/l 36 hours after surgery. Why?

The preoperative plasma [Na^+] was low because of water intake in the presence of ADH actions. The anxiety and anticipation of surgery were sufficient to drive the patient to drink coffee and release ADH. This response probably accounts for the wide normal range of plasma [Na^+] in hospitalized patients.

The plasma [Na^+] fell in the postoperative period because of water ingestion and ADH actions. Although the doctor was smart enough to avoid routine D_5W, the antibiotic was given in this solution and the patient was encouraged to drink postoperatively not only for his lungs, but also because his throat was sore and his mouth was dry. He did not think from a physiologic perspective.

7·14 What properties of glycine make this compound unique in the pathogenesis of the post-TURP hyponatremia syndrome?

The unique features of glycine are that it is is an amino acid and a neurotransmitter. It seems to distribute more rapidly into muscle cells than into the CNS. In the post-TURP syndrome, glycine could cause a delayed hyperammonemia, a urea-induced osmotic load, and perhaps behavioral or functional changes in the CNS.

7·15 Which is more important in causing a rapid "correction" of hyponatremia when hypertonic saline is infused, Na^+ gain or water loss? (Assume that when a 70-kg patient with hyponatremia (100 mmol/l) and a near-normal ECF volume received 150 mmol of NaCl, the plasma [Na^+] rose to 115 mmol/l.)

This question highlights the relative importance of Na^+ gain and water loss in rapid correction of hyponatremia. The concept is that as Na^+

are retained, water will shift from the ICF, so one must use total body water as the denominator of the Na^+:H_2O ratio in this case. Also, if the total body water is 50 liters, for easy math, the administered 150 mmol of Na^+ can only raise the plasma [Na^+] by 3 mmol/l. Therefore, if a much larger rise in [Na^+] is noted, there may be another reason for this change (e.g., a water diuresis caused by a suppressed release of ADH that is caused by reexpansion of the ECF volume).

7·16 What volume of an IV solution containing 152 mmol/l Na^+ (isotonic saline) or 856 mmol/l Na^+ (5% saline) must must be infused to raise the plasma [Na^+] by 1 mmol/l in a 70-kg patient? (Ignore the water infused for simplicity.)

Because a 70-kg person has 45 liters of water, and water moves rapidly across cell membranes, one must give 1 mmol of Na^+ for each liter of total body water even though Na^+ remain in the ECF. To give 45 mmol of Na^+ from isotonic saline, 300 ml the must be given; in contrast, 53 ml of the hypertonic saline will give this quantity of Na^+ (and with much less water, a desired feature).

7·17 If a 60-kg woman has a plasma [Na^+] of 110 mmol/l, how quickly will her [Na^+] rise with a restriction of water intake to 1 liter/day in fluids and food? (Assume that the urine osmolality is 600 mosm/kg H_2O and that she has 600 mosmoles to excrete.)

The key data required are the urine osmolality and the osmole excretion rate. She excretes 600 mosmoles at 600 mosm/kg H_2O; hence, her urine volume must be 1 liter of water, the quantity she ingested. The only negative water balance is her loss via the skin because water production from metabolism equals water loss via respiration. If she loses a few hundred milliliters per day, her plasma [Na^+] will rise close to 1 mmol/l.

8

HYPERNATREMIA

PART D REVIEW 310

Objectives

- To indicate that hypernatremia represents an increase in the amount of Na^+ relative to water in the ECF and signals generalized ICF volume depletion.
- To emphasize that hypernatremia rarely results from Na^+ gain; it is almost always due to water loss. The excessive water loss is almost always via the urine and is associated with weight loss. Urinary water loss is determined by the combination of the urine osmolality and osmole excretion rate. A high osmole excretion rate and low urine osmolality provide the potential for a high urinary water loss.
- To provide a diagnostic approach to the causes of hypernatremia based on an evaluation of the ECF volume, weight, and the physiological responses to a water deficit or Na^+ gain.
- To provide a diagnostic approach to patients with polyuria.
- To provide the basis for rational decision-making in the treatment of patients with hypernatremia.

Outline of Major Principles

1. Hypernatremia is not a specific disease: look for its cause and treat the underlying disease.
2. In all patients with hypernatremia, the ICF volume is contracted. The brain is most susceptible, and a CNS hemorrhage is more likely to ensue if hypernatremia is acute and/or severe.
3. Hypernatremia is almost always due to net water loss. The two major expected responses to water loss are thirst and the excretion of the minimum volume of maximally concentrated urine; at least one of these responses will usually be defective in a patient with a water deficit. The urine osmolality helps in differentiating the three major causes of water loss: diabetes insipidus (large volume of hypoosmolar urine), osmotic or pharmacologic diuresis (large volume of slightly hyperosmolar urine), and nonrenal water loss without water intake (minimum volume of maximally hyperosmolar urine).
4. A gain of Na^+ is rarely responsible for hypernatremia. Detect a Na^+ gain by finding an expanded ECF volume and then calculate the quantity of Na^+ retained in the ECF.
5. In terms of salt and water, treatment of a patient with hypernatremia has two components: first, stop electrolyte-free water loss, if possible; second, administer a hypotonic solution relative to the patient if oliguria is present or relative to the urine if polyuria is present. Hypotonic saline and glucose in water are IV solutions that contain electrolyte-free water. Do not give glucose faster than the patient can metabolize it, and do not

Abbreviations:
ADH = antidiuretic hormone.
CCD = cortical collecting duct.
CNS = central nervous system.
DI = diabetes insipidus.
D_5W = 50 g of glucose in 1 liter of water.
ECF = extracellular fluid.
ICF = intracellular fluid.
MCD = medullary collecting duct.

Importance of the [Na^+]:
In this instance, one uses a ratio (Na^+:water) to gain insights on the denominator (i.e., water) rather than the numerator of that ratio (i.e., Na^+) to reveal the ICF volume.

correct chronic hypernatremia too quickly. The best way to administer electrolyte-free water is by the oral route.

Introductory Case
Treat the Patient Before He (DI)es

(Case discussed on page 310)

A patient developed acute meningitis and was confused by the time of admission. The plasma [Na^+] was 140 mmol/l, the ECF volume was slightly low, and the urine volume was low. Shortly thereafter, a convulsion occurred; it was treated with a large dose of the anticonvulsant phenytoin (Dilantin). Over the next eight hours, polyuria developed, but thirst was not present. Physical exam revealed a modest degree of ECF volume contraction and a weight loss of 5 kg. The plasma [Na^+] rose to 157 mmol/l and urine osmolality was 100 mosm/kg H_2O.

Of what significance is the modest degree of contraction of the ECF volume?
Why was thirst absent?
Of what significance is the loss of body weight?
What is the significance of the polyuria?
Can any inference be made from the urine osmolality and the acuteness of the onset?
What should the therapy be?

PART A
ETIOLOGY OF HYPERNATREMIA

Background

[Na^+]

The actual plasma [Na^+] is 152 mmol/kg H_2O. If measured per liter of plasma, however, the plasma [Na^+] is 140 mmol/l because each liter contains 6–7% nonaqueous volume (lipids, proteins), and Na^+ are distributed only in the aqueous phase. The normal range of plasma [Na^+] is said to be 136–144 mmol/l. If blood lipids or proteins are excessively high, the laboratory values for the plasma [Na^+] may be much lower than the actual $Na:H_2O$ ratio if certain methods are used to measure the [Na^+] (see the margin).

Methods to measure the [Na^+]: If the plasma [Na^+] is measured using an Na^+-selective electrode or a conductance method (i.e., measuring the Na^+ to water ratio) on an undiluted sample, the normal value is 152 mmol/l; notwithstanding, the lab will "back-calculate" this value and report it as 140 mmol/l. There is no "correction factor" needed for hyperlipidemia or hyperproteinemia in these cases. If the method used determines the [Na^+] per volume of plasma (flame photometry) a "factitious" hyponatremia will result from an increased nonaqueous phase (e.g., hyperlipidemia).

URINE OSMOLALITY (OR SPECIFIC GRAVITY)

There is no normal value for urine volume and osmolality! The kidneys respond to a change in tonicity of body fluids by excreting the difference between what the body needs and what is consumed. If no water was taken in, the kidneys should excrete the minimum volume (400–800 ml/day) with the maximum osmolality (1200 mosm/kg H_2O, specific gravity 1.030); a lower osmolality is observed in the presence of renal disease. If the *osmole excretion rate* is low, much lower urine volumes should be expected for any given urine osmolality.

Osmole excretion rate: Urine volume × urine osmolality.

or

$$\text{Urine volume} = \frac{\text{osmole excretion rate}}{\text{urine osmolality}}$$

MINIMUM URINE VOLUME

The minimum urine volume is determined by the number of osmoles that the patient must excrete and the maximum urine osmolality that the patient can achieve. If the patient has 600 mosmoles to excrete and can achieve a urine osmolality of 1200 mosm/kg H_2O, the minimum urine volume is 500 ml. Urine volumes less than 400 ml/day are considered oliguric because they will not allow excretion of the usual osmole load.

THIRST

Thirst is initiated by a rise in the plasma [Na^+] of 2 mmol/l. Look for this response in all hypernatremic patients.

Clinical Approach

- Hypernatremia is almost always due to water loss.

By answering the following four questions, the etiology of hypernatremia can usually be deduced (Figure 8·1).

1. What is the ECF volume?
2. Has the body weight changed?
3. Is the thirst response normal?
4. Is the renal response normal?

WHAT IS THE ECF VOLUME?

A gain of Na^+ is characterized by ECF volume expansion and is rarely the sole cause of hypernatremia. All other causes of hypernatremia are due primarily to water loss (i.e., there is no ECF volume expansion).

Figure 8·1
Approach to the patient with hypernatremia
The final diagnoses appear in the shaded boxes.

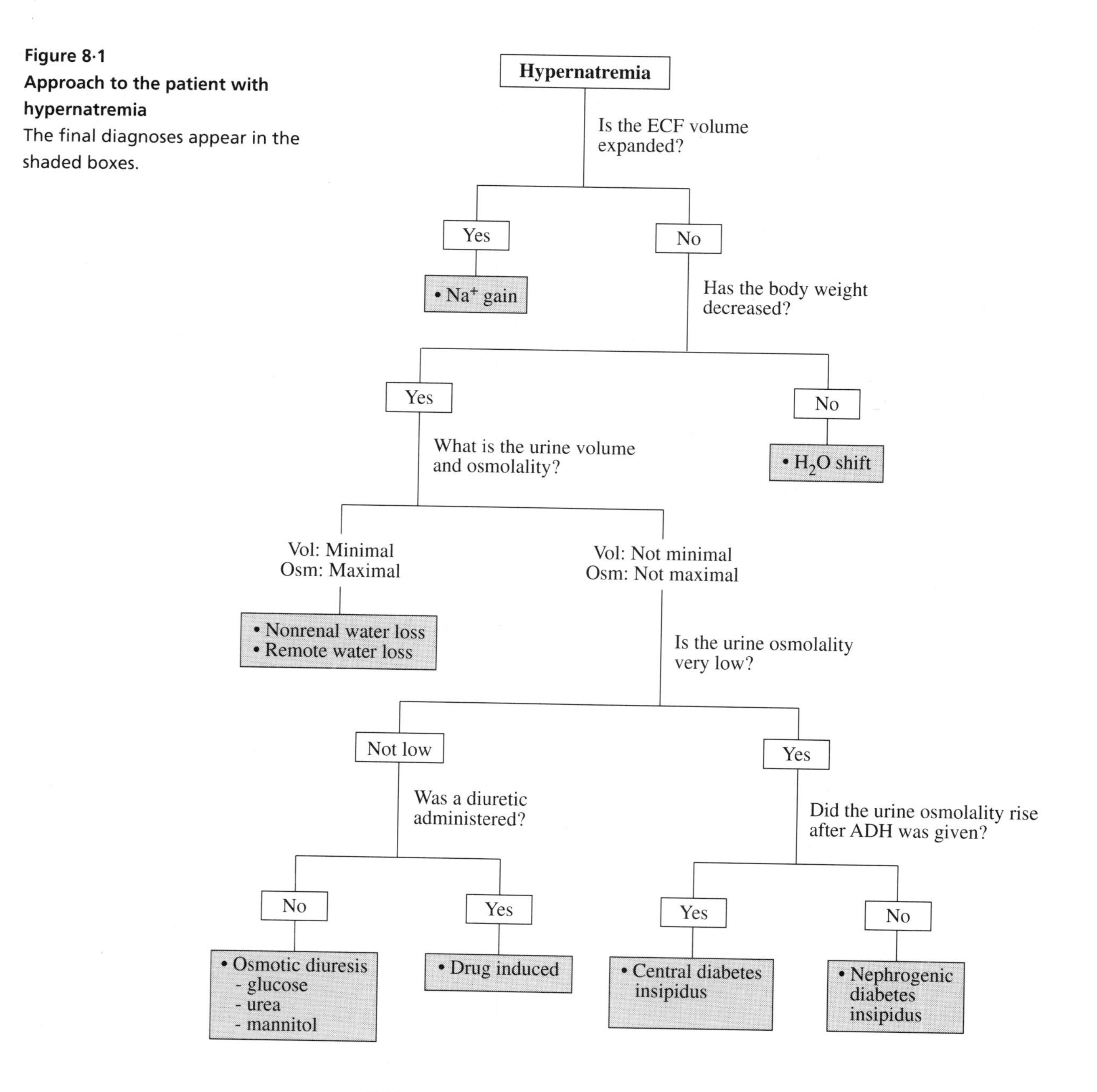

Questions

(Discussions on pages 315–16)

8·1 You are confronted with three patients; each has hypernatremia (154 mmol/l). One patient drank sea water, another lost pure water, and the third lost hypotonic saline (from sweating or via a diuretic). What parameter on physical examination would distinguish these three patients? What is (are) the major threat(s) to life in each case?

8·2 Two 70-kg patients present with hypernatremia (150 mmol/l). The first patient continues to excrete a very large amount of water; his ECF volume is normal (15 liters). The second patient has excreted a large urine volume; the [Na^+] in the urine is close to 50 mmol/l and the ECF volume is low (13 liters). What is the mechanism of hypernatremia in each patient?

HAS THE BODY WEIGHT CHANGED?

Very rarely, water shifts from the ECF into the ICF (e.g., with a convulsion or rhabdomyolysis). In this case, hypernatremia is accompanied by a decrease in the ECF volume and no loss of body weight.

Question

(Discussion on page 316)

8·3 Why does water shift into cells with a convulsion or rhabdomyolysis?

IS THE THIRST RESPONSE TO HYPERNATREMIA NORMAL?

A 2% rise in plasma tonicity provokes a powerful urge to drink; therefore, hypernatremia should be accompanied by thirst. Failure to drink could occur if the patient was unable to access water (was in a desert, was paralyzed, etc.). The absence of thirst should prompt the clinician to suspect a generalized or localized CNS lesion.

IS THE RENAL RESPONSE TO HYPERNATREMIA NORMAL?

The appropriate renal response to hypertonicity is the excretion of urine with the highest osmolality that the kidneys can achieve (not just above isotonicity, but more than 1000 mosm/kg H_2O) and a urine volume that is the minimum value that can be excreted (about 0.5 liters/day). Deviations from this response signal an ADH or renal problem (a disorder involving the renal medulla or an osmotic load). Table 8·1 shows the usual osmoles excreted in an adult.

Renal response to hypernatremia:
- Urine volume = 20 ml/hr (0.5 liters/day).
- Urine osmolality > 1000 mosm/kg H_2O.

Table 8·1
MAJOR CONSTITUENTS IN THE 24-HOUR URINE

Data are from a 70-kg adult eating a typical Western diet. The average urine volume is 1.5 liters.

Constituent	Content mmol	Concentration mmol/l
Urea	400	275
Creatinine	15	10
Na^+	150	100
K^+	75	50
Cl^-	150	100
NH_4^+	40	26
Phosphate	30	20
Sulfate	15	10
Osmolality	–	600

Questions

(Discussions on pages 316–17)

8·4 An elderly patient has not had enough water to drink. She is hypernatremic (150 mmol/l) and her maximum urine osmolality is 600 mosm/kg H_2O. How much of a deficit does she have in ability to concentrate her urine?

8·5 A patient with cirrhosis of the liver and ascites has a urine output of 0.4 liters/day (see the margin). Is this patient oliguric?

Note:
With ascites, there is a low "effective" circulating volume because of low levels of albumin in plasma; hence, high levels of ADH should be present.

Hypernatremia That Is Due To Water Loss

NONRENAL WATER LOSS

Water loss via the respiratory tract and skin (a hypotonic solution with an even lower [Na^+] when its volume is high) is, on average, just under 1 liter/day. Losses in perspiration can increase dramatically in hot environments and with exercise. Similarly, water losses in patients who are febrile and hyperventilating are higher. If the patient has a generalized or localized CNS lesion involving the thirst mechanism or is unable to obtain water, hypernatremia may develop because water loss is not matched by water intake. Infants may also experience hypernatremia from nonrenal water loss because they cannot specifically complain of thirst.

RENAL WATER LOSS

- Renal water loss is the most common cause of hypernatremia. It is usually accompanied by polyuria. The usual causes of polyuria with hypernatremia are diabetes insipidus and osmotic diuresis.

It is well known that patients with hypernatremia who excrete large volumes of hypoosmolar urine suffer from diabetes insipidus (DI) (see the margin). If the urine volume can be lowered and the urine osmolality raised appreciably following the administration of ADH, the defect is in the synthesis of ADH or its release from the brain; this disorder is termed "central DI" (Table 8·2). Alternatively, if the urine remains hypoosmolar or close to isosmolar and the volume is large after biologically active ADH administration, nephrogenic DI is present.

Notes:

- The urine osmolality is useful in assessing ADH actions and the osmolality of the renal medulla.
- The urine $[Na^+] + [K^+]$ is useful in detecting the kidney's influence on the tonicity of the body (cell size).

Central DI

- Central DI is due to lack of ADH. The major symptoms are polydipsia and polyuria. The plasma $[Na^+]$ will only be very high if there is a thirst defect or limited access to water. With ADH administration, the urine volume declines and osmolality rises.

A rise in the plasma $[Na^+]$ (but not hyperosmolality from hyperglycemia or a high urea concentration) stimulates the hypothalamic osmoreceptor and leads to an augmented synthesis of ADH in the paraventricular and supraoptic nuclei. This ADH is then transported by axonal flow to the posterior pituitary. A lesion at any of these sites will produce central DI. The common causes for a lesion are trauma (especially basal skull fractures), infections, space-occupying lesions, and neurosurgical procedures (Table 8·2). In close to 50% of cases of central DI, no specific cause is identified (called "idiopathic central DI").

Table 8·2
ETIOLOGY OF CENTRAL DI

Trauma (especially basal skull fractures)
Neurosurgery (hypophysectomy, other)
Space-occupying lesions
- Neoplasm
- Primary (craniopharyngioma, pineal cyst, pituitary tumors)
- Secondary (metastatic)
- Granuloma
- Sarcoid, histiocytosis X
Infection (meningitis, encephalitis)
Vascular (aneurysm)
Post-hypoxia
Drugs interfering with ADH release (e.g., phenytoin)
Idiopathic central DI (may be familial)

Selective removal of the posterior pituitary usually produces only a transient central DI (it seems that ADH can be released from the hypothalamic neurons that synthesize it). In almost every case, the onset of symptoms of polyuria and polydipsia are abrupt (see the margin). In central DI caused by

An enigma:
Why should slowly progressive lesions result in the abrupt onset of polyuria?

Of interest:
The volume of water reabsorbed in the MCD may be higher in DI than in antidiuresis mainly because of the very large volume of delivery to the MCD. This increased reabsorption can contribute to a medullary interstitial "wash-out."

Urine osmolality in DI:
For complete DI, the urine osmolality should be less than 50 mosm/kg H_2O. Greater values suggest partial central DI or renal influences (e.g., low GFR).

surgery or trauma, there may be an initial polyuria for up to a couple of days (because of inhibition of ADH release). Antidiuresis for up to a few days may ensue (stored ADH is released from the degenerating gland), but permanent central DI often follows.

The diagnosis of central DI is usually easy to confirm. The patient has a CNS problem together with a history of polyuria and polydipsia (cold liquids are usually preferred). Physical examination may help in identifying the underlying disorder. The ECF volume is usually normal or not appreciably reduced. Laboratory exam should show a normal or, more likely, a slightly high plasma [Na^+]; a very high plasma [Na^+] will only be present if thirst or access to water is compromised. The urine volume will be high (3–20 liters/day) unless the GFR is low, and the urine osmolality will be less than 150 mosm/kg H_2O. The diagnosis is confirmed when hypernatremia and polyuria occur with judicious water restriction and when ADH administration results in a prompt rise in the urine osmolality (to above that of plasma but not necessarily to maximum values because a period of time is required for the osmolality in the medullary interstitium to become elevated).

Anything that diminishes the delivery of filtrate to the distal nephron can lower electrolyte-free water excretion. Hence, polyuria is less dramatic in the presence of ECF volume contraction or anterior pituitary resection, which involves a loss of the hemodynamic benefits of cortisol and thyroid hormone.

Questions

(Discussions on pages 317–18)

8·6 A patient had a stroke that resulted in aphasia. All lab tests were normal. She was transferred to a chronic care facility, where tube feeding was instituted; no drugs were given. She was readmitted to the hospital several weeks later, at which time her ECF volume was contracted. The plasma [Na^+] was 160 mmol/l, the urine osmolality was 450 mosm/kg H_2O, and the urine volume was 3–4 liters per day. The urine glucose was negative. What is the diagnosis?

8·7 How may hypokalemia and hypercalcemia influence the urine volume?

A note of caution:
If your patient fails to develop an increase in urine osmolality in response to the ADH that was administered, ensure that the preparation of ADH contained biologically active material before making the diagnosis of nephrogenic DI. This precaution provides an ideal opportunity to demonstrate renal physiology to a willing colleague by administering an ADH analogue in the course of water diuresis.

Nephrogenic DI

This disorder can be divided into two major categories:

1. In some instances, ADH fails to increase the water permeability of the collecting duct, and thus osmotic equilibrium does not occur between the interstitial fluid and the hypoosmolal luminal fluid.
2. There is a group of diseases in which a loss of medullary hypertonicity occurs in response to a major medullary interstitial defect or infiltrate (Table 8·3).

In the latter category, the typical urine output is only 3 liters per day, and the urine osmolality is generally close to that of plasma. These values differ from the extreme polyuria and maximally dilute urine of central DI and from the values encountered when nephrogenic DI is characterized by a failure to respond to ADH. By definition, patients with nephrogenic DI do not have changes in urine volume or osmolality when ADH is given.

Table 8·3
ETIOLOGY OF NEPHROGENIC DI

Compromised ADH-induced rise in cyclic AMP in the CCD (and MCD)
- Lithium
- Demethylchlortetracycline
- Congenital nephrogenic DI
- Hyperkalemia

Loss of medullary hypertonicity
- Renal medullary pathology
  - Infiltrations (amyloid, etc.)
  - Infections (pyelonephritis)
  - Drug-induced (analgesics)
  - Hypoxic damage (sickle-cell anemia)
  - Obstructive uropathy
- Compromised medullary hypertonicity
  - Loop diuretics, transient phenomenon
- Generalized kidney disease
  - Polycystic disease
  - Hypokalemia

Idiopathic nephrogenic DI

Question

(Discussion on page 318)

8·8 If a person has nephrogenic DI, drinks a large volume of water, has a normal diet, and has a urine osmolality that is consistently close to 300 mosm/kg H_2O, what will his urine volume be?

Hypernatremia That Is Due To Na^+ Gain

- Hypernatremia is very rarely due to a gain of Na^+.

A gain of Na^+ is observed in several clinical situations (Table 8·4): when the patient receives a hypertonic Na^+ salt intravenously (e.g., $NaHCO_3$ in a cardiac arrest), when sea water is ingested, when sugar is replaced with salt in the pediatric formula, and, most commonly, when hypotonic Na^+ loss is replaced with isotonic saline (i.e., during the treatment of diabetic ketoacidosis). In the latter case, the urine in a patient with an uncomplicated

Limited rate of glucose metabolism: The maximal rate of glucose oxidation in an ill patient is close to 0.25 g/kg/hr, which is equivalent to 0.3 liters of D_5W in a 70-kg person. Therefore, administration of more D_5W could result in hyperglycemia and an osmotic diuresis, which could aggravate the degree of hypernatremia.

osmotic diuresis should have a [Na^+] of close to 50 mmol/l; a gain of Na^+ occurs when isotonic (152 mmol/l) or half-normal (76 mmol/l) saline is infused.

If hypernatremia is severe, confusion or convulsions may be present. In all cases, the ECF volume is expanded. Treatment in these examples is to increase Na^+ loss with a diuretic and to give water by mouth, if possible. Recall that there is a limited amount of glucose in water that can safely be administered intravenously (see the margin).

Table 8·4
CAUSES OF HYPERNATREMIA RESULTING FROM A GAIN OF NaCl

1. Treatment of polyuria with an infusion that contains a higher [Na^+] than in the urine
   - Half-normal saline in lithium-induced nephrogenic diabetes insipidus
   - Isotonic saline in a patient with a glucose-induced osmotic diuresis
2. Treatment of cardiac arrest with large volumes of hypertonic $NaHCO_3$
3. Salt-poisoning in infants
4. Ingestion of sea water
5. Dialysis error (use of hypertonic dialysate)
6. Combination of a low capacity to excrete NaCl and a thirst center defect

Question

(Discussion on pages 318–19)

8·9 What is the cause of hypernatremia in the following case? The plasma [Na^+] rose to 180 mmol/l overnight in a confused patient. He received no medications. His body weight had not changed, but his ECF volume was contracted. His blood sugar was 90 mg/dl (5 mmol/l), the urine volume was extremely low, and the urine osmolality was close to 1200 mosm/kg H_2O.

PART B
SYMPTOMS OF HYPERNATREMIA

Common Symptoms

The only symptom directly related to a modest degree of hypernatremia is mild confusion (and thirst if the thirst mechanism is intact); however, during severe hypernatremia, major CNS dysfunction can occur and can ultimately lead to coma and hemorrhages (subarachnoid or intracerebral). The exact $[Na^+]$ that can produce symptoms is lower if the onset of hypernatremia is acute (160 mmol/l seems to be close to the value at which symptoms are common). In contrast, symptoms may be absent if the hypernatremia develops gradually.

The polydipsia of central DI is associated with a strong preference for ice-cold liquids, a preference that is not as common in other polyuric states. Some patients can suppress frequency of voiding and thus urinate very large volumes (close to 1 liter) less often; they may develop a dilated bladder, hydroureter, and even hydronephrosis. The onset of polyuria in central DI is often a sudden one.

Polyuria

DEFINITIONS OF POLYURIA

- Polyuria is the excretion of too much urine for a given clinical setting.
- Interpret polyuria by considering each component of the osmole excretion formula: urine volume = osmole excretion/urine osmolality.

There are two definitions of polyuria. The first is driven by the urine flow rate and is an arbitrary value (greater than 1.5 to 2 ml/min, or 2.5 liters/day, in an adult). The second definition, the one we prefer, is based on pathophysiology. In essence, it is a greater rate of excretion of water than one would expect in that clinical setting.

Polyuria as a Function of the Osmole Excretion Rate

Normally, we excrete close to 900 mosmoles each day. If we presume, for simplicity, that a patient can achieve a urine osmolality of only 900 mosm/kg H_2O today, the 24-hour urine volume will be 1 liter (900 mosmoles at 900 mosm/kg H_2O). If this patient had an osmotic diuresis with 1800 mosmoles to excrete per day at this same urine osmolality, the urine volume would be 2 liters in 24 hours, and polyuria would almost be present.

Two further points need to be emphasized. First, it is extremely difficult to find an extra 900 mosmoles to excrete in a day. For example, an extra

Calculation:
- One kg of muscle contains 200 g of protein (80% water).
- Protein is 16% nitrogen.
- 0.16×200 g = 32 g of nitrogen.
- The molecular weight of nitrogen is 14, and urea has two nitrogens. Therefore, 32000 mg/28 = 1143 mmol of urea.

Note:
One liter of blood has:
- 140 g of hemoglobin,
- 48 g of protein (80 g per liter, but 0.6 liters of plasma).

Therefore:
For large urine volumes in an osmotic diuresis, the urine osmolality must not be very high.

Note:
The same volume of electrolyte-free water (e.g., 10 liters) can be excreted at two different urine osmolalities (e.g., 90 vs 45 mosm/kg H_2O) by having two different osmole excretion rates (e.g., 900 vs 450 mosm/day, respectively).

900 mmol of urea is equivalent to the catabolism of almost 1 kg of muscle or the digestion of 1 liter of blood (see the margin). The only other osmoles of endogenous origin to consider are glucose and salts—neither can be lost at this rate for any sustained period using endogenous supplies. The only way to excrete an extra 900 mosmoles is to have an exogenous input, and the solute will almost always be glucose (see the "drinker" subtype of hyperglycemic hyperosmolar syndrome, Chapter 12, pages 444–45).

The second point, and a very important one, is that during an osmotic diuresis, the urinary osmolality is unlikely to remain at 900 mosm/l as in the above example. Although one can achieve a very high urine osmolality (1200 mosm/kg H_2O) at low urine flow rates, the renal medullary osmolality falls at very high urine flow rates. It is therefore not surprising to find that the urine osmolality is closer to 450 mosm/kg H_2O during an osmotic diuresis; the urine volume will be at least 4 liters per day when a patient excretes these 1800 mosmoles daily (1800 mosmoles/450 mosm/kg H_2O).

Polyuria Based on an Unexpectedly Low Urine Osmolality

There are two major reasons to have an unexpectedly low urine osmolality. First, the renal medulla can be damaged by a number of diseases. If this damage occurs, the urine osmolality will be close to that of plasma when ADH acts (i.e., close to 300 mosm/kg H_2O for easy math). Accordingly, while one excretes the usual 900 mosmoles per day, the daily urine volume will be 3 liters (900 mosmoles/300 mosm/kg H_2O). Most clinical nephrologists consider this urine volume to be a form of nephrogenic DI.

A second basis for an unexpectedly low urine osmolality is a lack of ADH (central DI) or a failure of this hormone to open water pores in the late distal convoluted tubule, the CCD, and the MCD (nephrogenic DI). The urine volume now depends on the extent of the lesion. For example, if the urine osmolality is 90 mosm/kg H_2O, the daily urine volume will be 10 liters, but, if the urine osmolality is 45 mosm/kg H_2O, the daily urine volume will be 20 liters (900 mosmoles ÷ 90 or 45 mosm/kg H_2O, respectively; see the margin).

Primary Polydipsia

The excretion of a very large volume of urine in a patient who drinks a large volume of water is called polyuria in the first definition (polyuria based on arbitrary large volumes of urine) but not in our second definition because the appropriate volume of urine is being excreted based on physiologic signals. A "slave to proper definitions" would call this type of excretion a secondary form of polyuria—secondary to polydipsia.

One final point merits emphasis. Using the strict definition of polyuria as a urine volume that is inappropriately high for the clinical setting, a patient with primary polydipsia and hyponatremia can be oliguric while excreting 9 liters of urine per day. A look at the numbers makes our point: with an excretion of 900 mosmoles per day and a urine volume of 9 liters per day, the urine osmolality is 100 mosm/kg H_2O. But, if the expected urine osmolality is 45 mosm/kg H_2O, the urine volume should be 20 liters/day.

Why do we make this seemingly trivial point? Well, an examination of the data in Case 7·3 (Hyponatremia and the Guru, pages 274–75 and 278–79) indicates that ADH is present and acting at some times (urine osmolality greater than 100 mosm/kg H_2O; see the margin) but possibly not at others (reexpansion of the ECF volume by water ingestion or administration of salt). This view has one major advantage. By recognizing that ADH is acting and suppressing maximal water excretion by close to 50% in Case 7·3 (daily urine volume was 10 instead of the maximum of 20 liters), one will not make the mistake of reexpanding the ECF volume (by administering saline) without observing and later limiting the rate of electrolyte-free water excretion (this rate can be lowered by giving ADH once the desired quantity of water is being excreted). This approach will yield the desired rate of rise in the plasma [Na^+].

Note:
It is not possible to deduce the lowest possible urine osmolality in the absence of ADH action without knowing the rate of excretion of osmoles.

CAUSES OF POLYURIA

Because polyuria is very commonly associated with hypernatremia, we feel that it is appropriate to introduce an approach to polyuria at this point, even though not all causes of polyuria are associated with hypernatremia. Three major causes of polyuria can be identified on the basis of urine osmolality: hyperosmolar, isosmolar, and hypoosmolar (Figure 8·2). A more detailed approach to polyuria is provided below.

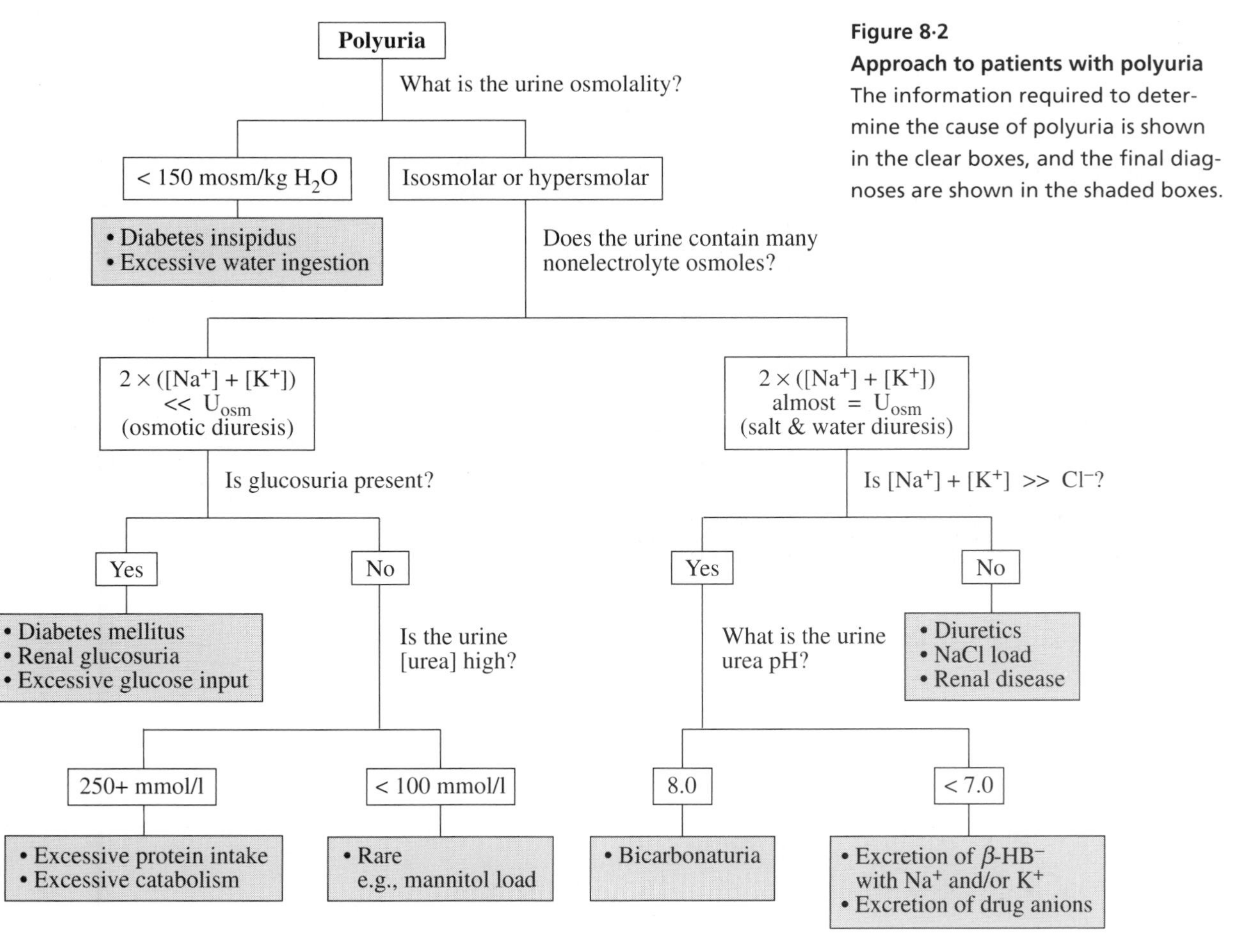

Figure 8·2
Approach to patients with polyuria
The information required to determine the cause of polyuria is shown in the clear boxes, and the final diagnoses are shown in the shaded boxes.

Hyperosmolar Urine

For a large volume of hyperosmolar urine to be excreted, there must be a source of these osmoles. If the urine osmolality is 400 mosm/kg H_2O, and 5 liters of urine are produced/day, 2000 mosmoles must be identified (normal is less than half this quantity). If the urine contains much less than 150 mmol/l of Na^+ plus K^+, then look for an osmotic diuretic (close to 300 mmol/l of glucose or urea). A high excretion of urea occurs with high-protein feeding, trauma, gastrointestinal hemorrhage, catabolic state (neoplasm or sepsis), or a high blood urea (recovery from obstructive uropathy or renal failure; other factors also operate to cause polyuria in these conditions). It is not necessary to administer ADH to establish the diagnosis in this group of patients.

It should be emphasized that each liter of urine excreted during an osmotic diuresis is expected to contain close to 50 mmol/l Na^+ and 25–50 mmol/l K^+. If the administration of Na^+ and K^+ is less than 150 mmol/day, deficiency states will ensue and lead to ECF volume contraction and hypokalemia.

Isosmolar Urine

The major differential diagnosis is summarized in the "loss of medullary hypertonicity" section of Table 8·3, page 299. These patients do not show an appreciable increase in urine osmolality following the administration of ADH. It is possible that patients with partial central DI can have isosmolar urine on presentation. In addition, a patient with DI and marked ECF volume contraction and a low urine flow rate can have urine that is close to isosmolal. In partial central DI, expect at least a significant rise in urine osmolality with ADH administration. The patient undergoing a "saline diuresis" as a result of NaCl administration or diuretics may also have isosmolar urine.

Hypoosmolar Urine

Almost all hypernatremic patients with very hypoosmolar urine have central DI; they excrete hyperosmolar urine when ADH is given. An exception to this rule is certain patients with nephrogenic DI (lithium or demethyltetracycline therapy, congenital nonresponsiveness to ADH); they do not have an appreciable increase in urine osmolality after ADH administration.

PART C
TREATMENT OF A WATER DEFICIT

- Stop the ongoing water loss.
- Replace the deficit slowly, by the oral route, if possible. If you must give electrolyte-free water by the intravenous route, use hypotonic saline plus a limited quantity of glucose in water.

The major objectives of therapy for patients with hypernatremia caused by water loss are to treat the underlying medical problems, to minimize ongoing water losses, and to replace the water deficit.

Stopping the Water Loss

The physician must curtail large ongoing water losses, if possible. The problem of ADH deficiency can easily be rectified with appropriate replacement therapy. If an osmotic diuresis is present, the addition of the osmotic agent must be stopped, if possible, and ongoing Na^+ and K^+ losses should be replaced.

Replacing the Water Deficit

The water deficit should be calculated, and only a portion (see the margin) should be replaced over the next 12 hours. Total correction should occur over several days. The oral route of water replacement is preferable if the patient is conscious and alert because it avoids the problems of hyperglycemia and glucosuria associated with intravenous glucose in water administration. The advantages and disadvantages of the intravenous solutions used for electrolyte-free water administration are discussed in the following paragraphs.

Clinical pearl:
Too rapid replacement of the water deficit should be avoided (> 1 mmol/l fall in $[Na^+]$/hr, or 12 mmol/l in a 24-hour period).

GLUCOSE IN WATER (D_5W)

D_5W appears to have ideal properties to deliver electrolyte-free water intravenously; it is isotonic (thus avoiding hemolysis), and the osmoles (glucose) disappear as a result of metabolism (see the discussion of Question 8·10 for more information).

Question

(Discussion on page 319)

8·10 How quickly can the glucose infused in one liter of D_5W be metabolized in a patient with hypernatremia?

HALF-NORMAL SALINE

Note:
The "effectiveness" of half-normal saline in the therapy of hypernatremia can be enhanced by increasing the $[Na^+]$ of the urine with a loop diuretic. The difference in $[Na^+]$ between the urine and the IV solution represents the electrolyte-free water administered.

As mentioned earlier, it is important to stop water losses early because it is difficult to administer large volumes of electrolyte-free water safely. Therapy with just "half-normal" (75 mmol/l) saline is not appropriate if polyuria is present and the $[Na^+]$ in the urine is less than that in the infusion (see the margin).

In patients with a degree of contraction of the ECF volume and no polyuria, the deficit of water should be replaced with half-normal saline and possibly small quantities of D_5W (a maximum of 0.3 liters/hr). Care should be taken to avoid a serious Na^+ overload. Since each liter of half-normal saline has 500 ml of electrolyte-free water that distributes throughout total body water, only 333 ml will enter the ICF. In quantitative terms, if the ICF volume is 30 liters, then 11 ml of electrolyte-free water will be added to each liter of ICF (the brain contains less than 1 liter of ICF). More electrolyte-free water can be administered if even more hypotonic saline solutions are infused (0.25% or 0.33%).

DISTILLED WATER

Note:
Do everything possible to avoid giving distilled water IV.

If a patient is severely hypertonic, is in congestive heart failure, is severely hyperglycemic, and cannot tolerate dialysis or oral water therapies, it may be necessary to administer sterile water intravenously. This water should be given by central vein, and several liters may be administered per day by this route. Extreme caution and frequent monitoring are required in this setting; the major risk is hemolysis if the water is given too quickly.

CALCULATION OF THE WATER DEFICIT

When calculating a water deficit, the simplest approach is to obtain quantitative estimates of ECF and ICF volume deficits.

Data

Imagine that a 70-kg patient lost enough water for the plasma $[Na^+]$ to rise from 140 to 160 mmol/l. The ECF volume, however, is normal on physical examination. The patient's normal values are an ICF volume of 30 liters and an ECF volume of 15 liters.

Long Form of the Calculation

The first step is to calculate body water balance by assessing the current vs the expected ICF and ECF volumes.

Usual "effective" osmoles in the ICF:
- Assume that a 70-kg adult has 30 liters of ICF and a plasma $[Na^+]$ of 140 mmol/l.
- 2 (140 mmol/l) × 30 liters = 8400 mosmoles.

Change in ICF volume: One can calculate this change by using the "effective" osmolality and by assuming a constant number of effective osmoles. Normally, the total number of effective osmoles in the ICF is the product of the ICF volume (30 liters) and 2 × plasma $[Na^+]$, or 8400 mosmoles. After the water loss, assume no change in number of effective osmoles. Thus, the new effective osmolality is 320 mosm/kg H_2O (2 × new plasma $[Na^+]$). Dividing 8400 by 320 yields the new ICF volume of 26.25 liters, a fall of 3.75 liters.

Change in ECF volume: The ECF volume is not revealed by the plasma [Na^+]. It is based on a clinical assessment of the vascular volume and, more importantly, the interstitial volume. This estimate is imprecise but still clinically useful.

Change in total body water: Add the new volumes of the ICF and ECF and then compare them to the normal value for total body water.

Change in Na^+ balance in the ECF: The second step is to calculate the change in content of Na^+ in the ECF. The normal content of Na^+ is the product of the [Na^+] in the ECF (for simplicity, assume 140 mmol/l, the same as that in plasma; see the margin) and the normal ECF volume (15 liters). The current content of Na^+ in the ECF is the measured [Na^+] in plasma multiplied by the clinical estimate of the ECF volume. If the current content of Na^+ exceeds the original value, Na^+ must have been gained; the converse is also true. An example is provided in Case 8·2 (see the margin).

[Na^+] in interstitial fluid:
- With a Donnan factor of 0.95 for cations, the [Na^+] in interstitial fluid would be lower than plasma.
- Since there is nonaqueous volume present (6%), the plasma [Na^+] is 94% of the actual value (150 mmol/kg H_2O).
- Overall, these two factors virtually cancel each other out, so the plasma and interstitial [Na^+] are almost equal.

Short Form of the Calculation

Assume total ICF osmoles do not change. Let ["e"] represent the concentration of effective osmoles, which is equal to 2 × [Na^+].

$$\text{ICF volume}_{normal} \times [\text{"e"}]_{normal} = \text{ICF volume}_{abnormal} \times [\text{"e"}]_{abnormal}$$

Sample calculation:
- [Na^+] = 160 mmol/l.
- Effective osmolality = 2 × 160 mmol/l = 320 mmol/l.
- Normal ICF volume = 30 liters.
- Therefore, 30 liters × 280 = new ICF volume × 320.
  New ICF volume = 26.25 liters.

Sample Na^+ balance:
- The infant in Case 8·2, pages 313–14, has a normal ECF volume of 1 liter and a [Na^+] of 140 mmol/l. Therefore, the expected content of Na^+ in the ECF of this infant is 140 mmol.
- Measured values were a [Na^+] of 180 mmol/l and an ECF volume of 0.7 liters.
- Calculate for yourself whether Na^+ balance is positive or negative.

Details of Therapy

- Objectives
  - To calculate mass balance for Na^+.
  - To calculate mass balance for water.
  - To estimate changes present in the ECF and ICF volumes.
  - To correct the ECF volume quickly if it is contracted and to correct the ICF volume slowly unless the patient is seriously symptomatic.

ESTIMATE MASS BALANCES FOR Na^+ AND WATER

- Replace the ECF volume quickly if it is contracted.
- Loss of K^+ without phosphate is matched for the most part by a shift of Na^+ into cells.

Na^+

Clinical pearls:

1. Replace the ECF volume deficit quickly.
2. The normal ECF volume is 20% of body weight. Shock occurs with a 30% deficit, but a 10% deficit is just detectable; iterate for intermediate values.
3. During therapy, administration of KCl is similar to administration of NaCl as far as the ECF is concerned (K^+ leave the ICF mainly in "exchange" for Na^+); the reverse occurs as K^+ are replaced.

The content of Na^+ is the product of the $[Na^+]$ in plasma and the ECF volume. The change in ECF volume must be estimated on clinical grounds.

A second consideration for total body Na^+ is the quantity of Na^+ that might have shifted into cells in conjunction with a negative balance for K^+ that is independent of the loss of intracellular anions (monovalent phosphate). Assume that most of this deficit of K^+ was matched by a shift of Na^+ into cells (see the discussions of Cases 7·1, pages 275–77, and 10·1, pages 379–80, for more information).

Ongoing losses of Na^+ will have to be replaced to achieve the desired change in Na^+ balance. To create a negative balance for Na^+, administer a loop diuretic, but reinfuse all losses of water and K^+.

Questions

(Discussions on pages 319–20)

8·11 If a 70-kg patient has a $[Na^+]$ of 154 mmol/l and a very marked degree of ECF volume contraction, what is the most likely pathophysiology for this lesion?

8·12 To expand the ECF volume of the patient described in Question 8·11 by 2 liters, but not change the ICF volume in this period, what would be the most appropriate $[Na^+]$ for the IV?

Water

- The asymptomatic patient requires a slow replacement of ICF volume.

There are two ways to expand the ICF volume: first, one can administer electrolyte-free water, two-thirds of which should distribute in the ICF in normal individuals; second, one can remove Na^+ but not water from the ECF. This second option should be considered only if the ECF volume is expanded. To lose Na^+ but not water, administer a loop diuretic to lose Na^+, K^+, Cl^-, plus water; then replace all the K^+, Cl^-, and water without replacing Na^+.

It is easiest to calculate the changes in the ECF and ICF volumes independently. The change in ECF volume is assessed clinically, and the change in ICF volume in liters is calculated as described on pages 306–07.

- During therapy, change the $[Na^+]$ by 0.5 to 1.0 mmol/l/hr if the patient is not seriously symptomatic (up to a total of 12 mmol/l/24 hours).

To achieve a positive water balance and a fall in the plasma $[Na^+]$, the $[Na^+]$ in the IV should be compared to the patient if the patient is oliguric and to the $[Na^+]$ plus the $[K^+]$ in the urine if the patient is polyuric (see the margin). Examples are provided in the Cases for Review section, pages 311–15.

If the patient is experiencing a convulsion, it is possible that the change in brain cell size has contributed to this emergency. Aside from the usual means of treating a patient who is convulsing, it is advisable to change the $[Na^+]$ to the value it was before the convulsion. In effect, the $[Na^+]$ should decrease quickly by less than 6 mmol/l. This change can be implemented with the administration of enough electrolyte-free water (D_5W, $D_{2.5}W$). The volume to infuse should be calculated as follows for a patient with a plasma $[Na^+]$ of 160 mmol/l:

- Assume no change in ICF osmoles.
- Let ["e"] represent the concentration of effective osmoles, which is approximately equal to $2 \times [Na^+]$.
- Initial ["e"] was 2×160 mmol/l = 320 mmol/l.
- Final ["e"] should be 2×156 mmol/l (i.e., a decline of 4 mmol/l) = 312 mmol/l.
- ICF volume at a $[Na^+]$ of 160 was 26.25 liters (see page 307).
- Therefore, $320 \times 26.25 = 312 \times$ new ICF volume.
  New ICF volume = 27 liters.

To keep the extra 0.75 liters of water in the ICF, one needs to administer 1.125 liters, given that approximately two-thirds of the water given remains in the ICF. One need not consider water excretion because it should be small over a very short time span.

Sample calculation:
In a 70-kg patient with hypernatremia (160 mmol/l), to lower the $[Na^+]$ by 1 mmol/l/hr, the positive water balance should be 250 ml/hr.
- Percent decline in $[Na^+] \times$ total body water = positive water balance.
- $((160 - 159)/160) \times 40$ liters = 250 ml.

Note:
If the patient is hyperglycemic, do not use D_5W; instead, use hypotonic saline (half the volume of half-normal saline is electrolyte-free water, or, better still, three-fourths of the volume of quarter-normal saline is close to electrolyte-free water).

Question

(Discussion on page 320)

8·13 A 70-kg patient has DI, hypernatremia (154 mmol/l), and a normal ECF volume. What is the most likely pathophysiology to explain this set of findings? Provide a quantitative answer. What are the objectives for therapy?

8·14 A patient has a clinical syndrome that resembles central DI. One unusual finding is present: the urine osmolality rises when DDAVP is given but not when ADH is given. How does this finding help in defining the etiology?

PART D
REVIEW

Discussion of Introductory Case
Treat the Patient Before He (DI)es

(Case presented on page 292)

Of what significance is the modest degree of contraction of the ECF volume?

Hypernatremia is due to Na^+ gain or water loss from the ECF. The reduced ECF volume indicates that the hypernatremia in this case is due primarily to loss of water.

Why was thirst absent?

Hypernatremia elicits a thirst response unless a CNS lesion is present (a generalized CNS problem such as confusion, a specific lesion involving the thirst center, or a communication problem). This patient was confused and thus had a thirst problem.

Of what significance is the loss of body weight?

The loss of weight indicates that the cause of hypernatremia is water loss from the body (not a shift of water into muscles).

What is the significance of the polyuria?

Polyuria suggests a problem of either DI or the presence of a diuretic. Since the urine is quite hypoosmolar in the face of hypernatremia, the patient has DI.

Can any inference be made from the urine osmolality and the acuteness of the onset?

The low osmolality and the acuteness of onset both suggest a diagnosis of central rather than nephrogenic DI. In this case, the meningitis and/or the phenytoin could have caused a low ADH release. This impression can be confirmed by an appropriate renal response to exogenous ADH (a rise in osmolality and a decrease in the volume of urine).

What should the therapy be?

Therapy has two major aims. First, stop water loss by giving ADH (see the margin); second, water must be given as half-normal saline (a danger is ECF volume expansion) or as glucose in water, provided that hyperglycemia is not present. Do not give more than 0.3 liters/hr (dangers are hyperglycemia and a glucose-induced osmotic diuresis, which will cause additional water loss).

Leverage for therapy:

- Increase water intake. A person who has 900 mosmoles to excrete and a urine osmolality of 100 mosm/kg H_2O needs more than 9 liters input per day. You just cannot infuse that volume of water safely.
- Stop water loss. This is where the leverage exists in this patient; give ADH.

Cases for Review

Case 8·1
The Shrink Shrank the Cells

(Case discussed on pages 312–13)

Manny, a 37-year-old, has been suffering from a bipolar affective disorder for the past five years. Lithium has helped him tremendously, but he drinks several liters of water each day. Elective surgery is planned for the near future.

To avoid lithium-induced nephrogenic DI, what preparations should be made for Manny before and during this operation?

SUPPLEMENTAL QUESTION

Postoperatively, Manny is passing 3 ml urine/min with an osmolality of 150 mosm/kg H_2O and a [Na^+] of 35 mmol/l; he is now hypernatremic (154 mmol/l). What advice would be appropriate now? Estimate his urine volume over the next 24 hours (assume no change in urine osmolality).

Case 8·2
Who Put the Na^+ in Mrs. Murphy's Breast Milk?

(Case discussed on pages 313–14)

Lois, aged 2 weeks, has had a rough time. She was normal at birth, weighing 3.5 kg (7.7 lb). She has deteriorated since then and has lost 1 kg (2.2 lb) of weight. She is breast-fed. Diarrhea and vomiting were not present and she has not taken medications. No other information was provided on history. On presentation, she has a very contracted ECF volume and is virtually anuric. The only laboratory data available are a plasma [Na^+] of 180 mmol/l, glucose of 1.1 mmol/l (20 mg/dl), and urea of 75 mmol/l (210 mg/dl). A brilliant intern sent a sample of breast milk for analysis, and the results were unexpected ([Na^+] was 107 mmol/l instead of the usual 7 mmol/l).

Is the hypernatremia due to the consumption of "hypernatric" breast milk?
What role might the high [Na^+] in breast milk play in this case?
What role might the hypoglycemia play?
What treatment should Lois receive?

Case 8·3
Steve Is Pee(d) Off

(Case discussed on pages 314–15)

After exams, the class partied. In the competition to see who could drink the most alcohol, Steve won! He went home and slept off his victory. He was very polyuric and was brought to the emergency room the next day.

His ECF volume was mildly contracted. His laboratory values are summarized below.

		Plasma	Urine
Na^+	mmol/l	152	5
K^+	mmol/l	4.0	10
Glucose	mmol/l (mg/dl)	4.0 (72)	0
Urea	mmol/l (mg/dl)	2.0 (5.6)	20
Osmolality	mosm/kg H_2O	420	287

How do you explain Steve's hypernatremia and polyuria?

Discussion of Cases

Discussion of Case 8·1 The Shrink Shrank the Cells

(Case presented on page 311)

To avoid lithium-induced nephrogenic DI, what preparations should be made for Manny before and during this operation?

Lithium-induced nephrogenic DI may or may not be reversed when lithium is discontinued. In consultation with Manny's psychiatrist, lithium was discontinued for three weeks. His 24-hour urine volume still remained at 5 liters/day. He is normonatremic, so a brief period of water restriction will distinguish between habitual polydipsia and nephrogenic DI. With water restriction, his [Na^+] rose, and he was still polyuric; there was no change in urine volume and osmolality following the administration of ADH. Hence, Manny has nephrogenic DI, which seems to be a permanent condition.

Manny should have hypotonic solutions infused in the postoperative period; the [Na^+] in the IV should be close to that in the urine. Hyperglycemia should be avoided and his plasma [Na^+] should be monitored closely.

SUPPLEMENTAL QUESTION

Postoperatively, Manny is passing 3 ml urine/min with an osmolality of 150 mosm/kg H_2O and a [Na^+] of 35 mmol/l; he is now hypernatremic (154 mmol/l). What advice would be appropriate now? Estimate his urine volume over the next 24 hours (assume no change in urine osmolality).

Manny was infused with half-normal saline to match his urine output throughout his surgery and postoperative period. This IV solution had a lower [Na^+] than his plasma. Hypernatremia developed because 75 mmol of Na^+ were infused for every liter of urine excreted (each liter of urine contained only 35 mmol of Na^+), and he was therefore in positive balance of 40 mmol of Na^+ per liter throughput. This balance should lead to the development of hypernatremia and an expanded ECF volume.

Advice concerning balance for Na^+: His ECF volume is normal, but each liter contains an extra 14 mmol of Na^+. Therefore, he must lose 210 mmol of Na^+ (15 liters × 14 mmol/l). To achieve this loss, stop the intake and promote the excretion of Na^+; it might be necessary to use a diuretic. This component of the analysis deals with balance for Na^+ but not water.

Advice concerning balance for water: His ECF volume is normal and his ICF volume is down 3 liters (10% of an ICF volume of 30 liters, see the margin); he therefore needs 3 liters of positive water balance; ongoing losses must also be replaced. Use the oral route to administer water, if possible.

Note:
A [Na^+] of 154 is 10% higher than the normal value of 140 mmol/l.

Practical considerations: If one gives Manny furosemide, induces a diuresis, and replaces the urine with D_5W, after 2 liters of urine output, Manny will lose 200 mmol of Na^+ (the [Na^+] is most likely to be 100 mmol/l). One should recall that D_5W should not be given at a rate that exceeds 300 ml/hr.

If Manny's ECF volume had been contracted, one would have replaced total body water with 3 liters of electrolyte-free water—2 would have gone to the ICF and 1 to the ECF. The ECF volume would have been coincidentally restored with isotonic saline.

Discussion of Case 8·2 Who Put the Na^+ in Mrs. Murphy's Breast Milk?

(Case presented on page 311)

Is the hypernatremia due to the consumption of "hypernatric" breast milk?

No. If one ingests NaCl with little water, hypernatremia may develop, but the ECF volume will be expanded, not contracted. With time and an expanded ECF volume, Na^+ will be excreted. Therefore, the ECF volume will return to normal or even become mildly contracted because of a natriuresis induced by hypernatremia. Nevertheless, shock is a very unlikely outcome.

If Lois' ECF volume was initially 1 liter (for easy math) and is now 0.7 liters, she will have a total body deficit of Na^+ (see the margin).

Deficit of Na^+:
Normal = 140 mmol (140 mmol/l × 1 liter)
Admission = 126 mmol (180 mmol/l × 0.7 liters)
Difference = 14 mmol deficit

What role might the high [Na^+] in breast milk play in this case?

To understand the pathogenesis of Lois' problem, one must appreciate that breast milk is isotonic to her mother's plasma (285 mosm/kg H_2O). Lactose normally constitutes the majority of osmoles (220 mosm/kg H_2O; electrolytes constitute 45 mosm/kg H_2O). In this case, with a [Na^+] of 107 mmol/l in breast milk, there is "room" for at most 45 mosm/kg H_2O lactose (285 – 2 × (107 + 13) mmol/l Na^+ and K^+, respectively), which is an insufficient nutritional supply. Therefore, Lois is carbohydrate-starved and must obtain the glucose for her brain from gluconeogenesis and glycogenolysis.

Calculation:
- Osmolality of breast milk equals the osmolality of the mother.
- Subtract 2 × ([Na^+] + [K^+]) to see the maximum concentration of lactose.

What role might the hypoglycemia play?

Lois' hypoglycemia led to the production of glucose from her body proteins because her stores of glycogen were depleted. For every millimole of glucose formed, 1.7 mmol of urea are formed (see the margin). Excretion

Calculation:
- 100 g of protein = 16 g of nitrogen.
  - 16 g of nitrogen = 572 mmol of urea (molecular weight of nitrogen = 14, and there are two nitrogens per urea).
- 100 g of protein = 60 g of glucose.
  - 60 g of glucose = 333 mmol (molecular weight of glucose = 180).
- 572/333 = 1.7

of this extra urea will lead to an osmotic diuresis in this setting because the immature kidneys do not reabsorb Na^+ as well as normal adult kidneys. Hence, Lois excreted large volumes of water with a $[Na^+]$ that is likely to be in the 50 mmol/l range. This excretion, driven by the urea load, could explain the low ECF volume and the hypernatremia.

What treatment should Lois receive?

Avoid hypoglycemia: Give enough glucose (5 mmol, 90 mg) to raise the concentration of glucose to normal acutely (ECF volume + 1/4 ICF volume); then give enough glucose to maintain euglycemia, but avoid hyperglycemia because it will lead to shrinking of cells and further osmotic diuresis.

Reexpand the ECF volume: Give 300 ml of saline with a $[Na^+]$ close to that of Lois, and replace urine losses when they occur.

Reexpand the ICF volume slowly: Give a positive water balance of close to 150 ml today to lower the $[Na^+]$ 12 mmol/l/day.

- Assume no change in ICF osmoles.
- Let ["e"] represent the concentration of effective osmoles, which is close to $2 \times [Na^+]$.
- Initial ["e"] was 2×180 mmol/l = 360 mmol/l.
- Final ["e"] should be $2 \times (180 - 12)$ mmol/l = 336 mmol/l.
- ICF volume at a $[Na^+]$ of 180 was close to 2 liters.
- Therefore, $360 \times 2 = 336 \times$ new ICF volume.
  New ICF volume = 2.15 liters, a positive balance of 0.15 liters, or 150 ml, is needed today.

Improve her nutritional state: Bottle and IV feeding should be sufficient.

Discussion of Case 8·3
Steve Is Pee(d) Off

(Case presented on pages 311–12)

How do you explain Steve's hypernatremia and polyuria?

Hypernatremia: Hypernatremia implies Na^+ gain or water loss. There was no Na^+ intake by history and his ECF volume was mildly contracted, so he had water loss from the ECF.

Water could have shifted into his cells, but there was no evidence of a convulsion or rhabdomyolysis. Subsequent investigations did not support this diagnosis either.

Water loss was occurring via the kidneys. Although he had excessive water intake with alcohol, the water loss was inappropriate because he was hypernatremic. Hypernatremia and excessive renal water loss imply very low levels of ADH. Although nephrogenic DI is possible, it does not develop this quickly. We speculate that the basis for the low levels of ADH was suppression of its release by ethanol. Note that he had a plasma osmolal gap of 110 mosm/kg H_2O $(420 - (2 \times 152) - 4 - 2)$, consistent with astronomic levels of ethanol in plasma (later confirmed by direct assay).

Two other points merit emphasis. First, the urine osmolality was very high considering the large loss of water that induced hypernatremia. The osmolal gap of the urine reveals that most of the osmoles were ethanol, an "ineffective" osmole with regard to water shifts. Hence, his nonethanol osmolality was close to 100 mosm/kg H_2O, consistent with central DI. Second, his natural history was interesting. As Steve metabolized ethanol in his body, his urine osmolality rose to 456 mosm/kg H_2O eight hours later without treatment. The ethanol level was 56 mmol/l in plasma at that time.

Polyuria: Even though his osmole excretion rate was above normal, most of the osmoles were ethanol, and ethanol does not cause an osmotic diuresis. He had ethanol-induced suppression of ADH release, not nephrogenic DI or an osmotic diuresis.

Summary of Main Points

- Hypernatremia is an increase in the Na^+ content relative to that of water. Although hypernatremia provides no insight into the ECF volume status, it does indicate that the ICF volume is contracted.
- Hypernatremia is most commonly due to net water loss and is rarely due to excess retention of exogenous Na^+.
- Hypernatremia indicates either a defect in water acquisition and/or excessive water excretion relative to the need to retain electrolyte-free water. Under normal circumstances, the hypernatremic patient should be thirsty and should excrete the minimum volume of maximally concentrated urine (> 1000 mosm/kg H_2O).
- Brain cells can regulate their ICF volume in hypernatremia by gaining particles. Therefore, the rate of correction of hypernatremia should be dictated by the clinical symptoms and not by the plasma [Na^+]. Generally, unless the CNS symptoms are severe, the [Na^+] should be lowered at a rate of 0.5–1 mmol/l/hr, up to a maximum of 12 mmol/l per day.

Discussion of Questions

8·1 You are confronted with three patients; each has hypernatremia (154 mmol/l). One patient drank sea water, another lost pure water, and the third lost hypotonic saline (from sweating or via a diuretic). What parameter on physical examination would distinguish these three patients?

The high Na^+:H_2O ratio of 154 mmol/l indicates that the ICF volume is contracted to the same degree in all cases. However, the patient who ingested sea water has ECF volume expansion. In contrast, the patient who lost hypotonic saline has a reduced ECF volume from loss of Na^+. The loss of water per se causes only a small degree of ECF volume contraction.

What is (are) the major threat(s) to life in each case?

In each case, the major organ at risk is the brain (hemorrhage could result). In addition, in the patient who drank sea water, there is the potential danger of congestive heart failure. In the patient who lost hypotonic saline, shock could ensue.

8·2 Two 70-kg patients present with hypernatremia (150 mmol/l). The first patient continues to excrete a very large amount of water; his ECF volume is normal (15 liters). The second patient has excreted a large urine volume; the [Na^+] in the urine is close to 50 mmol/l and the ECF volume is low (13 liters). What is the mechanism of hypernatremia in each patient?

Basis of hypernatremia: The basis of hypernatremia is water loss and Na^+ gain in the first patient and water loss with some Na^+ loss in the second patient. Apart from the renal water loss, both patients should have a powerful urge to drink; water intake should prevent the development of hypernatremia. Hence, each has a problem with thirst appreciation, the ability to communicate thirst, or a problem with access to water.

Balance for Na^+: In the first patient, when a negative balance for water occurs, the ECF and ICF volumes tend to contract. As a result of the lower ECF volume, there is a signal for the reabsorption of Na^+ by the kidney. Therefore, the ingestion of Na^+ will result in a positive balance for Na^+. Looked at in another way, each liter of ECF has retained an extra 10 mmol of Na^+. With a normal ECF volume (15 liters), the patient has a positive balance of 150 mmol of Na^+ (15 liters × 10 mmol/l) and will need to lose this quantity of Na^+ during therapy.

The second patient has lost Na^+ and water. This loss is a major cause of the marked degree of contraction of the ECF volume (see the margin). Because the ECF volume is 13 vs the normal 15 liters, as suggested, the content of Na^+ is 1950 mmol (13 liters × 150 mmol/l); the normal content of Na^+ is 2100 mmol (15 liters × 140 mmol/l). Hence, this patient will need a positive balance for Na^+ of 150 mmol during therapy.

Note:
When the ECF volume is contracted from water (+ Na^+) loss, the majority of this loss is in the interstitial fluid because the rise in the colloid osmotic pressure plus a fall in hydrostatic pressure tends to defend the intravascular volume.

8·3 Why does water shift into cells with a convulsion or rhabdomyolysis?

Water shifts into cells when cells gain particles or when the ECF gains water or loses particles (Na^+, Cl^-). In the examples cited, muscle cells gain particles because macromolecules break down to smaller molecules (lactic acid in the former example and polypeptides in the latter example).

8·4 An elderly patient has not had enough water to drink. She is hypernatremic (150 mmol/l) and her maximum urine osmolality is 600 mosm/kg H_2O. How much of a deficit does she have in ability to concentrate her urine?

The answer is a deficit of 67%. For round numbers, the normal plasma osmolality is 300 and the medullary interstitial osmolality is 1200 mosm/kg H_2O. Hence, the rise in osmolality from filtrate to final urine in the patient is 300 mosm/kg H_2O (600–300) vs a maximum of 900 mosm/kg H_2O (1200–300) normally.

8·5 A patient with cirrhosis of the liver and ascites has a urine output of 0.4 liters/day. Is this patient oliguric?

To determine if oliguria is present, assess the maximum urine osmolality and the osmole excretion rate. For simplicity, let us say the maximum urine osmolality that this patient can achieve is 600 mosm/kg H_2O. If the patient is on a low-protein diet and the urine is electrolyte-free, this patient could be excreting 100 mosmoles per day. If oliguria is present, the urine volume would be 0.167 liters per day, less than half the current urine output (see the margin). Hence, this patient is not really oliguric because the osmole excretion rate is so low.

Calculation:

- Urine osmolality = 600 mosm/kg H_2O.
- Excretion = 100 mosmoles.
- Urine volume

$$= \frac{\text{osmoles}}{U_{osm}} = \frac{100}{600}$$

$= 0.167$ liters

- With a stimulus for release of ADH (the low "effective" arterial volume), any urine volume > 0.167 liters/day is in fact a form of polyuria.

8·6 A patient had a stroke that resulted in aphasia. All lab tests were normal. She was transferred to a chronic care facility, where tube feeding was instituted; no drugs were given. She was readmitted to the hospital several weeks later, at which time her ECF volume was contracted. The plasma [Na^+] was 160 mmol/l, the urine osmolality was 450 mosm/kg H_2O, and the urine volume was 3–4 liters per day. The urine glucose was negative. What is the diagnosis?

Because hypernatremia was associated with ECF volume contraction, it was due to water loss (and Na^+ loss). The water loss was renal because the patient was polyuric; the urine osmolality of 450 mosm/kg H_2O indicates that that the water loss was not a simple water diuresis. Therefore, an osmotic diuresis is the most likely cause (no drugs were given). In the absence of glucosuria, a urea-induced diuresis should be suspected (to confirm, measure the urea in urine; it should be at least 300 mmol/l). The urea load was probably due to the high-protein feeding.

Note the absence of (or failure to communicate) thirst, a consequence of the previous stroke with resulting aphasia.

In addition, an osmotic diuresis leads to Na^+ loss—hence, the prominent degree of ECF volume contraction.

8·7 How may hypokalemia and hypercalcemia influence the urine volume?

Thirst: Primary polydipsia, which contributes to polyuria, has been shown with both hypokalemia and hypercalcemia. This derangement in thirst does not cause nephrogenic DI but does contribute to polyuria.

The renal aspects are considered below.

Hypokalemia: The mechanisms whereby hypokalemia produces polyuria are multiple. Circulating levels of ADH are normal, and although ADH-induced cyclic AMP generation is reduced in vitro, the physiologic response to ADH appears to be normal. Hypokalemia decreases medullary hyperosmolality and results in impaired NaCl reabsorption in the thick ascending limb. These effects require a large (200+ mmol) K^+ deficit; reversal with K^+ therapy may take many weeks because structural changes (e.g., cyst formation) occur.

Hypercalcemia: With hypercalcemia, levels of at least 11 mg/dl (2.75 mmol/l) are required to induce polyuria. Again, the mechanisms that produce nephrogenic DI are multiple. Circulating levels of ADH are normal during dehydration; however, there is evidence that the renal response to ADH is impaired. Hypercalcemia stimulates prostaglandin E_2 production, and this compound inhibits both the hydroosmotic effects of ADH and the stimulating effect of ADH on NaCl reabsorption in the medullary thick ascending limb of the loop of Henle. Hypercalcemia also causes a reversible decline in the GFR, but the major effect seems to involve a lower medullary hyperosmolality.

Osmole-free water:
This is an antiquated term because we are interested in the quantity of water excreted that influenced body tonicity.

Electrolyte-free water:
Divide the urine into two parts:
1. electrolyte excretion that is isotonic to the patient ((urine $[Na^+]$ + $[K^+]$)/$[Na^+]$ in plasma);
2. pure water loss (ignore urea).

8·8 If a person has nephrogenic DI, drinks a large volume of water, has a normal diet, and has a urine osmolality that is consistently close to 300 mosm/kg H_2O, what will his urine volume be?

On a normal diet, one can expect an osmole excretion rate that is close to 900 mosmoles per day. If these osmoles are excreted at 300 mosm/kg H_2O, the urine volume will be 3 liters (see the margin).

8·9 What is the cause of hypernatremia in the following case? The plasma $[Na^+]$ rose to 180 mmol/l overnight in a confused patient. He received no medications. His body weight had not changed, but his ECF volume was contracted. His blood sugar was 90 mg/dl (5 mmol/l), the urine volume was extremely low, and the urine osmolality was close to 1200 mosm/kg H_2O.

The patient did not receive Na^+ salts, and the ECF volume was contracted; therefore, hypernatremia was not due to Na^+ gain. Although gastrointestinal secretions may have a $[Na^+]$ that is less than 140 mmol/l, there was no clinical evidence to support the accumulation of very large volumes of these fluids (marked contraction of the ECF volume and an acid-base disorder). Because there was no change in body weight, hypernatremia was not the result of water loss from the body (water loss should have caused a weight loss of more than 5 kg), and the renal response to hypernatremia was normal (low volume of maximally concentrated urine). Hence, the rise in $[Na^+]$ was due to a shift of water from the ECF to another compartment.

Because the plasma $[Na^+]$ increased by 40 mmol/l (140 to 180 mmol/l), the osmolality rose by almost 30%; this increase could occur if close to 3.3 liters of water shifted from the ECF (plus an additional water shift from the ICF of other organs that were not involved in the pathologic condition) to the ICF of the diseased organ. Judging by the relative masses of individual organs, water must have accumulated in muscle.

For water to shift into cells, the number of particles in these cells must have increased (and Na^+ did not accumulate in these cells). Rhabdomyolysis, which involves a breakdown of intracellular muscle macromolecules into a larger number of smaller particles that are retained in the ICF of muscle, could have resulted in an increased tonicity of these cells.

From a clinical viewpoint, it is essential to realize that although water shifted into muscle cells, it shifted out of other normal cells (e.g., brain cells). Shrinkage of these cells could be life-threatening. Therefore, therapy should defend both the ECF volume and the ICF volume of nonmuscle tissues (e.g., brain cells); half-normal saline is the best solution for these purposes. A further expansion of the ICF of muscle should be anticipated with therapy, and specific steps may be necessary to cope with this expansion (i.e., decompression).

8·10 How quickly can the glucose infused in one liter of D_5W be metabolized in a patient with hypernatremia?

One liter of D_5W contains 278 mmol (50 g) of glucose. The total glucose content of the body (5 mmol/l × 15 liters ECF + 4 liters ICF) is only 95 mmol (17.1 g) when the blood glucose concentration is 5 mmol/l (90 mg/dl). Without metabolism, 1 liter of D_5W can increase the blood glucose concentration threefold, and, because several liters of D_5W are required for water replacement during severe hypernatremia, this addition could cause a severe degree of hyperglycemia in a hyperosmolar state.

Approximately 0.5 g of glucose can be metabolized per kilogram of body weight per hour, providing that a fat-derived fuel is not available for oxidation. However, during fasting, when ketoacids are oxidized, less than 0.1 g of glucose/kg of body weight is metabolized to CO_2. Therefore, oxidative metabolism of glucose in major organs is close to 7 g/hr. Because the liver does not oxidize large quantities of circulating glucose to CO_2, and because the conversion of glucose to glycogen in the liver and muscle is rather limited in acutely ill patients, the maximum rate of glucose metabolism could be 10 g/hr. Taken together, these factors should alert the clinician to the life-threatening danger of profound hyperglycemia consequent to large D_5W infusions. We therefore advise that the rate of administration of D_5W not exceed 0.3 liters/hr in a 70-kg adult. Monitor the blood sugar and slow the rate of glucose infusion when this value approaches 10 mmol/l (180 mg/dl).

8·11 If a 70-kg patient has a [Na^+] of 154 mmol/l and a very marked degree of ECF volume contraction, what is the most likely pathophysiology for this lesion?

Hypernatremia implies the presence of a thirst defect. If the hypernatremia were due to water loss, the ECF volume would be contracted by only 10%. Hence, the pathophysiology must reflect a thirst defect, a loss of water, and a loss of Na^+. These three aspects could occur together in a patient with a stroke who is fed a high-protein diet (urea-induced osmotic diuresis).

8·12 To expand the ECF volume of the patient described in Question 8·11 by 2 liters, but not change the ICF volume in this period, what would be the most appropriate [Na^+] for the IV?

The solution must be isotonic to the patient. Give 2 liters of saline with the same Na^+:water ratio as the patient.

8·13 A 70-kg patient has DI, hypernatremia (154 mmol/l), and a normal ECF volume. What is the most likely pathophysiology to explain this set of findings? Provide a quantitative answer.

The pathophysiology includes a thirst defect, a loss of water (3 liters), and a gain of close to 210 mmol of Na^+ (14 mmol extra Na^+ in each of the 15 liters of ECF). One example of such a defect can be seen in Case 8·1, The Shrink Shrank the Cells, pages, 311 and 312–13.

What are the objectives for therapy?
The objectives for therapy are to lose 210 mmol of Na^+ while gaining 3 liters of water. These objectives can be achieved by giving 3 liters of water orally and by replacing renal and nonrenal losses of water and electrolytes (except for the 210 mmol of Na^+).

The reason for the thirst problem must be explored.

8·14 A patient has a clinical syndrome that resembles central DI. One unusual finding is present: the urine osmolality rises when DDAVP is given but not when ADH is given. How does this finding help in defining the etiology?

The response to DDAVP and the lack of response to ADH suggest that something is present that destroys the latter compound but not the former. The differences chemically are the absence of an amino terminal (desamino) and the presence of D- instead of L-arginine in DDAVP. The answer to the problem is the presence of an enzyme (a "vasopressinase" produced by the placenta of the patient) that destroys only ADH because it has an amino-terminal.

SECTION THREE

Potassium

9

POTASSIUM PHYSIOLOGY

Concepts in Potassium Physiology:

Objectives

- To emphasize that K^+ are the primary intracellular cations and to elucidate the mechanisms responsible for the distribution of K^+.
- To demonstrate that there are two components of K^+ excretion, the $[K^+]$ in the urine and the rate of urine flow, and to explain the regulation of each.

Outline of Major Principles

1. The major clinical concern in a patient with hyperkalemia or hypokalemia is a cardiac arrhythmia.
2. The vast majority (98%) of K^+ are in cells, held there by an electrical force (inside negative). This charge separation across cell membranes is called the resting membrane potential (RMP).
3. There are two important factors leading to the RMP: first, the electrogenic Na^+K^+ATPase creates a high intracellular $[K^+]$ (3 Na^+ are pumped out and 2 K^+ enter); second, and more importantly, K^+ diffuse out of cells, down their concentration gradient. The entry of Na^+ into cells is limited because the permeability of these membranes to Na^+ is so much lower than the permeability to K^+. Most anions do not diffuse out of cells down their electrical gradient because the quantitatively important ones are impermeable macromolecular compounds.
4. Excretion of K^+ is the product of the $[K^+]$ in the urine times the urine flow rate. The $[K^+]$ in the lumen of the CCD is determined by the lumen-negative transepithelial potential difference (TEPD). This TEPD is generated by the reabsorption of Na^+ at a faster rate than the reabsorption of Cl^- (called *electrogenic reabsorption of* Na^+). The major factors that adjust this TEPD are aldosterone and possibly the $[HCO_3^-]$ in the luminal fluid; the delivery of Na^+ is rarely a limiting factor. The other major influence on K^+ excretion is the volume of fluid delivered to the terminal CCD (this volume determines the number of K^+ that enter the lumen to reach electrochemical equilibrium).

Abbreviations:
CCD = cortical collecting duct.
ECF = extracellular fluid.
ICF = intracellular fluid.
MCD = medullary collecting duct.
RMP = resting membrane potential.
TEPD = transepithelial potential difference.
TTKG = transtubular $[K^+]$ gradient.

Electrogenic reabsorption of Na^+: When Na^+ are reabsorbed without Cl^-, a lumen-negative TEPD to drive the countermovement of K^+ and/or H^+ is created.

Electroneutral reabsorption of Na^+: When Na^+ are reabsorbed equimolarly with Cl^-, a TEPD to drive the countermovement of K^+ and/or H^+ is not created.

Introductory Case
Lee's [K^+] (K)rashed

(Case discussed on page 343)

Lee, whom you met in Chapters 1, 2, 3 and 6, returns. She has been unwell for two weeks with an upper respiratory tract infection. She presented to the emergency room and was diagnosed as having diabetic ketoacidosis. She was treated with isotonic saline and insulin.

		Admission	**4 hours later**
Na^+	mmol/l	130	134
K^+	mmol/l	5.6	3.2
Cl^-	mmol/l	93	98
HCO_3^-	mmol/l	10	12
Glucose	mmol/l (mg/dl)	25 (450)	20 (360)

What was the total body K^+ content at the time of presentation and why?
What factors led to her hyperkalemia?
Why did hypokalemia develop four hours later?

PART A
DISTRIBUTION OF K^+ IN THE ICF

Overview of K^+ Distribution

- The ICF contains 98% of K^+ in the body.
- When faced with a load of K^+, entry of K^+ into cells is rapid. This shift is the body's initial defense mechanism.

Concepts:
1. Potassium ions (K^+) play a key role in the generation of the resting membrane potential, which, in turn, influences many important biologic events.
2. The clinical importance of K^+ is that surpluses or deficits of K^+ in the ECF may predispose the patient to cardiac arrhythmias.

Most (98%) of the K^+ in the body is in cells (40–50 mmol/kg body weight). In contrast, the total content of K^+ in the ECF is quite low, less than 1 mmol/kg body weight, a quantity similar to the dietary intake of an adult on a typical Western diet. A steady state is maintained with a [K^+] in the ICF that is close to 35-fold greater than that in the ECF. To avoid important changes in this [K^+] gradient, regulatory mechanisms must be rapid and extremely sensitive to small variations in the input and output of K^+ (see the discussion of Question 9·1). Ultimately, all the K^+ ingested must be excreted to maintain balance for K^+.

Factors influencing the [K^+] in the ICF:
- There are several ways to change the [K^+] in the ICF. One can change the RMP by changing the activity of the Na^+K^+ATPase and/or by increasing the permeability of the cell membrane to Na^+. One can also lower the permeability of this membrane to K^+.
- K^+ movement across cell membranes requires either the countermovement of a cation (Na^+ or H^+) or a parallel movement of an anion (phosphate in most cells).
- The number of K^+ that must move to generate the RMP is very small.

Question
(Discussion on page 344)

9·1 What quantity of K^+ of dietary origin (70 mmol) will be distributed in cells if there is no excretion of K^+ nor a change in the RMP? What will the plasma [K^+] be?

The Resting Membrane Potential

There are two major factors that generate the RMP, an active one, the Na^+K^+ATPase, and a passive one, the diffusion of K^+ from the ICF where the [K^+] is high (150 mmol/l) to the ECF where the [K^+] is low (4.1 mmol/l). Further diffusion of K^+ is retarded by an electrical force (negative charge inside cells, the RMP). Quantitatively, the Na^+K^+ATPase creates a minor component of the RMP by pumping 3 Na^+ out of cells in exchange for 2 K^+ entering cells. The passive diffusion of K^+ out of cells is responsible for the majority of the RMP.

In addition to the two factors described above, low permeability of the cell membrane to Na^+ and the fact that the intracellular anions are mostly macromolecular helps maintain the RMP (Figure 9·1).

In the paragraphs to follow, the factors influencing the distribution of K^+ between the ECF and ICF will be considered in more detail (Table 9·1).

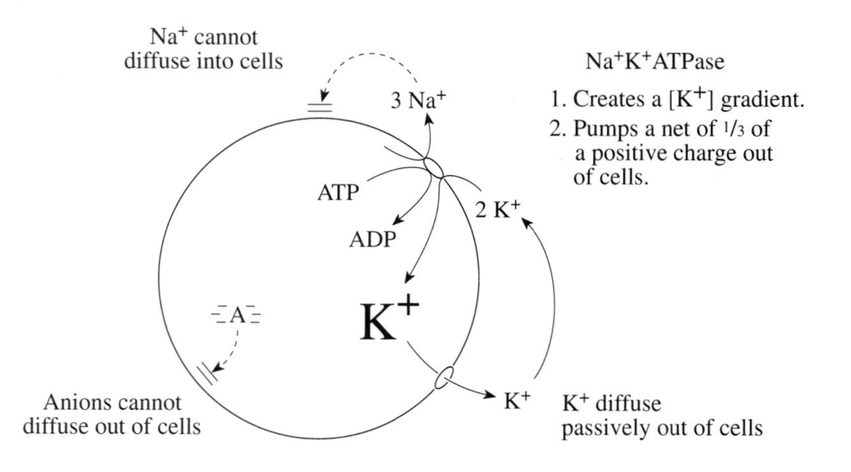

Figure 9·1
Generation of the RMP
The RMP is largely due to the activity of the $Na^{+}K^{+}$ATPase and the passive diffusion of K^{+}, but not Na^{+} or anions, across cell membranes.

Table 9·1
FACTORS INFLUENCING A K^{+} SHIFT FROM THE ICF TO THE ECF

Hormones
Lack of insulin, β_2-adrenergic antagonists, lack of aldosterone
Acid-base disturbances
Acute HCl gain or $NaHCO_3$ loss (other acid-base disturbances have little direct effect on K^{+} distribution)
ICF anion change
Catabolism, loss of organic phosphates
Cell necrosis
Rare factors
Cellular depolarization (e.g., succinylcholine)
Decrease in K^{+} permeability (barium poisoning)
Unusual cation accumulations in the ICF (lysine/arginine toxicity)
Hypertonicity, which decreases the ICF volume (an overrated factor; see the discussion of Question 9·5, page 345)
Hyperkalemic periodic paralysis

$Na^{+}K^{+}$ATPASE

- Hydrolysis of 1 molecule of ATP pumps 3 Na^{+} out of cells for every 2 K^{+} that enter; therefore, this pump is electrogenic.

Note:
Some cells do not have the $Na^{+}K^{+}$ATPase as their principal ion translocating pump. For example, the H^{+}ATPase is the principal ion pump in α- and β-intercalated cells in the cortical and medullary collecting ducts.

Almost every cell contains $Na^{+}K^{+}$ATPase, an ion translocating pump that pumps 3 Na^{+} out of cells for every 2 K^{+} that enter these cells (see the margin). The net result of having a normal activity of the $Na^{+}K^{+}$ATPase is a large transmembrane cation gradient with a very low intracellular [Na^{+}] of close to 10 mmol/l (vs 150 mmol/l in interstitial fluid); the converse applies

GENERATORS

1. K^+ diffuse from the ICF to the ECF.
2. Na^+K^+ATPase pumps 1/3 of a charge from the ICF to the ECF.

EFFECTS

1. Na^+/H^+ antiport
2. Na^+/Ca^{2+} exchange
3. Na^+/nutrient cotransport
4. K^+ secretion
5. Lower RMP

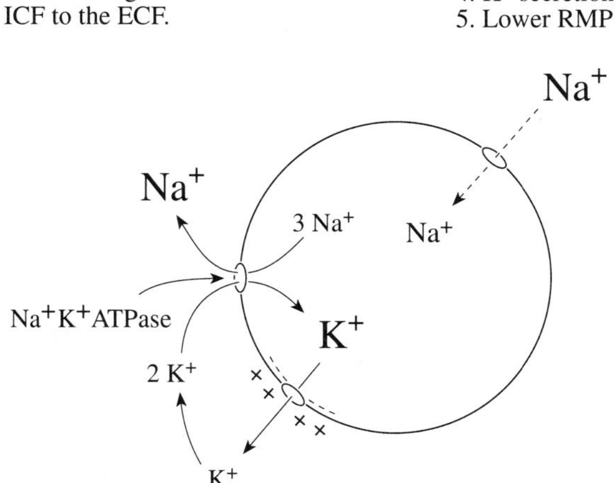

Figure 9·2
The role of the [Na^+] gradient across cell membranes
The generators of the [Na^+] gradient are shown on the left side of the figure, and the effects of this gradient are shown on the right side of the figure.

Note:

- The intracellular [Na^+] is close to 10 mmol/l.
- The cell membrane has ion channels for Na^+, but they remain closed unless there is a signal for depolarization of the cell.

to K^+ (150 mmol/l inside most cells and 4.3 mmol/l in interstitial fluid). These [Na^+] and [K^+] gradients have implications for many vital functions of the cell (Figure 9·2).

Questions

(Discussions on pages 344–45)

9·2 What is the intracellular cation composition in the red blood cells of a dog? These cells lack Na^+K^+ATPase pumps.

9·3 Why does the entry of Na^+ into cells increase the RMP when augmented by insulin but decrease the magnitude of the RMP in either hyperkalemic or hypokalemic periodic paralysis?

[K^+] in the ECF:
The [K^+] in interstitial fluid is several fold higher than the K_m for K^+ of the Na^+K^+ATPase. Hence, this pump activity will rarely be influenced by this [K^+].

Factors Influencing the Activity of the Na^+K^+ATPase Pump

The major factors influencing the activity of the Na^+K^+ATPase pump are the [Na^+] in the ICF and the number and/or activity of individual pump units.

Intracellular [Na^+]: In general, the activity of the Na^+K^+ATPase is enormous relative to the quantity of Na^+ that need to be pumped because cell membranes have an extremely low permeability to Na^+. The Na^+K^+ATPase extrudes Na^+ rapidly when the [Na^+] rises transiently in the ICF. Hence, it follows that the principal regulator of pump activity is the [Na^+] in the cytosol in immediate proximity to this pump (see the margin). Given this pump-leak system, if the activity of the Na^+K^+ATPase changes, the [Na^+] in the ICF will have to change in a reciprocal way because this pump does not seem to be involved in an equilibrium system.

Nonequilibrium systems:
Flux in systems that are at equilibrium depend only on the concentrations of substrates and products; flux in systems that are not at equilibrium depend on the substrate concentration and the activity of the regulatory enzyme involved. For the Na^+K^+ATPase, more Na^+ are pumped if the [Na^+] in cells rises or if there is activation or insertion of Na^+K^+ATPase units in the basolateral membrane of principal cells.

Activity of the Na^+K^+ATPase: The Na^+K^+ATPase has α and β subunits. The assembly of these inactive precursors permits rapid changes in activity to occur. Phosphorylation and dephosphorylation also modulate the activity of the Na^+K^+ATPase. Longer-term control is exerted by synthesis of new α and β subunits. A commonly used inhibitor of this pump is ouabain, a compound in the digitalis family of drugs. There is evidence of the existence of endogenous compounds with ouabain-like actions.

FACTORS INFLUENCING A SHIFT OF K^+ ACROSS CELL MEMBRANES

Hormones

Hormones can influence the quantity of Na^+ pumped by the Na^+K^+ATPase in three general ways (Table 9·2). First, they can increase the electroneutral entry of Na^+ into cells (activate the Na^+/H^+ antiporter), which will lead to more pumping of Na^+ out of cells via this electrogenic pump. Second, hormones may activate existing Na^+K^+ATPase enzymes in cell membranes via phosphorylation. Third, hormones may induce the net synthesis of more Na^+K^+ATPase enzyme molecules.

- The most important hormones are insulin and the β-adrenergics.

Other actions of insulin:
Insulin causes the synthesis of phosphate esters in the ICF (hexose phosphates, RNA, etc.). An increase in intracellular anions could "attract" K^+ into cells. This effect is usually slow and small.

Insulin: Insulin causes K^+ to shift into cells. The most important mechanism seems to be the activation of the electroneutral Na^+/H^+ antiporter. For full expression of this effect, supraphysiologic doses of insulin may be required. Once Na^+ have entered cells, the $[Na^+]$ in the ICF will rise transiently. When these Na^+ are pumped out of cells, there will be an increased flux through the electrogenic Na^+K^+ATPase (Figure 9·3). Quantitatively, if the basal insulin concentration halves, the $[K^+]$ in the ECF rises by about 0.5 mM within 30 minutes.

On a clinical note, administration of insulin causes a significant fall in the $[K^+]$ in the ECF; this fall is especially important during treatment for diabetic ketoacidosis and for hyperkalemia with ECG changes.

Other actions of catecholamines:
β_1-Adrenergics, which stimulate renin release from the kidneys, increase aldosterone production and, as a result, promote the excretion of K^+.

Catecholamines: The effects of catecholamines on the $[K^+]$ in the ECF are complex. β_2-Adrenergics stimulate the movement of K^+ into cells, possibly via effects on the Na^+K^+ATPase. The mechanism may involve a rise in cyclic AMP and, as a result, phosphorylation and activation of the Na^+K^+ATPase.

Catecholamines can also act in an indirect fashion. For example, they stimulate glycogenolysis, which leads to hyperglycemia and the release of insulin from β cells of the pancreas. Insulin causes K^+ to move into the ICF as described above.

α-Adrenergic actions lead to a direct shift of K^+ out of cells. Hormones with this action may also influence the $[K^+]$ in plasma indirectly. In addition, because they inhibit release of insulin from pancreatic β cells, their actions may also lead to development of hyperkalemia.

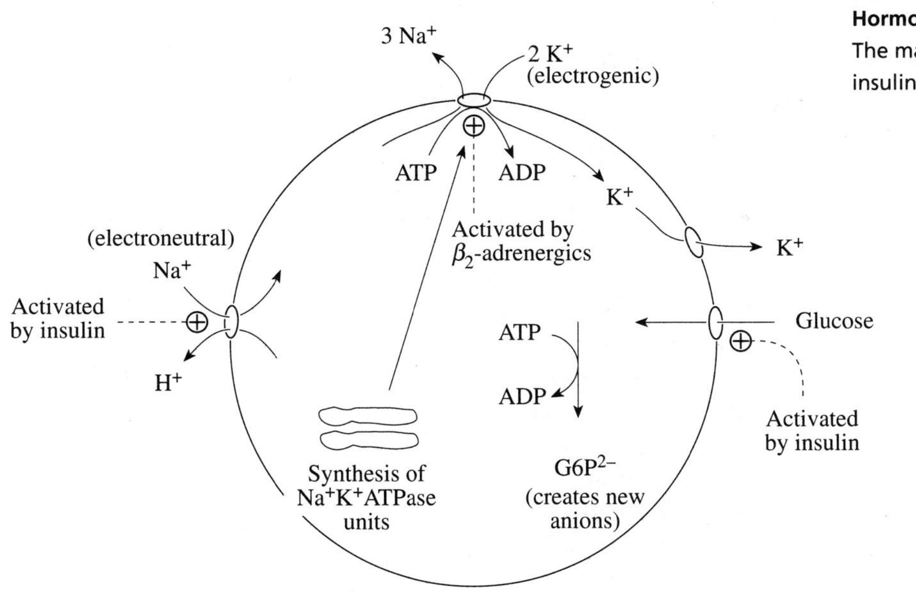

Figure 9·3
Hormones cause K^+ to shift into cells
The major hormones involved are insulin and β_2-adrenergics.

Aldosterone: If a patient lacking aldosterone is hyperkalemic and is given aldosterone, the plasma [K^+] falls promptly before excretion of K^+ rises appreciably. Therefore, aldosterone seems to permit the entry of K^+ into cells in this setting. In contrast, when aldosterone is given to normal individuals, there is no acute fall in the plasma [K^+]. Hence, aldosterone does not promote K^+ entry if its level was not extremely low initially.

Table 9·2
EFFECTS OF SOME HORMONES ON DISTRIBUTION OF K^+ BETWEEN ICF AND ECF

Hormone	Possible mechanisms	Quantitative role
Insulin	• Increased entry of Na^+ via activation of the Na^+/H^+ antiporter	• A lack of insulin leads to an acute rise in the [K^+] in the ECF of close to 0.5 mmol/l.
β_2-Adrenergics	• Activation of Na^+K^+ATPase	• Pharmacologic doses can bring the [K^+] in the ECF below 3.0 mmol/l.
Aldosterone	?	• If aldosterone is given to a patient who lacks this hormone, the [K^+] in the ECF can fall by about 0.5 mmol/l.

Question

(Discussion on page 345)

Note:
Question 9·4 is for the more curious.

9·4 What property of the $Na^+K^+ATPase$ might allow it to bind K^+ at a $[K^+]$ of 4 mmol/l in the ECF yet also permit K^+ to dissociate from this pump at a $[K^+]$ of 150 mmol/l in the ICF?

Acid-Base Changes

- Only metabolic acidosis caused by loss of $NaHCO_3$ (or gain of HCl) leads to a shift of K^+ across cell membranes and hyperkalemia.
  - With time, normal kidneys and aldosterone response return the plasma $[K^+]$ to normal.
- A gain of an organic acid does not cause hyperkalemia.
  - Lack of insulin or hypoxia may lead to hyperkalemia.
- There is little K^+ shift across cell membranes during acute respiratory acid-base disorders.

Data in animals: When HCl was infused into nephrectomized dogs, there was a rise in the $[H^+]$ in the ECF and a shift of H^+ into the ICF; some K^+ moved out of cells for charge balance. In contrast to HCl, infusion of organic acids, L-lactic acid, and ketoacids did not cause a net shift of K^+ into cells as a result of ICF buffering because L-lactate and ketoacid anions entered the ICF to almost the same degree as protons (Figure 9·4). Therefore, when metabolic acidosis is due to the accumulation of these organic acids, factors other than the acidemia per se cause hyperkalemia (if it is present).

Respiratory acid-base disorders cause only small changes in the plasma $[K^+]$. Therefore, if the $[K^+]$ changes with a Pco_2 change, look for a cause other than the simple acid-base disturbance (see the margin).

Shift of K^+ in acute respiratory acid-base disorders:
During acute respiratory acidosis or alkalosis, there is little change in the $[HCO_3^-]$ in the ECF and only a very small movement of K^+ across the cell membrane. In acute experiments, no appreciable change in the $[K^+]$ in the ECF was detected when the change in P_aco_2 was in the range of 20–80 mm Hg.

Intracellular Anions

- The $[K^+]$ in the ICF is "electrically balanced" mainly by intracellular, macromolecular anions, largely organic phosphates; these anions are restricted to the ICF compartment.

The positive and negative charges in a compartment are virtually equal in number. In the ICF, many of the negatively charged molecules do not cross

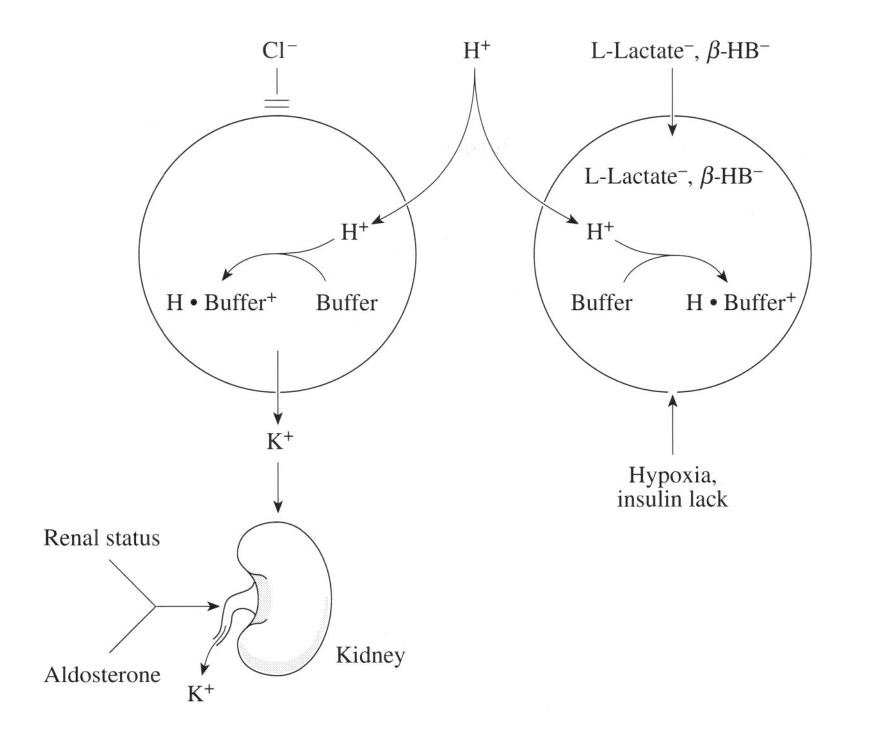

Figure 9·4
Buffering of H^+ and the consequent K^+ shift
The circles represent the cell membrane. Although Cl^- remain largely in the ECF (left-hand circle), organic anions enter the ICF (right-hand circle). Hence, hyperkalemia is due to a shift of K^+ with HCl only. Hyperkalemia should stimulate aldosterone release and result in excretion of the extra K^+. Therefore, a steady-state hyperkalemia in hyperchloremic metabolic acidosis should only be seen if there is low excretion of K^+. A decrease in excretion of K^+ may occur during renal insufficiency or low aldosterone bioactivity.

cell membranes (mainly the organic phosphates, DNA, RNA, ATP, creatine phosphate; intracellular proteins do not make a major net contribution to this anionic charge). The intracellular anionic composition is normally relatively constant (although the specific phosphate compounds differ from organ to organ). However, loss of organic phosphates (largely RNA), as can occur in specific disease states (e.g., diabetic ketoacidosis), leads to the parallel loss of K^+. The reason for the fall in RNA is the lack of the actions of insulin (i.e., degradation of ribosomal RNA needed for protein synthesis—the catabolic state of a relative lack of insulin); this topic will be explored further in the discussion of Question 9·7. The converse will occur during states with anabolism (i.e., recovery from DKA).

Net charge on intracellular proteins: At the ICF pH, the anionic charges on proteins are on the C-terminal carboxyl group and on the anionic amino acids glutamate and aspartate. The positive charges on ICF proteins are on the N-terminal amino group and on the cationic amino acids lysine, arginine, and close to half of the histidines. Simple arithmetic shows that the anionic groups do not exceed the cationic residues in most proteins. Hence, proteins are not likely to be major ICF anions unless very marked changes in the pK of amino acids occur in specific ICF environments.

Questions

(Discussions on pages 345–46)

9·5 How much will the plasma $[K^+]$ change when 1 liter of pure water is shifted from the ICF to the ECF? Assume a constant resting membrane potential.

9·6 Will a drug that depolarizes cell membranes cause a change in the plasma $[K^+]$?

9·7 Why does the loss of intracellular phosphate anions cause hyperkalemia in DKA but not in poliomyelitis?

9·8 During K^+ depletion, Na^+ replace K^+ in cells. How does this exchange affect the Na^+K^+ATPase?

PART B
RENAL REGULATION OF K^+ EXCRETION

Physiology of K^+ Excretion

Concept:

3. The kidneys adjust overall K^+ homeostasis by increasing or decreasing the rate of excretion of K^+.

- Events controlling K^+ excretion have their greatest impact in the CCD.
- To analyze K^+ excretion, assess both the $[K^+]$ in urine and the urine volume.
  K^+ excretion
  = Urine $[K^+]$ × Urine volume
  = Aldosterone and HCO_3^- in the lumen × Na^+ and H_2O intake, diuretics, and urea

K^+ delivery to the CCD:
A reasonable estimate of the volume of fluid delivered to the early distal convoluted tubule in humans is 20–30 liters. With a $[K^+]$ of 3 mmol/l, 60–90 mmol of K^+ could be delivered here. This amount is equivalent to the quantity of K^+ consumed and excreted daily by most adults on a typical Western diet.

Each day, the kidneys must excrete virtually all the K^+ that were absorbed from the GI tract; in adults ingesting a typical Western diet, this amounts to about 1 mmol/kg body weight. Excretion of K^+ requires two major events: the $[K^+]$ in each liter of fluid traversing the CCD must be increased, and there must be a sufficient number of liters traversing this portion of the nephron.

Other nephron segments also influence the excretion of K^+ (Figure 9·5). A very important point to emphasize is that the amount of K^+ delivered out of the loop of Henle is similar to that ingested and excreted per 24 hours (see the margin).

EVENTS IN THE CCD

- Virtually all regulation of urinary K^+ occurs in the CCD.

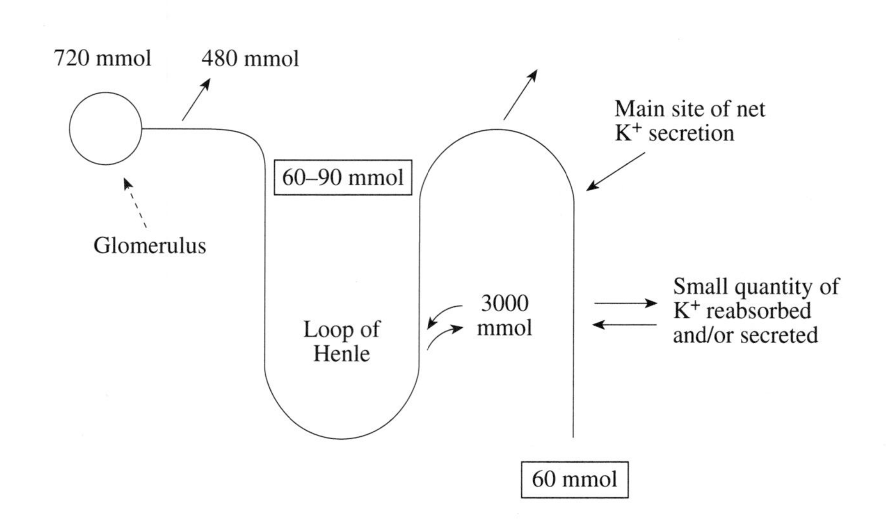

Figure 9·5
Handling of K^+ in segments of the nephron
Most of the filtered K^+ (720 mmol/day in an adult) is reabsorbed in the proximal convoluted tubule (480 mmol); the $[K^+]$ in the luminal fluid is not very different from that in the ECF. In the loop of Henle (LOH), close to 3000 mmol of K^+ are secreted and then reabsorbed in the thick ascending limb via the Na^+, K^+, 2 Cl^- cotransporter. This cycle permits Na^+ and Cl^- to be reabsorbed. The $[K^+]$ in the luminal fluid that exits the LOH is close to 3 mmol/l.

In the CCD, the major type of cell involved is the principal cell. For secretion of K^+, Na^+ are reabsorbed through an ion-specific channel (induced or activated by aldosterone and inhibited by K^+-sparing diuretics such as amiloride).

In this nephron segment, reabsorption of Na^+ is either electrogenic or electroneutral (Figure 9·6). If more Na^+ than Cl^- are reabsorbed, there is electrogenic reabsorption, which creates the electrical driving force that augments the net secretion of K^+. In contrast, if equal quantities of Na^+ and Cl^- are reabsorbed in the CCD, the reabsorption is electroneutral and does not "drive" the net secretion of K^+. In the following section, the components responsible for the net secretion of K^+ are considered.

[Na^+] in the lumen of the CCD:
The luminal [Na^+] required for half-maximal rates of K^+ secretion in the CCD is 10–15 mmol/l in the rat. No comparable data are available in humans, so we shall assume a similar value for the human CCD. Because the osmolality of fluid in the terminal CCD is equal to that in plasma when ADH acts, and because the maximum concentration of urea in the lumen of the CCD is close to 100 mmol/l in normal individuals, electrolytes represent close to 200 mosm/l. Half of these mosmoles are cations, the bulk of which are Na^+. Thus, the [Na^+] can be less than 15 mmol/l only when the concentration of urea exceeds 200 mmol/l in this fluid (i.e., with a marked degree of contraction of the ECF volume).

Delivery of Na^+ and Their Absorption in the CCD

The delivery of Na^+ to the CCD rarely influences net secretion of K^+. This delivery falls to limiting values if the ECF volume is markedly contracted (see the margin for further information). Na^+ are reabsorbed through a specific ion channel that is regulated in part by the availability of aldosterone (aldosterone opens ion channels). The Na^+ channel is inhibited by the diuretic amiloride. Entry of Na^+ into cells of the CCD has two effects:

1. It makes the membrane potential across the luminal membrane more electronegative; this action favors the secretion of K^+ into the lumen.
2. The transient small elevation in the [Na^+] in the cell increases flux through the basolateral Na^+K^+ATPase and thereby brings K^+ from the fluid adjacent to the basolateral membrane into principal cells. The net effect is movement of K^+ from the ECF to the lumen. Said another way, the actions of aldosterone lead to an elevated luminal [K^+] that is approximately 10-fold higher than that in the ECF (Figures 9·7 and 9·8).

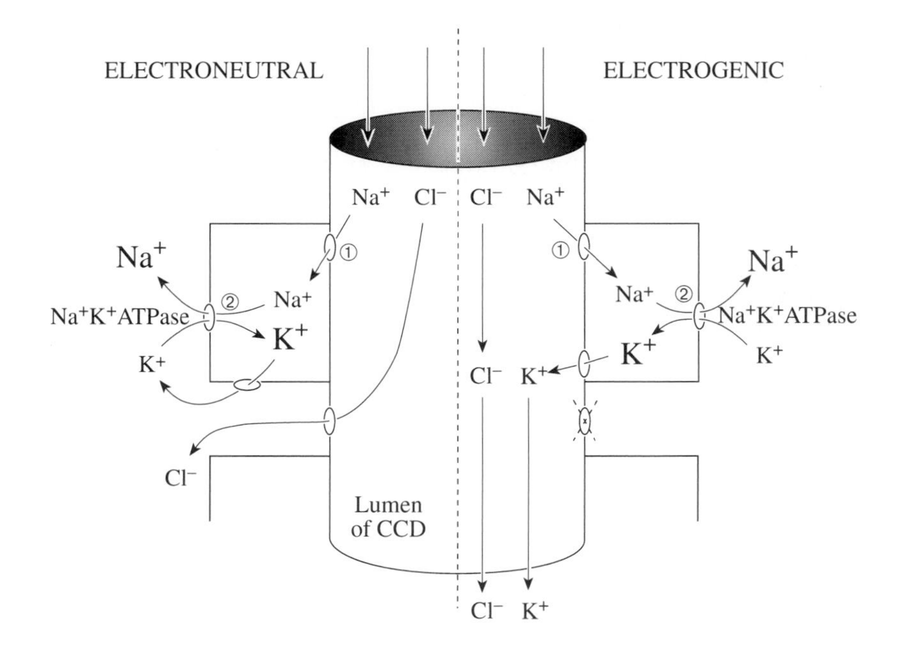

Figure 9·6
Electrogenic and electroneutral reabsorption of Na^+
The barrel-shaped structure is the CCD. When Na^+ and Cl^- are reabsorbed in equimolar amounts, there is no TEPD to favor net secretion of K^+; this reabsorption, termed electroneutral, is shown on the left side of the figure. In contrast, reabsorption of Na^+ without Cl^- is electrogenic and generates a TEPD (lumen negative), which drives the net secretion of K^+ (depicted on the right side of the figure). There are two components to Na^+ reabsorption: (1) the Na^+ channel in the luminal membrane, and (2) the Na^+K^+ATPase on the basolateral membrane.

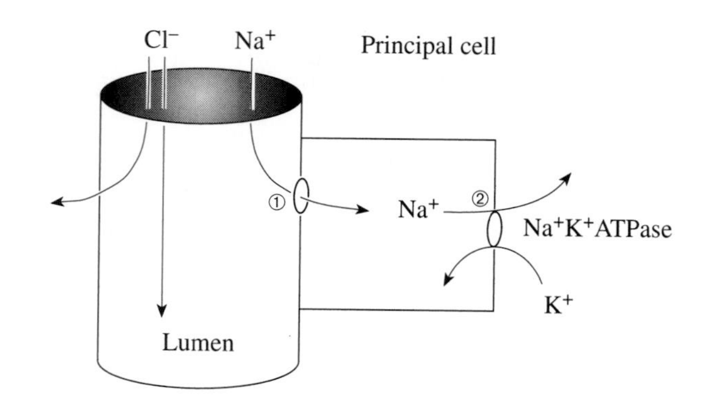

Figure 9·7
Physiology of aldosterone-induced K$^+$ secretion
The barrel-shaped structure represents the lumen of the CCD, and the square represents a principal cell. The actions of aldosterone open the Na$^+$ channel (1) and cause more Na$^+$K$^+$ATPase units to be inserted in the basolateral membrane (2). Reabsorption of Na$^+$ in the CCD can be electroneutral or electrogenic, depending on whether Cl$^-$ are more or less permeable in the CCD luminal membrane. If the flux of Cl$^-$ is less than that of Na$^+$, a TEPD is generated and K$^+$ secretion rises.

Cl$^-$ Reabsorption in the CCD

The mechanisms that regulate the reabsorption of Cl$^-$ in the CCD are relatively unclear. The bulk of Cl$^-$ movement is believed to occur via a paracellular rather than transcellular route, driven in part by the lumen-negative TEPD and limited by the permeability of the luminal membrane to Cl$^-$ (Figure 9·6). Intracellular and luminal events may also regulate this permeability to Cl$^-$. The influence of HCO_3^- is considered below.

Role of Anions Other Than Cl$^-$ in the Lumen of the CCD

Among anions that influence the excretion of K$^+$, HCO_3^- seem to be unique; they lead to a very high rate of excretion of K$^+$ (provided that aldosterone is also present) by permitting a very high [K$^+$] in the luminal fluid of the CCD. There are two possible mechanisms of action:

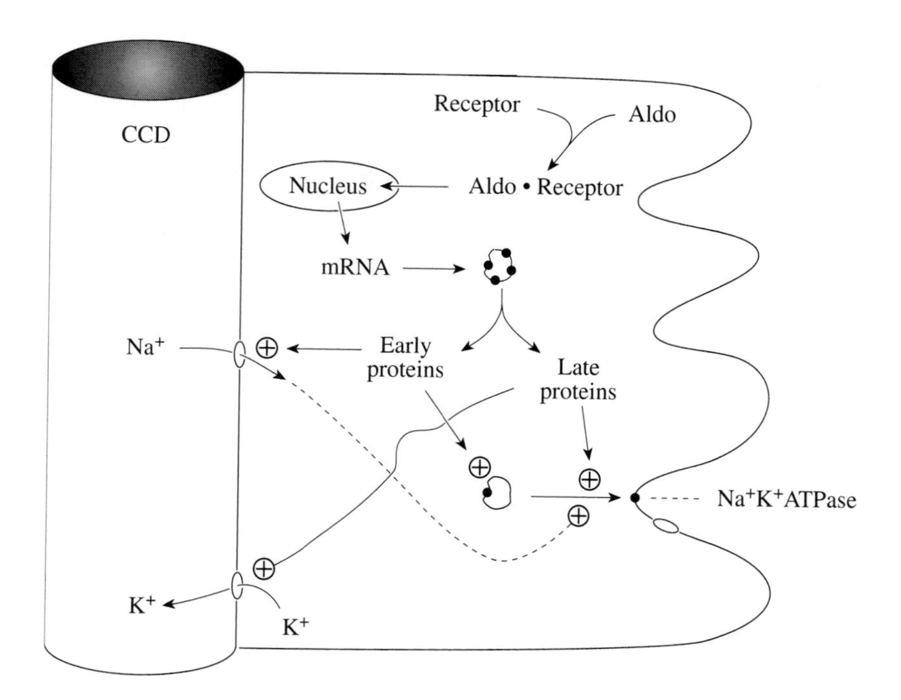

Figure 9·8
Summary of actions of aldosterone on principal cells of the CCD
Aldosterone (Aldo) enters the principal cell via the basolateral membrane and binds to a specific receptor. This hormone receptor complex enters the nucleus and causes the synthesis of new proteins. The early action of aldosterone is the opening of luminal Na$^+$ channels and, possibly, the synthesis of the precursor of the Na$^+$K$^+$ATPase. The later actions of aldosterone include the insertion of more Na$^+$K$^+$ATPase units into the basolateral membrane and K$^+$ channels into the luminal membrane.

1. **HCO_3^- may act simply as poorly reabsorbed anions in the CCD (like SO_4^{2-}):**
   A high TEPD will be generated if Na^+ reabsorption is electrogenic (without Cl^-). For this electrogenic reabsorption, Cl^- should be absent from the luminal fluid. Because HCO_3^- cause a high TTKG even if Cl^- are present in the urine, the mechanism of action cannot simply be as a poorly reabsorbed anion (see the margin).

Note:
SO_4^{2-} causes a high TTKG when given with aldosterone only if the [Cl^-] in the urine is very low.

2. **HCO_3^- inhibit the reabsorption of Cl^- in the CCD:**
   We do not know the mechanism, but some data are consistent with it. Clinically, conditions associated with enhanced delivery of HCO_3^- to the CCD (e.g., vomiting, treatment of proximal RTA with $NaHCO_3$, the use of carbonic-anhydrase-inhibitor type of diuretics such as acetazolamide, and the subgroup of distal RTA that results from a defect in distal nephron H^+ secretion) have in common hypokalemia with an inappropriately high [K^+] in the urine. Our speculation on the ways in which HCO_3^- might enhance net secretion of K^+ is considered in the discussion of Question 9·12, pages 347–48.

K^+ Transport in the CCD

- There are three major pathways for K^+ movement: antiporters, specific K^+ channels, and cotransporters.

Antiporters: K^+ may be reabsorbed in the CCD and MCD. Reabsorption is, in general, smaller in magnitude and mediated by the major antiport system, the K^+/H^+ antiporter. This antiporter leads to the reabsorption of K^+ and the secretion of H^+ (providing that there is a suitable H^+ acceptor such as HCO_3^-, NH_3, or divalent phosphate in the lumen). The quantity of these H^+ acceptors in the lumen is not large.

Specific K^+ channels: The major transport system involved in net K^+ secretion seems to be K^+ channels. There are two major types of K^+ channels in the luminal membrane of the CCD: one that is activated by Ca^{2+} or depolarization, and another that is almost permanently open and apparently makes up the bulk of the luminal K^+ conductance. These latter channels can be inhibited by intracellular acidification and ATP. They are also inactivated by protein kinase C and arachidonic acid; nevertheless, they are usually active enough so that conductance of K^+ does not generally impose a major limit to net secretion of K^+.

Cotransporters: There is also a K^+, Cl^- cotransporter for secretion of K^+, but it seems to operate only when the luminal [Cl^-] is exceedingly low (less than 10–15 mmol/l, a rare event); this transporter may therefore be important only on rare occasions.

The Adrenal Gland and Aldosterone Release

- Hyperkalemia and ECF volume contraction (via angiotensin II) are the two major stimuli for aldosterone release.

The release of aldosterone is stimulated primarily by hyperkalemia and angiotensin II; these two stimuli act in a fashion that is more than additive. The level of angiotensin II rises when renin is released from the juxtaglomerular apparatus of the kidneys. Renin release is stimulated by ECF volume contraction, renal artery stenosis, and β_1-adrenergic stimulation; in contrast, it is inhibited by ECF volume expansion and adrenergic β_1-blockers. Destruction of the juxtaglomerular apparatus by interstitial renal disease also results in lower levels of renin release. In some cases, angiotensin II levels may not be elevated despite high renin levels. Such cases can occur when drugs inhibit the conversion of angiotensin I to its active form, angiotensin II (Figure 9·9).

The release of aldosterone from zona glomerulosa cells is diminished by atrial natriuretic factor.

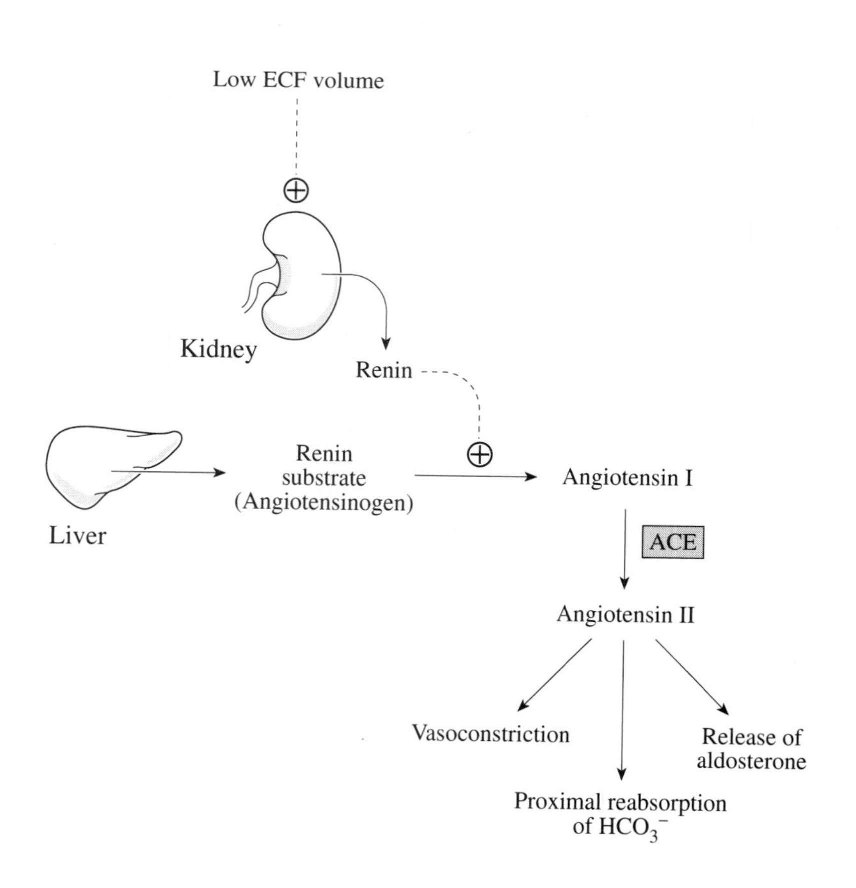

Figure 9·9
Site of action of angiotensin-converting enzyme (ACE) inhibitors
An important class of antihypertensive drugs that influence angiotensin II levels is the ACE inhibitors.

ACE:
Angiotensin
Converting
Enzyme

Questions

(Discussions on pages 346–48)

9·9 What is the maximum rate of K^+ excretion in 24 hours? (Given: plasma $[K^+]$ is 4 mmol/l, TTKG is 10, 5 liters of filtrate reach the CCD, and there is no medullary K^+ secretion or reabsorption.)

9·10 Can you determine whether a high urine $[K^+]$ is due to K^+ addition to the CCD or to water reabsorption in the MCD?

9·11 Does the excretion of K^+ conserve Na^+ and thereby defend the ECF volume? This question requires one to consider the origin of the K^+ that is excreted.

9·12 A very high $[K^+]$ in the lumen of the CCD occurs in a patient who vomits. A similarly high $[K^+]$ occurs when acetazolamide (a carbonic anhydrase inhibitor) acts. Common to both situations is bicarbonaturia. What might the mechanism be?

9·13 Does a normal person excrete more K^+ than a patient with adrenal insufficiency (assume they eat the same diet)?

9·14 Is the volume delivered to the terminal CCD lower in a patient with a normal or reduced GFR (lower by 50%)?

Pathophysiologic Approach to Disturbances in the Plasma $[K^+]$

Analysis of the rate of excretion of K^+ in patients with hypokalemia or hyperkalemia can indicate why the renal response is not adequate. This rate should be compared to the expected values for an otherwise normal person with a deficit of K^+ (excretion less than 10–15 mmol per day) or a surplus of K^+ (excretion close to 400 mmol per day). If the rate of excretion of K^+ is abnormal, the next step is to determine whether the fault is with the $[K^+]$ in the luminal fluid in the CCD or with the flow rate in the CCD. The levels of aldosterone and renin often help in clarifying the basis of the problem. In

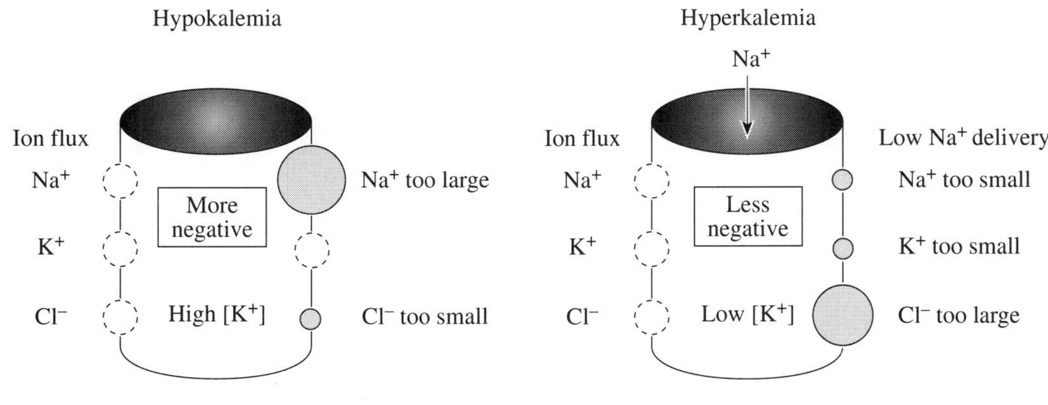

Figure 9·10
Changes in ion channel flux that may be the renal basis of hypokalemia or hyperkalemia
The barrel-shaped structures represent the CCD and the dashed ovals represent pathways for ion movement. The potential lesions causing hypokalemia are shown in the left figure and those causing hyperkalemia are depicted in the right figure; possible lesions are depicted in the right-hand margin of each figure. More detailed information is provided in Table 9·3.

Note:
A decreased conductance or permeability to K^+ in the luminal membrane of the CCD could theoretically lead to hyperkalemia. In this setting, there would be little augmentation of K^+ excretion when a diuretic increases the flow rate.

Notes:
- 9αF = fludrocortisone, a drug with mineralocorticoid action.
- We switched terminology from "greater open probability" to "faster" for economy of space.
- With a very low flow rate, one can achieve a high $[K^+]$ in the lumen of the CCD, but only a small quantity of K^+ will be excreted.

these analyses, the components leading to a high or low net secretion of K^+ should be deduced (Figure 9·10, Table 9·3, and the review flow charts on pages 355 and 394).

Table 9·3
POTENTIAL LESIONS IN THE CCD THAT MAY CAUSE HYPOKALEMIA OR HYPERKALEMIA

The lesions described are speculative and cause a change in K^+ homeostasis by changing the $[K^+]$ in the lumen of the CCD (see Figure 9·10 as well). The other way to change the rate of K^+ excretion is to change the flow rate in the CCD.

Lesion in CCD	ECF volume	Renin	Urine $[Na^+]$ and $[Cl^-]$ with low ECF volume	TTKG + 9αF
Hypokalemia caused by:				
- Faster Na^+ channel	High	Low	Absent	> 6
- Lower Cl^- permeability	Low	High*	Not < 20 mmol/l	> 15
Hyperkalemia caused by:				
- Low Na^+ delivery	Low	High*	Absent	Low or high
- Slower Na^+ channel	Low	High*	Not < 20 mmol/l	Low
- Higher Cl^- permeability	High	Low	Absent	Low

* The high level of renin can be suppressed by expansion of the ECF volume.

PART C
DIAGNOSTIC TESTS

The clinician may employ any or all of the following tests to evaluate the K^+ secretory process: the rate of excretion of K^+, $[K^+]$ in a random urine sample, the ratio of K^+ to Na^+ in the urine, the fractional excretion of K^+, and the TTKG. Each test has its advantages and shortcomings (Table 9·4).

Table 9·4
TESTS USED TO MONITOR THE K^+ EXCRETION PROCESS
For a description, see the text.

Test	Strengths	Weaknesses	Expected value HypoK	Expected value HyperK
Measure of K^+ excretion				
- 24-hour K^+ excretion (mmol/day)	• Valuable	• Does not indicate pathophysiology	< 15	> 200
- K^+ per creatinine (mmol/mmol)	• Can use random urine	• Must know expected rate of creatinine excretion	< 1	> 20
TTKG	• Physiologic basis • "Translates" urine to CCD • Separates $[K^+]$ from urine flow rate	• Many unverified assumptions	< 2	> 10
Less useful tests				
- Random urine $[K^+]$	• Simple	• Does not consider MCD water reabsorption	None	None
- K^+/Na^+	• None	• Depends on dietary Na^+	?	?
- Fractional excretion of K^+	• None	• Expected values depend on GFR • Is not based on the physiology of K^+	Need nomogram	

24-Hour K^+ Excretion Rate

This test is the most important to perform because it indicates whether or not a renal disorder is present. Expected values are less than 15 mmol/day during hypokalemia and greater than 200 mmol/day if hyperkalemia is present. Because there is a diurnal pattern for K^+ excretion, a 24-hour specimen is ideal, but it is not really necessary given the wide ranges of expected values.

$[K^+]$ in a Random Urine Sample

One major advantage of this test is its simplicity. Nevertheless, on its own, the urine $[K^+]$ may be frankly misleading because reabsorption of water in the MCD has a great influence on the $[K^+]$ in the urine (see the margin). The other major disadvantage of a simple random urine sample is that the sample might be nonrepresentative (not usually a major concern).

With either of two manipulations, the $[K^+]$ in a random urine sample can be very useful. First, the daily rate of excretion of K^+ can be deduced by dividing the $[K^+]$ in the random urine sample by the concentration of creatinine in that sample multiplied by an estimate of the 24-hour creatinine excretion (the daily rate of excretion of creatinine is both constant and predictable; see the margin). This calculation will provide results that are satisfactory at the bedside. Second, the TTKG can be calculated (see below). It will permit a separate analysis of the $[K^+]$ and the volume components of the K^+ excretion formula, as well as provide a way to "translate" data from the urine to the CCD.

Rise in $[K^+]$ in the MCD:
Three-fourths of the water content can be reabsorbed from MCD fluid (rise of urine osmolality from 300 to 1200 mosm/kg H_2O) in the MCD.
- This reabsorption will elevate the $[K^+]$ fourfold, from a maximum of 40 to 160 mmol/l.
- Therefore, a 120 mmol/l rise in the $[K^+]$ can be due to water reabsorption.

Use of creatinine excretion:
In any individual, the rate of excretion of creatinine is relatively constant throughout the day. It depends on muscle mass, and the daily rate is close to 20 mg (0.2 mmol)/kg body weight. If a person excretes 10 mmol of creatinine per day and the urine has a $[K^+]$ of 40 mmol/l and a creatinine concentration of 5 mmol/l, the 24-hour K^+ excretion rate should be close to 80 mmol (40 mmol/l K^+/5 mmol/l creatinine = 80 mmol of K^+/10 mmol of creatinine).

Ratio of K^+ to Na^+ in the Urine

This ratio has been suggested to give some insights into aldosterone action. In our opinion, it is not useful because the denominator, the $[Na^+]$ in urine, depends primarily on dietary intake of NaCl and/or diuretics. Therefore, unless the patient is on a fixed diet, the test is useless, and, even on a fixed diet, the test has little merit.

Fractional Excretion of K^+

This frequently performed calculation relates the quantity of K^+ excreted to that filtered. It does not provide any insights as to the mechanisms in the CCD, and it depends heavily on the rate of filtration of K^+. One needs to carry a nomogram to interpret the fractional excretion. For these reasons, we do not recommend it (see the discussion of Question 9·15 for more detail).

Question

(Discussion on page 348)

9·15 In what way does the calculation of the fractional excretion of K^+ differ from that of the TTKG?

Transtubular $[K^+]$ Gradient

The TTKG is a test designed to reflect the driving force for K^+ secretion. Before describing the calculation, we shall first present the premise and then consider the weaknesses and assumptions required for this calculation. One should recall that the excretion of K^+ is a function of the $[K^+]$ in the lumen of the CCD and the distal flow rate.

PREMISE

To calculate the $[K^+]$ in the lumen of the CCD, one must adjust the $[K^+]$ in the urine to "correct" for changes in the $[K^+]$ caused by the reabsorption of water in the MCD.

ASSUMPTIONS

1. **Reabsorption of water in the MCD can be estimated:**
   The quantity of water reabsorbed in the MCD is calculated by comparing the rise in the osmolality of the fluid from the terminal CCD (same as that of plasma when ADH is acting) with that in the final urine. A twofold rise in urine osmolality implies that half the water was reabsorbed in the MCD, provided that no particles were reabsorbed in the MCD. This assumption is true for urea, NH_4^+, and K^+ for the most part, but it is not true for Na^+ (see Figure 9·11 and the discussion of Question 9·16).

Question

(Discussion on page 349)

9·16 If NaCl is reabsorbed in the MCD, in what direction will the TTKG change?

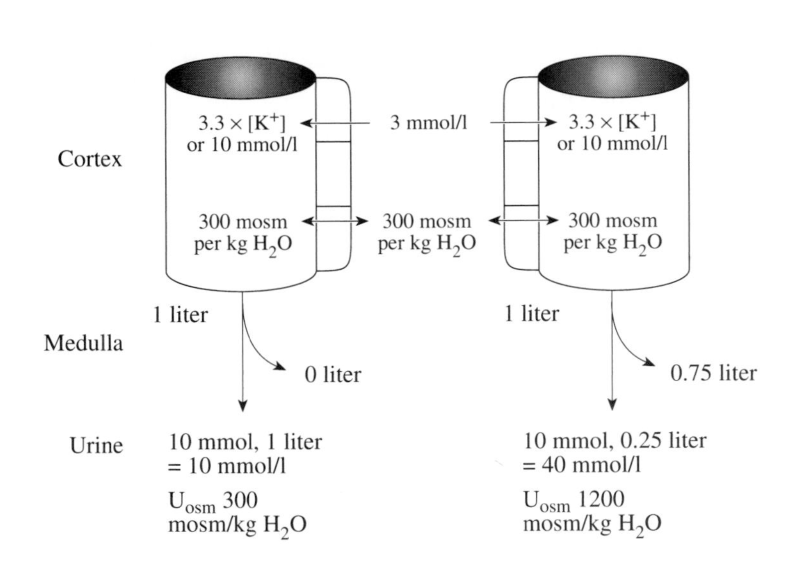

Figure 9·11
Effect of medullary water reabsorption on the urine $[K^+]$
The barrel-shaped structures represent the CCD and the arrows below them the MCD. In both examples, there is a TTKG of 3.3. Consider what happens when 1 liter of fluid traverses the MCD. In the left-hand example, there is no medullary water reabsorption, but, in the right-hand example, 75% of the water is reabsorbed. In both cases, no K^+ are reabsorbed or secreted in the medulla. Therefore, although the excretions of K^+ are equal in both cases, the $[K^+]$ in urine is fourfold higher (40 mmol/l) when water is reabsorbed in the medulla (right side). This increase must be taken into account when assessing the urine $[K^+]$.

2. **K^+ are not reabsorbed or secreted in the MCD:**
   In the rat, K^+ are not usually secreted or reabsorbed. Nevertheless, with profound K^+ depletion, there is net K^+ reabsorption, and when "subindustrial" doses of K^+ are given, there is net secretion of K^+ in the MCD. We assume that humans do have this capacity, but perhaps only in these extremes. Because this assumption is impossible to verify, one must have "faith" and a sense of caution when using the TTKG.

3. **The osmolality is known in the terminal CCD:**
   Knowing the osmolality of the tubular fluid in the CCD is a critical component of the calculation of the TTKG. If the urine osmolality is not greater than the plasma osmolality, the TTKG cannot be used because one does not know how much water was reabsorbed in the MCD. In contrast, with a higher urine osmolality, it is safe to assume that the fluid in the CCD has the same osmolality as that in the plasma.

4. **The $[K^+]$ in plasma reflects that in the peritubular fluid around the CCD:**
   This assumption cannot be verified, but the $[K^+]$ in plasma should be representative unless K^+ excretion rates are enormous (see the discussion of Question 9·17).

Question

(Discussion on page 349)

9·17 If all the filtered K^+ were excreted (100% fractional excretion), how would the calculated value for the TTKG change?

CALCULATION OF THE TTKG

- TTKG = $[K^+]_{urine}/(urine/plasma)_{osm}/[K^+]_{plasma}$

If one divides the $[K^+]$ in urine by the urine-to-plasma osmolality ratio ($[K^+]_{urine}/(urine:plasma)_{osm}$), a reasonable approximation of the $[K^+]$ at the end of the CCD can be deduced. This value will be high (7 or more times the plasma $[K^+]$) if mineralocorticoids are acting and low (less than 2 times the plasma $[K^+]$) if they are not. It is critical to ensure that the urine is not hypoosmolal to plasma before doing this calculation.

PART D
REVIEW

Discussion of Introductory Case
Lee's [K^+] (K)rashed

(Case presented on page 324)

What was the total body K^+ content at the time of presentation and why?

At the time of presentation, most patients with DKA have a reduced total body K^+ content primarily because of renal loss of K^+ subsequent to the glucose-induced osmotic diuresis. The degree of K^+ depletion depends on the antecedent intake of K^+ and more importantly on prior loss of K^+. In this regard, a history of vomiting or diuretic use might suggest excessive K^+ loss. On the other hand, an underlying kidney problem might limit the degree of K^+ depletion.

What factors led to her hyperkalemia?

The primary factor is insulin deficiency. A contributing factor may be that hyperglycemia caused a shift of water from K^+-rich ICF; this shift raised the [K^+] in the ICF (not a very important factor; see the discussion of Question 9·5).

Several other factors could influence the degree of hyperkalemia.

1. Acidemia is not an important factor.
2. Although the rate of excretion of K^+ is not high now, it might have been higher when her GFR was not so low (before the development of a markedly contracted ECF volume caused by the glucose-induced osmotic diuresis).
3. Intrinsic renal disease caused by longstanding diabetes mellitus could have compromised her renal excretion of K^+.
4. Certain drugs often given to diabetics may impair the excretion of K^+ (e.g., angiotensin-converting enzyme inhibitors).

Clinical pearls:

- Changes that occur between a deficiency of a hormone and a level that is low but in the normal range have a larger effect on the [K^+] than do changes from the high end of normal to very high levels.
- Combinations of hormone deficiencies can have large effects on the [K^+] in the ECF (e.g., the combination of a lack of both insulin and aldosterone may lead to a very severe degree of hyperkalemia, especially in the presence of hyperglycemia).

Why did hypokalemia develop four hours later?

Patients with DKA have a large deficit of K^+ but are hyperkalemic because of a lack of insulin. Hypokalemia can develop after two hours when insulin promotes the entry of K^+ into cells. To a very minor extent, dilution will occur when the ECF is reexpanded with saline that contains no K^+.

In DKA there is little renal excretion of K^+ on admission despite hyperkalemia (the mechanism is unclear, but it prevents further K^+ depletion; for our speculation, see the margin).

Hyperglycemia and K^+ excretion:
Diabetics in very poor control have a low TTKG despite high aldosterone levels and adequate distal delivery of Na^+. A possible explanation is that persistent glucosuria led to glycation and inhibition of the Na^+ channels in the CCD (similar to the actions of amiloride).

Summary of Main Points

- Potassium, principally an intracellular cation, is electrically balanced primarily by large macromolecular anions that are not free to leave the ICF.
- The ratio of the [K^+] in the ICF to that in the ECF (35–40:1) is due to the electrogenic Na^+K^+ATPase and diffusion of K^+ from the ICF to the ECF. This distribution, which is primarily influenced by hormones (insulin and β-adrenergics), generates the resting membrane potential across cell membranes, an important determinant of cell function.
- Renal excretion of K^+ is a function of the [K^+] in the urine and the urine flow rate. The major determinant of the [K^+] in the urine is the electrogenic reabsorption of Na^+ in the CCD; the major determinant of the flow rate in the CCD is the rate of excretion of osmoles.

Discussion of Questions

9·1 What quantity of K^+ of dietary origin (70 mmol) will be distributed in cells if there is no excretion of K^+ nor a change in the RMP?

The calculation is based on the following facts: the distribution of K^+ between the ICF and the ECF is 35–40:1 at the usual RMP. Because the RMP did not change, the ingested K^+ will distribute at a 35–40:1 ratio. Since 70 mmol of K^+ were ingested, approximately 2 mmol will remain in the ECF and 68 mmol will enter the ICF.

What will the plasma [K^+] be?

The ECF volume is 15 liters, so the rise in [K^+] is 2/15, or 0.075 mmol/l.

9·2 What is the intracellular cation composition in the red blood cells of a dog? These cells lack Na^+K^+ATPase pumps.

If there is no Na^+K^+ATPase, the [Na^+] will be very high (close to 140 mmol/l) and the [K^+] will be very low (close to 9 mmol/l) in these cells. Perhaps this is why we refer to dogs as canines.

9·3 Why does the entry of Na^+ into cells increase the RMP when augmented by insulin but decrease the magnitude of the RMP in either hyperkalemic or hypokalemic periodic paralysis?

The answer becomes obvious when one counts the number of positive charges that move.

Insulin: The entry of Na^+ into cells is electroneutral because it occurs in conjunction with the export of H^+ from cells; flux is via the Na^+/H^+ antiporter. When Na^+ are pumped out of cells via the electrogenic

Na^+K^+ATPase, a net of one-third of a positive charge exits the cell per Na^+ that entered via the Na^+/H^+ antiporter; this movement increases the electronegativity of the RMP (see the margin).

Periodic paralysis: The entry of Na^+ is electrogenic; Na^+ enter via their ion-specific channels, so one positive charge enters the cell per Na^+. As discussed above, only a net of one-third of a charge exits the cell per Na^+ pumped via the Na^+K^+ATPase. Hence, the net effect is net positive charge entry and a diminished magnitude of the RMP.

Na^+K^+ATPase:
One-third of a positive charge is exported per Na^+ pumped (3 Na^+ are pumped out and 2 K^+ enter).

9·4 What property of the Na^+K^+ATPase might allow it to bind K^+ at a $[K^+]$ of 4 mmol/l in the ECF yet also permit K^+ to dissociate from this pump at a $[K^+]$ of 150 mmol/l in the ICF?

Note:
Question 9·4 is for the more curious.

This question is challenging and the answer is unknown. The affinity for K^+ has to be considerably less in cells than in the ECF (smaller by orders of magnitude). We offer the following suggestions: a conformational change in the K^+-binding subunit of the Na^+K^+ATPase that occurs once it is in the intracellular domain and/or the presence of a nearby positive charge in the cell that helps "repel" K^+ from this ion pump.

9·5 How much will the plasma $[K^+]$ change when 1 liter of pure water is shifted from the ICF to the ECF? Assume a constant resting membrane potential.

The assumption requires that the ratio of the $[K^+]$ in the ICF to the $[K^+]$ in the ECF remain constant. If 1 liter leaves the ICF, its volume declines by 1/30, or 3.33%. This decline raises the $[K^+]$ in the ICF (and ECF) by 3.33% because the ratio of the $[K^+]$ in the ICF to the $[K^+]$ in the ECF remains constant. Therefore, the $[K^+]$ in plasma will rise from 4 to 4.13 mmol/l at steady state, a trivial rise. If hyperosmolal states cause hyperkalemia, a simple shift of water is not the primary mechanism involved. Rather, hyperosmolality leads to a lowering of the RMP, or it is associated with decreased excretion of K^+.

9·6 Will a drug that depolarizes cell membranes cause a change in the plasma $[K^+]$?

Yes. With depolarization, the RMP falls; thus, the ratio of the $[K^+]$ in the ICF to the $[K^+]$ in the ECF falls, and hyperkalemia will be present.

9·7 Why does the loss of intracellular phosphate anions cause hyperkalemia in DKA but not in poliomyelitis?

The answer hinges on three facts: first, the major anions in the ICF are organic phosphates; second, the major cation is K^+; third, electroneutrality must be preserved.

Poliomyelitis: In this disorder, there is a wasting of muscles and the loss of phosphate, water, and K^+ in the same proportion as they exist in cells. There is no change in the RMP. Thus, there is just a smaller (but otherwise normal) ICF in this disorder. There is ample time to excrete the K^+ and phosphate.

DKA: In this acute catabolic setting, there is a loss of organic phosphates (RNA, etc.), K^+, and water from the ICF. The key to hyperkalemia is the fall in the RMP associated with a lack of insulin; this fall means that more K^+ will be distributed into the ECF. The degree of hyperkalemia will also depend on the rate of excretion of K^+ (osmotic diuresis) in DKA. Usually in DKA there is a large total body deficit of K^+ and a redistribution of K^+ from the ICF to the ECF.

9·8 During K^+ depletion, Na^+ replace K^+ in cells. How does this exchange affect the Na^+K^+ATPase?

A rise in the intracellular [Na^+] should make the Na^+K^+ATPase pump more quickly. Faster pumping should, in turn, lower the [Na^+] in cells back to normal. If the elevated [Na^+] persists, something else must have happened. Perhaps there was also a fall in the number of Na^+K^+ATPase units or their activity. There may have also been a lower affinity of the Na^+K^+ATPase pump for Na^+. At least one of these events must have occurred for a higher intracellular [Na^+] to be present in steady state.

9·9 What is the maximum rate of K^+ excretion in 24 hours? (Given: plasma [K^+] is 4 mmol/l, transtubular [K^+] ratio is 10, 5 liters of filtrate reach the CCD, and there is no medullary K^+ secretion or reabsorption.)

Note:
K^+ excretion = urine [K^+] $\times$ volume.

The luminal [K^+] increases to almost 10-fold above the plasma [K^+] when stimulated by aldosterone. If the plasma [K^+] is 4 mmol/l, then the luminal [K^+] is 40 mmol/l (i.e., 10×4 mmol/l). If distal delivery is 5 liters, then 200 mmol of K^+ could be excreted; alternatively, a 10-liter delivery could result in the excretion of 400 mmol at the same TTKG. Hence, the importance of the volume of filtrate delivered to this nephron site can be appreciated. In addition, it has been reported that the [K^+] in luminal fluid can be 20 times the plasma [K^+] if bicarbonaturia is present. Therefore, there is the potential to augment the excretion of K^+ considerably; the highest excretion rates reported are close to 400 mmol/day.

9·10 Can you determine whether a high urine [K^+] is due to K^+ addition to the CCD or to water reabsorption in the MCD?

The [K^+] is the K^+:H_2O ratio. If the [K^+] were to rise from the reabsorption of water in the MCD, the particle:H_2O ratio (osmolality) would also rise in fluid traversing the MCD if the particles themselves were not reabsorbed (half of the particles are normally urea, and urea is not reabsorbed in the MCD to an appreciable extent). Hence, if the urine osmolality has not risen, the rise in [K^+] is due to the net secretion of K^+, a cortical action. In contrast, a rise in urine osmolality implies a role for water reabsorption in the MCD. This issue is examined quantitatively in the section addressing the transtubular [K^+] gradient (TTKG), pages 340–42.

9·11 Does the excretion of K^+ conserve Na^+ and thereby defend the ECF volume? This question requires one to consider the origin of the K^+ that is excreted.

To answer this question, one must consider the actions of aldosterone in terms of Na^+ and K^+ physiology, as depicted in Figure 9·12. The first step in aldosterone action is the reabsorption of Na^+ in the CCD. This reabsorption does not necessarily conserve Na^+ for the ECF. Consider Pathway 1a, for example: to reabsorb Na^+, K^+ are transported from the ECF to the urine. Because the ECF has a small quantity of K^+, the source of the K^+ is really the ICF. To transport most of the K^+ out of the ICF, Na^+ enter this compartment. Therefore, the net effect is the excretion of intracellular K^+ in the urine and the placement of Na^+ from the ECF into the ICF. These actions do not appear to defend either the ECF volume (Na^+ content) or the ICF composition (K^+); however, in a different context, the reabsorption of Na^+ may indeed defend the ECF volume.

If the source of the urinary K^+ was dietary K^+ (Pathway 1b), if Na^+ reabsorption was linked to the excretion of NH_4^+ (hypokalemia stimulates ammoniagenesis), or if a Na^+ and Cl^- were reabsorbed, the ECF volume would be defended. Hence, it is necessary to obtain a broad overview to integrate Na^+, K^+, and acid-base physiology.

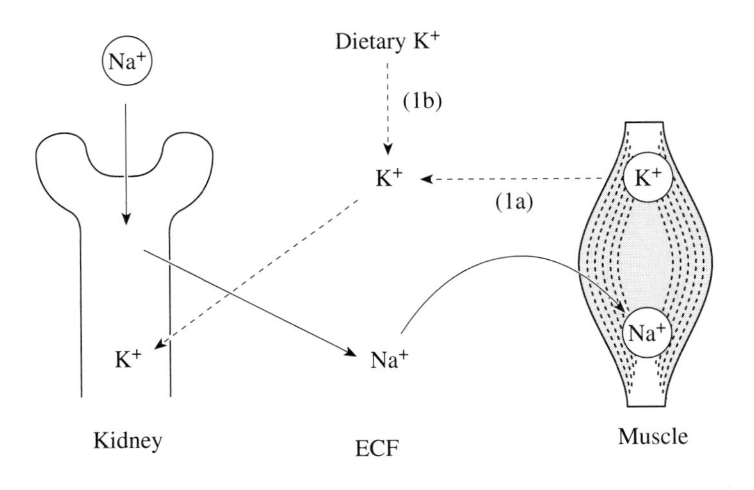

Figure 9·12
The source of excreted K^+ and the defense of the ECF volume
The reabsorption of filtered Na^+ with K^+ excretion will result in the preservation of the ECF volume only when the source of the K^+ was dietary.

9·12 A very high $[K^+]$ in the lumen of the CCD occurs in a patient who vomits. A similarly high $[K^+]$ occurs when acetazolamide (a carbonic anhydrase inhibitor) acts. Common to both situations is bicarbonaturia. What might the mechanism be?

The mechanism has not yet been defined, but we offer the following speculation. For HCO_3^- to cause such a high TTKG, it seems that HCO_3^- inhibit the reabsorption of Cl^- in the CCD. Assume that the mouth of the Cl^- channel contains a positively charged group, such as the epsilon amino group in lysine (pK close to 9). These positive charges "attract" Cl^- to the mouth of the channel, where they are per-

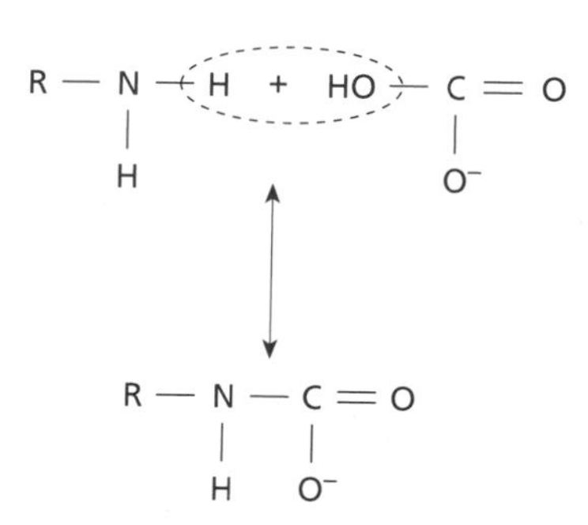

Carbamate

mitted or encouraged to enter this channel. In contrast, if HCO_3^- enter this environment, alkalization occurs. The epsilon $R\text{-}NH_3^+$ group dissociates (the equation below is driven to the right by the fall in the $[H^+]$).

$$R\text{-}NH_3^+ \longleftrightarrow RNH_2 + H^+$$

The uncharged RNH_2 group is very reactive and will combine with CO_2 to form a carbamino compound that bears a negative charge (the equation in the margin is drawn with HCO_3^- for convenience).

This anionic carbamino compound may face the lumen, repel Cl^- from the mouth of the Cl^- channel, and diminish the reabsorption of Cl^-. This explanation is highly speculative.

9·13 Does a normal person excrete more K^+ than a patient with adrenal insufficiency (assume they eat the same diet)?

Adrenal insufficiency is a chronic condition, so the patient is in balance; all the K^+ ingested and absorbed are excreted. Thus, both people eat and excrete the same quantity of K^+. The difference is that the normal person uses aldosterone to "drive" K^+ excretion, but the patient who lacks aldosterone requires hyperkalemia to achieve excretion of K^+ and K^+ balance. Nevertheless, K^+ excretion would be much higher in a hyperkalemic individual who is otherwise normal (i.e., has aldosterone).

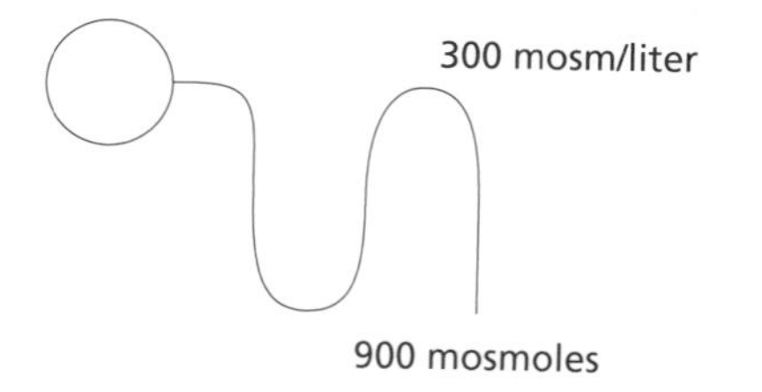

9·14 Is the volume delivered to the terminal CCD lower in a patient with a normal or reduced GFR (lower by 50%)?

The volume delivered to the terminal CCD depends on the osmole excretion rate if ADH is acting. When ADH acts, the osmolality in this fluid is the same as in plasma (say 300 mosm/kg H_2O). If 900 mosmoles are excreted per day with a normal or 50% reduction in GFR, the volume delivered to the terminal CCD is the same in both examples, 3 liters/day (see the margin).

9·15 In what way does the calculation of the fractional excretion of K^+ differ from that of the TTKG?

The TTKG "adjusts" for water reabsorption in the MCD, and the fractional excretion adjusts for water reabsorption throughout the nephron. They differ in the methods used to monitor water reabsorption—osmolality vs creatinine.

The formulae for these two calculations are as follows (U = urine and P = plasma):

$$\text{Fractional excretion} = (U/P)_{[K^+]}/(U/P)_{\text{creatinine}}$$

$$\text{TTKG} = (U/P)_{[K^+]}/(U/P)_{\text{osmolality}}$$

9·16 If NaCl is reabsorbed in the MCD, in what direction will the TTKG change?

If more particles are reabsorbed in the MCD, for the same urine osmolality, one will underestimate the volume of water reabsorbed in the MCD. With more water reabsorption, the $[K^+]$ or the K^+:urine volume will be higher. Hence, the value back-calculated for the $[K^+]$ in the CCD will be higher than it should be, and thereby the TTKG will overestimate the ratio of tubular fluid to plasma $[K^+]$.

9·17 If all the filtered K^+ were excreted (100% fractional excretion), how would the calculated value for the TTKG change?

Because close to one-fourth of renal plasma flow is filtered, one-fourth of the K^+ is removed and excreted when the fractional excretion is 100%. Therefore, the $[K^+]$ in the peritubular fluid (and renal vein plasma) is 3 vs 4 mmol/l. Hence, the TTKG using systemic venous blood will underestimate the TTKG, and the result will be 75% of the real value.

10

HYPOKALEMIA

PART C REVIEW

Objectives

- To provide a clinical approach to the patient with *hypokalemia* based on an assessment of K^+ intake, shift of K^+ into cells, and increased loss of K^+, with an appreciation that most clinical problems fall into the renal loss category.
- To describe how to evaluate both components of renal K^+ loss—the $[K^+]$ in the luminal fluid of the CCD and the volume of fluid delivered to the terminal CCD.
- To illustrate how an assessment of the urine electrolytes (Na^+, Cl^-, and K^+) helps in elucidating the basis of excessive excretion of K^+.
- To provide a plan of treatment based on physiologic principles.

Abbreviations:
CAI = carbonic anhydrase inhibitor.
CCD = cortical collecting duct.
CHF = congestive heart failure.
ECF = extracellular fluid.
ECG = electrocardiogram.
11β-HSDH = 11β-hydroxysteroid dehydrogenase.
ICF = intracellular fluid.
JVP = jugular venous pressure.
MCD = medullary collecting duct.
RMP = resting membrane potential.
TEPD = transepithelial potential difference.
TTKG = transtubular potassium gradient.

Outline of Major Principles

1. **General considerations:** Because the ECF contains only 2% of K^+ (60 mmol), physicians monitor total body K^+ through a tiny and inadequate window. The ratio of the $[K^+]$ across the cell membrane is 150 mmol/l/4.3 mmol/l, or close to 35:1. If the plasma $[K^+]$ falls from 4 to 3 mmol/l and this ratio of $[K^+]$ remains constant, the fall in the content of K^+ in the ICF would be 25% of the total, or close to 1125 mmol (0.25×4500 mmol). Because the deficit of K^+ is only half of this value at best in clinical studies (100–400 mmol), the transcellular ratio of $[K^+]$ must rise (i.e., the ratio of $[K^+]$ in the ICF to that in the ECF is higher; see the margin). A rise in this ratio implies that the RMP has hyperpolarized (i.e., more positive charges (K^+) diffused out of cells down their higher concentration difference).

2. **Etiology of hypokalemia:** Loss of K^+ occurs most often in patients who vomit, have nasogastric suction, diarrhea, or who use (abuse) diuretics, but in each case the major route of K^+ loss is renal. The clues are found in the history (vomiting, laxative abuse, diuretics—all of which the patient may deny) and the physical finding of ECF volume contraction. If there is no ECF volume contraction, look for a primary reason for high levels of aldosterone. The presence of hypertension is a useful clue when deducing the cause of hyperaldosteronism. A shift of K^+ into cells can occur under the influence of hormones, metabolic alkalosis, or recent anabolism; however, these causes are uncommon for chronic and severe hypokalemia.

3. **Diagnosis of hypokalemia:** Clinical suspicion is essential because weakness is a late symptom and is not present until K^+ depletion is quite severe; hypokalemia is most often found on "routine" analysis of electrolytes, as a result of ECG changes, or in patients with a history of disorders associated with renal K^+ loss. Urine electrolyte levels are most

Hypokalemia:
- Plasma $[K^+] < 3.5$ mmol/l.
- Threats are:
  - cardiac arrhythmias,
  - respiratory failure,
  - hepatic encephalopathy.

Note:
For the cell to lose 400 mmol of K^+, there must be a loss of 400 mEq of intracellular anions or a gain of 400 mmol of Na^+ and/or H^+ in the ICF. Hence, major changes in the ICF environment should be expected.

helpful in establishing the etiology, and a useful approach is to interpret the urine [K^+] by adjusting it for water reabsorption in the renal medulla to reflect the [K^+] in the lumen of the CCD (i.e., the TTKG; see Chapter 9, pages 340–42).

Digitalis:
A glycoside that is given to improve myocardial contractility or to slow atrioventricular conduction.

Note:
The major danger of hypokalemia and hyperkalemia is cardiac arrhythmias.

Summary:
After being consumed, K^+ are stored in cells temporarily and are then excreted several hours later.

4. **Dangers of hypokalemia:** The major danger of hypokalemia is a cardiac arrhythmia, especially in the presence of *digitalis* and/or metabolic alkalosis. A second danger is hypoventilation in the patient with a severe degree of metabolic acidosis because high values for alveolar ventilation are required to maintain a very low P_aco_2; hypoventilation is often due to muscle weakness when associated with hypokalemia. A third setting in which hypokalemia may be detrimental is hepatic encephalopathy. The hypokalemia may aggravate toxicity by permitting more NH_4^+ to accumulate in cells (hypotheses mentioned include enhanced production of NH_4^+ in the kidney and their distribution in cells because of acidosis of the ICF).
5. **Treatment:** Large quantities of K^+ must be given when the total body deficit is great. In most cases, KCl is needed; $KHCO_3$ should be given in patients with hypokalemia and a severe degree of metabolic acidosis. In the patient receiving digitalis and in the patient with diabetic ketoacidosis in whom K^+ will shift into cells 1–2 hours following insulin administration, K^+ must be replaced more aggressively. When a large quantity of KCl is given, there may be a significant expansion of the ECF volume because Na^+ will move from the ICF to the ECF when K^+ enter cells (Figure 10·1).

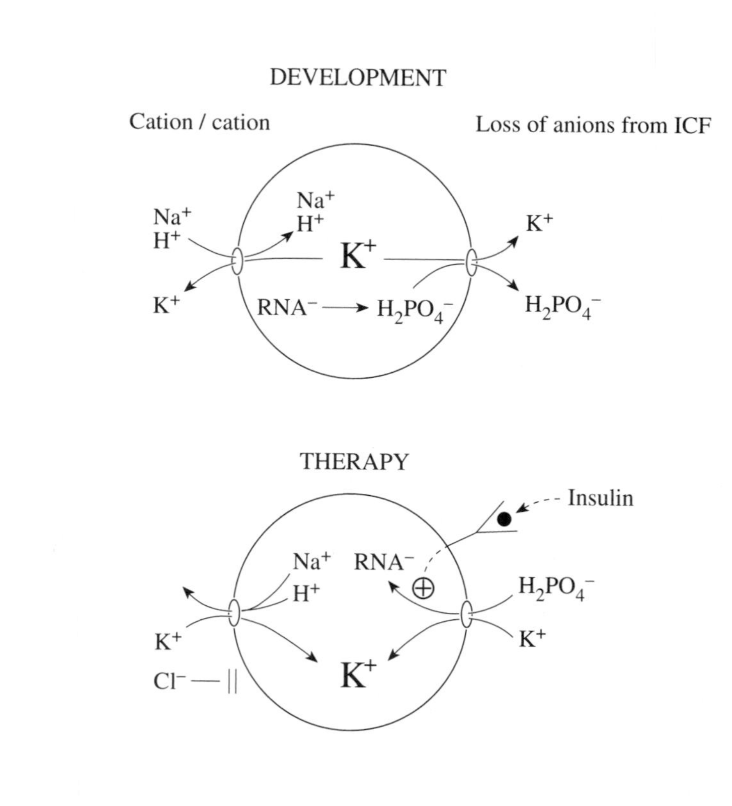

Figure 10·1
The effect of KCl administration in a patient with K^+ deficiency
As shown on the top portion, K^+ can be lost from the ICF by two mechanisms: a cation-cation pathway on the left and a loss of K^+ plus an ICF anion (phosphate) on the right. During therapy (bottom portion), if the entire K^+ deficit was not produced by the cation-cation exchange, KCl alone will not correct the ICF deficit of K^+ because Cl^- is predominantly an ECF anion. To replace the entire deficit of K^+ in the ICF, some K^+ must be given along with phosphate once anabolism occurs.

Introductory Case
Toby Is Losing More Than Weight

(Case discussed on page 376)

Toby, the ballerina, returns (pages 148, 168–69). She is on the verge of landing a job with a prestigious ballet company. She believes that a possible impediment is a gain of weight. She tries measures that worked in the past during her adolescence. Although she loses weight, her stamina is not the same as it used to be. She sees a physician for advice. She denies taking medications and vomiting. On physical examination, a degree of contraction of the ECF volume is detected. Laboratory results are provided below.

		Plasma	Urine (random sample)
Na^+	mmol/l	136	7
K^+	mmol/l	3.1	22
Cl^-	mmol/l	108	84
HCO_3^-	mmol/l	19	0
pH		7.35	5.6
Osmolality	mosm/kg H_2O	278	589
Glucose	mmol/l (mg/dl)	5 (90)	0

What is her total body K^+ deficit?
Was a shift of K^+ into cells a major cause of hypokalemia?
What factors could have contributed to a renal loss of K^+?
What tests might be helpful in this regard?

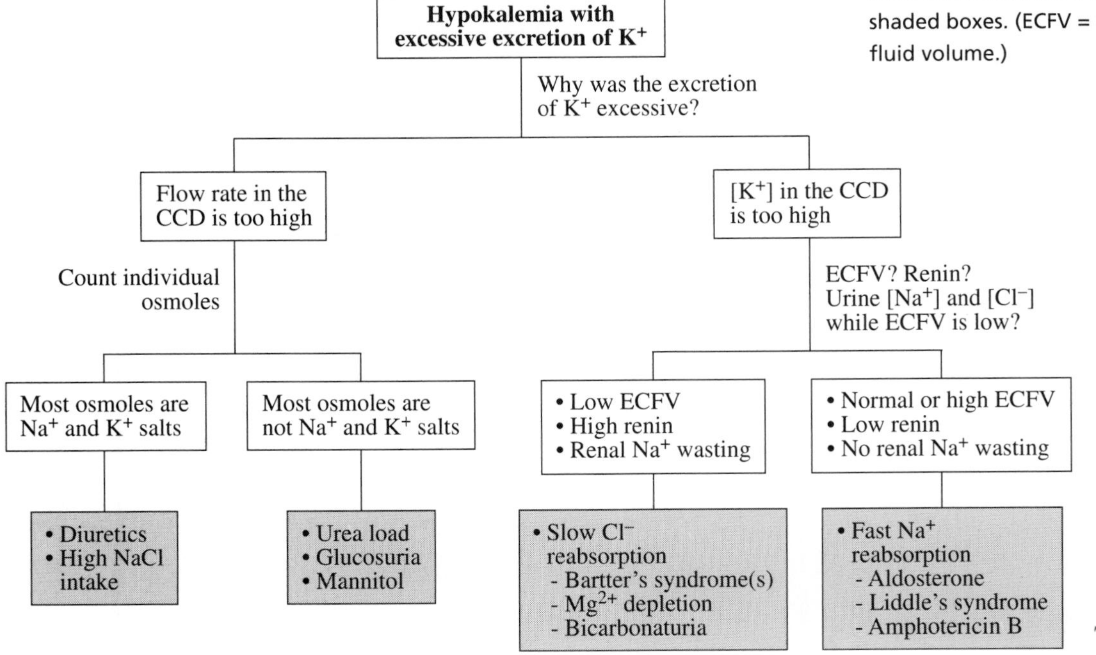

Review flow chart
Causes of hypokalemia with excessive excretion of K^+
The causes of excessive excretion of K^+ are too high a flow rate in the CCD (left limb) and too high a $[K^+]$ in the lumen of the CCD (right limb). Both flow rate and $[K^+]$ in the CCD should be evaluated in each patient. Final considerations are shown in the shaded boxes. (ECFV = extracellular fluid volume.)

PART A
DIAGNOSIS

Approach to the Patient with Hypokalemia

Dangers:

1. Cardiac arrhythmias
2. Respiratory weakness
3. Hepatic encephalopathy

Because hypokalemia can lead to a life-threatening event, one might have to embark on therapy before the investigation can be completed (Figure 10·2). If there is an important change in the ECG, if digitalis is being used in a patient with a cardiac problem, or if respiratory failure or hepatic encephalopathy is imminent, proceed directly with treatment (see pages 372–75); otherwise, use the steps outlined in Figure 10·2 to make a diagnosis. The questions to ask are as follows:

Excretion of K^+:
K^+ excretion =
Urine $[K^+]$ × Urine volume

1. Is the rate of excretion of K^+ excessive?

A rate of excretion of K^+ that is less than 15 mmol/day in an adult with hypokalemia indicates two possibilities: first, the hypokalemia may not be due to excessive excretion of K^+; second, renal K^+ loss may have occurred at an earlier time, but the provocative agent or event is not present now (e.g., remote vomiting or diuretic use). Nevertheless, in most patients with hypokalemia, there is excessive renal excretion of K^+ (greater than 15 mmol/day).

2. Why is the rate of excretion of K^+ high?

The first step to take to understand why the rate of excretion of K^+ is too high is to determine whether a high $[K^+]$ or a high flow rate in the CCD

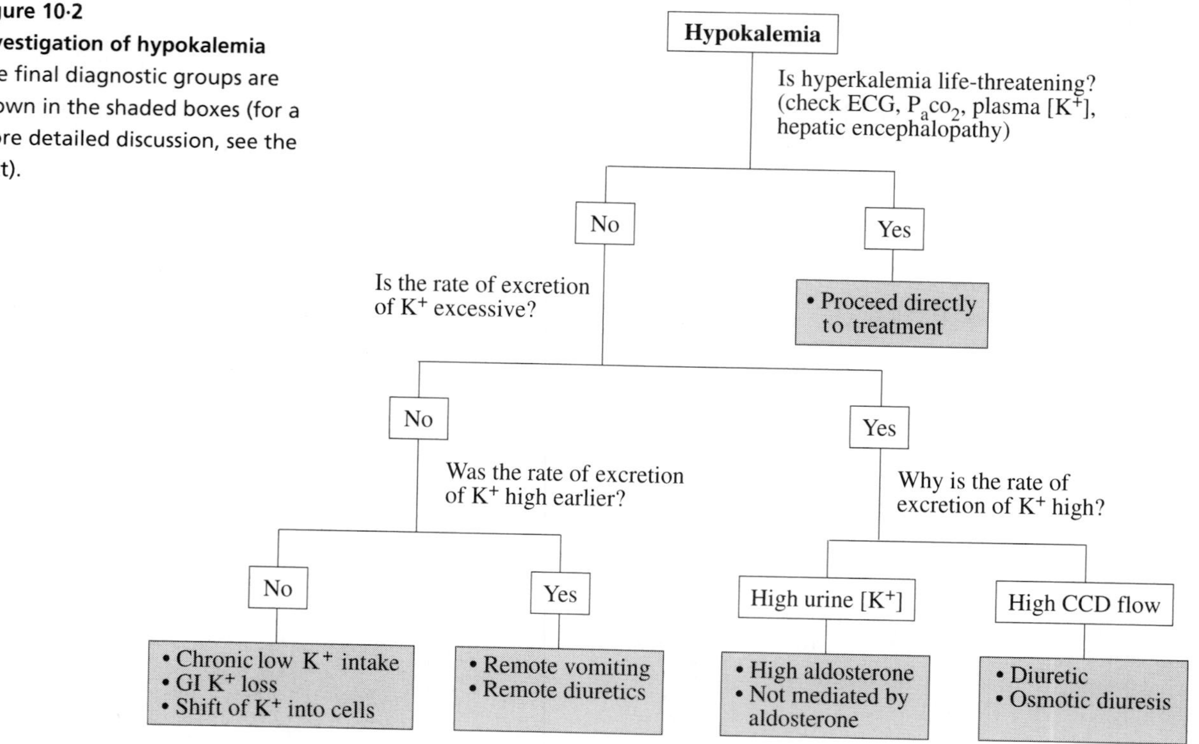

Figure 10·2
Investigation of hypokalemia
The final diagnostic groups are shown in the shaded boxes (for a more detailed discussion, see the text).

is responsible. First, consider the $[K^+]$ in the urine: if the TTKG is greater than 4, one cause of the high rate of excretion of K^+ is a high $[K^+]$ in the terminal CCD. The question at this point is to determine whether or not the high $[K^+]$ is due to actions of aldosterone. This aspect will be considered in detail later.

The other major contributor to a high rate of excretion of K^+ is a large flow rate in the CCD, driven by an osmole excretion load. The osmoles can be electrolytes (usually caused by a diuretic; Table 10·1) or nonelectrolytes (usually from an osmotic diuresis caused by glucose or urea).

Table 10·1
URINE ELECTROLYTE LEVELS DURING CHRONIC SEVERE HYPOKALEMIA OF RENAL ORIGIN

Electrolyte	Vomiting	Diuretics
Na^+	• High if vomiting was recent despite ECF volume contraction; otherwise < 10 mmol/l.	• Less than 10 mmol/l unless use of diuretics was recent.
K^+	• High relative to plasma value; absolute value depends on medullary water abstraction.	• Can be greater than 20 mmol/l even if no diuretics were used recently.
Cl^-	• Always less than 10–15 mmol/l.	• Less than 10 mmol/l if ECF volume is low and there was no recent use of diuretics.
HCO_3^-	• Abundant with recent vomiting.	• Zero unless CAI type of diuretics.

CAI = carbonic anhydrase inhibitor.

Questions

(Discussions on pages 383–84)

10·1 What is the most likely cause of K^+ loss in patients A, B, C, and D? They all have hypokalemia, ECF volume contraction, and metabolic alkalosis.

		Patient			
Urine electrolyte		A	B	C	D
Na^+	mmol/l	25	25	25	0
K^+	mmol/l	60	60	60	10
Cl^-	mmol/l	0	85	0	0
HCO_3^-	mmol/l	85	0	0	0

10·2 Will a patient consuming a very high NaCl load become hypokalemic as a result of enhanced renal K^+ excretion? With respect to K^+ homeostasis, how does such a patient differ from one who receives a loop diuretic? Assume equal Na^+ excretion rates.

Etiology of Hypokalemia

Three factors must be examined to determine the most likely basis for hypokalemia: intake of K^+, distribution of K^+ in the ICF, and the excretion of K^+. The most important of these factors is excessive renal excretion of K^+ in a patient with chronic hypokalemia.

DECREASED INTAKE OF K^+

- Hypokalemia is almost never solely due to low intake of K^+.

Usual K^+ intake:

1. In an adult on a Western diet, the intake of K^+ is 60–90 mmol per day, on average.
2. Most of the K^+ ingested is absorbed. Certain materials, such as resins, bind K^+ so that K^+ cannot be absorbed. This binding can be considered a decrease in the "net intake" of K^+.

   In therapy of hyperkalemia, gastrointestinal loss of K^+ can be promoted with K^+-binding resins.

A deficiency of K^+ develops slowly if its basis is low intake of K^+ because renal excretion of K^+ falls to very low levels (10–15 mmol/day) in K^+-depleted subjects. Two major populations are at risk for low dietary K^+ intake: first, the urban poor because they do not consume the meat and vegetables that contain abundant K^+ (Table 10·2); second, populations who live in rural, near-equatorial regions, such as the northeast area of Thailand, because their diet consists mostly of polished rice (high in carbohydrates, low in K^+) and they do not ingest other foods rich in K^+.

Given the ability of the kidneys to diminish the excretion of K^+ so efficiently and so promptly, it should take weeks to months to lower the overall K^+ content by 100 mmol. Notwithstanding, a less severe restriction in K^+ intake can contribute to hypokalemia if there is a greater loss of K^+. This tendency becomes important when people lose K^+ by sweating excessively and when diuretics are given to people who eat a low quantity of foods that contain K^+ (e.g., the elderly who are hypertensive, see the discussion of Case 10·1, pages 379–80).

Table 10·2
EXAMPLES OF FOODS THAT SUPPLY 60 mmol OF K^+

Foods	Weight (g)
Vegetables	
Potatoes or beans	500
Peas	5000
Fruits	
Bananas, canteloupe	800
Oranges	1200
Meats	
Beef or chicken	600

Question

(Discussion on pages 384–85)

10·3 A patient takes a diuretic, which causes the excretion of an extra 60 mmol of K^+ per day. The doctor recommended that this patient consume bananas to replace the K^+ lost each day. If all the kcal were retained as stored fat (9 kcal/g), what would be the weight gain at the end of one year?

INCREASED SHIFT OF K^+ INTO CELLS

The factors promoting a shift of K^+ into cells are listed in Table 10·3. In general, these factors, when acting individually, lead to a fall in serum $[K^+]$ of less than 1 mmol/l, but when acting in combination, with very high doses of drugs, or in the right clinical setting (e.g., recovery from DKA), a much larger fall in the plasma $[K^+]$ can occur.

In the paragraphs to follow, the factors we shall discuss are the influences of the acid-base state, the actions of hormones, the role of anions in the ICF, and the loss of K^+ by nonrenal and renal routes.

Table 10·3
CAUSES OF A SHIFT OF K^+ INTO CELLS

Acid-base disorder
- Metabolic alkalosis

Acting via hormones
- Insulin, β_2-adrenergic agonists, possibly aldosterone, α-adrenergic antagonists

Anabolism
- Growth, recovery from diabetic ketoacidosis, total parenteral nutrition, red blood cell synthesis (e.g., recovery from pernicious anemia)

Rare disorders
- Hypokalemic periodic paralysis

Other
- Anaesthesia? (see the margin)

Anaesthesia:
In animals, the induction of anesthesia leads to a significant fall in the plasma $[K^+]$ that is independent of the acid-base status. It may be due to catecholamine release and/or a direct membrane effect of anaesthetics. No comparable data are available in humans.

Influence of the Acid-Base States

- In metabolic alkalosis, K^+ shift into cells; no shift of K^+ occurs in respiratory alkalosis.

A shift of K^+ from the ECF into cells should occur as a result of the movement of H^+ in the opposite direction (Figure 10·3). Since the magnitude of this H^+ shift is much smaller with respiratory acid-base disorders, hypokalemia in a patient with respiratory alkalosis is probably due to causes other than this acid-base disorder. In contrast, K^+ do shift into cells during metabolic alkalosis.

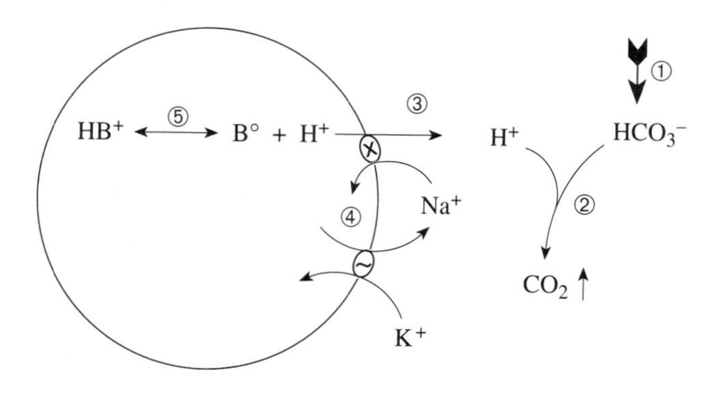

Figure 10·3
The shift of K^+ from the ECF to the ICF
The addition of HCO_3^- (1) causes the $[H^+]$ in the ECF to fall (2). This fall causes Na^+ to enter the ICF in conjunction with H^+ export on the Na^+/H^+ antiporter (3). The $Na^+K^+ATPase$ (4) pumps the Na^+ out of cells while pumping K^+ into cells. Continued flux requires back-titration of buffers in the ICF (5). There is no way to predict whether Na^+ or K^+ will remain in the ICF when H^+ exit; the "decision" is made by the K_m for Na^+ of the $Na^+K^+ATPase$ (i.e., does this K_m change?).

Actions of Hormones

- Hormone-induced K^+ shifts are due mainly to insulin and β_2-adrenergics.

Insulin: Insulin promotes the entry of K^+ into the ICF by making the RMP more electronegative. A more negative voltage in cells is the result of the electroneutral entry of Na^+ and the electrogenic export of Na^+ via the $Na^+K^+ATPase$ (see Figure 9·2, page 327, and the margin); actions of insulin also lead to an increase in the content of anions in the ICF (phosphate esters such as RNA). The major clinical setting in which this action is observed is following the administration of insulin to treat patients with diabetes mellitus in poor control. The fall in plasma $[K^+]$ can be quite large (approximately 90 minutes after insulin administration).

Note:
By activating the electroneutral Na^+/H^+ antiporter, insulin will cause a greater flux of Na^+ through the electrogenic $Na^+K^+ATPase$. Very large doses of insulin can activate the Na^+/H^+ antiporter.

β_2-Adrenergic activity: β_2-Adrenergics lead to the entry of K^+ into cells. Elevated levels of these hormones are seen during the response to hypotension, stress, exercise, hypoglycemia, and in the clinical setting of delirium tremens, among others. Drugs with these actions are commonly given to alleviate bronchospasm. With the usual dose of β_2-agonist, the plasma $[K^+]$ can fall by 0.2–0.3 mmol/l, whereas massive overdoses cause a fall of almost 1.5–2.0 mmol/l. In contrast, β_2-antagonists cause a much smaller rise in the plasma $[K^+]$ (0.3 mmol/l).

Aldosterone: Hypokalemia in patients with high aldosterone activity is due primarily to renal K^+ loss rather than a shift of K^+ into cells. Nevertheless, in patients with a deficiency of aldosterone, there may also be a diminished capacity to retain K^+ in the ICF (mechanism unknown).

Gain in ICF anions: During acute anabolic states, such as recovery from diabetic ketoacidosis or when malnourished subjects are given a nutritional supplement, organic phosphate anions accumulate in cells (e.g., RNA, DNA, phospholipids); for electrical balance, K^+ will also return to cells, and hypokalemia will develop if the intake of K^+ is not increased appropriately.

Questions

(Discussions on pages 385–86)

10·4 Elite athletes often have a $[K^+]$ in plasma that is close to 3.0 mmol/l at rest. What mechanisms might be involved?

10·5 If a subject is K^+-depleted and has lost 300 mmol of K^+, what is the most likely change in ionic composition of the ICF and what adjustments were needed to permit this change?

10·6 The lesion in hypokalemic periodic paralysis is a higher open probability of Na^+ channels in the plasma membrane. How might this lesion cause hypokalemia? In what way is this lesion different from that in hyperkalemic periodic paralysis, in which Na^+ channels also have a more open probability?

Note:
These questions are for the more curious.

ENHANCED LOSS OF K^+

- Renal excretion of K^+ is the major type of K^+ loss.
- Certain types of diarrhea cause hypokalemia via actual GI K^+ loss.

Obligate loss of K^+:

- Nonrenal
  - sweat (close to 10 mmol/l, but volume can vary from 0.2 to 12 liters/day)
  - stool loss (100 mmol/l × 0.1 liters/day)
  - diarrhea ($[K^+]$ close to 40–50 mmol/l)
- Renal
  - minimum of 10 mmol of K^+ per day

Gastrointestinal Tract

The $[K^+]$ of upper gastrointestinal secretions is usually 15 mmol/l or less (Table 10·4). Since gastric secretions have such a low $[K^+]$, hypokalemia due to vomiting or nasogastric suction actually results from K^+ loss in the urine (also see the discussion of Question 10·7). In contrast, loss of colonic contents can lead directly to hypokalemia. When diarrhea is due to more distal colon involvement (villous adenoma of the rectum), the K^+ content might be two or three times higher. Suspect that diarrhea is the cause of hypokalemia when metabolic acidosis with a normal plasma anion gap is present and the rate of excretion of NH_4^+ is not low (greater than 100 mmol/day).

Note:
In cholera, for example, 6 liters of water containing 750 mmol of Na^+ and 100 mmol of K^+ can be lost in a 24-hour period.

Laxative abuse:
The loss of K^+ is much larger when a phenolphthalein type of laxative is used instead of an osmotic or bulk type of laxative.

Table 10·4
REPRESENTATIVE ELECTROLYTE CONTENTS AND VOLUMES OF UPPER GASTROINTESTINAL SECRETIONS

Site	Volume liters/day	$[Na^+]$ mmol/l	$[K^+]$ mmol/l	$[Cl^-]$ mmol/l	$[HCO_3^-]$ mmol/l
Gastric	1.5	20	10	130	0
Duodenal	3–8	110	15	115	10
Pancreas	0.5	140	5	30	115
Bile duct	0.5	140	5	100	25
Jejunal	3.0	140	5	100	20
Ileal	0.5	80	10	60	75

Table 10·4 Legend
Values are representative for a typical 70-kg adult.

Questions

(Discussions on pages 386–87)

10·7 What is the cause of hypokalemia in patients who vomit and have metabolic alkalosis?

10·8 When a subject starves for three weeks, there is a negative balance for K^+ of close to 300 mmol, but the subject is normokalemic. What changes may have occurred in cells to cause this intracellular K^+ deficit?

10·9 Should the deficit of K^+ in a starved subject be replaced during the fast?

Urinary K^+ Loss

- Excretion of K^+ is high when there is either a higher $[K^+]$ in the CCD or higher volume of fluid traversing the terminal CCD.
- Although hypokalemia can result from decreased intake of K^+, shift of K^+ into cells, or K^+ loss, virtually all cases with chronic hypokalemia have increased renal K^+ loss.

Note:
This mechanism requires release of aldosterone.

From a clinical perspective, patients with hypokalemia on a chronic basis usually have hypokalemia caused by excessive renal excretion of K^+. This form of hypokalemia should be suspected if patients were treated with diuretics (usually for hypertension or edema states) or if vomiting or nasogastric suction was prominent in the clinical picture (see the margin). As outlined earlier, the two components of the K^+ excretion formula, $[K^+]$ and the flow rate in the CCD, should be examined independently.

A high $[K^+]$ in the CCD is seen with:
- mineralocorticoids,
- bicarbonaturia,
- low urine $[Cl^-]$.

High $[K^+]$ in the urine: A TTKG that is greater than 4 in a patient with hypokalemia indicates that certain factors have raised the $[K^+]$ in the terminal CCD. These factors form part (or all) of the basis for the hypokalemia. In this setting, we look for stimulators of this process: first, aldosterone or compounds with aldosterone-like actions (Table 10·5 and page 370); second, bicarbonaturia; and third, a very low $[Cl^-]$ in the luminal fluid in the CCD.

1. **Plasma aldosterone:**
   In the patient with excessive renal loss of K^+, hypertension, and a normal ECF volume, the plasma renin and aldosterone levels help one to differentiate adrenal from nonadrenal causes of hyperaldosteronism (Table 10·5). The diagnostic efficacy of these hormone assays is improved by expanding the ECF volume with saline to suppress endogenous renin release; one should make these measurements while the patient is recumbent (when the patient is in the upright position, the ECF volume redistributes and thereby stimulates the release of renin).

2. **Bicarbonaturia:**
   Bicarbonate in the lumen of the CCD causes the $[K^+]$ to rise markedly at this location if aldosterone is also present (e.g., vomiting). The mechanism is not clear but could involve inhibition of reabsorption of Cl^- at this nephron site (see Chapter 9, page 335).

3. **A low $[Cl^-]$ in the luminal fluid of the CCD:**
   When Na^+ are delivered to the CCD with an anion other than Cl^-, reabsorption of Na^+ will be electrogenic. This electrogenic reabsorption makes the lumen more negative and thereby increases the $[K^+]$ in the lumen of the CCD (provided that Cl^- are not present in appreciable amounts).

Table 10·5
CAUSES OF A HYPERMINERALOCORTICOID STATE

Endogenous release of aldosterone
Associated with increased renin levels
- Low effective circulating volume usually caused by diuretics or vomiting, less commonly the result of intrinsic renal disorders such as Bartter's syndrome
- Renal artery stenosis
- Juxtaglomerular apparatus hypertrophy or tumor

Associated with low renin levels
- Primary adrenal cortical hyperplasia or adenoma
- Defects in enzymes of the adrenal gland

Nonaldosterone-mediated (associated with low renin levels)
Endogenous compounds with mineralocorticoid effects
- Cushing's syndrome or tumors that secrete excessive ACTH
- Dexamethasone-suppressible release of of aldosterone
- Defects in enzymes of the adrenal gland
- Apparent mineralocorticoid excess syndrome
- Etiology unclear (e.g., magnesium deficiency)

Exogenous compounds
- Administered steroids, such as high doses of hydrocortisone
- Licorice, swallowed chewing tobacco, carbenoxolone, which block 11β-hydroxysteroid dehydrogenase (see page 370)
- Amphotericin B

Questions

(Discussions on page 387)

10·10 Can excessive delivery of Na^+ to the CCD augment the rate of excretion of K^+?

10·11 How do poorly reabsorbed anions lead to an augmented excretion of K^+?

Clinical Features

- Unless the plasma [K^+] is less than 3 mmol/l, few symptoms will be present.

SYMPTOMS

Symptoms have a high degree of variability from patient to patient. A classification of these symptoms with respect to the organ affected is shown in Table 10·6.

Table 10·6
SYMPTOMS OF HYPOKALEMIA

Muscular
- Cardiac muscle
  - Arrhythmias (especially if digitalis or heart disease is present)
- Skeletal muscle
  - Weakness, rhabdomyolysis with extreme K^+ depletion, cramps, myalgias
- Smooth muscle
  - Intestinal disturbances (constipation, ileus)

Renal
- Concentrating defect (polyuria, nocturia)
- Medullary interstitial disease (a direct consequence of K^+ deficiency and a result of enhanced NH_3 availability)

Neurologic
- Thirst, decreased deep tendon reflexes, paresthesias

Cardiac Muscle

Severely hypokalemic patients are likely to develop a number of cardiac arrhythmias. This tendency is important if the patient is on digitalis or has underlying myocardial disease.

Note:
Some individuals with marked degrees of hypokalemia do not have weakness; we do not know how to explain this difference.

Skeletal Muscle

The most common complaint related to hypokalemia is weakness or easy fatigability. Symptoms become prominent at a plasma [K^+] below 3.0 mmol/l. Low levels can ultimately lead to paralysis. This consequence is especially important during metabolic acidosis because respiratory muscle weakness can compromise ventilation and result in a life-threatening fall in pH in cells. The pattern of muscle weakness caused by K^+ deficiency seems to be most evident in the lower extremities. With time and/or continuing K^+ loss, the trunk and upper extremities become involved. Rarely, patients may also complain of cramps, muscle tenderness, and tetany.

With a severe degree of hypokalemia, the patient is at greater risk of developing *rhabdomyolysis*. If hypokalemia remains untreated for a prolonged time, muscle atrophy could occur.

Rhabdomyolysis:
A disorder in which macromolecules in skeletal muscle are hydrolyzed and retained; they may also be released. This disorder occurs more frequently in patients with hypokalemia and hypophosphatemia. Interestingly, it is not a common cause of severe hyperkalemia.

Smooth Muscle

With hypokalemia, GI motility might decline and lead to constipation and, in the extreme, an ileus (see the margin).

Clinical pearl:
If patients with hypokalemia and low GI motility are given K^+ supplements by the oral route, K^+ may accumulate in the GI tract. Later, these K^+ may be reabsorbed and provide a sudden K^+ load with a resulting hyperkalemia. Chronic hypokalemia retards subsequent excretion of K^+ for 24 hours or so.

Renal Problems

Polydipsia is a common complaint during severe hypokalemia and may be due to a more direct effect on the thirst center. Because of a reduction in renal concentrating ability during severe hypokalemia, polyuria and nocturia may develop. These symptoms occur when the K^+ deficiency is chronic and severe (greater than 200 mmol). The renal lesion may ultimately become irreversible because of interstitial fibrosis and tubular damage. In contrast, the ability to excrete a dilute urine is relatively preserved during hypokalemia.

Other Factors

The central nervous system seems to be spared in hypokalemia. The $[K^+]$ in the cerebrospinal fluid is constant despite wide swings in the plasma $[K^+]$.

ECG CHANGES DURING HYPOKALEMIA

Hypokalemia is associated with a more negative RMP. Nevertheless, although ECG changes occur, they correlate poorly with the plasma $[K^+]$. In general, the earlier changes include T-wave flattening or even inversion. The S-T segment becomes depressed and may have a "trough-like" appearance. A "U" wave may also appear. In addition, severe hypokalemia tends to promote arrhythmias and conduction disturbances. The Q-T interval is normal or it may be prolonged during hypokalemia, and this difference may help distinguish the ECG effects of hypokalemia from those seen with digitalis toxicity. The ECG abnormalities of hypokalemia are exaggerated if hypercalcemia is also present.

Note:
Hypokalemia is often associated with hypomagnesemia. This combination of abnormalities may lead to polymorphic ventricular tachycardia (Torsade de pointe).

IMPORTANCE OF ECF VOLUME STATUS AND/OR HYPERTENSION

From a clinical perspective, it is helpful to subdivide patients presenting with hypokalemia based on their ECF volume and blood pressure (Figure 10·4, page 366).

Contracted ECF Volume and Lower Blood Pressure

Those patients with a contracted ECF volume will commonly have vomiting or diuretic use (abuse) as the most likely basis (see the discussion of Question 10·1, pages 383–84). Less often, Bartter's syndrome may be present.

Normal or Expanded ECF Volume and Hypertension

In those patients who do not have a contracted ECF volume, the presence of hypertension suggests that hyperaldosteronism is present; its basis should be sought.

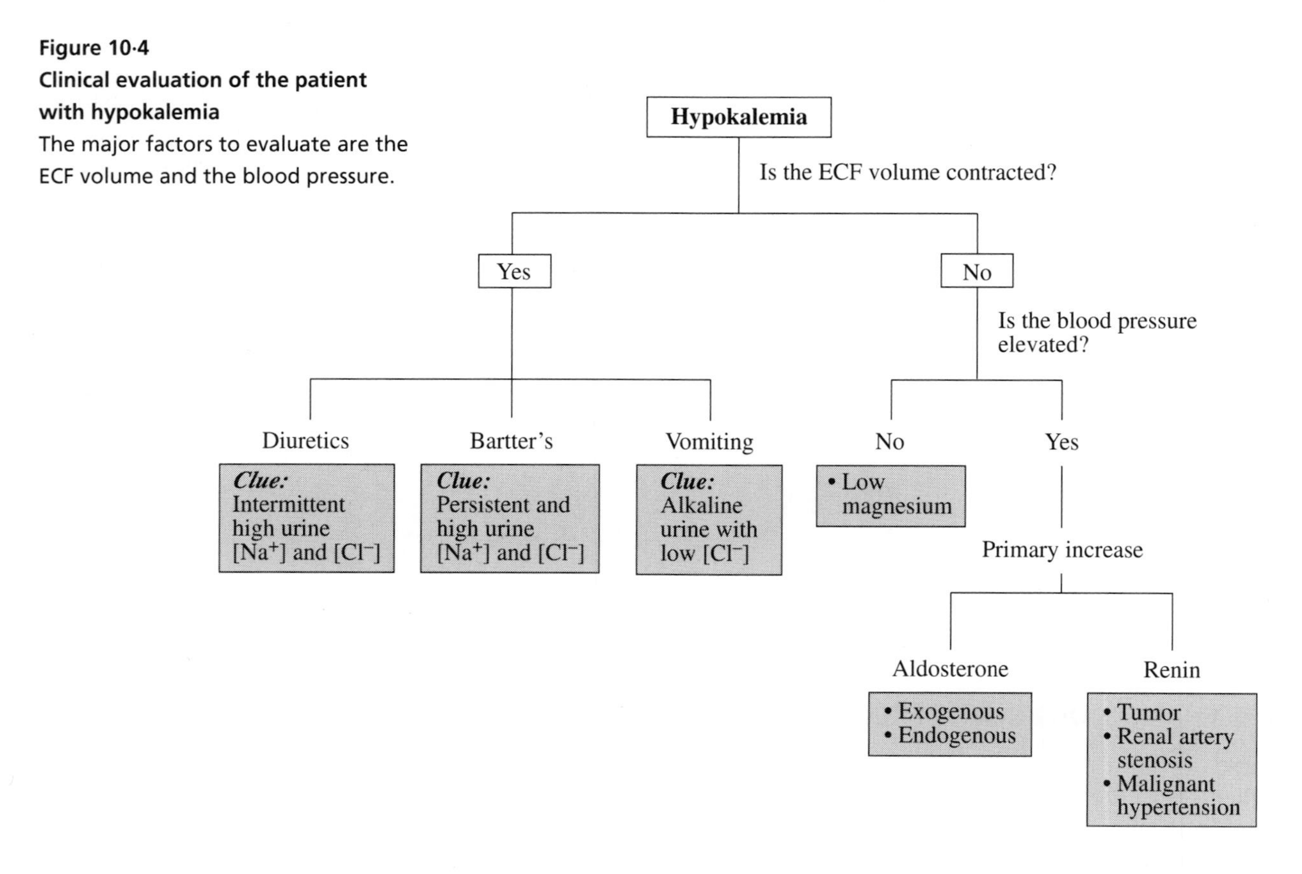

Figure 10·4
Clinical evaluation of the patient with hypokalemia
The major factors to evaluate are the ECF volume and the blood pressure.

Patients with Mg^{2+} depletion have high aldosterone levels despite hypokalemia. The mechanism for aldosterone release is not clear. Suspect Mg^{2+} depletion if there is hypocalcemia, alcoholism, and/or a GI problem, or if the patient was treated with drugs that cause Mg^{2+} wasting (large doses of loop diuretics, cisplatin, or aminoglycosides). Mg^{2+} depletion also occurs frequently in Bartter's syndrome.

[K^+] in the CCD:
This concentration is the [K^+] in urine divided by the urine/plasma osmolality, provided that the urine osmolality equals or exceeds that of plasma.

Note:
To deliver a small quantity of Cl^- to the CCD, the stimulus of a low effective circulating volume must be present, together with the presence of a large delivery of Na^+ with anions other than Cl^- (e.g., carbenicillin, hippurate, or sulfate). Another way to reduce the reabsorption of Cl^- is if permeability to Cl^- in the CCD is diminished.

Application of Basic Physiology to the Bedside

- This section is speculative and requires "faith" in possible physiologic mechanisms.

A HIGH [K^+] IN THE LUMEN OF THE CCD

For there to be an unduly high [K^+] in the lumen of the CCD, the electrogenic reabsorption of Na^+ must be augmented (a higher TEPD). This augmentation, from an ion movement point of view, means that there was either more reabsorption of Na^+ or less reabsorption of Cl^- in the CCD (see the margin, Figure 10·5, and the review flow chart on page 355).

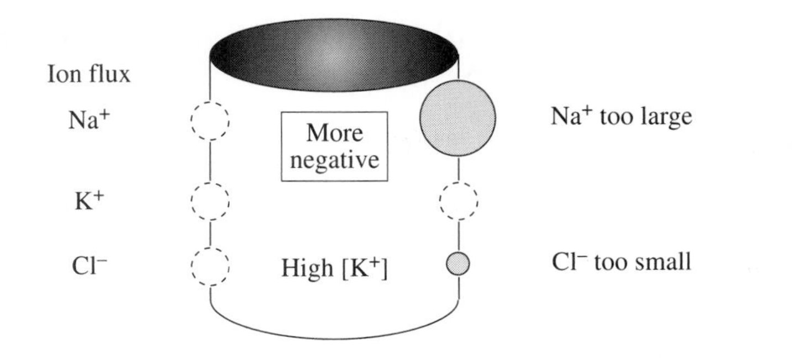

Figure 10·5
Causes of increased [K+] in the lumen of the CCD
The "normal size" of the ion channels or conductive pathways responsible for the net secretion of K+ are represented on the left side of the figure. The major factors that lead to a high [K+] in the CCD are increased Na+ reabsorption and decreased Cl− reabsorption (shaded circles shown on the right side). Both lead to an increase in the lumen-negative TEPD. Plasma renin levels and urine electrolytes help in confirming clinical suspicions.

High Conductance for Na^+ in the CCD

High mineralocorticoid activity causes the Na^+ channel to have an open probability that is greater than normal (Table 10·5). Liddle's syndrome is a rare disorder that results from a greater open probability of Na^+ channels in the CCD. Another form of excessive Na^+ permeability is the insertion of a nonspecific cation channel, as occurs in amphotericin B-induced nephrotoxicity; this channel enables Na^+ reabsorption independent of factors that regulate the normal Na^+ channel; it is thus insensitive to amiloride.

As a result of the increased reabsorption of Na^+, there should be a tendency for ECF volume expansion, low levels of renin, and perhaps even hypertension to go along with the hypokalemia and the high TTKG. The other finding is the ability to excrete a Na^+-poor and Cl^--poor urine when the ECF volume is contracted. With ECF volume contraction, renin levels should rise.

Criteria:
1. Hypokalemia, high TTKG
2. Increased ECF volume and low renin
3. Urine can be Na^+-poor if ECF volume decreases

Amphotericin B:
A drug used to treat certain fungal infections.

Lower Reabsorption of Cl^- in the CCD

An increase in the electrochemical driving force for K^+ excretion can occur if there is reabsorption of Na^+ without enough Cl^-; this lower reabsorption of Cl^- may be due to a reduced delivery of Cl^- or a reduced permeability of the luminal membrane of the CCD to Cl^-. Reduced delivery of Cl^- occurs in rare circumstances (marked ECF volume contraction and a high rate of excretion of urea). With normal Na^+, but a relatively low reabsorption of Cl^- in the CCD, there should be hypokalemia and an excessively high $[K^+]$ in the urine (a very high TTKG) along with Na^+ and Cl^- in the urine even if the patient has ECF volume contraction. There is a tendency to ECF volume contraction and hyperreninemia, and, because of the hypokalemia and ECF volume contraction, metabolic alkalosis. Because of their hypovolemia, these patients should not be hypertensive despite high levels of angiotensin II. In many ways, these features resemble Bartter's syndrome.

Criteria:
1. Hypokalemia, high TTKG
2. Low ECF volume and high renin
3. Urine $[Na^+]$ high when ECF volume decreases

HIGH FLOW RATE IN THE CCD

The two major reasons for a high flow rate in the CCD are a water diuresis and an osmotic diuresis (see the margin). The main factor controlling flow rate in the CCD in the presence of ADH is the rate of delivery of osmoles. The two major groups of osmoles are urea and electrolytes. The former primarily reflects the load of protein ingested, and the latter largely reflects the ingestion of salt or the use of diuretics. It is possible that both a high flow rate and a high $[K^+]$ in the CCD may be present in a given patient.

Note:
A water diuresis does not cause a high rate of excretion of K^+ even if aldosterone levels are high; perhaps this feature reflects the requirement of ADH for normal K^+ conductance in the luminal membrane of the CCD.

Specific Disorders in which Hypokalemia Is a Prominent Feature

BARTTER'S SYNDROME

The cardinal features of Bartter's syndrome are hypokalemia, a variable degree of renal Na^+ wasting, a contracted ECF volume, metabolic alkalosis, hyperreninemia, and, in many cases, a deficiency of Mg^{2+} caused by excessive excretion of this cation. Secondary features include hypertrophy of the juxtaglomerular apparatus and a resistance to the vasoconstrictor actions of angiotensin II.

The pathophysiology of Bartter's syndrome has not yet been clarified. The following hypothetical lesions have been proposed:

1. Indirect techniques have suggested impaired reabsorption of NaCl in earlier nephron segments, the most important of which is in the thick ascending limb of the loop of Henle. This possibility is attractive because it could account for the wasting of K^+, Na^+, and Mg^{2+} (Mg^{2+} is reabsorbed in this area). Nevertheless, the components of K^+ excretion differ in Bartter's syndrome and in chronic loop diuretic administration (a model of reduced Na^+ and Cl^- reabsorption in the loop of Henle). With loop diuretics, the kaliuresis is due primarily to an increased flow in the CCD plus a small increase in the $[K^+]$ in the CCD (TTKG close to 6) when aldosterone acts in this setting (stimulated by the high levels of renin that occur in response to ECF volume contraction). In Bartter's syndrome, on the other hand, kaliuresis is due to a high $[K^+]$ in the CCD and does not require aldosterone (it is not affected by adrenalectomy); the TTKG is extremely high, in the 20–30 range. This high TTKG is noteworthy because it appears in the presence of Cl^- in the urine, may occur in the absence of bicarbonaturia, and implies a more negative voltage in the lumen of the CCD.

2. Our speculation to relate the lesion in the loop of Henle and in the CCD is as follows: the lesion in the loop of Henle leads to enhanced delivery of electrolytes to the CCD. In addition, there is hypomagnesemia. In the CCD, it appears from preliminary data that hypomagnesemia causes a high TTKG. This speculation is based on the fact that the TTKG and $[K^+]$ in the CCD fall markedly, as does K^+ excretion, when enough Mg^{2+} are infused to cause a high normal level of Mg^{2+} in plasma. Thus, the link between the loop of Henle and the CCD in Bartter's syndrome could be hypomagnesemia.

   From a pathophysiologic viewpoint, an exceedingly high TTKG could be the result of either a Na^+ channel in the CCD with a higher open probability or a reduced permeability to Cl^- in the luminal membrane of the CCD (Figure 10·5 and the review flow chart on page 355). If the former is the case, one would expect to see hypokalemia, an ECF volume that is normal or high, renin levels that are not elevated, and the ability to excrete a urine that is devoid of Na^+ and Cl^- with ECF volume contraction. None of these features reflects the situation in Bartter's syndrome; therefore, we favor a reduced permeability to Cl^- in the CCD because it could lead to hypokalemia, Na^+ wasting, a high level of renin, and the obligate excretion of Na^+ and Cl^- when the ECF volume is contracted.

Therapy for patients with Bartter's syndrome is not satisfactory. Usually, high intake of KCl provides only partial benefit. A high-salt diet is beneficial for the low ECF volume but may be detrimental for K^+ balance (see the discussion of Question 10·12). Other measures advised, such as *nonsteroidal anti-inflammatory drugs* (NSAIDS), may lead to long-term problems (interstitial nephritis), so the minor short-term benefit must be weighed against this theoretical detrimental effect. Although dietary supplements of Mg^{2+} make sense, absorption is not sufficient to achieve a therapeutic benefit; high doses cause diarrhea. The other therapeutic option is the use of the drug amiloride, but this K^+-sparing diuretic may only be of modest benefit.

Nonsteroidal anti-inflammatory drugs (NSAIDS):
The rationale for NSAIDS is that they depress the synthesis of prostaglandins of the E_2 family, and PGE_2 levels are raised in states with marked deficiency of K^+.

Questions

(Discussions on page 388)

10·12 In a patient with Bartter's syndrome, in what ways will a greater salt intake increase the rate of excretion of K^+? What are the dangers of giving a small load of NaCl to this patient?

10·13 When a patient presents with all features characteristic of Bartter's syndrome, a clinician must rule out surreptitious use (abuse) of diuretics, vomiting, and laxative abuse. What features should a clinician rely on in each of these settings?

HYPOKALEMIA FOLLOWING VOMITING AND DIURETICS

Hypokalemia as a result of vomiting and the use of diuretics was considered in detail in Chapter 4. During vomiting, the excessive loss of K^+ is due to a very high $[K^+]$ in the CCD that results from both the high level of aldosterone and, of special importance, the presence of bicarbonaturia.

With diuretics, the high levels of aldosterone are secondary to ECF volume contraction; it is associated with a high volume delivery to the CCD. These factors lead to an augmented rate of excretion of K^+. Nevertheless, the degree of hypokalemia is generally quite mild (3.0–3.5 mmol/l). Larger degrees of hypokalemia can occur if the dietary intake of K^+ is very low (e.g., in the aged; see the discussion of Case 10·1, page 379–80) or if hyperaldosteronism is present for other reasons (e.g., primary hyperaldosteronism).

PRIMARY HYPERALDOSTERONISM

A tumor of the adrenal cortex leads to high levels of aldosterone, ECF volume expansion, hyporeninemia, and hypertension. The characteristic electrolyte finding is hypokalemia in the range of 3.0 mmol/l. These findings suggest that the lesion in the CCD is a Na^+ channel with a greater open probability. The patients are in K^+ balance because they excrete the usual K^+ content of the diet. Typical values for the TTKG are close to 6 and perhaps reflect the influence of hypokalemia and K^+ depletion. Treatment is surgical removal, if feasible. Medical therapy includes aldosterone antagonists such as spironolactone.

11β-HYDROXYSTEROID DEHYDROGENASE AND HYPOKALEMIA

11β-Hydroxysteroid dehydrogenase (11β-HSDH) exists in cells that respond to mineralocorticoids. This enzyme converts cortisol to an inactive product so that cortisol will not occupy the aldosterone receptor illicitly and mimic the actions of aldosterone in these cells. Cortisol, if permitted to enter and accumulate in principal cells of the CCD, would bind to the aldosterone receptor and lead to activation of the K^+ secretory process.

Cortisol is present in much higher levels than is aldosterone. Therefore, the enzyme 11β-HSDH exists in principal cells to inactivate cortisol before it can "contaminate" the aldosterone message (see the margin). There are three circumstances in which cortisol might have aldosterone-like actions. First, if the enzyme 11β-HSDH has a low activity or is absent on a congenital basis, principal cells will behave as if they were continuously exposed to aldosterone because of the high intracellular levels of cortisol (Figure 10·6). This condition is called the "apparent mineralocorticoid excess syndrome." Second, at excessively high levels of cortisol (as may occur with ACTH-producing tumors), this hormone may "overwhelm" the activity of 11β-HSDH and activate the aldosterone receptor because not all of the intracellular cortisol has been destroyed. Third, 11β-HSDH may be inhibited, as occurs when licorice is consumed, when chewing tobacco is swallowed, or when the drug carbenoxolone is ingested.

Note:
A receptor has a much higher affinity for a ligand than does an enzyme. Therefore, 11β-HSDH must act before cortisol enters the cytoplasm.

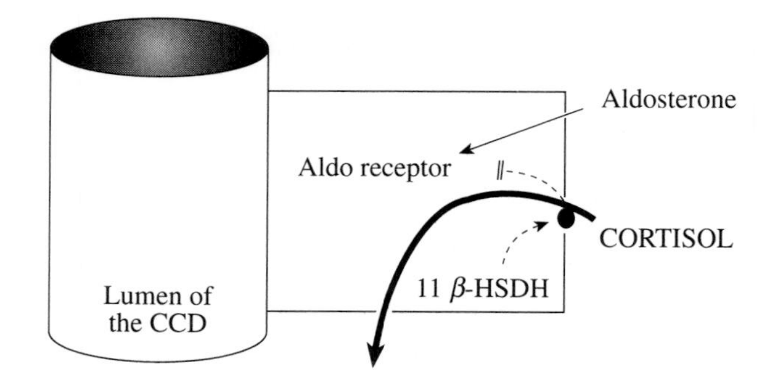

Figure 10·6
Influence of 11β-hydroxysteroid dehydrogenase on "aldosterone-like" actions in the CCD
Cortisol has a very high affinity to the aldosterone receptor. As cortisol enters principal cells of the CCD, the enzyme 11β-HSDH inactivates it before it can bind to the aldosterone receptor. There are three circumstances in which cortisol will successfully bind to the aldosterone receptor: when there is a deficiency of 11β-HSDH (apparent mineralocorticoid excess syndrome), when an inhibitor of 11β-HSDH is present (e.g., licorice), and possibly when there is an excess supply of cortisol relative to the normal activity of 11β-HSDH (perhaps ectopic production of ACTH by a tumor).

DEXAMETHASONE-SENSITIVE HYPERALDOSTERONISM

A chromosomal crossover has been recently documented: ACTH now promotes the synthesis of aldosterone and other steroids in addition to the usual glucocorticoids. Hypokalemia is often absent. Suppressing ACTH with dexamethasone ameliorates the hypertension and the hypokalemia, if present. It has been suggested that steroids other than aldosterone may contribute to the hypertension, which may be unduly severe.

LIDDLE'S SYNDROME

Note:
See page 355 for a review flow chart on the causes of hypokalemia with excessive excretion of K^+.

In 1963, Liddle described a patient with chronic hypokalemia, excessive excretion of K^+, metabolic alkalosis, and a severe degree of hypertension associated with hyporeninemia and near-absent levels of aldosterone. There were no abnormal levels of steroid hormones. The TTKG was in the range of 10, and the flow rate in the CCD was normal, as judged by the osmole excretion rate. The urine contained a small quantity of Na^+ and Cl^- when the ECF volume was contracted. Taken together with the hyporeninemia, the lesion behaved as if there was more open probability of the Na^+ channel in the CCD. Moreover, the above abnormalities responded to blockers of the Na^+ channel in the CCD but not to aldosterone antagonists. Of extreme interest, renal transplantation relieved all of these abnormalities.

HYPOKALEMIC PERIODIC PARALYSIS

Hypokalemic periodic paralysis is a rare disorder that results in a net shift of K^+ from the ECF to the ICF. It occurs more frequently in people of Asian descent, and, in many of these cases, it occurs in conjunction with hyperthyroidism. Affected individuals suffer abrupt attacks of mild to severe degrees of hypokalemia; these attacks often last 6–24 hours. The major symptoms relate to weakness, which might be severe enough in degree to cause paralysis. Some patients may develop a cardiac arrhythmia during an attack.

The etiology is unknown. Given the fact that meals rich in carbohydrate may precipitate an attack, an early hypotheses to explain the lesion was an enhanced sensitivity of the Na^+K^+ATPase to insulin, perhaps via a higher Na^+/H^+ antiporter flux. This sensitivity should make the RMP more electronegative, but the RMP is less electronegative, so this hypothesis is unlikely.

In other patients, the association of precipitation in response to stress, anxiety, exercise, cold, and hyperthyroidism has suggested a role for enhanced β-adrenergic sensitivity. Again, the less negative RMP makes this hypothesis unattractive.

More recent electrophysiologic studies have identified an abnormality of a Na^+ channel in the cell membrane. This Na^+ channel differs from that in hyperkalemic periodic paralysis in that it is not sensitive to tetradotoxin. The lesion involves the majority of Na^+ channels, which do not close appropriately with depolarization. Accordingly, many more Na^+ enter and are pumped out by the electrogenic Na^+K^+ATPase. The RMP lowers as a result (see the margin). If this action occurred in conjunction with a lower K^+ conductance, hypokalemia could be maintained for the period of time that the Na^+ channel had a greater open probability.

Note:
When a Na^+ enters via its ion-specific channel, one positive charge enters. In contrast, only one-third of a positive charge exits per Na^+ pumped out of cells via the Na^+K^+ATPase. Thus, net entry of a positive charge depolarizes the RMP.

Therapy for this disorder is largely symptomatic or empiric. The patients are advised to avoid carbohydrate-rich meals and take β-blockers if symptoms are related to sympathetic activity. Hyperthyroidism is treated in the usual fashion. If this treatment fails, patients should take 250–750 mg of acetazolamide daily. Acetazolamide acts via unknown mechanisms; its benefits are based on empiric observations.

PART B
TREATMENT

Because hypokalemia may have many causes, there is no universal therapy. In this section, we shall consider only the replacement of the deficit of K^+. The urgency for this replacement is dictated by a number of factors (Table 10·7):

1. the presence of drugs (e.g., digitalis) or heart disease, which may increase the likelihood of a cardiac arrhythmia;
2. the possibility that K^+ will shift into cells (e.g., during recovery from diabetic ketoacidosis or in β_2-adrenergic administration);
3. the development severe weakness in the patient who must have maximum hyperventilation because of metabolic acidosis;
4. the severity of the deficit and the continuing K^+ losses.

At the bedside, it is helpful to monitor the ECG and assess the possibility of hypoventilation and the degree of weakness.

Table 10·7
INDICATIONS FOR INITIATING K^+ THERAPY DURING HYPOKALEMIA

Absolute indications
Digitalis therapy
Therapy for diabetic ketoacidosis
Presence of symptoms (e.g., respiratory muscle weakness causing hypoventilation)
Severe hypokalemia (< 2.0 mmol/l)
Strong indications
Myocardial disease
Anticipated hepatic encephalopathy
Anticipated increase in another factor that causes a shift of K^+ into the ICF (e.g., β_2-adrenergics)
Modest indications
Development of glucose intolerance
Mild hypokalemia ([K^+] closer to 3.5 mmol/l)
Need for better antihypertensive control

Quantity of K^+ to Administer

When considering the quantity of K^+ to administer, the aim should be to get the patient out of danger quickly but to replace the entire K^+ deficit more slowly (see the margin). How quickly can K^+ be administered? The consideration here is to deliver enough K^+ to the cell membrane of the heart before a life-threatening arrhythmia occurs. This approach requires an assessment of the dose required and proper mixing of K^+ in the venous blood (administration should be through a large central vein because K^+ are very irritating to smaller veins where the concentration must be raised considerably). Therefore, the emergency replacement of K^+ is by a central vein with the patient on a cardiac monitor.

Rise in [K^+] with therapy:
During chronic hypokalemia, renal mechanisms are called into play to minimize the aldosterone-induced rise in the TTKG; this effect may persist for 24–36 hours. Hence, aggressive therapy with K^+ can lead to retention of K^+ and an undue but transient degree of hyperkalemia.

CALCULATION OF THE BOLUS OF KCl TO ADMINISTER

If a 70-kg adult has a very severe degree of hypokalemia (1.5 mmol/l) and an abnormal ECG, the aim will be to raise the plasma [K^+] from 1.5 to 3.0 mmol/l in one minute. To do so, one should infuse 4.5 mmol of KCl over one minute into a central vein, a minimum estimate. The basis for this decision is as follows: the total blood volume (5 liters) circulates each minute (cardiac output is 5 liters/min). Because 60% of this blood volume is plasma (3 liters) and the target for the plasma [K^+] might be 3.0 mmol/l, then 1.5 mmol/l $\times$ 3 liters must be given over 1 minute. The infused K^+ will mix with interstitial fluid (4 $\times$ the plasma volume) before reaching the cell membrane, so there will be a much smaller rise in [K^+] next to cell membranes. Following this bolus, the rate of infusion of K^+ should be reduced to 1 mmol/min, and measurement of the plasma [K^+] should be repeated in 5 minutes. If the plasma [K^+] is still much less than 3.0, the above procedure can be repeated. The rate of infusion of K^+ should be much slower once the plasma [K^+] approaches 3.0 mmol/l.

REPLACEMENT OF THE DEFICIT OF K^+

- To correct severe hypokalemia, several hundred mmol of K^+ may be required.

It is difficult to relate accurately the degree of hypokalemia to the extent of total body K^+ deficit. A fall in the plasma [K^+] from 4 to 3 mmol/l is usually associated with a total body K^+ deficit of 100–400 mmol; a much larger deficit is required to lower the plasma [K^+] from 3 to 2 mmol/l. Nevertheless, if the patient also has a shift of K^+ into or out of cells, the magnitude of this deficiency is not accurately reflected by these numbers (see the margin).

Clinical pearl:
Although one has rough guidelines for relating the deficit of K^+ and the degree of hypokalemia, these are very unreliable in an individual patient. The "clinical trial" in a patient is critical and requires frequent observations.

Methods of K^+ Administration

ROUTE OF K^+ ADMINISTRATION

The safest route to give K^+ is by mouth. The conditions demanding intravenous therapy are gastrointestinal problems that limit intake or absorption of K^+, a severe degree of hypokalemia with either respiratory muscle weakness or cardiac arrhythmias (especially if digitalis is present), and an anticipated shift of K^+ into the ICF (recovery from diabetic ketoacidosis).

RATE OF ADMINISTRATION

The rate of K^+ infusion should not exceed 40–60 mmol/hr in most circumstances. The [K^+] in intravenous solution should be less than 60 mmol/l for use in peripheral veins because a higher [K^+] leads to local discomfort, venous spasm, or sclerosis. In general, do not administer K^+ in glucose-containing solutions because the glucose (via insulin) can lead to an initial lowering of the plasma [K^+]. As discussed above, when there is an acute medical emergency (cardiac arrhythmia, extreme weakness especially if it involves respiratory muscles), K^+ will have to be given more rapidly intravenously via a central vein. This situation requires extreme caution, cardiac monitoring, and frequent determination of the [K^+] in plasma.

PREPARATION OF K^+

The anions accompanying K^+ fall into three major classes. The choice depends on the clinical situation.

KCl

- The most common K^+ preparation used is KCl.

In all cases in which hypokalemia is associated with ECF volume contraction, Cl^- are essential to correct the K^+ deficit. Diuretics and vomiting (the two most common clinical causes of K^+ deficiency) are associated with a Cl^- deficit and ECF volume deficit. Because Cl^- permit Na^+ to be reabsorbed, the ECF volume can now be restored.

One can obtain KCl as a liquid or in solid form. Enteric-coated KCl can lead to the development of small intestinal ulcerations that are due to local irritation by hypertonic KCl. Because the liquid form of KCl is relatively unpalatable, KCl in a wax matrix or coated microspheres may be preferable.

$KHCO_3$

- Give $KHCO_3$ only if the patient needs the HCO_3^- load.

Ingesting K^+ salts of organic acids (acetate, citrate, gluconate, etc.) is equivalent to ingesting $KHCO_3$ because HCO_3^- are produced when these organic anions are metabolized. Administer these salts when the patient has lost $KHCO_3$ (e.g., in diarrhea). Obviously, administration of $KHCO_3$ is a poor choice for K^+ replacement if the patient has a high plasma [HCO_3^-] (metabolic alkalosis) or if the patient is unlikely to metabolize the accompanying anion (hypoxia, etc.).

K^+ Phosphate

- Give phosphate (< 6 mmol/hr) to ensure that K^+ will stay in the ICF during anabolism.

If the K^+ deficit is due to or associated with a loss of intracellular anions (phosphate), the K^+ deficit will be corrected only when phosphate is administered. Clinically, the need for K^+ phosphate is most evident in ICF anabolism during TPN or in the recovery from diabetic ketoacidosis. Nevertheless, clinicians may satisfy this need indirectly by giving KCl, while relying on the patient to eat the phosphate (plus Mg^{2+}, etc.) required for ICF replacement. Recall that a large phosphate load has the danger of inducing hypocalcemia and metastatic calcification.

Dietary K^+

The best way to replace K^+ deficits or to keep up with ongoing losses is to use the oral route. The diet should contain the phosphate, Cl^-, Mg^{2+}, and HCO_3^- needed. Foods rich in K^+ are fresh fruits, juices, and meats (Table 10·2). However, a disadvantage may be that these foods contain a significant number of calories—a problem for obese patients.

Controversy Concerning Hypokalemia and K^+ Therapy

The clinician must decide whether a hypokalemic patient needs to receive K^+ supplements. Apart from the replacement of existing deficits (to avoid more serious degrees of K^+ depletion), the basis of K^+ therapy is to prevent serious cardiac arrhythmias. On the opposite side of the argument, there is a significant cost: the real danger of hyperkalemia when patients are taking certain drugs (K^+-sparing diuretics, β-blockers), when an ileus is intermittent, or when renal function is poor. There is lack of proof that treating modest hypokalemia (3.0–3.5 mmol/l) really does diminish the cardiac arrhythmias and improve the patient's well-being. If the patient is taking a digitalis preparation, however, KCl treatment should be given. Furthermore, in other acute clinical situations in which there are factors that will cause a shift of K^+ into cells (treatment of severe metabolic acidosis), K^+ therapy for hypokalemia is mandatory. Therefore, at present, prudent patient selection should avoid most of the dangers of KCl therapy.

Question

(Discussion on page 389)

10·14 What is the time course and the expected degree of hypokalemia with diuretic therapy? At the usual doses, is the hypokalemia particularly severe with one class of diuretics?

PART C
REVIEW

Discussion of Introductory Case
Toby Is Losing More Than Weight

(Case presented on page 355)

What is her total body K^+ deficit?

Without a major shift of K^+ into the ICF, a 1 mmol/l fall in the [K^+] would imply that the deficit of K^+ is 100–400 mmol. In patients with high levels of catecholamines, a shift of K^+ into cells can occur without a deficit of K^+. Therefore, it is not possible to be sure of her actual deficit of K^+ at present; it may be as large as 400 mmol, but its magnitude will become clear during therapy.

Was a shift of K^+ into cells a major cause of hypokalemia?

A shift of K^+ into cells rarely causes a large change in the plasma [K^+]. Metabolic acidosis will not cause a fall in the [K^+] in the ECF. Of the hormones, insulin is not likely to be important because the blood sugar is normal; β_2-adrenergic activity secondary to hypotension could have caused some K^+ to shift into cells, but it is difficult to be quantitative here.

What factors could have contributed to a renal loss of K^+?

Aldosterone, bicarbonaturia, and a high rate of urine flow cause high rates of K^+ loss in the urine. ECF volume contraction should lead to aldosterone release and contribute to the high [K^+] in the random urine sample. Bicarbonaturia is not present, but it might have been present earlier in her history if she had induced vomiting to lose weight.

What tests might be helpful in this regard?

To establish the basis for renal K^+ loss, measure the urine flow rate, osmole excretion rate, and [K^+]. Because the TTKG is high, it seems that aldosterone is acting. The low plasma [HCO_3^-] in conjunction with a high urine [Cl^-] and a negative urine net charge (metabolic acidosis with increased renal NH_4^+ excretion) makes laxative abuse more likely. Alternatively, vomiting of fluid rich in intestinal secretion with or without the presence of achlorhydria could produce these results (see the margin). At this point, it will be necessary to have a chat with Toby to see if she needs more serious counselling regarding a self-induced metabolic disorder (long-term K^+ depletion can lead to progressive renal interstitial disease).

Note:
The blood pH may be different in young children who vomit if pyloric stenosis is present. If present, metabolic alkalosis is found (loss of HCl); metabolic acidosis is common (loss of $NaHCO_3$) with a patent pylorus.

Cases for Review

Case 10·1
Diuretics in the Elderly

(Case discussed on pages 379–80)

Emily is 73 years old; her usual lifestyle involves entertaining visitors in her home. She enjoys toast with jam along with her traditional cup of tea (a large cup). Visitors come because she is such a fun lady.

On her annual checkup, her doctor noted that her blood pressure was higher (170/95 mm Hg), so he prescribed a thiazide diuretic. Emily has not felt as well since she began taking this medication. She has become light-headed when she stands up, she feels weak, and she is less able to perform at her high intellectual level.

On physical examination, her blood pressure is now 150/90 mm Hg and she now has a postural fall in blood pressure of 15 mm Hg. No other abnormalities are observed.

The results of blood and urine tests are provided below.

		Plasma	Urine
Na^+	mmol/l	107	3
K^+	mmol/l	1.9	12
Cl^-	mmol/l	67	7
HCO_3^-	mmol/l	30	0
pH		7.47	5.1
Urea	mmol/l (mg/dl)	1.5 (4.2)	-
Glucose	mmol/l (mg/dl)	6.0 (108)	0
Osmolality	mosm/kg H_2O	220	405

What is the basis of the hyponatremia?
What is her anticipated K^+ deficit (her weight is 50 kg)?
What is her Na^+ balance?
What is her total body water balance?
What dangers should be anticipated during acute therapy?
What therapy should be recommended initially?
Compare her results with her son, Jerry, Case 7·1, pages 273, 275–77.

Case 10·2
I Told You I Do Not Vomit

(Case discussed on page 381)

Alicia has always been concerned about her body image but claims that she is more realistic now. She has vomited in the past but vigorously denies vomiting recently. Her food intake is now normal. She has no specific complaints.

When Alicia appeared for her routine physical exam, she was hypokalemic (2.7 mmol/l). Upon more detailed investigation, she denied vomiting and the use of diuretics or laxatives. Her blood pressure was on the low side (90/55 mm Hg) and she had clinical signs suggesting that her ECF volume was indeed contracted (low JVP, low blood pressure).

Laboratory values are provided below.

		Plasma	Urine
Na^+	mmol/l	138	36
K^+	mmol/l	2.7	61
Cl^-	mmol/l	99	57
HCO_3^-	mmol/l	28	0
Mg^{2+}	mmol/l	0.5	4
Creatinine	μmol/l (mg/dl)	88 (1.0)	-
Osmolality	mosm/kg H_2O	287	563

Do these urine values suggest that Alicia is a surreptitious vomiter or diuretic abuser?
In what way might the hypomagnesemia help with the diagnosis?
Why is her excretion of K^+ not low?
What is the final diagnosis?

Case 10·3
Hypokalemia in a Patient with Malignancy

(Case discussed on page 382)

Alfie, aged 57 years, has cancer of the lung with hepatic metastases. He has not been doing well and was admitted to the hospital. He is not taking any medications. The physical examination was not very helpful. The only recent new finding was an expanded ECF volume.

Laboratory investigation on admission revealed the following results.

		Plasma	Urine
Na^+	mmol/l	141	38
K^+	mmol/l	1.9	22
Cl^-	mmol/l	96	27
HCO_3^-	mmol/l	31	0
pH		7.43	6.0
Creatinine	μmol/l (mg/dl)	89 (1.0)	ND
Urea	mmol/l (mg/dl)	5.9 (16)	ND
Magnesium	mmol/l	0.7	ND
Osmolality		288	306

ND = Not done

Was the rate of excretion of K^+ high and, if so, what is the most likely reason for excessive excretion of K^+?
What could be responsible for the lesion?
What should be done for therapy? Focus on the hypokalemia.

Discussion of Cases

Discussion of Case 10·1 Diuretics in the Elderly

(Case presented on page 377)

What is the basis of the hyponatremia?

Emily has three components contributing to her hyponatremia.

1. **Water intake:**
   Emily always drank a relatively large volume by habit (her cup of tea). Now she may drink even more because of thirst secondary to her ECF volume contraction.

2. **Less water excretion:**
   There are two factors to consider in this regard. First, she has a higher urine osmolality than expected because of ADH actions. ADH was released as a result of low ECF volume. Second, she has a low osmole excretion rate, largely because of her low-protein diet. Her urine osmolality is 405 mosm/kg H_2O, and she is excreting only 200 mosmoles, so she can excrete only 0.5 liter of water in urine. In contrast, if she excreted the usual 800 mosmoles, she could excrete close to 2 liters of water per day. Her diet which is low in protein ("tea and toast") provides little urea for excretion (note the low concentration of urea in plasma). This low urea excretion limits her ability to excrete water by limiting the delivery of fluid to the CCD.

3. **Loss of Na^+:**
   As a result of the actions of the diuretic, she will lose Na^+ from her ECF; this loss will decrease the Na^+:H_2O ratio in plasma. She will also shift Na^+ from her ECF to her ICF as K^+ exit the ICF.

What is her anticipated K^+ deficit?

This deficit is difficult to assess. Without a major reason for a sudden shift of K^+ into cells, the K^+ deficit could easily be 200–400 mmol, perhaps even more! She is unlikely to lose a vast quantity of intracellular anions (largely phosphate esters such as DNA, RNA, phospholipids, ATP, creatine phosphate) in the absence of a major catabolic disease. For K^+ to leave the ICF and electroneutrality to be maintained, a large number of cations must enter cells. Since the degree of metabolic alkalosis is mild, she probably has a very large gain of Na^+ in her ICF (see the margin).

Note:
To have a higher [Na^+] in the ICF, the Na^+K^+ATPase activity must be low or, more likely, this enzyme must have a higher K_m for Na^+ in the ICF.

What is her Na^+ balance?

If her ECF volume is nearly normal (physical exam) and is normally 10 liters, she should have lost close to 330 mmol of Na^+ from her ECF (see the

Calculation of Na^+ in the ECF:

- Fall in $[Na^+]$ is 33 mmol/l (140 – 107 mmol/l).
- Original and current volume of ECF is 10 liters.
- Fall in content of Na^+ = 33 mmol/l × 10 liters = 330 mmol.

Calculation:

- Assume the content of osmoles in the ICF does not change.
- 2 $[Na^+]$ × normal ICF volume of 20 liters = number of particles normally in the ICF.
- 2 [140] × 20 liters = 2 [107] × new ICF volume.
  New ICF volume = 26 liters.

margin). Now, if her ICF gained 200–400 mmol of Na^+ during the K^+ deficit, she could be virtually in Na^+ balance but have a large maldistribution of Na^+—more in the ICF and less in the ECF.

What is her total body water balance?

Her ECF volume is near normal, but the plasma $[Na^+]$ indicates an expanded ICF volume. If we apply the formula described in the margin to calculate her ICF volume, we find that her ICF volume should be expanded by close to 6 liters.

What dangers should be anticipated during acute therapy?

The major dangers to anticipate are cardiac arrhythmias (from the hypokalemia) and excessively rapid water diuresis once ADH levels are suppressed (via reexpansion of her ECF volume); the latter could lead to a rapid loss of water in the urine and subsequently to osmotic demyelination with a rapid rise in the $[Na^+]$. Reexpansion of the ECF volume will come about equally well if NaCl is given or if KCl is given (K^+ enter cells and Na^+ exit; see Figure 10·1, page 354). Emily is elderly and her heart may not tolerate rapid expansion of her ECF volume. Therefore, a final risk is the development of congestive heart failure (CHF) subsequent to KCl therapy, which will probably be equivalent to NaCl administration as far as the ECF is concerned.

What therapy should be recommended initially?

If we keep in mind the above balance data, the dangers to anticipate, and the fact that Emily is not symptomatic from her hyponatremia, we may decide that she is not likely to need Na^+ now. Therefore, the initial recommendation should be the slow, oral replacement of her K^+ deficit with KCl. The dose of KCl to choose depends on the desired rate of correction of her hyponatremia. Because the initial aims are to replace the K^+ deficit and to raise the $[Na^+]$ in plasma by only 1 mmol/l/hr, 1 mmol of KCl should be given per liter of total body water for each of the first several hours. Emily has close to 35 liters of total body water, so 30–40 mmol of KCl should be given per hour as a starting therapy. We are looking for a positive balance of K^+, so urine losses should be monitored. In total, her plasma $[Na^+]$ should rise by 12 mmol/l in a 24-hour period, and her plasma $[K^+]$ should not be permitted to rise above 3.5 mmol/l on the first day. Her ECG should be monitored closely (for arrhythmias), as should her plasma $[K^+]$ (for hyperkalemia), her urine volume (for rapid water diuresis), and her plasma $[Na^+]$ (to see the expected rise in her $[Na^+]$). ADH should be given if her urine volume increases suddenly; such an increase could lead to too rapid a loss in water and ultimately to too rapid a rise in her plasma $[Na^+]$.

One should watch for the early signs of CHF (increasing JVP, crackles in the chest) and be prepared to use a diuretic. If the urine $[Na^+]$ is lower than her plasma $[Na^+]$, and she has a substantial diuresis, be careful of the rate at which the plasma $[Na^+]$ rises.

Discussion of Case 10·2
I Told You I Do Not Vomit

(Case presented on pages 377–78)

Do these urine values suggest that Alicia is a surreptitious vomiter or diuretic abuser?

Vomiting: These values do not suggest vomiting. The hallmark for a diagnosis of vomiting is near-absent excretions of Cl^- in the urine.

Diuretics: With a recent intake of diuretics, the urine [Na^+] and [Cl^-] may be high. Because the concentrations of Na^+ and Cl^- were both high in a patient with a low ECF volume, these values in the urine are typical for diuretic abuse. Diuretic abuse was the initial clinical diagnosis, but two facts made this diagnosis unlikely. First, every random urine sample contained a high [Na^+] and [Cl^-]. Second, urine assays for diuretics were consistently negative. Therefore, a renal salt-wasting syndrome is present and the nephron site responsible for the salt-wasting is not the CCD because of the very high TTKG (> 10).

In what way might the hypomagnesemia help with the diagnosis?

The principal nephron site that regulates the renal excretion of Mg^{2+} is the thick ascending limb of the loop of Henle. Because this patient has hypomagnesemia, renal excretion of Mg^{2+}, a low ECF volume, and renal excretion of Na^+ plus Cl^-, the thick ascending loop of Henle seems to be involved. The high TTKG is also consistent with this interpretation. Hypomagnesemia is associated with use of loop diuretics, GI disease, chemotherapy, and Bartter's syndrome.

Why is her excretion of K^+ not low?

During hypokalemia, the excretion of K^+ should be low. Alicia has a high rate of excretion of K^+ because the [K^+] in the urine is too high (TTKG is close to 11); it does not appear that her osmole excretion rate is excessive, but more data will be needed (urine flow rate and osmolality) to confirm this suspicion.

The cause of the high TTKG should be sought. Perhaps it is due to the presence of a high level of aldosterone. This hormone could be released as a result of ECF volume contraction. She has a high delivery of Na^+ with Cl^- to the CCD, and perhaps this increased delivery contributes to her unexpectedly high rate of excretion of K^+ and the high [K^+] in the urine. She behaves as if the luminal membrane of the CCD has a low permeability to Cl^-.

Note:
See page 355 for a review flow chart on the causes of hypokalemia with excessive excretion of K^+.

What is the final diagnosis?

We believe that she has adult-acquired Bartter's syndrome.

Discussion of Case 10·3
Hypokalemia in a Patient with Malignancy

(Case presented on pages 378–79)

Was the rate of excretion of K^+ high and, if so, what is the most likely reason for excessive excretion of K^+?

Rate of excretion of K^+: The helpful data provided are the urine [K^+] (22 mmol/l) and the urine osmolality (306 mosm/kg H_2O). Because normal adults excrete 600–900 mosmoles per day at a relatively consistent rate and the random urine osmolality is 306 mosm/kg H_2O, one could extrapolate (at some risk) that his 24-hour K^+ excretion is two to three times 22 mmol, or close to 50 mmol/day. Hence, the rate of excretion of K^+ is much greater than 15 mmol per day, so it is excessive for a patient with hypokalemia.

Reasons for a high excretion of K^+: A high excretion of K^+ is due to a high volume and/or a high [K^+] in the CCD. The former cannot be assessed (we do not know the osmole excretion rate), so we can only comment on the latter. The TTKG is very high, close to 10. A high TTKG implies electrogenic reabsorption of Na^+ in the CCD. Two possibilities are an increased conductance for Na^+ or a decreased Cl^- reabsorptive rate (Figure 10·5, page 367, and the review flow chart on page 355). Given the expanded ECF volume and the high TTKG, the results favor a higher conductance for Na^+. This view would be reinforced if he could elaborate a Na^+-poor and Cl^--poor urine with a contracted ECF volume (he was able to do so).

TTKG in subjects with a deficit of K^+ and hyperaldosteronism:
The TTKG falls to close to 6 in these subjects and does not rise with exogenous mineralocorticoids unless the K^+ deficit is repaired or bicarbonaturia is induced.

What could be responsible for the lesion?

High levels of aldosterone could be a possibility, but the [K^+] in plasma is too low for the diagnosis of primary hyperaldosteronism alone. Nevertheless, aldosterone levels in plasma should be measured. When measured, the level of this hormone was not elevated; hence, another explanation should be sought. Our hypothesis is illustrated in Figure 10·6, page 370. We speculate that high levels of cortisol have resulted from ACTH production by the tumor. This cortisol could be "acting like aldosterone" in the principal cells of his CCD because the 11β-HSDH activity is not high enough to deal with such a high load of cortisol. Both ACTH and cortisol were measured and found to be extremely high on several occasions during the day. At this time, an ACTH-producing tumor (along with little intake of K^+) is our best guess for the principal cause of hypokalemia.

What should be done for therapy? Focus on the hypokalemia.

His dietary NaCl should be restricted, and he needs KCl supplements. Perhaps amiloride will be needed if these measures do not cause his plasma [K^+] to rise to 3.0 mmol/l.

One must monitor the plasma [K^+] carefully when using the combination of amiloride and KCl therapy because of the risk of developing hyperkalemia later when the GI tract absorbs the K^+.

Summary of Main Points

- Potassium is primarily an intracellular cation, and its concentration in the ICF relative to the ECF is the major determinant of the resting membrane potential (RMP). As hypokalemia develops, the RMP becomes more electronegative (i.e., the ratio of K^+ in the ICF to K^+ in the ECF increases). The magnitude of a K^+ deficit is difficult to determine from the plasma $[K^+]$.
- Renal K^+ excretion has two components: the $[K^+]$ in CCD fluid and the flow rate through the CCD. The former is determined primarily by mineralocorticoid activity and bicarbonaturia, the latter by osmole delivery to the CCD.
- The most prevalent cause of hypokalemia is renal K^+ loss, the most likely basis being diuretic use and vomiting.
- In some circumstances, replacement of large K^+ deficits constitutes an emergency. Cardiac arrhythmias may occur (more likely in patients receiving a digitalis preparation), hypoventilation may result from hypokalemic muscle weakness, and some patients may have a shift of K^+ into the ICF (e.g., recovery from diabetic ketoacidosis).

Discussion of Questions

10·1 What is the most likely cause of K^+ loss in patients A, B, C, and D? They all have hypokalemia, ECF volume contraction, and metabolic alkalosis.

		Patient			
		A	B	C	D
Urine electrolyte					
Na^+	mmol/l	25	25	25	0
K^+	mmol/l	60	60	60	10
Cl^-	mmol/l	0	85	0	0
HCO_3^-	mmol/l	85	0	0	0

A. The major clue is the anion present in the urine. In this case, it is HCO_3^-. Because the patient has a high content of HCO_3^- (metabolic alkalosis) and is losing HCO_3^- in the urine, there must be an additional source of HCO_3^-. In the absence of HCO_3^- intake, loss of HCl (vomiting) is the most likely cause. The low $[Cl^-]$, which indicates a normal renal response to ECF volume contraction in the presence of Na^+ and HCO_3^- excretion, is further supportive evidence of vomiting as the basis of the findings.

B. The major clue is the high [Cl^-] in urine. Loop diuretics cause the excretion of NaCl, which leads to ECF volume contraction, and thereby, aldosterone release; aldosterone leads to a high [K^+] in the urine. A high urine [Na^+] (and [Cl^-]) in the face of ECF volume contraction suggests that the diuretic was acting on the kidney at the time the urine was collected. Alternatively, if no diuretics were taken, the patient could have Bartter's syndrome.

C. The major clue here is that the sum of measured cations (Na^+ and K^+) greatly exceeds that of the usual anions (Cl^- and HCO_3^-). Therefore, another anion (e.g., β-HB^-, hippurate, or a drug) is excreted initially as its Na^+ salt. This excretion leads to ECF volume contraction, aldosterone release, and loss of K^+ in the urine.

D. There is little renal K^+ loss at present; the kidney is behaving normally by conserving Na^+ and Cl^-. If the delivery of Na^+ and volume to the lumen of the distal nephron is low, there is a small quantity of Na^+ for aldosterone to act on. Thus, the diagnostic features of urine electrolytes disappear when the provocative stimulus (vomiting/diuretics) is not a recent event. One can suspect their presence in the past if treatment with NaCl plus KCl readily returns the plasma electrolyte values to normal. Furthermore, frequent random urine electrolyte studies should be done to see if a sudden rise in the concentration of Na^+, K^+, Cl^- and/or HCO_3^- ensues. If it does, either the patient vomited ([HCO_3^-] high in urine) or took diuretics ([Cl^-] high in urine). In the latter case, appropriate urine studies to measure the diuretic at this time may confirm a suspicion of occult diuretic abuse.

10·2 Will a patient consuming a very high NaCl load become hypokalemic as a result of enhanced renal K^+ excretion? With respect to K^+ homeostasis, how does such a patient differ from one who receives a loop diuretic? Assume equal Na^+ excretion rates.

There must be a stimulus for aldosterone release plus a high distal flow rate to get very high rates of K^+ excretion. The patient who has a high intake of NaCl that increases distal flow rate but does not have ECF volume contraction to stimulate aldosterone release will not become hypokalemic. Thus, he differs from the patient who takes a diuretic that causes ECF volume contraction, aldosterone release, and an increase in the distal tubular flow rate.

10·3 A patient takes a diuretic, which causes the excretion of an extra 60 mmol of K^+ per day. The doctor recommended that this patient consume bananas to replace the K^+ lost each day. If all the kcal were retained as stored fat (9 kcal/g), what would be the weight gain at the end of one year?

A banana (120 g) contains close to 10 mmol of K^+; therefore, the patient will have to consume 720 g (1.6 lb) of bananas daily to supply the needed 60 mmol of K^+. This amount is equivalent to 30 cubic inches of banana. The caloric content of bananas depends on their dry weight and whether the major constituent is carbohydrate (4 kcal/g) or fat (9 kcal/g). Assuming that bananas contain 67% water and that the remaining dry weight has 4 kcal/g, the number of extra kcal ingested per day is close to 1000, the equivalent of 1/4 lb of stored fat (see the margin). Thus, the overall weight gain would be approximately 90 lb/year!

Quantities:

- Daily surplus of kcal: If the wet weight is 720 g, then the dry weight (33%) = 240 g. 240 g × 4 kcal/g = close to 1000 kcal.
- Stored fat has 9 kcal/g. If 111 g of fat is stored each day (1000 kcal/9 kcal/kg), there will be a weight gain of 1/4 lb per day (1 lb = 454 g and close to 4000 kcal/lb).
- Overall: 365 days × 1/4 lb/day = 91 lb/year.

10·4 Elite athletes often have a [K^+] in plasma that is close to 3.0 mmol/l at rest. What mechanisms might be involved?

We are not sure. Either there is a total body deficit of K^+ or there is a hyperpolarized RMP that maintains more K^+ in the ICF.

Deficit of K^+: K^+ can be lost in sweat or the urine.

1. **Sweat:**
   Although the [K^+] in each liter of sweat is low (close to 10 mmol/l), the number of liters lost can be large (> 10/day). Hence, nonrenal loss of K^+ can be significant. If this loss were the only effect, the urinary excretion of K^+ would be less than 15 mmol/day, but it is higher and therefore suggests renal mechanisms as well.

2. **Renal excretion of K^+:**
   A higher rate of excretion of K^+ suggests hyperaldosteronism. Perhaps the high levels of aldosterone reflect the state induced by chronic exercise. In response to the threats to the ECF volume during exercise (sweating and shift of water into cells induced by the higher osmolality in muscle cells when muscles contract), the body retains extra Na^+ to expand the ECF volume. This expansion provides the "reserve" that the athlete needs to fill his or her blood volume (high stroke volume, dilated capillary beds in exercising muscle and skin).

Hence, loss of K^+ may occur by both renal and nonrenal routes.

Shift of K^+ into cells: Perhaps higher catecholamine levels or increased sensitivity to β_2-adrenergics can lead to a higher RMP.

10·5 If a subject is K^+-depleted and has lost 300 mmol of K^+, what is the most likely change in ionic composition of the ICF and what adjustments were needed to permit this change?

To lose 300 mmol of K^+ from the ICF, either 300 mEq of anions must be lost (organic phosphates) or 300 mEq of cations (H^+ or Na^+) must enter cells (Figure 10·1, page 354). Thus, in the former case, cells would have to lose RNA, DNA, ATP, creatine phosphate, or phospholipids; in the latter case, cells would have to become more acidic and/or have a higher [Na^+]. To maintain this higher [Na^+], the Na^+K^+ATPase would have to be less active (fewer pumps, inhibited pumps, and/or pumps with a higher K_m for Na^+).

10·6 The lesion in hypokalemic periodic paralysis is a higher open probability of Na^+ channels in the plasma membrane. How might this lesion cause hypokalemia? In what way is this lesion different from that in hyperkalemic periodic paralysis, in which the Na^+ channels also have a more open probability?

In hypokalemic periodic paralysis, most of the Na^+ channels are more open. Thus, there is a large flux of Na^+ into cells. Because a higher $[Na^+]$ in cells could drive the Na^+K^+ATPase, some of the positive charges that enter are exported. In quantitative terms, one positive charge enters per Na^+ entry but only one-third of a charge exits per Na^+ pumped via the Na^+K^+ATPase; this net entry of positive charges depolarizes the RMP. The fall in RMP should cause K^+ to exit and result in hyperkalemia. Given the fact that hypokalemia is present in hypokalemic periodic paralysis, a possible explanation is that the K^+ channels have a reduced open probability in this disorder and, as a result, more of the K^+ that enter cells could remain in the ICF. This explanation is just our guess as to what might happen.

In the hyperkalemic variety, different mutant Na^+ channels have a consistent higher open probability. Thus, the RMP is also depolarized. If, in this setting, the K^+ channels maintained their high open probability, K^+ would exit from cells because of the fall in RMP.

10·7 What is the cause of hypokalemia in patients who vomit and have metabolic alkalosis?

Vomiting is an example of K^+ loss by more than one route. As outlined in Figure 4·1, page 150, the plasma $[HCO_3^-]$ rises and $[Cl^-]$ falls with each loss of gastric HCl. The consequent increase in the filtered load of HCO_3^- exceeds its renal reabsorption and leads to the loss of Na^+ from the body. The resultant decrease in ECF volume stimulates aldosterone release. Aldosterone action, together with a large distal nephron delivery of Na^+-rich fluid that contains HCO_3^-, produces a large renal loss of K^+.

To a lesser extent, there is a shift of K^+ into the ICF (Figure 10·3, page 360).

The loss of K^+ by the GI tract is quite small; the $[K^+]$ of gastric fluid is usually less than 15 mmol/l.

10·8 When a subject starves for three weeks, there is a negative balance for K^+ of close to 300 mmol, but the subject is normokalemic. What changes may have occurred in cells to cause this intracellular K^+ deficit?

The loss of K^+ from the ICF could result from three major reasons.

1. There might be a shift of cations such as H^+ into cells for buffering, but this loss is trivial in ketoacidosis of starvation.
2. There might be a lower quantity of Na^+ pumped by the Na^+K^+ATPase as a result of the low level of insulin.

3. A loss of intracellular anions (phosphate esters such as RNA) occurs during fasting and requires a similar loss of K^+ for electroneutrality. The $[K^+]$ in the ICF need not change because water will be lost as well. This type of K^+ loss is important during chronic fasting.

10·9 Should the deficit of K^+ in a starved subject be replaced during the fast?

Although there is a large negative balance for K^+, much of the intracellular stores of K^+ cannot be replaced before the net synthesis of RNA and phospholipids occurs (insulin acts to stimulate anabolism) and phosphate is ingested. Accordingly, only small supplements of K^+ are advisable during a fast, and larger supplements are needed upon refeeding. One can imagine that the renal response that led to the negative K^+ balance was appropriate during fasting because it prevented the development of hyperkalemia.

10·10 Can excessive delivery of Na^+ to the CCD augment the rate of excretion of K^+?

To determine whether the delivery of Na^+ to the CCD actually limits its net secretion of K^+, one needs to know the $[Na^+]$ in the luminal fluid and the $[Na^+]$ needed for the net secretion of K^+ (a $[Na^+]$ of 15 mmol/l can support half the maximal rate of K^+ transport in the CCD in rats).

$[Na^+]$ in the luminal fluid in the CCD: To calculate this $[Na^+]$, one first needs to know the osmolality of luminal fluid. For easy math, when ADH acts (i.e., most of the day), the osmolality of the luminal fluid is the same as that of plasma, say 300 mosm/kg H_2O. The next step is to decide how many of these osmoles are urea because the rest will be largely due to electrolytes. The likely concentration of urea is 100 mosm/liter (see the margin). Therefore, the concentration of electrolytes is close to 200 mosm/liter, half of which are Na^+ plus K^+. Because the $[Na^+]$ exceeds the K_m for the Na^+ needed to augment K^+ secretion by 5- to 10-fold, delivering more Na^+ per se will not drive more K^+ secretion unless volume delivery is increased.

Calculation:
Because 400 mmol of urea are excreted per day and close to 4 liters of GFR will reach the terminal CCD, the concentration of urea in the lumen of the CCD is 100 mmol/l.

10·11 How do poorly reabsorbed anions lead to an augmented excretion of K^+?

A poorly reabsorbed anion can augment K^+ secretion by two mechanisms. First, it can augment the delivery of volume to the CCD by increasing the osmolar load. This contribution is usually minor. The second way is to deliver more Na^+ without Cl^- to the CCD (i.e., during ECF volume contraction). In experiments in humans, a load of poorly reabsorbed anions did not cause a significant kaliuresis in normovolemic subjects because the $[Cl^-]$ in their urine was not less than 10–15 mmol/l.

10·12 In a patient with Bartter's syndrome, in what ways will a greater salt intake increase the rate of excretion of K^+?

The excretion of K^+ can rise if the $[K^+]$ in the CCD or the flow rate in this nephron segment rises. Because the $[Na^+]$ in luminal fluid does not limit net secretion of K^+ in the CCD, a higher delivery of Na^+ and Cl^- can increase the flow rate in the CCD by increasing the rate of excretion of osmoles. In a patient with an obligatory high $[K^+]$ in the CCD, as in Bartter's syndrome, this high flow rate could lead to much higher rates of excretion of K^+.

What are the dangers of giving a small load of NaCl to this patient?

If too small a load of NaCl is given, a smaller quantity of K^+ will be excreted, but the patient might develop a deficit of Na^+, and the ECF volume will be contracted to a greater extent.

10·13 When a patient presents with all features characteristic of Bartter's syndrome, a clinician must rule out surreptitious use (abuse) of diuretics, vomiting, and laxative abuse. What features should a clinician rely on in each of these settings?

Diuretic abuse: The patient probably will not admit to the abuse of diuretics. Therefore, the findings of hypokalemia, low ECF volume (high renin), and a urine with abundant quantities of Na^+ and Cl^- provide the common database. At this point we would, without prior notice, ask for many random urine samples. If at least one is Cl^--poor or Na^+-poor, we would suspect diuretic abuse; assays should be carried out in multiple urine samples. Assays for diuretics would be diagnostic in urine that contains Na^+ and Cl^-. Finally, hypomagnesemia often is very significant in the patient with Bartter's syndrome.

Vomiting: Again, the urine is very helpful. A urine that contains very little Cl^- makes loss of HCl a very likely diagnosis. A urine that contains Na^+ and is alkaline should literally clinch the diagnosis.

Laxative abuse: If the subject is ECF volume contracted, the urine should have a small quantity of Na^+. If metabolic acidosis is present, more NH_4^+ will be excreted (along with Cl^-). Therefore, in this setting, the quantity of Na^+ (not Cl^-) in urine reflects the ECF volume status. Urine with a low $[Na^+]$ and a high $[Cl^-]$ suggests that NH_4^+ are being excreted in response to acidosis (or hypokalemia). On the other hand, a urine that contains Na^+ but not Cl^- or HCO_3^- suggests that organic anions are being excreted.

As a final caution, patients with a PhD in deception may utilize multiple abusive techniques simultaneously and increase the diagnostic challenge. A high degree of suspicion, careful observation, an enjoyment of subterfuge, and a combination of the above analyses can help one unravel these problems.

10·14 What is the time course and the expected degree of hypokalemia with diuretic therapy?

The plasma [K^+] falls to its steady-state value within one week of diuretic therapy. The degree of hypokalemia is modest—only 5% of patients have values less than 3.0 mmol/l (if lower than 3.0 mmol/l, look for another cause for K^+ loss, including reasons for high aldosterone and/or urine flow rate). Elderly Caucasian women seem to be the most susceptible to developing hypokalemia, perhaps because their diets have a much lower quantity of K^+.

At the usual doses, is the hypokalemia particularly severe with one class of diuretics?

The degree of hypokalemia is not really different with the various classes of diuretics, with the possible exception of acetazolamide, which can cause a greater loss of K^+. This difference is, at best, however, very modest.

11

HYPERKALEMIA

Objectives

- To provide the physiologic background so that a classification of *hyperkalemia* can be developed. Although intake and shift of K^+ from the ICF may contribute to hyperkalemia, reduced excretion of K^+ in the urine is virtually always present in chronic hyperkalemia.
- To emphasize that the components of renal K^+ excretion are the $[K^+]$ in the urine and the urine volume.
- To elucidate the factors that lead to a rise or fall in the urine $[K^+]$ and the urine volume and to recognize their impact on maintaining hyperkalemia.
- To provide an approach to the therapy of hyperkalemia.

Abbreviations:
CCD = cortical collecting duct.
ECF = extracellular fluid.
ECG = electrocardiogram.
$9\alpha F$ = fludrocortisone, a drug with mineralocorticoid action.
GFR = glomerular filtration rate.
ICF = intracellular fluid.
MCD = medullary collecting duct.
NSAIDS = nonsteroidal anti-inflammatory drugs
P_{osm} = plasma osmolality
RMP = resting membrane potential.
TEPD = transepithelial potential difference.
TTKG = transtubular $[K^+]$ gradient.
U_{osm} = urine osmolality

Outline of Major Principles

1. **General considerations:** Hyperkalemia may produce serious side effects when the plasma $[K^+]$ is greater than 6 mmol/l. The correlation between the plasma $[K^+]$ and untoward effects (arrhythmias) is not reliable. In general, the rate at which hyperkalemia develops is as important as its degree in determining its overall threat to the body.

2. **Etiology of hyperkalemia:** Chronic hyperkalemia implies that the rate of excretion of K^+ is lower than expected in almost every case. In the absence of renal failure, suspect low aldosterone bioactivity or diminished distal nephron flow.

   A shift of K^+ out of cells is a less common cause for severe hyperkalemia; it occurs with hormone deficiency, certain types of metabolic acidosis, catabolism, or a metabolic abnormality such as hyperkalemic periodic paralysis.

3. **Diagnosis of hyperkalemia:** Hyperkalemia may be suspected clinically and confirmed by measuring the plasma $[K^+]$. Urine electrolytes help in establishing the etiology of chronic hyperkalemia.

4. **Treatment:** Eliminate K^+ intake. Antagonize the cardiac effects of K^+ with calcium (a very rapid effect). Shift K^+ into cells with insulin and possibly $NaHCO_3$ (requires 0.5 to 2 hours). If hypoaldosteronism or a low urine flow rate is present, promote K^+ loss in the urine. Induce K^+ loss in the GI tract with resins (a slow process). Be prepared to promote loss of K^+ by dialysis if hyperkalemia is severe or rapidly progressive (i.e., if there is an ongoing K^+ load).

Hyperkalemia:

- Definition: plasma $[K^+] > 5.0$ mmol/l.
- Expected renal responses:
  - > 200 mmol of K^+ excreted/day,
  - TTKG > 7.
- The main danger is a cardiac arrhythmia.
- The diagnosis is usually made by the routine determination of plasma electrolytes supported by clinical suspicions.

Introductory Case
Lee's [K$^+$] Is on a High

(Case discussed on pages 414–15)

Lee, our noninsulin-dependent diabetic, returns yet again. Now glycemic control is relatively poor (blood glucose is 360 mg/dl, or 20 mmol/l) and she is hyperkalemic (5.4 mmol/l). On physical examination, there is evidence of a mild degree of ECF volume contraction.

Pertinent laboratory values are presented below.

		Plasma	Urine (random sample)
Na^+	mmol/l	130	53
K^+	mmol/l	5.4	20
Cl^-	mmol/l	99	51
HCO_3^-	mmol/l	19	0
pH		7.35	5.4
Glucose	mmol/l (mg/dl)	20 (360)	302
Osmolality	mosm/kg H_2O	285	600
Creatinine	mmol/l (mg/dl)	180 (2.0)	–
Urea	mmol/l (mg/dl)	10 (28)	135

Review flow chart

Causes of hyperkalemia with low excretion of K$^+$

The causes of a low rate of excretion of K$^+$ are too low a flow rate in the CCD (left limb) and too low a [K$^+$] in the lumen of the CCD (right limb). Both flow rate and [K$^+$] in the CCD should be evaluated in each patient. Final considerations are shown in the shaded boxes. In renal failure, the excessive flow rate per nephron limits electrogenic reabsorption of Na$^+$ and thereby causes the kidneys to behave as if there is slow reabsorption of Na$^+$. (ECFV = extracellular fluid volume.)

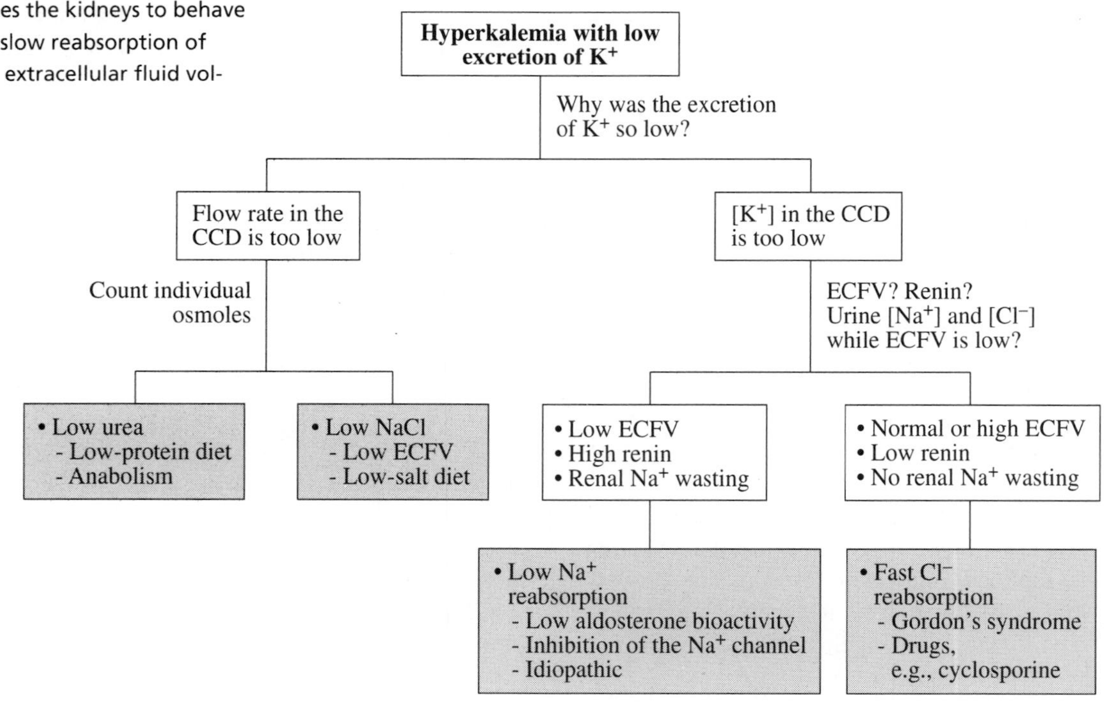

What factors contributed to hyperkalemia?

What additional information is needed to understand the relative importance of each factor?

What initial therapy would be most appropriate?

PART A
PATIENTS WITH HYPERKALEMIA

Clinical Approach

The clinical approach we recommend is illustrated in Figures 11·1A and 11·1B and outlined in Table 11·1. It requires answers to the following questions:

1. Is hyperkalemia life-threatening?

Hyperkalemia endangers patients by provoking cardiac arrhythmias. There is no specific plasma [K^+] at which an arrhythmia occurs. In general, hyperkalemia is better tolerated in the very young and if the hyperkalemia is chronic. The factors that suggest the possibility of an arrhythmia are changes in the ECG (see page 407), a sudden rise in the degree of hyperkalemia, the likelihood that K^+ will be released into the ECF (cell lysis and recovery from an ileus, especially if oral K^+ were given), and the presence of medications that may permit hyperkalemia to worsen (e.g., β-blockers). The absence of these factors does not mean that an arrhythmia will not occur.

If a patient is in danger from hyperkalemia, address therapy first (Figure 11·1B). The steps here are to antagonize the effects of hyperkalemia, a benefit that occurs in minutes (with Ca^{2+}), to promote progressive entry of K^+ into cells (with insulin and possibly $NaHCO_3$), and, as more remote therapy, to promote the excretion of K^+. Details will be provided later in this chapter.

On the other hand, if the danger of an arrhythmia is judged to be small, we recommend limiting K^+ intake and proceeding towards a diagnosis, as shown on the left side of Figure 11·1A.

Table 11·1
APPROACH TO A PATIENT WITH HYPERKALEMIA

Rule out a life-threatening condition in which therapy for hyperkalemia is the first consideration.
- Antagonize effects of hyperkalemia with Ca^{2+} if there are important ECG changes.
- Allow no K^+ intake.
- Shift K^+ into cells with insulin and possibly $NaHCO_3$.
- Promote K^+ loss in the urine, GI tract, or via dialysis.

Rule out pseudohyperkalemia.
- Determine if the technique for drawing blood was improper.
  - Excessive muscular contraction
- Did blood cell lysis occur?
  - Red blood cells, fragile tumor cells in blood, enhanced platelet volume

Assess the cause for reduced excretion of K^+ in the urine.
- Is the [K^+] in the urine low?
- Is the distal flow rate low?

Assess K^+ intake.
- A high intake will aggravate hyperkalemia if there is compromised renal K^+ loss.

Table 11·1 is continued on the next page.

Assess a shift of K^+ from the ICF to the ECF.
- What was the cause of the shift?
  - Metabolic acidosis (nonorganic)
  - Hormonal causes
    - Insulin deficiency
    - β_2-Adrenergic blockade
    - Aldosterone deficiency
    - More than one hormone deficiency can lead to a severe degree of hyperkalemia, aggravated by hyperglycemia.
  - Necrosis or depolarization
  - Rare disorders (e.g., hyperkalemic periodic paralysis)

2. Is the hyperkalemia real?

Pseudohyperkalemia:
An elevated [K^+] in plasma that does not represent events in vivo. It is usually due to cell rupture or depolarization of cell membranes during withdrawal of blood.

A useful clue is that there are no ECG changes.

At times (see below), hyperkalemia may be reported by the laboratory but not represent values in the patient (called "*pseudohyperkalemia*"). The common causes for this error are improper techniques in drawing blood and lysis of blood cells. A relatively common way to achieve an artificial elevation in the [K^+] in plasma is to draw blood from a patient who clenches his fist excessively; the muscular contraction leads to depolarization of cell membranes and, as a result, release of K^+ from

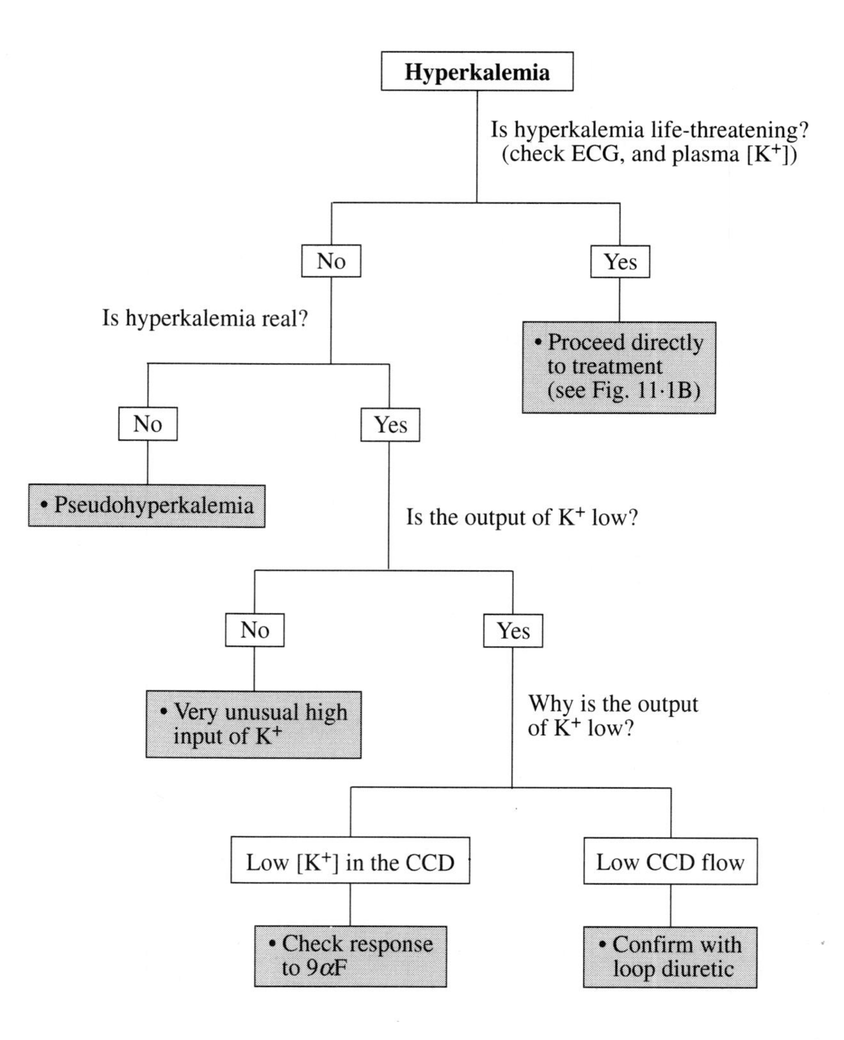

Figure 11·1A
Diagnostic approach to hyperkalemia
The final diagnostic groups are shown in the shaded boxes. For details, see the text.

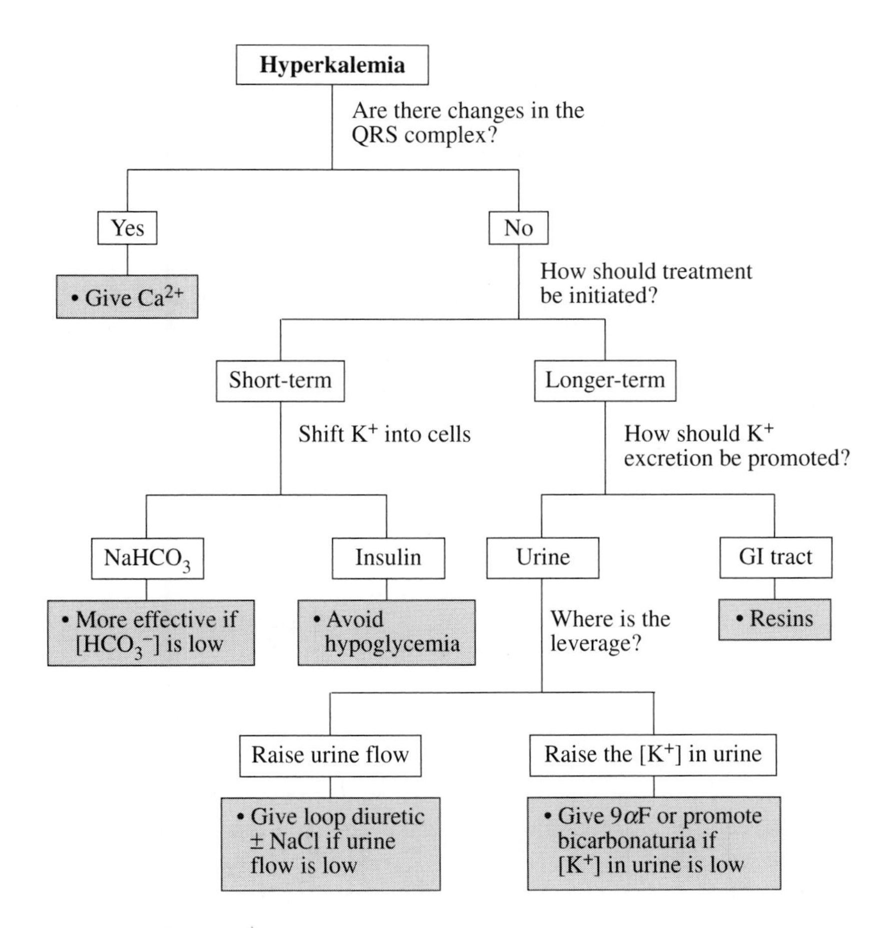

Figure 11·1B
Therapeutic approach to hyperkalemia
If treatment must be given promptly, the steps to take are as follows. In a patient with hyperkalemia, multiple treatment options are usually required.

cells. This release of K^+, however, does not usually go into the general pool of interstitial fluid but, rather, enters *T-tubules*. Certain patients, especially those with *cachexia* and/or those whose veins are difficult to puncture (more fist clenching), are more prone to release K^+ into their venous blood.

Alternatively, K^+ may be released from red cells that have lysed. In this case, the plasma is red. Platelets can rupture during clotting or if they contact glass; nevertheless, significant hyperkalemia is only evident if the platelets are abnormally large in diameter. Given the very small size of platelets, an abundant number (thrombocytosis) does not correlate well with pseudohyperkalemia (see the discussion of Question 11·1, page 422). In rare instances, circulating tumor cells (leukemias) can be unduly fragile. To minimize these errors, blood should be drawn slowly into a silicone-coated tube that contains heparin to avoid the lysis that is caused by a glass surface or that occurs during clotting. To rule out pseudohyperkalemia, one could draw two blood samples simultaneously—one from the arm using the usual technique and another from the femoral vein or an artery using a heparinized, silicone-treated tube.

Once pseudohyperkalemia has been ruled out or proven not to be the sole cause of hyperkalemia, the next stage in the approach is to address the renal response. Chronic hyperkalemia implies a defect in the excretion of K^+ in almost every case.

T-tubules:
Invaginations in the plasma membrane of skeletal muscle where most of the K^+ are released during depolarization of muscle cells. The K^+ released are taken up rapidly before they can mix with the general interstitial fluid.

Cachexia:
A term used to describe a patient with a marked degree of muscle wasting. Disturbed local architecture (T-tubules) may permit K^+ to enter interstitial fluid during depolarization.

Note:

If the rate of excretion of K^+ exceeds 200 mmol/day or 140 μmol/min, look for a very high intake of K^+; a high K^+ intake is an extremely unlikely sole cause of hyperkalemia.

3. Is the rate of excretion of K^+ low? If so, why?

In almost every case, the excretion of K^+ will be low in a patient with chronic hyperkalemia. Since the excretion of K^+ is the product of the $[K^+]$ in the urine and the urine flow rate, each component should be examined independently.

$[K^+]$ in urine: Two factors lead to a high $[K^+]$ in the urine. First, there must be electrogenic reabsorption of Na^+ in the CCD to generate the driving force for the net secretion of K^+. Second, the luminal membrane must be permeable to K^+ for this secretion to occur. Both components of the K^+ secretory process can be evaluated by examining concentrations of electrolytes and the osmolality of the urine (see pages 340–42). In practice, calculating the TTKG is very helpful.

If the TTKG is lower than expected (less than 7 in a patient with hyperkalemia), one can question whether the level of aldosterone is low or whether there is a lesion in the CCD so that it fails to respond to aldosterone. To answer these questions, blood should be drawn for renin and aldosterone levels (there is a long delay before results are available) and an acute challenge with 100 μg of *9αfludrocortisone* (9αF) should be attempted. If the TTKG rises to greater than 8, suspect a problem with aldosterone synthesis; a problem in the CCD is present if the TTKG does not rise to greater than 8 two hours after 9αF is given.

9αfludrocortisone (9αF):

A drug with mineralocorticoid actions. It is active when given orally.

Volume delivery to the CCD: Low volume delivery to the CCD is uncommon as the sole cause of hyperkalemia. In essence, volume delivery is assessed by measuring the osmole excretion rate; details are provided later in this chapter (Figure 11·3, page 403). To confirm that a low volume is the reason for the low rate of excretion of K^+, a loop diuretic should be given. Expect the rate of excretion of K^+ to rise to greater than 140 μmol/min at peak diuresis (flow 10 ml/min). If the patient has ECF volume contraction, be sure to replace the Na^+ and water losses.

4. Has either a high intake of K^+ or a shift of K^+ from the ICF to the ECF contributed to the hyperkalemia?

The intake of K^+ was discussed above. The only point to add is that resolution of an ileus might provide an "occult" load of K^+ (see the margin).

A shift of K^+ from cells occurs if there is necrosis, depolarization, hyperchloremic metabolic acidosis, and/or a deficiency of insulin and/or β_2-adrenergics. If present, these disturbances should be addressed.

Occult load of K^+:

If a patient with hypokalemia and an ileus receives an oral K^+ load, the K^+ may remain in the GI tract. Once GI motility recovers, there will be a sudden load of K^+ that the kidneys cannot excrete promptly. It may take 24 hours of normokalemia to restore K^+ excretion to normal.

Question

(Discussion on page 422)

11·1 What is the total K^+ content in the platelets of 1 liter of blood? (Assume 400,000 cells/mm^3, a 150 mmol/l $[K^+]$ in the ICF, and a platelet volume of 1 μ^3.)

Etiology of Hyperkalemia

ROLE OF EXCESSIVE K^+ INTAKE

- Chronic severe hyperkalemia occurs with increased K^+ intake only if the excretion of K^+ is compromised.

After appropriate redistribution, the retention of 70 mmol of K^+ produces only an 0.1 mmol/l rise in the plasma [K^+] in normal individuals (see the discussion of Question 9·1, page 344). Notwithstanding, should a defect in K^+ movement into cells be present, severe hyperkalemia could ensue. Extremely high K^+ intake leads to chronic hyperkalemia only in patients with compromised K^+ excretion. In certain cultures, K^+ intake can be 10 or 400 mmol/day.

Under special conditions, an unusually large load of K^+ can be administered IV (e.g., when large volumes of whole blood are transfused); the degree of this danger depends on the preservative, the temperature of the blood, cell lysis, the duration of blood storage, and whether or not the blood has sedimented (plasma may deliver a large K^+ load because it has a high [K^+] and will be infused faster as a result of a lower viscosity). If the citrate that is used as an anticoagulant acutely lowers the plasma ionized calcium concentration, the biologic response to hyperkalemia might be more severe. The cardioplegic solution used in cardiac surgery also may provide a substantial K^+ load to the patient.

Very high K^+ intake:

1. A very high K^+ intake is rare; examples include the use of large quantities of salt substitutes that contain 10–13 mmol of K^+ per gram and the intake of K^+ salts of organic acids (citrate) given to alkalinize the urine.
2. In patients undergoing aortic surgery, rapid reperfusion after clamp removal can result in a sudden bolus of K^+ from the previously ischemic limbs to the heart. An analogous situation may occur following renal transplantation if the transplanted kidney was perfused with a K^+-rich solution during preservation.

EXCESSIVE SHIFT OF K^+ FROM CELLS

- K^+ will shift from cells with hormone deficiencies, acidosis, or cell damage.

Hormones

Insulin and β_2-adrenergic agonists are the major hormones that cause movement of K^+ into cells. Hyperkalemia is seen if these hormones are relatively inactive.

Insulin: Insulin promotes the entry of K^+ into cells by hyperpolarizing the cell membrane (more flux through the Na^+/H^+ antiporter leads to more pumping of Na^+ via the Na^+K^+ATPase and, as a result, a more negative resting membrane potential). Insulin also increases the net anionic charge (organic phosphates) in cells.

The shift of K^+ out of cells during diabetic ketoacidosis is not due to a pH change but is due to the lack of insulin. This cause can be deduced from direct experiments and the time course of events (i.e., the plasma [K^+] falls 1–2 hours after insulin is given; at this time there is little change in the acid-base status; see the margin).

Quantities:

In acute experiments in which the basal insulin level was halved, the plasma [K^+] rose 0.5 mmol/l.

Note:
The plasma [K^+] rises by only a few tenths of a millimole per liter in normal subjects on β-blockers.

β_2-Adrenergics: β_2-Adrenergics lead to a shift of K^+ into cells; hence, β-blockers may cause hyperkalemia. β_1-Blockers diminish renin release, and this action can reduce aldosterone levels and thereby K^+ excretion. In contrast, α-adrenergics have the opposite effect; they tend to increase the severity of hyperkalemia if there is already a stimulus to move K^+ out of cells.

Aldosterone: Low levels of aldosterone contribute to the serious hyperkalemia that is due primarily to a reduced renal excretion of K^+.

Combined hormone deficiency: If a patient has both hypoaldosteronism and low insulin levels, a much greater degree of hyperkalemia can occur. If hyperglycemia develops in this hormonal setting, it can markedly exacerbate the degree of hyperkalemia by mechanisms that are not entirely clear.

Questions

(Discussions on pages 422–23)

11·2 What sets the limit on K^+ uptake into cells?

11·3 How might β-blockers cause a severe degree of hyperkalemia?

11·4 What happens to the plasma [K^+] during exercise?

11·5 Why do patients with a problem regenerating ATP in muscle cells suffer from weakness but not hyperkalemia?

11·6 How much of a K^+ load might a patient receive if one liter of blood is digested in the GI tract in the course of a GI bleed?

Acid-Base Factors

- Hyperkalemia is not caused directly by respiratory acidosis, ketoacidosis, and L-lactic acidosis.
- If hyperkalemia accompanies chronic acidosis, look for a problem with K^+ excretion or insulin deficiency.

Acute respiratory acidosis per se does not cause an appreciable rise in the plasma [K^+]. Similarly, metabolic acidosis caused by the accumulation of lactic acid or ketoacids does not cause an appreciable exit of K^+ from cells because H^+ enter the ICF with their conjugate base. In contrast, if a patient has metabolic acidosis from $NaHCO_3$ loss, hyperkalemia may ensue. Notwithstanding, if the kidneys and adrenal glands are normal with respect to K^+ handling, hyperkalemia will not persist (see Figure 9·4, page 331).

Cell Damage

Because the [K^+] in the ICF is close to 35-fold higher than that in the ECF, if an appreciable number of cells are damaged, hyperkalemia may result.

This type of hyperkalemia is seen in patients suffering from acute organ necrosis, trauma, or those given cytotoxic treatment for neoplasms. The degree of hyperkalemia will be more severe if the patient also has a reduced rate of excretion of K^+ or deficiencies of insulin or β_2-adrenergics.

Although not technically an example of cell damage, hyperkalemia can be caused in special circumstances by a cellular K^+ shift in response to "destruction" of the resting membrane potential. In large doses, succinylcholine may depolarize the entire muscle and lead to a severe degree of hyperkalemia (instead of the less than 1 mmol/l transient rise in the plasma $[K^+]$ when the usual dose of succinylcholine is used during anaesthesia).

A Rise in "Effective" Osmolality

A rise in plasma effective osmolality is associated with hyperkalemia in a number of settings. Nevertheless, for a shift of K^+ from the ICF to the ECF to cause chronic hyperkalemia, either the resting membrane potential across cell membranes must become less negative (depolarization) or permeability of cell membranes to K^+ must have decreased so that K^+ cannot reenter cells. There are no compelling data to support such a role for hyperosmolality. For example, in the clinical setting of diabetic ketoacidosis, there are other factors, such as deficiency of insulin, that cause the hyperkalemia. Similarly, in experiments in animals in which a rise in the plasma $[K^+]$ was associated with a rise in osmolality, enormous infusions were given, and hyperkalemia was not an isolated finding.

REDUCED EXCRETION OF K^+ IN THE URINE

To excrete K^+, there must be electrogenic reabsorption of Na^+ in the CCD. It is difficult to establish just what constitutes a reduced rate of excretion of K^+ in a hyperkalemic patient. Because a normal person can excrete as much as 450 mmol of K^+ per day when K^+-loaded, anything less than this amount should constitute a low rate of excretion of K^+. If the 24-hour K^+ excretion rate is less than 200 mmol, the rate of excretion of K^+ is clearly lower than expected in the presence of hyperkalemia. The causes of reduced excretion of K^+ can be an unexpectedly low $[K^+]$ in the luminal fluid in the CCD and/or a low delivery of volume to the CCD. In the latter circumstance, the pathophysiology will include a low rate of excretion of osmoles (see the margin).

A low $[K^+]$ in the fluid in the lumen of the CCD usually reflects a reduced aldosterone bioactivity (Table 11·2) and/or an intrinsic renal lesion. The role of the adrenal gland can be assessed by hormone measurements and by the kaliuretic response to exogenously administered mineralocorticoid (100 μg 9αF).

Clinical pearls:
- Assess the rate of K^+ excretion first.
- If this rate is low, assess the $[K^+]$ and volume of urine separately.
- The TTKG reflects the $[K^+]$ in the CCD relative to that in plasma.
- The osmole excretion rate reflects the volume delivered to the terminal CCD (osmole excretion rate/P_{osm} provides a minimum estimate of the number of liters delivered to the CCD if ADH is acting).

Table 11·2
FACTORS TO EVALUATE IN PATIENTS WITH HYPERKALEMIA AND LOW ALDOSTERONE BIOACTIVITY

Plasma renin concentration (while ECF volume is low)
Low
• Problem with juxtaglomerular apparatus
High
• Converting enzyme inhibitor, angiotensin II receptor blocker
• Adrenal gland problem
• Low renal response to aldosterone

Table 11·2 is continued on the next page.

Plasma aldosterone concentration

Low

- Low renin
- Converting enzyme inhibitor, angiotensin II receptor blocker
- Adrenal gland problem

High

- Renal problem (interstitial nephritis, Cl^- shunt, etc.)
- Aldosterone antagonists
- K^+-sparing diuretics or drugs such as trimethoprim that mimic K^+-sparing diuretics

Renal response to mineralocorticoids (physiologic dose)

Urine [K^+] not high enough

- Interstitial nephritis or tubular damage
- Very low distal Na^+ delivery
- K^+-sparing diuretics
- Cl^- shunt in the CCD
- Low Na^+ conductance in the CCD

High urine [K^+]

- Low aldosterone levels because of an adrenal problem

In the following paragraphs, we shall present a speculative way to assess why the [K^+] in the lumen of the CCD may not be high enough in a patient with hyperkalemia. We shall apply newer discoveries in physiology to the bedside (Figure 11·2 and the review flow chart on page 394).

Causes of an Unexpectedly Low [K^+] in the CCD

- This approach is speculative but based on physiologic principles, and we find it useful.

Five potential lesions may cause hyperkalemia via a low [K^+] in the CCD (Figure 11·2 and Tables 11·3 and 11·4).

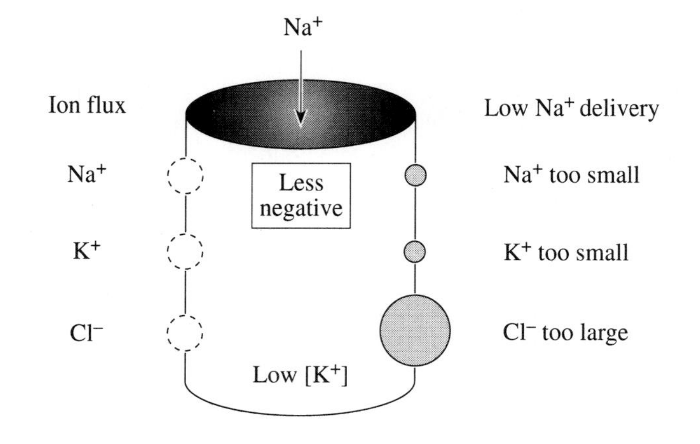

Figure 11·2
Factors that may cause a low [K^+] in the CCD
The barrel-like structure is the lumen of the CCD. For details, see the text. The potential lesions are a low delivery of Na^+ to the CCD, fewer open Na^+ channels, fewer open K^+ channels and/or more Cl^- permeability in the CCD.

Table 11·3
PATHOPHYSIOLOGIC CLASSIFICATION OF LOW EXCRETION OF K^+

For a description, see the text.

Low [K^+] in the lumen of the CCD
- Low delivery of Na^+ (rare)
- Low Na^+ channel open probability
  - Low aldosterone bioactivity, amiloride, trimethoprim, congenital disorders
- High Cl^- permeability
  - Possibly in some patients taking cyclosporine, in those with diabetes mellitus, in those with Gordon's syndrome
- Increased anion permeability
  - In patients with a low TTKG that fails to rise with bicarbonaturia
- Low K^+ conductance
  - Not yet described

Low volume delivery to the CCD
- Low urea excretion rate plus ECF volume contraction (Figure 11·3)

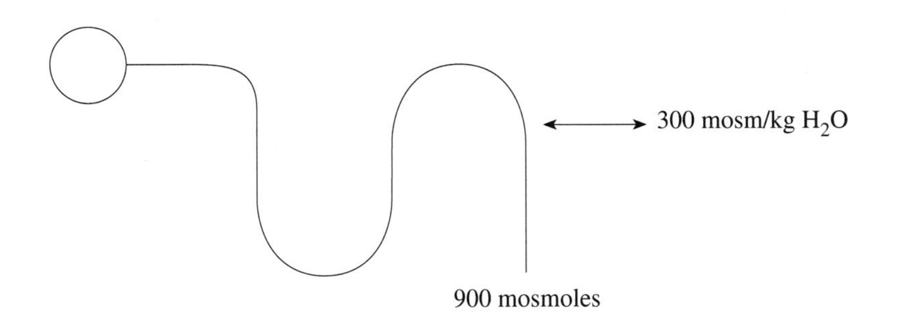

Figure 11·3
Causes of a low flow rate in the CCD
When ADH acts, the osmolality in the lumen of the CCD is the same as that in plasma, say 300 mosm/kg H_2O. If 900 mosmoles are excreted in a period of time, three liters of urine traversed the terminal CCD.

Table 11·4
CAUSES OF AN UNEXPECTEDLY LOW [K^+] IN THE CCD

The major findings are hyperkalemia and a low rate of excretion of K^+. The lesions are listed below and described in the text.

Suspected lesion	Major features	Other comments
Low delivery of Na^+	• Na^+-free urine • Very low effective arterial volume	• Rare cause of hyperkalemia
Low open probability of the Na^+ channel	• Low ECF volume and a high [Na^+] and [Cl^-] in the urine • High renin	• Adrenal gland may be the problem • Also suspect K^+-sparing diuretics or trimethoprim
High permeability to Cl^-	• ECF volume tends to be expanded • Low renin • Can achieve Na^+-free and Cl^--free urine with ECF volume contraction	• Hypertension may be present • May be called "hyporeninemic hypoaldosteronism" • Can achieve high TTKG with bicarbonaturia

Quantitative analysis:

1. Close to 1000 mmol of Na^+ are delivered to the CCD, and the majority of Na^+ (700 mmol or so) are reabsorbed in an electroneutral fashion with Cl^-. Hence, only 50–70 mmol of Na^+ are reabsorbed daily in an electrogenic manner, and this reabsorption is responsible for ensuring the net secretion of 50–70 mmol of K^+.
2. Close to 90 mmol of K^+ exit the loop of Henle because the volume of fluid that exits from the loop is close to 30 liters per day in an adult and the $[K^+]$ in this fluid is close to 3 mmol/l. A smaller quantity is actually delivered to the CCD because some K^+ are reabsorbed in the distal convoluted tubule.

1. Low delivery of Na^+ to the terminal CCD:

There are many instances that suggest a low delivery of Na^+, but it is not clear whether the delivery is low enough to limit the net secretion of K^+. We recommend the following approach. When Na^+ are excreted in the urine, one can assume that Na^+ were present in the CCD because Na^+ are not secreted into the lumen of the MCD.

How can one deduce what the $[Na^+]$ was in the CCD when Na^+ were not present in the urine? The following data permit an analysis:

(a) The $[Na^+]$ in the CCD required for half-maximal net secretion of K^+ is 10–15 mmol/l in rats.

(b) When ADH acts, the osmolality in the luminal fluid in the CCD is the same as that in plasma (say 300 mosm/kg H_2O for easy math).

(c) If the volume delivered to the terminal CCD is 5 liters/day in a 70-kg adult and if the quantity of urea excreted is 400 mmol/day, the concentration of urea is 80 mmol/l.

(d) Hence, the concentration of electrolytes is close to 200 mmol/l, half of which is Na^+, so delivery of Na^+ will only limit the net secretion of K^+ in very rare circumstances (e.g., marked degree of contraction of the ECF volume with high urea excretion rates).

(e) If one suspects that a low TTKG is due to a low delivery of Na^+ to the CCD, one should administer a loop diuretic and expect to see an increase in the TTKG and K^+ excretion.

2. Low open probability of the Na^+ channel in the luminal membrane of the CCD:

Criteria:

1. Hyperkalemia and low TTKG ± 9αF
2. Somewhat low ECF volume and high renin
3. Na^+ and Cl^- wasting while the ECF volume is contracted

The Na^+ channel in the luminal membrane of the CCD is normally regulated by aldosterone (which leads to a greater open probability) and is inhibited by K^+-sparing diuretics, such as amiloride, and by drugs bearing a positive charge, such as the antimicrobial trimethoprim.

There are inborn or possibly acquired errors in Na^+ reabsorption in the CCD; a possible example is discussed in Case 11·2. The cardinal features of a Na^+ channel with a reduced open probability are tendencies for hyperkalemia and ECF volume contraction. The latter leads to high renin levels and the inability to achieve a Na^+-free and Cl^--free urine when the ECF volume is contracted. Finally, the TTKG should be on the low side when a physiological supplement of a mineralocorticoid is given (100 μg 9αF). A tendency to hypotension could be present in some patients, especially when they consume a low-salt diet.

3. An increased permeability to Cl^- in the luminal membrane of the CCD:

Criteria:

1. Hyperkalemia and low TTKG ± 9αF
2. Somewhat high ECF volume and low renin
3. Can excrete Na^+-poor and Cl^--poor urine

If the luminal membrane of the CCD had an increased permeability to Cl^-, there would be an enhanced electroneutral rather than electrogenic absorption of Na^+ because there would be little restriction to the movement of Cl^- across the luminal membrane. Hence, the cardinal features would be hyperkalemia, a tendency to an expanded ECF volume with low levels of renin, and possibly hypertension in those subjects who are particularly sensitive to an expanded ECF volume despite the low level of vasoconstrictors (i.e., low angiotensin II). Possible clinical examples of this variety of hyperkalemia are found in patients who receive cyclosporine, those with Gordon's syndrome (also called a Cl^- shunt or pseudohypoaldosteronism type II), or even those with diabetes mellitus given the label of *hyporeninemic hypoaldosteronism*. In these patients, bicarbonaturia may lead to a marked rise in the TTKG if aldosterone is present; we use this observation to support the speculation that there is a "Cl^- shunt" in the CCD.

4. **Nonspecific anion permeability in the CCD:**
   Nonspecific anion permeability in the CCD is a theoretical lesion. It would be characterized by a low TTKG, but if the urine volume were to rise (after a loop diuretic is given), one might expect to see a rise in the rate of excretion of K^+. These subjects should be able to have a Na^+-poor and Cl^--poor urine when salt-restricted. Bicarbonaturia would not augment kaliuresis or the TTKG with this type of lesion. There would be a tendency to ECF volume expansion and low renin levels.

5. **A low open probability of the luminal K^+ channel:**
   As far as we know, there are no reports that specifically characterize a low open probability of the luminal K^+ channel. We would suspect it if there were hyperkalemia and low excretion of K^+ (low TTKG) and if the TTKG would not rise in the presence of aldosterone and bicarbonaturia. Perhaps the most distinguishing feature would be a lack of an appreciable rise in the rate of excretion of K^+ when the flow rate in the CCD is increased following the administration of a loop diuretic; in this setting the TTKG would actually fall.

Hyporeninemic hypoaldosteronism:
We do not like to use this descriptive term because it represents a heterogeneous pathophysiology. In the Cl^- shunt, the hyporeninemia is due to ECF volume expansion; with primary destruction of the juxtaglomerular apparatus, however, there is low renin and secondary ECF volume contraction.

Causes of a Low Delivery of Volume to the CCD

Patients with a low delivery of volume to the CCD may have a high [K^+] and a high TTKG in the urine while the rate of excretion of K^+ is low. Thus, these patients are distinct from those whose hyperkalemia is caused by a low excretion of K^+ in response to a low delivery of Na^+.

A low rate of excretion of K^+ will occur in patients with a very low rate of excretion of urea (low-protein diet) and a degree of ECF volume contraction (low delivery of Na^+ and Cl^-). The rate of excretion of K^+ should rise appreciably in response to a loop diuretic if it is accompanied by an increased delivery of osmoles to the CCD.

PLASMA ALDOSTERONE AND RENIN

In the hyperkalemic patient, failure to find a high aldosterone concentration suggests an adrenal or a renin problem because hyperkalemia should stimulate the release of aldosterone. If the renin level is high, there are two possibilities: either the problem is in the adrenal gland or there is inhibition of the angiotensin-converting enzyme (see the margin). In contrast, if the patient's ECF volume is low or on the low side of normal and the renin level is not elevated, then the problem is renal in origin (Tables 11·2 and 11·5 and Figure 11·4). In some patients, there may be two renal problems: low renin release and low renal response to aldosterone.

Note:
Some patients who are severely ill behave as if they cannot generate high angiotensin II levels.

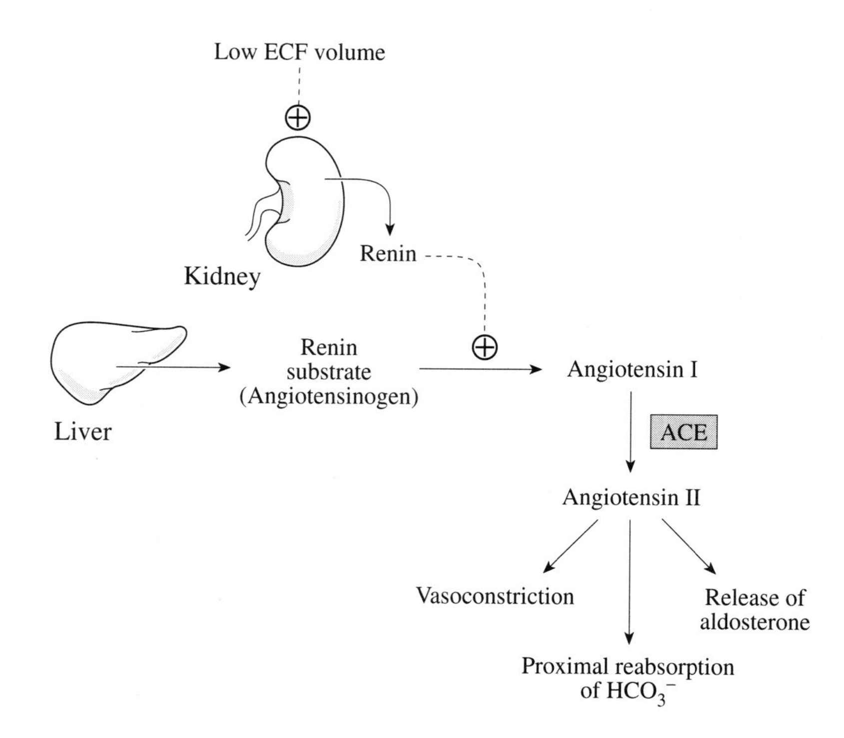

Figure 11·4
The adrenal-renal axis
Renin, which is produced in the kidney, acts on renin substrate from the liver to yield angiotensin I; this latter compound is converted into the biologically active angiotensin II by the angiotensin-converting enzyme (ACE). Angiotensin II is also a stimulator of thirst.

Table 11·5
CAUSES OF HYPORENINEMIA IN PATIENTS WITH HYPERKALEMIA

Destruction of juxtaglomerular apparatus
- Interstitial nephritis or infiltration
  - Infection
  - Drugs such as nonsteroidal anti-inflammatory agents (NSAIDS) and certain antibiotics, including methicillin
  - Depositions such as urate or amyloid
- Diabetes mellitus

Pharmacologic blockade with β_1-adrenergic blockers

Unknown causes (see the margin)

Unknown causes:
Hyporeninemia may be caused by excessive Na^+ retention and ECF volume expansion (e.g., a Cl^--shunt disorder).

Clinical Assessment

SYMPTOMS AND SIGNS

Note:
When hyperkalemia is present, the resting membrane potential is less negative than normal (partial depolarization).

The major symptom of hyperkalemia is weakness; ultimately, paralysis may occur. These symptoms appear only with a very severe degree of hyperkalemia. This weakness is first evident in the lower extremities. Often, the patient is asymptomatic and the diagnosis is "stumbled upon" when the electrolytes are "routinely" determined.

DRUGS THAT MAY CAUSE OR AGGRAVATE THE DEGREE OF HYPERKALEMIA

Table 11·6 contains a list of drugs that can cause hyperkalemia; these drugs are classified according to mechanism of action.

Table 11·6
DRUGS THAT MAY CAUSE HYPERKALEMIA

Preparations such as KCl and other K^+ salts that contain K^+ (only if renal function is compromised)

Drugs that cause a shift of K^+ from ICF to ECF
- β_2-Adrenergic blockers, α-adrenergic agonists
- Drugs that impair insulin release from β cells of the pancreas
- Drugs that depolarize cell membranes (e.g., succinylcholine, massive digitalis overdose)
- Drugs that cause cell necrosis (e.g., in cancer chemotherapy)

Drugs that interfere with excretion of K^+ in the urine
- Drugs that cause acute renal failure
- Drugs that cause interstitial nephritis (e.g., NSAIDS)
- Drugs that interfere with aldosterone bioactivity
  - Disrupted release from the adrenal gland
    - β_1-Adrenergic blockers, which diminish renin release
    - Converting enzyme inhibitors, angiotensin II receptor blockers
    - Heparin
  - Blocked binding of aldosterone to its renal receptor
    - Spironolactone
  - Presence of postreceptor blockers
    - Na^+ channel blockers in the CCD (e.g., amiloride)
    - Increased Cl^- permeability in the CCD (possibly cyclosporine)

ECG

The margin contains a list of the changes that occur in the ECG during hyperkalemia. The plasma [K^+] at which the ECG changes is variable. These changes may be seen at a lower [K^+] if hypocalcemia, hyponatremia, or acidosis is present; they are more dramatic if the changes in plasma [K^+] are acute. Early changes include an increase in T-wave amplitude that leads to tall, narrow, peaked, symmetrical T-waves. With a greater degree of hyperkalemia, the R-wave amplitude decreases, the S wave increases, and the S-T segment becomes depressed. The P-R, QRS, and Q-T intervals all become prolonged. The P-wave duration increases, yet its amplitude decreases. With extreme hyperkalemia, there is progressive QRS and T-wave widening. Terminally, ventricular tachyarrhythmias may be observed.

ECG changes during hyperkalemia:
1. Increased T-wave amplitude
2. Decreased R-wave amplitude
3. S-T segment depression
4. Decreased P-wave amplitude
5. Prolonged P-R, QRS, and Q-T
6. Absent P waves
7. Bradycardia
8. Sine-wave pattern of QRST
9. Ventricular arrhythmias

Specific Clinical Examples

ADDISON'S DISEASE

- K^+ excretion = [K^+] in urine × urine volume
  - [K^+] in urine: (aldosterone) (bicarbonaturia)
  - urine volume: (osmole excretion) (diuretics)

A simple examination of the natural history of patients with aldosterone deficiency illustrates the problems associated with the clinical assessment of reduced excretion of K^+. If the patient has chronic hyperkalemia and is

Clinical pearl:
The cardinal feature of Addison's disease is a low rate of excretion of K^+ relative to the degree of hyperkalemia.

in a steady state, the 24-hour urine K^+ excretion equals the amount absorbed from the diet. Therefore, to be diagnostic, the 24-hour urine K^+ excretion must be interpreted in conjunction with the plasma [K^+]. As a generalization, hyperkalemia and aldosterone produce many of the same results in the CCD, but hyperkalemia is less effective. Hence, the K^+ excretion rate matches the diet (60–80 mmol/day) but does not reach the expected rate of several hundred mmol/day, which a normal person would excrete if rendered hyperkalemic by K^+-loading. The TTKG is close to 5–6 vs the expected value of 10. Given these constraints, patients with Addison's disease have a plasma [K^+] in the mid-5 range, day in and day out. A higher plasma [K^+] occurs in two circumstances: first, if there is a very low GFR (low urine volume); second, if there is a very high intake of K^+ or a shift of K^+ from the ICF.

Treatment of hyperkalemia will have two features:

1. saline to restore the ECF volume and the volume of fluid delivered to the CCD;
2. a mineralocorticoid to augment Na^+ reabsorption and K^+ secretion in the CCD.

HYPERKALEMIC PERIODIC PARALYSIS

Hyperkalemic periodic paralysis is a rare inborn error of metabolism that has undergone remarkable clarification in the past several years. The clinical picture is dominated by weakness or even paralysis. Symptoms are precipitated by an event that leads to hyperkalemia (e.g., after exercise). At times, patients may suffer from a chronic increase in motor activity (clonus).

From a biochemical perspective, the lesion seems to be in the regulation of specific Na^+ channels in the cell membrane (3–15% of Na^+ channels, those sensitive to tetrodotoxin, a toxin from uncooked spiny fish from seas near Japan; see the margin). Normally, when a stimulus arrives for muscular contraction, Na^+ channels open; as flux of Na^+ increases, these cells depolarize. When the RMP approaches –50 mvolts, normal Na^+ channels close. In hyperkalemic periodic paralysis with hyperkalemia, the mutant channels remain open. Hence, cells have a persistently less negative RMP. Depending on the absolute voltage, lesser changes result in myotonia, and larger changes lead to paralysis. Hyperkalemia develops because of the less negative RMP together with relatively open K^+ channels.

Molecular biology in hyperkalemic periodic paralysis:
Recent studies have suggested a similar gene location for growth hormone and the Na^+ channel.

Treatment, which is directed at measures that avoid hyperkalemia, involves the use of certain drugs, such as acetazolamide, whose mechanisms of action are unclear but empirically effective.

CHRONIC RENAL INSUFFICIENCY

Patients with renal failure develop hyperkalemia when their GFR falls towards 20% of normal. This decline results in a lower rate of excretion of K^+. Examining the components of this excretion individually (volume, [K^+]) suggests that the basis for the low rate of excretion of K^+ is not a fall in the volume delivered to the terminal CCD but rather the very rapid rate of flow in the remaining CCDs. In more detail, if a patient with renal failure is excreting the usual 900 mosmoles each day, the volume delivered to the terminal CCD is close to the same as in normal individuals, but this volume is traversing many fewer CCDs (Figure 11·3, page 403).

The second component of the analysis is the $[K^+]$ in the CCD. Because of an extremely rapid flow rate per CCD, the $[K^+]$ in each CCD cannot be raised to 10 times that in the plasma. Therefore, the TTKG is always less than maximal, even if aldosterone is acting.

In a normal individual, 30 liters of dilute urine exit from the loop of Henle, and each liter contains 3 mmol of K^+ per liter. When 80% of this volume is reabsorbed, the $[K^+]$ can rise to 15 mmol/l without additional secretion of K^+. In contrast, in renal insufficiency, many fewer liters of fluid leave the loop, so to have a $[K^+]$ of 15 mmol/l in fluid entering the CCD, more K^+ must be secreted.

PART B
TREATMENT OF PATIENTS WITH HYPERKALEMIA

- Urgency of treatment depends on ECG changes and the anticipated future rise in plasma [K^+].

Because hyperkalemia can arise from many causes, there is no universal therapy for this electrolyte abnormality. We shall assume for simplicity that the cause of hyperkalemia is known, all diagnostic tests have been performed, and specific therapy, where applicable, was initiated. Treatment is now dictated by the degree of ECF volume expansion or contraction and the degree of elevation of the plasma [K^+] together with the anticipated rate of K^+ release from the ICF. The degree of abnormality in the ECG can help the clinician decide on the urgency of therapy because the major aim initially is to prevent cardiac arrhythmia. A normal ECG, however, does not imply that a casual approach should be taken because the ECG can change quickly. A list of the treatment modes and their time frame for effect is presented in Table 11·7. Table 11·8 summarizes therapy for hyperkalemia.

Table 11·7
TIME FRAME FOR THE EFFECT OF THERAPEUTIC AGENTS FOR HYPERKALEMIA

1. **Seconds to minutes**
   - Antagonist to cardiac effects of hyperkalemia (intravenous Ca^{2+} salts)
2. **30 minutes–1 hr**
   - $NaHCO_3$ (isotonic), but the effect is small
3. **1–4 hr**
   - Insulin
   - Aldosterone agonists
   - Rectally administered K^+-binding resins
4. **Greater than 6 hr**
   - Orally administered K^+-binding resins
5. **Immediate once instituted**
   - Dialysis

Table 11·8
SUMMARY OF THERAPY FOR HYPERKALEMIA

Stop K^+ intake.

Shift K^+ into cells.
- Hormones
  - Insulin
- Acid-base factors
  - Bicarbonate
- Gain of ICF anions
  - Replace phosphate deficits.

Note:
Table 11·8 is continued on page 411.

Increase K^+ loss from the ECF.
- Bind K^+ in the GI tract.
  - K^+ exchange resins
    - Oral resins have a slower onset of action.
    - Resins given by enema have a faster onset of action.
- **Promote urinary K^+ loss.**
  - Ensure adequate distal Na^+ and volume delivery to the CCD (e.g., give diuretics, restore the ECF volume if it is contracted).
  - Remove mineralocorticoid antagonists, K^+-sparing diuretics, or other Na^+ channel blockers (e.g., trimethoprim).
  - Administer mineralocorticoids and acetazolamide, if appropriate.

Preventing a Further Rise in Plasma [K^+]

INTAKE OF K^+

Intake of K^+ should be as low as possible. Do not overlook the fact that certain medications are K^+ salts (certain penicillin preparations, alkalinizing salts).

PREVENTING A SHIFT OF K^+ FROM THE ICF

In certain cases, preventing a shift of K^+ from the ICF is an important form of therapy. If a patient is in a catabolic state, this catabolism should be arrested. Among the drugs to consider discontinuing are those that cause cell lysis and those that promote catabolism (anticancer medications, drugs causing hemolysis, etc.). Another potential for cell lysis that should be considered is bleeding into the gastrointestinal tract. Nevertheless, because 1 liter of blood has only 0.4 liters of RBC (the [K^+] in the ICF = 150 mmol/l), the quantity of K^+ released is only 60 mmol per liter of blood digested.

Although it may be difficult to stop all of the causes of cell lysis, their presence should dictate a more active role in promoting K^+ loss on a longer-term basis. Without a rapid reversal in etiology, a patient who cannot excrete the K^+ released should be considered for dialysis before life-threatening hyperkalemia ensues.

Release of K^+ from tumors:
If one lb (454 g) of tumor is lysed, there would be a release of close to 60 mmol of K^+ (150 mmol/l in the IC 80% of the weight is water, and o lb of tumor contains 400 ml of water).

PREVENTING THE ABSORPTION OF DIETARY K^+

Certain resins bind K^+ avidly enough to diminish the net absorption of K^+. In the process, the cations Na^+, Ca^{2+}, or H^+ (depending on the nature of the resin) are displaced and absorbed. The major resin used is Na^+ polystyrene sulphonate; each gram of resin may bind 1 mmol of K^+. Usually 30 g of resin are given every two to four hours with 70% sorbitol to hasten transit to the colon because the colon is the major site of K^+ binding (where K^+ are secreted). Oral resins are therapeutically effective only after many hours; in contrast, if the resins are administered by enema, they are effective much sooner. For administration by enema, give 100 g of resin in as little water as needed to dissolve it (200 ml); keep it in the colon for as long as possible.

Note:
If diarrhea is provoked by phe thalein rather than sorbitol, t loss from the GI tract will be enhanced.

PROMOTING K^+ LOSS IN THE URINE

Two events are required for K^+ loss in the urine: an increase in the $[K^+]$ in the lumen of the CCD (aldosterone) and an increased flow through the CCD. If the patient has a low aldosterone level, administer a physiologic dose of 9αF (100 μg). If the urine $[K^+]$ (adjusted for medullary water reabsorption) does not rise in 4 hours to at least 6 × plasma $[K^+]$, the dose may be quadrupled and the same parameters may be followed (see the margin). Aldosterone action requires a lag period of up to two hours, and the maximum biological effect may take days to develop. If the patient is taking drugs that interfere with aldosterone action (K^+-sparing diuretics, trimethoprim, competitive inhibitors), they should be discontinued if the hyperkalemia cannot be controlled by other measures. Reduced K^+ excretion may occur if the volume of fluid reaching the CCD is small. Suspect this occurrence in patients with ECF volume contraction or a low cardiac output. In the former case, saline administration is appropriate. Digitalis and/or a loop diuretic may help patients in the second category. Even if the $[K^+]$ in the CCD is lower than one might hope for, raising the flow rate in the CCD with a loop diuretic can increase the rate of excretion of K^+.

Bicarbonaturia:
The most effective agent to increase the $[K^+]$ in the CCD is HCO_3^-; the excretion of HCO_3^- can be enhanced by acetazolamide.

$NaHCO_3$ THERAPY

If the patient had metabolic acidosis caused by loss of $NaHCO_3$, the acidemia could have led to a shift of K^+ into the ECF. This shift can be minimized by administering sufficient $NaHCO_3$ to bring the blood pH towards normal. Even in patients who have a normal blood pH, therapy with $NaHCO_3$ (50 mmol over 5–10 minutes, repeat 30 minutes later if necessary) can be effective in lowering the plasma $[K^+]$. Nevertheless, in clinical trials, $NaHCO_3$ is relatively ineffective unless it augments renal excretion of K^+. The dangers of this therapy are related to the Na^+ load given (ECF volume expansion) and to potential problems of alkalemia (tetany, etc.).

Emergency treatment of hyperkalemia:

- If the QRS abnormality is significant, give 10 ml of 10% Ca^{2+} gluconate.
- Begin an infusion of 500 ml $D_{10}W$ that contains 10 units of regular insulin. If there is no heart failure, add 50 mmol of $NaHCO_3$ and run the infusion at 150–200 ml/hour.
- If acidemia is present ($[HCO_3^-] < 15$ mmol/l and no heart failure), give 50 mmol of $NaHCO_3$ IV over 5 minutes.
- Give 100 g of K^+-binding resin by enema.
- If there is renal failure, make plans for dialysis.
- If the patient has heart failure and is acidemic, consider phlebotomy to allow $NaHCO_3$ administration. HCl loss from the stomach can also be promoted (see page 150).

HORMONAL THERAPY

Two major hormones need to be considered—insulin and β_2-adrenergics. If the patient is severely hyperglycemic, insulin can have a dramatic effect in lowering the plasma $[K^+]$. Insulin therapy can also be dramatic in diabetic patients with aldosterone deficiency.

To lower the plasma $[K^+]$, large doses of insulin may be necessary, but do not rely on this administration as a long-term therapy. As an example, giving 500 ml of 10% glucose in water stimulates endogenous insulin release; if diabetes mellitus is present, give a bolus of 10 units of insulin. A much higher dose of insulin may be required to activate the Na^+/H^+ antiporter.

β_2-Adrenergics shift K^+ into cells. Therefore, β-blockers should be discontinued. There is not much therapeutic leverage in administering β_2-adrenergics to lower the plasma $[K^+]$.

Although a lack of an essential ICF constituent such as phosphate or magnesium is not directly related to hormonal or acid-base effects, any deficits should be replaced to promote anabolism and a subsequent shift of K^+ into cells.

ANTAGONIZING THE EFFECTS OF K^+

The administration of calcium salts can decrease membrane excitability and thereby protect against the effects of hyperkalemia. This benefit begins within minutes but is relatively short-lived; it "buys 60 minutes" for other forms of therapy (insulin, $NaHCO_3$) to lower the plasma [K^+]. The usual dose is 10 ml of 10% Ca^{2+} salt infused over 2–3 minutes. This dose can be repeated in 5–10 minutes if ECG changes persist (note the danger of hypercalcemia-induced digitalis toxicity if cardiac glycosides are being given).

REMOVING K^+ BY DIALYSIS

When renal function cannot eliminate sufficient K^+, dialysis therapy should be considered. The major indications for dialysis are ECG changes or a plasma [K^+] that is very high and the anticipation of an ongoing shift of K^+ from cells. Hemodialysis is preferred over peritoneal dialysis because it is more efficient at removing K^+. Although there will be some variation because of characteristics of dialyzer membranes, differences in the [K^+], and blood flow rates, close to 30–50 mmol of K^+ may be removed per hour by hemodialysis and 15–25 mmol of K^+ per hour by peritoneal dialysis.

Apart from the dialysis membrane and the flow rate past this membrane, the major factor influencing the dialysis of K^+ is the [K^+] gradient: the higher the plasma [K^+] and the lower the bath [K^+], the greater the rate of K^+ removal. Therefore, once dialysis is started, one should stop the glucose and insulin drip and reduce the bath glucose to minimize those forces moving K^+ into the ICF.

Questions

(Discussions on pages 423–24)

11·7 Does aldosterone act only on the kidney and the ICF-ECF interface to lower the plasma [K^+]?

11·8 An otherwise healthy patient presents with hyperkalemia associated with mild ECF volume expansion and hypertension, metabolic acidosis with a normal plasma anion gap, and an unexpectedly low urine [K^+]. No drugs were taken. What is the lesion in this patient?

11·9 A patient has a severe degree of hyperkalemia and ECF volume contraction. A random urine sample has the following values: [Na^+] = 3 mmol/l, [K^+] = 80 mmol/l, [Cl^-] = 23 mmol/l. What should be done to increase the rate of excretion of K^+ promptly?

PART C
REVIEW

Discussion of Introductory Case Lee's [K^+] Is on a High

(Case presented on page 394)

What factors contributed to hyperkalemia?

First, pseudohyperkalemia caused by a hemolyzed sample, thrombocytosis, or leukemia should be ruled out (and was); because there is no indication that K^+ intake was excessive, hyperkalemia is probably due to a K^+ shift from the ICF and/or reduced K^+ excretion.

K^+ shift from the ICF: There is no evidence of excessive destruction of cells, so a single or multiple hormone deficiency should be suspected if there is a major shift of K^+ from cells. Insulin deficiency is present; aldosterone deficiency is discussed below. In the setting of insulin plus aldosterone deficiencies, hyperglycemia exacerbates hyperkalemia. Because the metabolic acidosis was not ketoacidosis, as judged from the normal value for the plasma anion gap, the acidemia may have caused some K^+ to shift out of cells.

Reduced K^+ excretion in the urine: The GFR is not markedly reduced and the urine volume is large, so aldosterone deficiency is possibly the cause of a low rate of K^+ excretion. All of the following could point to this diagnosis: Lee has ECF volume contraction but is losing Na^+ in the urine; the urine [K^+] is low (especially when correction is made for water abstraction in the MCD); and the TTKG is in the range that suggests a low aldosterone bioactivity (see the margin). To determine if this low TTKG is due to low levels of aldosterone, a physiologic replacement dose of aldosterone (100 μg 9αF) should be given and the TTKG measured. If the TTKG rises toward 10, an adrenal problem is likely; it if remains low, there is a problem with the K^+ secretory process in Lee's CCD.

TTKG:

Assume that her plasma osmolality is 300 mosm/kg H_2O for easy math. Her TTKG is 20/(600/300)/5.4 = close to 2; the expected value would be close to 10 with "perfect" kidneys and adrenals.

Osmotic diuresis:

The glucose-induced osmotic diuresis could explain the natriuresis and the low TTKG.

What additional information is needed to understand the relative importance of each factor?

When Lee received insulin, her blood glucose concentration fell to less than 180 mg % (10 mmol/l), yet her plasma [K^+] fell to only 4.8 mmol/l.

When Lee received sufficient $NaHCO_3$ to bring the plasma [HCO_3^-] to normal, the plasma [K^+] fell to 4.3 mmol/l; nonetheless, the Na^+ load could have led to an increased renal K^+ excretion.

Blood samples were drawn to measure aldosterone and renin. Because the results require several days, a physiologic dose of mineralocorticoids was administered. There was little change in the plasma and urine [K^+]. Thus, the major problem seems to be reduced K^+ excretion in the presence of normal concentrations of mineralocorticoids. The problem therefore lies in the kidney (the plasma renin and aldosterone levels were found to be high), and the final diagnosis is end-organ (kidney) hyporesponsiveness to aldosterone. One would then further characterize this defect by individually assessing the Na^+ and K^+ channels and the Cl^- permeability in the CCD (see pages 332–35).

What initial therapy would be most appropriate?

Initial therapy should consist of a low K^+ intake, insulin for better glycemic control (and to shift K^+ into the ICF), $NaHCO_3$ to correct the metabolic acidosis (and to shift K^+ into the ICF), and saline to maintain a normal ECF volume. Because end-organ hyporesponsiveness to aldosterone may be transient, Lee needs to be observed closely.

Cases for Review

Case 11·1
Ewe Too Can Avoid Hyperkalemia

(Case discussed on pages 417–18)

Ewe, a female sheep, weighs 70 kg. She eats 800 mmol of K^+ daily but does not become hyperkalemic because she excretes 800 mmol of K^+ in her urine each day.

What quantity of K^+ can Ewe store in her cells if she does not change her acid-base status?
What renal mechanisms permit such a large excretion of K^+?
How might a gastric H^+/K^+ATPase offer Ewe an advantage in excreting her dietary K^+ load?
What factors lead to the delivery of HCO_3^- to her CCD?

Case 11·2
Channel Your Thoughts

(Case discussed on pages 418–19)

Shirley, aged 30 years, was referred by her family physician because the basis of her hypertension (150/110 mm Hg) could not be detected after an extensive work-up, including renal arteriography. The only value that was considered abnormal was a very mild degree of ECF volume contraction on physical examination and hyperkalemia (5.1 mmol/l). Laboratory data from the investigations are provided in the table below. Samples were obtained when Shirley was off all medications. The excretions of Na^+ and K^+ in a 24-hour urine sample (volume 1 liter) were 150 and 53 mmol, respectively.

		Plasma	**Random urine**
Na^+	mmol/l	138	160
K^+	mmol/l	5.2	46
Cl^-	mmol/l	106	161
HCO_3^-	mmol/l	18	0
Urea	mmol/l (mg/dl)	4.3 (12)	-
Creatinine	μmol/l (mg/dl)	84 (0.9)	-
Albumin	g/l (g/dl)	40 (4.0)	-
Osmolality	mosm/kg H_2O	285	606

SPECIAL STUDIES

1. The TTKG did not rise after the administration of 9αF.
2. Plasma renin was very elevated; it fell to normal with salt-loading. Plasma aldosterone levels moved in parallel.
3. In response to ECF volume contraction, the [Na^+] and [Cl^-] in the urine were 46 and 76 mmol/l, respectively.

What might be the pathophysiology of her hyperkalemia?
How might it be related to the hypertension?
What specific therapy might help ameliorate her hypertension and hyperkalemia?

Case 11·3
Hyperkalemia After Renal Transplantation

(Case discussed on page 420)

Andrea, aged 24 years, had a successful renal transplant. Postoperative medications included cyclosporine. On routine visit six months later, several abnormalities were detected. First, a mild degree of hypertension developed (a new finding). On laboratory investigation, there was a mild degree of hyperkalemia and metabolic acidosis with a normal plasma anion gap. The only other findings were a very low renin level and a normal value for aldosterone in plasma. There were no changes in K^+ excretion (or the TTKG) when aldosterone was given.

		Plasma	Random urine sample
Na^+	mmol/l	137	103
K^+	mmol/l	5.2	46
Cl^-	mmol/l	108	111
HCO_3^-	mmol/l	19	0
pH		7.35	5.1
Creatinine	μmol/l (mg/dl)	123 (1.4)	-
Osmolality	mosm/kg H_2O	285	606

What is the most likely basis for the hyperkalemia?
What additional studies might help in clarifying this pathophysiology?
Why was metabolic acidosis present?

Case 11·4
Treatment Threatens More Than the Protozoa

(Case discussed on pages 420–21)

Pat has had AIDS for several years. He recently developed a nonproductive cough. Because the cough persisted and he developed chills and shortness of breath, he went to the emergency room. His blood pressure was 110/70 and his pulse was 86. The physical examination was normal except that he had crackles at both lung bases. His lab results were all normal (K^+ = 4.3 mmol/l and creatinine = 86 μmol/l). The chest X-ray showed bilateral interstitial infiltrates. The presumptive diagnosis was Pneumocystis pneumonia, so he was started on trimethoprim (240 mg) and sulfamethoxazole (1200 mg) every eight hours. Three days later, although his pneumonia was stable, his blood pressure fell to 90/70 when he was standing. His jugular venous pulse was not visible. His lab results on day three are listed below. Because his 24-hour urine volume was 0.8 liters, his osmole excretion rate was somewhat low, 480 mosmoles per day.

		Plasma	Random urine sample
Na^+	mmol/l	131	82
K^+	mmol/l	5.6	18
Cl^-	mmol/l	99	77
HCO_3^-	mmol/l	21	0
pH		7.35	5.1
Creatinine	μmol/l (mg/dl)	111 (1.3)	-
Osmolality	mosm/kg H_2O	268	577

Why has hyperkalemia developed?
What is the most likely intrarenal defect?

Discussion of Cases

Discussion of Case 11·1 Ewe Too Can Avoid Hyperkalemia

(Case presented on page 415)

What quantity of K^+ can Ewe store in her cells if she does not change her acid-base status?

If Ewe wanted to shift K^+ into cells and could neither generate new organic phosphate esters (RNA, DNA, etc.) nor change the $[H^+]$, the only way for K^+ to enter her cells is if Na^+ exit. Because the ICF of a 70-kg adult contains close to 300 mmol of Na^+ (10 mmol/l × 30 liters), this amount sets the maximum for net entry of K^+ into cells (one meal's worth). This limit underscores the need for Ewe to have a rapid rate of excretion of K^+.

What renal mechanisms permit such a large excretion of K^+?

To have a rapid rate of excretion of K^+, there must be a large volume delivered to the terminal CCD (many osmoles) and a very high $[K^+]$ in the lumen of the CCD. With regard to the former, sheep excrete threefold to fourfold more osmoles than do humans (half are urea). With regard to the latter, the $[K^+]$ in the CCD is higher because aldosterone is present from the K^+ load, and bicarbonaturia is always present (the major anions accompanying K^+ in the diet are organic anions, which are ultimately metabolized and yield HCO_3^- as a product; see below as well).

How might a gastric H^+/K^+ATPase offer Ewe an advantage in excreting her dietary K^+ load?

To excrete K^+ rapidly, one must ensure that HCO_3^- are delivered to the lumen of the terminal CCD and that aldosterone is present. If H^+ are secreted in conjunction with gastric reabsorption of luminal K^+, the net effect will be the addition of K^+ and HCO_3^- to the venous blood leaving the stomach (Figure 4·1, page 150). Thus, the sheep need not depend on very rapid metabolism of the dietary organic anions to yield HCO_3^-. An advantage of bicarbonaturia is the initiation of a high rate of kaliuresis.

What factors lead to the delivery of HCO_3^- to her CCD?

For a high [K^+] in the lumen of the CCD, one needs a high [HCO_3^-] in the luminal fluid and a high level of aldosterone in plasma. To have a high [HCO_3^-] in the CCD, more HCO_3^- must be filtered and/or less must be reabsorbed. Reabsorption of HCO_3^- in the proximal tubule is inhibited mainly by ECF volume expansion (low angiotensin II) and hyperkalemia (possibly via ICF alkalemia). The overall effect of providing K^+ and HCO_3^- to the body is that together they augment the delivery of HCO_3^- to the CCD by the following mechanisms: the alkalemia should depress the indirect reabsorption of filtered HCO_3^- in the PCT; a higher plasma [K^+] also depresses this reabsorption and ensures delivery of Na^+ and HCO_3^- to the CCD. In the presence of aldosterone actions, there will be a very high TTKG (close to 20). HCO_3^- can still be reabsorbed in the MCD if stimulated by a fall in the plasma [HCO_3^-] because there are many H^+ATPase pumps in this nephron segment.

Discussion of Case 11·2 Channel Your Thoughts

(Case presented on pages 415–16)

What might be the pathophysiology of her hyperkalemia?

Note:
See page 394 for a review flow chart on causes of hyperkalemia with low excretion of K^+.

Because long-term homeostasis for K^+ is maintained by regulation of renal excretion, the first step is to assess the renal response to hyperkalemia. The expected value for the rate of excretion of K^+ with hyperkalemia and normal kidneys is the excretion of several hundred mmol/day; 24-hour K^+ excretion in Shirley was only 53 mmol/day. Because K^+ excretion is the product of the [K^+] and the volume of urine, there are two major components that could be responsible for her low rate of excretion of K^+: a low volume of urine or a decreased ability to raise the [K^+] (reduced K^+ secretion in the CCD).

Volume in the CCD: The volume of urine and osmole excretion rate were not particularly low; therefore, the problem must be a low [K^+] in the CCD.

TTKG = 46/(606/285)/5.2 = 4.2

TTKG: The measured value of the TTKG was 4·2 (see the margin), much lower than the expected value of 10. This low TTKG may indicate the lack of a stimulator, the presence of an inhibitor, or an intrinsic defect in the K^+ secretory process. The latter seems to be the case because the TTKG failed to rise in response to 9αF. Moreover, the levels of renin and aldosterone in her plasma were elevated. Therefore, lack of mineralocorticoids was not the cause of hyperkalemia with a low TTKG.

Our approach: The subnormal kaliuretic response to hyperkalemia and the tubular insensitivity to aldosterone may result from low delivery of Na^+ to the CCD, low open probability of the Na^+ or K^+ conductive pathways, or high permeability to Cl^- in the CCD (Figure 11·2, page 402).

Less availability of Na^+: In the rat, the $[Na^+]$ required for half-maximal rates of K^+ secretion in the CCD is 10–15 mmol/l. Because Shirley had a much higher $[Na^+]$ in her urine and because some Na^+ are normally reabsorbed in the MCD, reduced delivery of Na^+ to the CCD was very unlikely to limit her K^+ secretory system.

High permeability to Cl^- in the CCD: Failure to generate a favorable TEPD in the CCD because of increased "permeability" to Cl^- in the CCD, the so-called "Cl^--shunt disorder" has been postulated as one lesion that could cause hypertension, hyperkalemia, low levels of renin, and a low rate of excretion of K^+ in a patient with a normal GFR. The shunting of Cl^- when Na^+ are reabsorbed (electroneutral reabsorption of NaCl) leads to ECF volume expansion and suppression of the renin-angiotensin-aldosterone system. These features were not present in Shirley.

Low Na^+ channel activity in the CCD: The mild degree of contraction of the ECF volume on clinical examination, the presence of high renin and aldosterone, and the high concentrations of Na^+ and Cl^- in urine suggest a different pathophysiology for the hyperkalemia and the tubular insensitivity to mineralocorticoids in Shirley. Failure to reabsorb Na^+ in the CCD could diminish the TEPD in the CCD and thereby reduce the driving force to secrete K^+. Moreover, this wasting of Na^+ could lead to a degree of contraction of the ECF volume, which would in turn stimulate the release of renin and aldosterone; all are consistent with Shirley's case.

How might it be related to the hypertension?

A second lesion is necessary to explain the presence of hypertension. Not only does Shirley need high levels of vasoconstrictors, but there must also be an increase in sensitivity to these agents to produce hypertension because the ECF volume is contracted.

If her hypertension is due to the ECF volume contraction, salt-loading would decrease the levels of vasoconstrictors and benefit her blood pressure.

What specific therapy might help ameliorate her hypertension and hyperkalemia?

There are two therapeutic options with respect to the hyperkalemia. One is to increase the urine flow rate; if the TTKG does not change, excretion of K^+ will rise. A diet high in NaCl will not only increase the flow rate, but will also reexpand the ECF volume. The other strategy is to try to increase the TTKG by causing bicarbonaturia (Na^+ load with $NaHCO_3$) to reduce Cl^- reabsorption.

One also can maintain her on a low K^+ diet.

Note:
See page 394 for a review flow chart on causes of hyperkalemia with low excretion of K^+.

TTKG:
$[K^+]$ in urine (46)
÷ $(U/P)_{osm}$ (606/285)
÷ $[K^+]$ in plasma (5.2)
= 4

Volume to the CCD:
= U_{osm} (606) × urine volume (1 liter)
÷ P_{osm} 285
= More than 2 liters to the terminal CCD per day.

Discussion of Case 11·3
Hyperkalemia After Renal Transplantation

(Case presented on page 416)

What is the most likely basis for the hyperkalemia?

Chronic hyperkalemia is due at least in part to a low excretion of K^+. The basis for the low excretion was an unexpectedly low TTKG (close to 4; see the margin). The components of K^+ excretion indicate that there was a reasonable urine volume and osmole excretion rate (see the margin), and Na^+ delivery was not rate-limiting for K^+ excretion, as judged by the urine $[Na^+]$. One possible explanation is a low open probability for the Na^+ channel, but this basis for the hyperkalemia is unlikely because renin level was low, not high. This constellation better reflects a so-called Cl^- shunt, at best a good guess.

What additional studies might help in clarifying this pathophysiology?

A Cl^- shunt seems to respond to bicarbonaturia or a protocol that lowers the $[Cl^-]$ in the CCD. The TTKG rose with bicarbonaturia; it also rose with Cl^--free urine. One might expect to see the excretion of urine with little Na^+ and Cl^- upon NaCl restriction.

Why was metabolic acidosis present?

The urine net charge was positive (more Na^+ + K^+ than Cl^-), which suggests a low rate of excretion of NH_4^+ (confirmed later by direct measurements). The acidosis and low NH_4^+ excretion were no longer evident when normokalemia was present, so the basis for the low NH_4^+ excretion was hyperkalemia (see the discussion of Question 1·24, page 42).

Discussion of Case 11·4
Treatment Threatens More Than the Protozoa

(Case presented on pages 416–17)

Why has hyperkalemia developed?

Because this development of hyperkalemia is acute, one should begin by considering two initial events: first, there may be a greater input of K^+ from the diet, cell lysis, or a shift of K^+ from the ICF (e.g., lack of insulin); or second, there may be a decreased rate of excretion of K^+. A low flow in the CCD raises the probability of a low osmolar load (e.g., anorexia and/or a very low GFR). The very low TTKG suggests that the low $[K^+]$ in the urine is important.

What is the most likely intrarenal defect?

1. **$[K^+]$ vs flow rate**: The excretion of K^+ is the product of the $[K^+]$ in the urine and the urine flow rate. Translating these events to the CCD reveals a somewhat reduced volume (excretion of 480 mosmoles) but a $[K^+]$ that is much lower than expected (TTKG of 1.5 vs the expected value of 10; see the margin).

TTKG:
$[K^+]$ in urine (18)
÷ $(U/P)_{osm}$ (577/268)
÷ $[K^+]$ in plasma (5.6)
= 1.5

2. **Renal vs prerenal cause for low TTKG:** Because hypoaldosteronism may be present (low ECF volume), a physiologic dose of aldosterone was given; two hours after receiving 100 μg of 9αF, the TTKG did not rise in freshly voided urine. Hypoaldosteronism, if present, is therefore not the sole cause of the low TTKG.
3. **Channel analysis in the CCD:** A low TTKG in this setting implies either an abnormally low flux via the Na^+ channel or increased permeability to Cl^- in the luminal membrane of the CCD. The low ECF volume with renal Na^+ wasting is more consistent with Na^+ channel problem than with a Cl^- permeability issue, which would be associated with ECF volume expansion.
4. **Inhibition of Na^+ channels:** Aside from low levels of aldosterone, drugs that are cationic, such as amiloride, inhibit Na^+ channel flux. Pat was taking trimethoprim, an antimicrobial that bears a positively charged amine that can act like amiloride. Later, his renin and aldosterone results were reported and both were elevated.

Summary of Main Points

- Hyperkalemia is an important electrolyte disorder because it predisposes the patient to potentially lethal cardiac arrhythmias.
- Although intake and shift of K^+ from the ICF may contribute to hyperkalemia, reduced excretion of K^+ in the urine is virtually always present in chronic hyperkalemia.
- An unexpectedly low rate of excretion of K^+ implies either a low flow rate in the CCD (low osmole excretion rate) or, more commonly, a low $[K^+]$ in the luminal fluid of the CCD.
- The major causes of a low $[K^+]$ in the luminal fluid of the CCD are a Na^+ channel with a low open probability (e.g., low aldosterone bioactivity or drugs such as amiloride or trimethoprim) or a Cl^--shunt type of disorder. Evaluating the ECF volume, the ability to excrete a Na^+-free and Cl^--free urine with salt restriction, and the TTKG after mineralocorticoid administration helps in the differential diagnosis.
- In an emergency, the adverse cardiac effects of hyperkalemia can be attenuated by infusing Ca^{2+} salts. This infusion can be followed by measures to shift K^+ into cells (insulin, $NaHCO_3$). Longer-term strategies include dietary restriction of K^+, binding of K^+ in the GI tract with resins, and/or promoting the excretion of K^+ by increasing the flow rate (loop diuretics) or $[K^+]$ (aldosterone, bicarbonaturia). These strategies depend on the type of lesion that is present.

Discussion of Questions

11·1 What is the total K^+ content in the platelets of 1 liter of blood? (Assume 400,000 cells/mm^3, a 150 mmol/l [K^+] in the ICF, and a platelet volume of 1 μ^3.)

The platelet count in normal blood is 4×10^5 cells/mm^3, which is equivalent to 4×10^{11} cells per liter (1 cm^3 = 1 ml).

If the volume of an individual platelet is 1 μ^3 (equal to 1×10^{-15} liters), the total platelet ICF volume is the product of these two numbers, or 4×10^{-4} liters.

The [K^+] in the ICF of a platelet was given as 150 mmol/l, so the total K^+ content of platelets is 150 mmol $\times 4 \times 10^{-4}$ liters, or 6×10^{-2} mmol. If lysed, the rise in the [K^+] would be only 0.06 mmol/l.

To get an appreciable rise in the plasma [K^+] from the platelet disruption, the platelets must be increased considerably in volume; a change in the platelet number is much less important.

11·2 What sets the limit on K^+ uptake into cells?

When a K^+ enters a cell, either an anion must enter or a cation (Na^+ or H^+) must exit to maintain electroneutrality. Because there is a limit to the synthesis of new intracellular anions, and because the major anion available to enter is a Cl^- and it primarily has an ECF distribution, an exit of Na^+ or H^+ sets the limit on K^+ entry in an acute setting. The total quantity of Na^+ that can exit is equal to the Na^+ content in the ICF (300 mmol in a 70-kg adult; i.e., 10 mmol/l $\times$ 30 liters of ICF). The [H^+] gradient across cells and the content of HCO_3^- in the ECF (375 mmol) set the upper limit on a countermovement of K^+ for H^+.

11·3 How might β-blockers cause a severe degree of hyperkalemia?

There are several settings in which it is important to have the "K^+-shift defense" operating at its greatest efficiency (for example, following heavy physical exercise, when trauma causes cell necrosis, after a large K^+ intake, during renal failure, in insulin plus aldosterone deficiencies, and after surgery); β-blockade at these times may cause severe hyperkalemia.

11·4 What happens to the plasma [K^+] during exercise?

The plasma [K^+] may rise by several mmol/l following exhausting exercise. Although all the mechanisms are not clearly defined, the most likely cause is depolarization of muscle cells, which yields a high interstitial [K^+] that is washed into the circulation (i.e., K^+ are not restricted to the T-tubules). Cell damage (muscle, red blood cells mechanically ruptured by persistent pressure on the soles of the feet) may make a small contribution to this rise. In exercise, the hyperkalemia is particularly well-tolerated by the otherwise normal subject, and its duration is short-lived. β_2-Adrenergic activity helps in minimizing the degree and the duration of the hyperkalemia.

11·5 Why do patients with a problem regenerating ATP in muscle cells suffer from weakness but not hyperkalemia?

We do not know, but hyperkalemia is not generally observed. If these patients lack ATP, the Na^+K^+ATPase should stop pumping ions. Because these patients do not become hyperkalemic, they probably have markedly reduced K^+ conductance in their cell membranes.

Note:
A deficiency in carnitine palmitoyl transferase (i.e., a diminished ability to oxidize fatty acids and therefore an increased sensitivity to carbohydrate deprivation) is an example of a disease with a problem regenerating ATP.

11·6 How much of a K^+ load might a patient receive if one liter of blood is digested in the GI tract in the course of a GI bleed?

One liter of blood has only 0.4 liters of red blood cells. The $[K^+]$ in the ICF is as high as 150 mmol/l; therefore, the quantity of K^+ released is only 60 mmol/l of blood ingested. The plasma contains another 2.4 mmol (0.6 liters × 4 mmol/l).

11·7 Does aldosterone act only on the kidney and the ICF-ECF interface to lower the plasma $[K^+]$?

No. Aldosterone also acts on the colon to promote fecal K^+ loss, but this loss is only quantitatively important in patients with chronic renal failure in whom almost half of the daily K^+ load is excreted in this manner.

The mechanisms involved in gastrointestinal K^+ loss are analogous to those in the cortical distal nephron. In addition, glucocorticoids promote K^+ loss via the colon.

Aldosterone also promotes the loss of K^+ in sweat. This loss can become important in athletes who train in hot environments.

11·8 An otherwise healthy patient presents with hyperkalemia associated with mild ECF volume expansion and hypertension, metabolic acidosis with a normal plasma anion gap, and an unexpectedly low urine $[K^+]$. No drugs were taken. What is the lesion in this patient?

Hyperkalemia should lead to aldosterone release and enhanced K^+ excretion. These changes did not occur. Therefore, either aldosterone is absent or not working. The absence of aldosterone as a primary lesion would lead to Na^+ loss and ECF volume contraction. This patient does not have a typical aldosterone deficiency because the ECF volume is expanded.

If aldosterone is present and is preventing Na^+ loss, why does it not lead to K^+ loss? Recall that a major action of aldosterone is to promote Na^+ reabsorption in the CCD (it opens the Na^+ channels). In this case, if the accompanying anions (Cl^-) were reabsorbed in parallel, there would not be a luminal negative voltage, and there would be little secretion of K^+ and H^+; the low secretion of H^+ (together with hyperkalemia-induced inhibition of NH_4^+ excretion) could cause the metabolic acidosis. The lesion is very likely a "Cl^--shunt disorder."

Note:
See page 394 for a review flow chart on causes of hyperkalemia with low excretion of K^+.

11·9 A patient has a severe degree of hyperkalemia and ECF volume contraction. A random urine sample has the following values: [Na^+] = 3 mmol/l, [K^+] = 80 mmol/l, [Cl^-] = 23 mmol/l. What should be done to increase the rate of excretion of K^+ promptly?

There is little leverage to raise the [K^+] in the urine further by giving mineralocorticoids (you may be able to raise it twofold if the urine osmolality is now 1200 mosm/kg H_2O). A better strategy is to raise the urine flow rate with a loop diuretic, which should increase the rate of excretion of K^+ considerably. Of course, NaCl must be infused more rapidly than it is excreted to result in reexpansion of the ECF volume.

SECTION FOUR

Hyperglycemia

12

HYPERGLYCEMIA

Objectives

- To explain why a severe degree of hyperglycemia develops.
- To explain how hyperglycemia causes a shift of water across cell membranes and an osmotic diuresis and why hyperglycemia is often associated with catabolism of lean body mass.
- To provide the rationale for designing appropriate intravenous therapy for a patient with a severe degree of hyperglycemia.
- To demonstrate in a quantitative fashion the factors that determine the fall in the concentration of glucose that occurs during treatment of a patient with a severe degree of hyperglycemia.

Abbreviations:
CCD = cortical collecting duct.
CNS = central nervous system.
ECF = extracellular fluid.
GFR = glomerular filtration rate.
HHNC = hyperglycemic hyperosmolar nonketotic coma.
HHS = hyperglycemic hyperosmolar syndrome.
ICF = intracellular fluid.
IDDM = insulin-dependent diabetes mellitus.
MCD = medullary collecting duct.
NIDDM = noninsulin-dependent diabetes mellitus.
PDH = pyruvate dehydrogenase.

Outline of Major Principles

1. Hyperglycemia requires a relative lack of insulin or a resistance to the actions of insulin.
2. A severe degree of hyperglycemia usually requires a low output of glucose, including a low GFR. On occasion, there may be an excessive intake of glucose and little reduction in the GFR.
3. The concentration of glucose in plasma will decline early during therapy, primarily as a result of dilution and glucosuria. Metabolism of glucose is a minor pathway for the removal of glucose at this time.
4. The actions of insulin become important in lowering the blood glucose level many hours after administration of this hormone.
5. The major complications of hyperglycemia are shifts of water across cell membranes and losses of Na^+ and K^+ via osmotic diuresis.

Note:
To describe the actions of low net insulin more completely, we must include both the low levels of insulin and high levels of glucagon, the normal hormonal responses to fasting with hypoglycemia. For full-blown ketoacid production, both a low insulin level and a high glucagon level are required.

Introductory Case
Terry Is Confused

(Case discussed on pages 447–48)

Terry, aged 72, has had NIDDM for the past 10 years; since she lost weight, her diabetes has been under reasonable control. Two weeks ago, her physician prescribed a thiazide diuretic to treat hypertension (160/95 mm Hg). Since she began taking this medication, Terry has not felt well, her urine output has increased, thirst has become prominent (she has been drinking a large quantity of apple juice) and, more recently, she has become lightheaded when standing. Today Terry's daughter found her quite confused. In the hospital, the two principal new findings are confusion and a marked degree of ECF volume contraction (her blood pressure is now 130/60 mm Hg).

Laboratory results (summarized in Table 12·1) reveal a marked degree of hyperglycemia, hyponatremia, and a high value for creatinine in plasma.

Table 12·1
TERRY'S VALUES ON ADMISSION
The urine output was very low on admission.

		Normal	Terry
Na^+	mmol/l	140	126
K^+	mmol/l	4.0	4.1
Cl^-	mmol/l	103	82
HCO_3^-	mmol/l	25	25
pH		7.40	7.40
$[H^+]$	nmol/l	40	40
P_aco_2	mm Hg	40	40
Glucose	mmol/l (mg/dl)	4 (72)	50 (900)
Creatinine	μmol/l (mg/dl)	80 (0.9)	200 (2.3)
Urea	mmol/l (mg/dl)	4 (11)	20 (56)

What is the basis for her severe degree of hyperglycemia?
Why does Terry have hyponatremia?
What is the appropriate therapy?
Why will the concentration of glucose fall during therapy?

PART A
GENERAL APPROACH TO A SEVERE DEGREE OF HYPERGLYCEMIA

- A very high concentration of glucose indicates a low GFR and possibly a high input of glucose.
- A large change in the concentration of glucose can occur rapidly because its pool size is small and the rate at which it is consumed can be very rapid.

Background

It is beyond the scope of this text to describe the normal metabolism of glucose in any detail. Nevertheless, the following points should serve as a review.

1. In normal metabolism, glucose is very important because it is the principal fuel for the CNS. Only when subjects do not eat might the supply of glucose be inadequate to meet this demand (Table 12·2); in this setting, the brain oxidizes ketoacids.

2. Outside the CNS (see the margin), there is a *hierarchy of fuel oxidation:* fatty acids, if present, are the preferred fuel to oxidize, then ketoacids; if neither fatty acids nor ketoacids are available, glucose can be oxidized. The basis for this hierarchy is that pyruvate dehydrogenase (PDH), the highly regulated enzyme that controls the oxidation of glucose, is inhibited by acetyl-CoA, NADH, and ATP, the products of the oxidation of fatty acids or ketoacids (Figure 12·1). A lack of insulin or a resistance to its actions makes fatty acids or ketoacids available for oxidation.

3. A mild degree of hyperglycemia is the result of a lack of insulin. A severe degree of hyperglycemia requires either an extraordinarily large intake of glucose or, much more commonly, a low excretion of glucose because the GFR is so low.

Note:
Because fatty acids cannot cross the blood-brain barrier at a significant rate, fatty acids are never an important fuel for the brain.

Fuel oxidation in the brain:
The brain can burn only two fuels: the first is ketoacids if they are present, and the second is glucose. When the levels of ketoacids and glucose fall (in response to the actions of insulin), the brain will have insufficient fuel for its needs.

Hierarchy of fuel oxidation:
The hierarchy depends on which fuel generates acetyl-CoA, the substrate for the TCA cycle. When acetyl-CoA is formed from fat-derived fuels, this intermediate cannot be formed from carbohydrates or proteins.

Table 12·2
DEGREE OF HYPERGLYCEMIA AND ITS CLINICAL SIGNIFICANCE

Concentration of glucose mmol/l	mg/dl	Clinical significance
4–8	70–140	Normal fed state
3–3.5	55–65	Normal fasted state
10–15	180–270	Relative insulin deficiency
> 25	> 390	Relative lack of insulin plus a low GFR, and/or possibly a very large intake of glucose

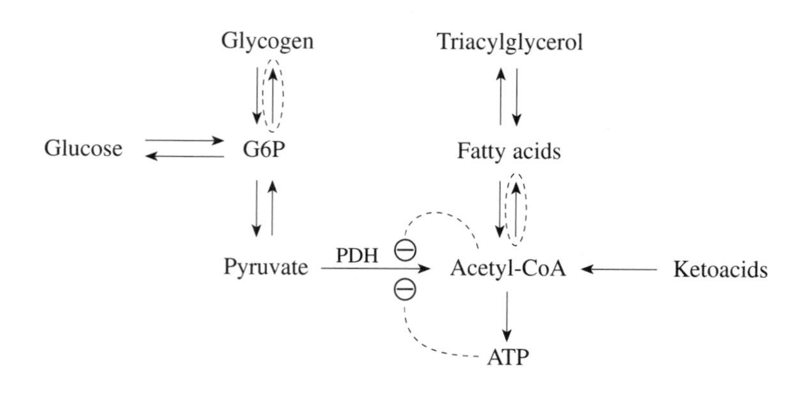

Figure 12·1
Hyperglycemia caused by limited metabolism of glucose
For oxidation, the critical fact is that acetyl-CoA, NADH (not shown), and ATP, the products of fatty acid or ketoacid oxidation, inhibit PDH. The enzyme pathways in the hatched ovals are active only when levels of insulin are high (levels of glucagon are low).

Note:
This figure is an oversimplification. It depicts events in the liver (synthesis of glucose, glycogen, and fatty acids) with events in nonhepatic organs (oxidation of ketoacids).

Clinical pearl:
Extremely sensitive controls maintain a tiny pool of glucose in the body despite a high flux of glucose through the body. If these controls do not operate properly, hyperglycemia will develop and may do so rapidly.

Quantitative Aspects

The approach to an abnormal concentration of a metabolite (glucose in this case) requires an analysis of two factors, its rate of input and its rate of output (Figure 12·2). It is important to recognize that the quantity of glucose in the body is very small relative to the amount of glucose consumed and oxidized each day (Figure 12·3).

POOL SIZE OF GLUCOSE

- Glucose is distributed in 50% of total body water.
- The pool of glucose is close to 100 mmol, or 18 g.

To calculate the amount of glucose in the body, one must consider the following facts: the volume of distribution of glucose is the ECF plus the ICF of organs that do not require insulin to transport glucose across their cell

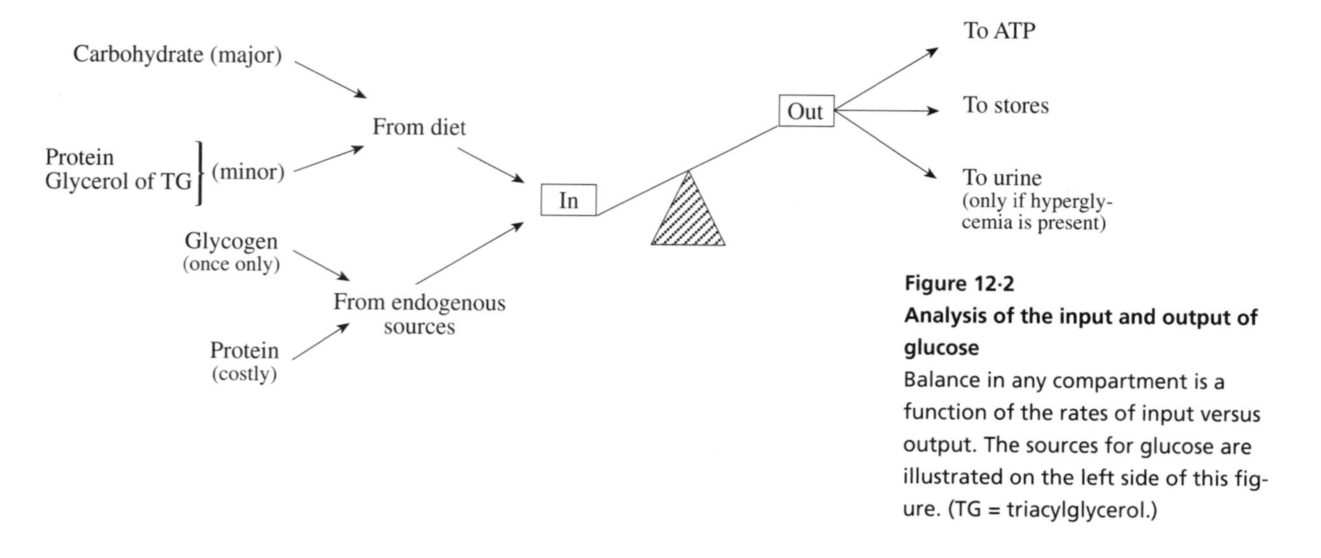

Figure 12·2
Analysis of the input and output of glucose
Balance in any compartment is a function of the rates of input versus output. The sources for glucose are illustrated on the left side of this figure. (TG = triacylglycerol.)

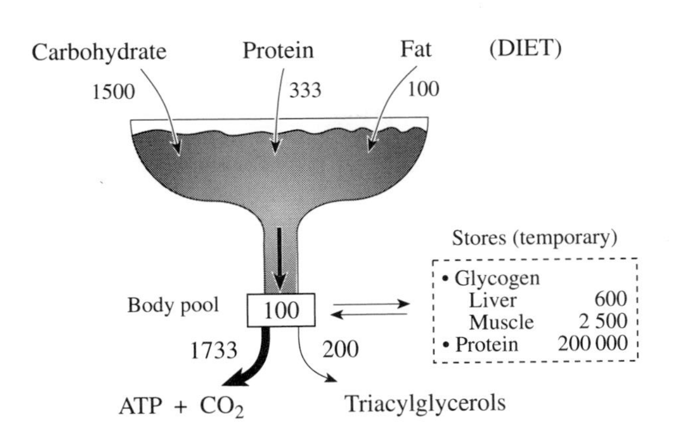

Figure 12·3
Glucose content vs turnover of glucose
The pool of glucose in the body (represented by the solid rectangle) is very small relative to the turnover of glucose each day. Conversion of endogenous protein to glucose is not rapid compared with the production of glucose from dietary sources. All the numbers are in millimoles of glucose; multiply by 0.18 to convert to grams.

membranes (most organs other than skeletal muscle; see the margin). In the ECF, the concentration of glucose is normally close to 5 mmol/l, or 100 mg/dl (1 g/liter), for easy math. Therefore, the body contains close to 100 mmol (18 g) of glucose (see the margin).

Volumes in a 70-kg adult:
ECF: 15 liters
ICF where glucose distributes: 4-5 liters
ICF of skeletal muscle: close to 25 liters
Total body water: close to 45 liters

Calculation:
- Concentration of glucose = 5 mmol/l.
- Volume of distribution is the ECF (15 liters) plus 5 liters of ICF.
- 5 mmol/l × 20 liters = 100 mmol.

QUANTITY OF GLUCOSE OXIDIZED EACH DAY

A normal adult consumes close to 1500 mmol (270 g) of carbohydrate each day (Figure 12·3). In addition, glucose is synthesized during the metabolism of proteins (333 mmol, or 60 g) and triglycerides (glycerol of dietary origin can yield close to 100 mmol, or about 18 g, of glucose). In this example, balance is maintained, so a net of 1933 mmol of glucose must be oxidized directly or indirectly each day.

Note:
The daily input (and output in balance) of glucose exceeds the pool size of glucose by close to 20-fold!

HOW MUCH GLUCOSE IS STORED AS GLYCOGEN?

- The size of the pool of glycogen in the liver is only 600 mmol (100 g).
- Liver glycogen can provide glucose on a "one time only" basis.
- The synthesis of glycogen will be slow when insulin is given to a patient with chronic, severe hyperglycemia.
- For the most part, glycogen in muscle (2500 mmol, 450 g) cannot be converted directly to glucose.

Liver

The liver in a 70-kg adult can store about 600 mmol (100 g) of glucose; this store can empty quickly under the catabolic signals of a lack of insulin and high levels of glucagon (called "*low net insulin*" hereafter).

Net insulin:
The combined effects of insulin and glucagon. "Low net insulin" refers to low levels of insulin and high levels of glucagon. "High net insulin" indicates high levels of insulin and low levels of glucagon.

Conversion of glucose to glycogen: Two major types of changes occur in the presence of insulin: rapid changes in the levels of activators and inhibitors of enzymes, and slower changes involving the number of enzyme molecules. This latter type of change requires the synthesis of new proteins, a process that may require a period of hours. This delay is the major reason why the administration of insulin to a patient with HHS does not lead to a prompt fall in glycemia.

Conversion of glucose to glycogen is rapid if a person is exposed chronically to high net insulin signals; if, however, the net insulin levels are low and then raised abruptly, as in a patient with diabetes mellitus in poor control, conversion to glycogen will be very slow.

Muscle

Large amounts of glucose (2500 mmol, or 450 g) are also stored as glycogen in skeletal muscle. Synthesis here requires high net insulin levels and, for rates to be rapid, there must be longer periods of exposure to high net insulin levels. Breakdown of glycogen in muscle does not result in the release of glucose (muscle lacks the enzyme glucose 6-phosphatase); rather, lactic acid will be released. The major stimuli for the breakdown of glycogen in muscle are exercise and adrenaline, not hypoglycemia in a direct way.

Question

(Discussion on page 452)

12·1 In a patient with hypoglycemia, glycogen in muscle breaks down. Why? How might this pathway be activated in a patient with hyperglycemia?

Analysis of the Rate of Removal of Glucose

- Metabolic removal of glucose is via oxidation or conversion to storage compounds. During hyperglycemia with insulin deficiency, renal excretion is the only major path for removal of glucose.

REMOVAL BY METABOLISM

There are two major means of removing glucose from the circulation via metabolism: oxidation to regenerate ATP and conversion of glucose to its storage form, glycogen. In the liver, glucose is also converted to stored fat, but this pathway usually proceeds at a slow rate.

Oxidation of Glucose

- Oxidation of fat-derived fuels prevents the complete oxidation of glucose.

For simplicity, the following generalization can be made: oxidation of fat-derived fuels prevents the oxidation of glucose because pyruvate dehydrogenase (PDH), the key regulatory enzyme, is inhibited by the products of fatty acid and ketoacid oxidation (refer to Figure 12·1, page 430).

If ketoacids are not present, the brain can oxidize close to 28 mmol (5 g) of glucose per hour. In contrast, if ketoacidosis is present, this oxidation is curtailed markedly and the only major option for removal of glucose is glucosuria.

Question

(Discussion on page 452)

12·2 In what organ(s) might glucose be oxidized to CO_2 at an appreciable rate in a patient with chronic hyperglycemia?

Conversion to Storage Forms

- There is virtually no conversion of glucose to stores when net insulin levels are low.

The pathway for the conversion of glucose to glycogen requires the hormonal setting of high net insulin, which leads to synthesis (induction) of relevant enzymes plus provides the signals required for conversion of glucose to glycogen. Conversely, the hormonal setting of low net insulin, which is typical of fasting with hypoglycemia, requires that the conversion of glucose to glycogen be slow. There is little or no conversion of glucose to fat with this hormonal milieu. Thus, because of the effects of a relative lack of insulin, one should expect the level of glucose to decline at a slow rate when insulin is given to a diabetic in poor glycemic control.

Clinical pearl:
In a metabolic setting in which glucose cannot be removed by metabolism, any input of glucose is too much glucose.

EXCRETION OF GLUCOSE IN THE URINE

- During severe hyperglycemia, the only important avenue for removal of glucose is excretion.
- Excretion of glucose falls when the GFR falls.
- The usual maximum rate of glucose reabsorption by the kidney is 325 g/day (1800 mmol/day).

Virtually the only pathways for removal of glucose in a patient with hyperglycemia and low net insulin levels are oxidation of glucose in the brain and urinary excretion of glucose. Valuable water-soluble nutrients such as glucose (referred to as "goodies" in Figure 12·4) must normally not be excreted. Because there are only small stores of glucose in the body, the excretion of glucose will result in the loss of precious fuels for vital organs (brain). Further, a loss of glucose in the urine will "drag" valuable ions (e.g., Na^+, K^+) and water in the urine (osmotic diuresis). In addition, if the body has to rely on high rates of production of glucose

Figure 12·4
Reabsorption of glucose by the kidney
Reabsorption occurs in the proximal convoluted tubule. Limits are usually set by the GFR; stimulation occurs with a low "effective" circulating volume contraction and metabolic acidosis.

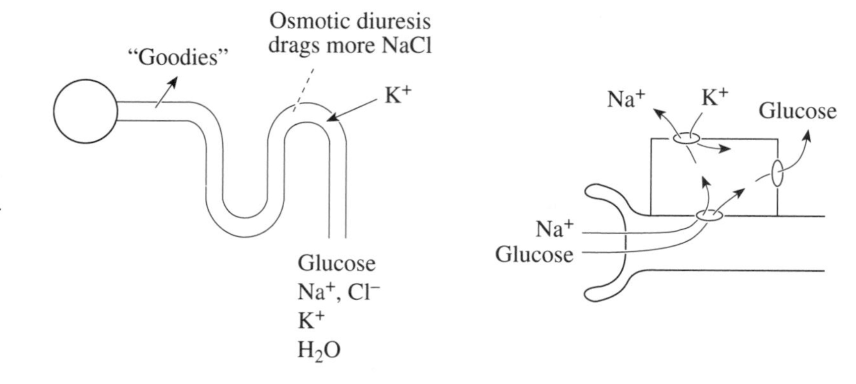

Calculation:
In a 70-kg adult, 1 kg of lean body mass can supply the brain with glucose for 24 hours:

- The brain needs 120 g of glucose (480 kcal) per day.
- 120 g of glucose is derived from 200 g of protein (60% of protein to glucose).
- 200 g of protein is derived from 1 kg of muscle (80% water).

Quantities:
In briefest terms, the plasma glucose in normal individuals after a meal is less than 8 mmol/l (140–150 mg/dl), and the normal kidney can reabsorb 1800 mmol (325 g) of glucose per day. Because the usual GFR in a 70-kg adult is 180 liters per day, the filtered load of glucose will not exceed 1800 mmol (325 g) as long as the plasma glucose is less than 10 mmol/l (180 mg/dl).

Excretion of 175 g of glucose per day:
This rate of excretion reflects the intake of 275 g of glucose and the metabolism of 100 g in organs that do not need insulin to promote the oxidation of glucose (e.g., the brain).

from endogenous compounds such as proteins to maintain the pool of glucose, the cost in terms of loss of lean body mass will be great (see the margin).

The general features of reabsorption of glucose in the proximal convoluted tubule reflect the general properties of this nephron segment—a high capacity relative to the normal filtered load, yet a "leaky" epithelium that prevents the development of large concentration gradients for osmolality and Na^+ (Figure 12·4). Quantitatively, the maximal reabsorption of glucose is linked to the GFR and exceeds the normal filtered load (Table 12·3). Quantitative considerations appear in the margin.

Table 12·3
EFFECT OF HYPERGLYCEMIA AND THE GFR ON THE EXCRETION OF GLUCOSE
Excretion of glucose (g/day) = [GFR (liters/day) × plasma glucose (g/l) – 325 g glucose reabsorbed/day]. Note the very high levels of glucosuria when there is both hyperglycemia and a near-normal GFR. A usual diet will supply 270 g of glucose each day, but in a patient with chronic steady-state hyperglycemia, expect an *excretion of 175 g of glucose per day*. Hence, one can predict the GFR required to maintain a given degree of hyperglycemia and the rate at which the concentration of glucose will fall once the GFR rises abruptly.

GFR (liters/day)	180	100	50	25
Glucose (mg/dl)	Glucose excreted (g/day)			
200	35	0	0	0
400	395	75	0	0
600	755	275	0	0
1000	1475	675	175	0
1500	2375	1175	425	50

The main reasons that the concentration of glucose might fall during therapy are reexpansion of the ECF volume (via dilution) and the consequent rise in the GFR, which markedly enhances glucosuria. Table 12·3 indicates just how much glucose will be excreted with a given rise in GFR.

Question

(Discussion on page 452)

12·3 A patient with NIDDM has hyperglycemia (50 mmol/l, 900 mg/dl). Assume that the GFR is normal. What are the implications for the intake of glucose and the electrolyte balance over the next 24 hours?

What Permits a Severe Degree of Hyperglycemia to Develop?

A lack of insulin or resistance to its actions is necessary for a modest degree of hyperglycemia to develop. For a modest degree of hyperglycemia to become severe, usually either the renal excretion of glucose will be low or the intake of glucose will be high.

Summary:
A combination of factors is required to develop a severe degree of hyperglycemia. A relative deficiency of insulin leads to poorly controlled diabetes mellitus. A severe degree of hyperglycemia requires excessive intake of glucose or, more commonly, a low excretion of glucose (a low GFR resulting from ECF volume contraction).

LOW EXCRETION OF GLUCOSE

As shown in Table 12·3, the excretion of glucose will be low during severe hyperglycemia only if the GFR is reduced; most commonly, this reduction is secondary to prerenal failure, which, in turn, is due to an osmotic diuresis with a resultant negative Na^+ balance (Table 12·4). The excretion of glucose can also be low in patients with chronic renal disease who cannot produce an osmotic diuresis because of their very low GFR.

Clinical pearl:
If the ECF volume is not low in a patient with a severe degree of hyperglycemia, look for chronic renal disease rather than a prerenal cause for the low GFR.

Table 12·4
COMPOSITION OF ONE LITER OF GLUCOSE-INDUCED OSMOTIC DIURESIS

The values are approximations. Later in time, there will be a marked contraction of the ECF volume and little caloric intake. The volume of urine will then become much smaller (see the margin).

Glucose		Urea	Na^+	K^+
g/l	mmol/l	mmol/l	mmol/l	mmol/l
54	300	100–200	50	30

Osmotic diuresis:
During an osmotic diuresis, the $[Na^+]$ in fluid exiting the proximal tubule is two-thirds that of plasma, but more liters exit. The net result is no change in absolute Na^+ reabsorption in the proximal convoluted tubule. A smaller quantity of Na^+ is reabsorbed in the loop of Henle, so delivery to the distal nephron is increased. If all the extra Na^+ delivered are not reabsorbed, a natriuresis occurs.

HIGH INTAKE OF GLUCOSE

The second component that may contribute to a severe degree of hyperglycemia is an extraordinary input of glucose, which can overwhelm the normal kidney's ability to excrete it. Most commonly, the source of glucose is excessive quantities of fruit juices or sweetened soft drinks consumed in response to thirst (Table 12·5). Given the very large capacity to excrete glu-

Note:
See the margin of page 439 for a calculation depicting the net catabolism of endogenous protein and the yield of glucose and urea.

cose, a degree of reduction in GFR is required for these patients to maintain hyperglycemia on a chronic basis. Very rarely, an increased endogenous input of glucose can result transiently from the breakdown of protein (e.g., with increased protein reabsorption from a major gastrointestinal blood loss).

K^+ loss in DKA or HHS:
- Early in the natural history there is a large K^+ loss in the urine.
- Later, even though there is distal Na^+ delivery and high aldosterone levels, the $[K^+]$ in urine is not high (Table 12·5). We speculate that chronic glucosuria may lower the open probability of the Na^+ channel in the luminal membrane of the CCD (possibly by glycation) and thereby yield a low TTKG.

Table 12·5
COMPOSITION OF FLUIDS CONTAINING GLUCOSE

All examples are based on one liter volume. Reprinted with permission from Halperin and Rolleston, *Clinical Detective Stories* (London: Portland Press, 1993).

Sample	Glucose		Sodium	Potassium
	g	mmol	mmol	mmol
Plasma (normal)	0.9	5	140	4
Urine (osmotic diuresis)	54	300	50	28
Intake				
- D_5W	50	275	0	0
- Apple juice	132	732	0	32
- Sweetened drinks	110	612	2	0
- Normal saline	0	0	152	0

PART B
IMPACT OF HYPERGLYCEMIA ON FLUID AND ELECTROLYTES

Hyperglycemia has two major influences on the ECF and ICF volumes. First, as a result of the increased number of osmoles (glucose) restricted to the ECF, severe hyperglycemia will lead to a shift of water from the ICF to the ECF, which will expand the ECF volume and lower the [Na^+] in the ECF. Second, water and electrolytes will be lost in the urine because hyperglycemia induces an osmotic diuresis.

Shift of Water Across Cell Membranes

- During severe hyperglycemia, myocytes will shrink and hepatocytes will swell. Of greater importance, brain cell size is probably close to normal.

There are two types of particles to consider with respect to shifts of water:

1. Certain particles (urea, ethanol) are "ineffective" in causing a shift of water because they always ultimately achieve an equal concentration in the ECF and ICF.
2. Particles that are restricted to one or other compartment (e.g., Na^+ in the ECF) will cause water to shift into or out of that compartment when their concentrations rise or fall, respectively.

Given these two categories of particles, it remains to be defined whether glucose is a particle with no influence on water shift or one that causes water to exit from the ICF.

Glucose appears to be an "ineffective" osmole in many cells where its transport is independent of insulin (e.g., hepatocytes), since its concentration will be equal in both the ECF and the ICF. In contrast, the concentration of glucose is always much higher outside cells that depend on insulin for the transport of glucose (e.g., myocytes), and the behavior of glucose is similar to Na^+ with respect to water shifts (Figure 12·5).

At any stage of hyperglycemia, it is difficult to be certain of the overall volume of individual brain cells. During severe hyperglycemia, water

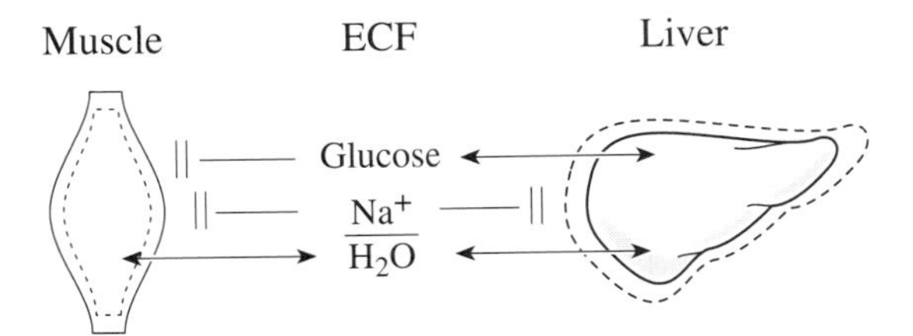

Figure 12·5
Hyperglycemia and the shift of water
Water will move across cell membranes if the concentration of particles restricted to the ECF or ICF changes. The concentration of glucose will always be much higher in the ECF than in the ICF of muscle. Hence, muscle shrinks during hyperglycemia (dashed line); as a result, the concentration of Na^+ in the ECF falls. In contrast, in liver cells, the concentration of glucose is equal to that in the ECF. Hence, hyperglycemia per se has no influence on water shifts in the liver. Hyponatremia, however, will cause water to enter hepatocytes and they will swell (dashed line). Overall, the volume of the ECF will rise if no excretion of water or electrolytes occurs because the volume of ICF in muscle is at least fourfold larger than that of cells that behave like hepatocytes.

Change in ECF volume when 1000 mmol of glucose are retained:

- Background:
  - Glucose distributes in ECF (15 liters) + 4 liters of ICF.
  - ECF contains 2100 mmol of Na^+ (140 mmol/l × 15 liters).
  - The body contains 12 600 "effective" mosmoles (2 × 140 mmol/l × 45 liters).
- Add 1000 mmol of glucose to its distribution volume (19 liters):
  - Total body osmolality = total osmoles (12 600 + 1000)/45 liters = 302.2 mosm/l.
  - Now glucose distribution volume has 6320 "effective" mosmoles (1000 mmol + (2 × 140 × 19 liters)).
- New ECF volume:
  - Divide osmoles in glucose distribution volume (6320) by new osmolality (302.2 mosm/l) = 20.9 liters
  - ECF volume = (15/19) × 20.9 liters = 16.5 liters
  - Therefore, ECF volume rises by 1.5 liters and the rise in concentration of glucose is 1000 mmol/20.9 liters = 47.8 mmol/l.

Replacement of K^+ deficit:

- That portion of intracellular K^+ loss associated with a countermovement of Na^+ or H^+ can be replaced with KCl.
- That portion of intracellular K^+ loss associated with a loss of phosphate cannot be replaced with KCl. Phosphate and time for the anabolic effects of insulin (days) are required to restore K^+ stores in the ICF.

moves out of myocytes and leads to hyponatremia because there is a gain of water without Na^+ in the ECF. In quantitative terms, muscle contains close to half of the water in the body (25 of 45 liters in a 70-kg adult). The volume of distribution of glucose will be 19 liters (15 liters of ECF and about 4 liters of ICF). One can calculate that if the concentration of glucose rises to almost 50 mmol/l, the volume of the ECF should increase by 1.5 liters (see the margin). The converse will also be true when the concentration of glucose falls from 50 to 5 mmol/l; 1.5 liters of water will shift back from the ECF into the ICF, and a decrease in the volume of the ECF will result.

The Impact of an Osmotic Diuresis on Body Fluid Compartments

- The major losses of Na^+, K^+, and water during hyperglycemia are the result of the osmotic diuresis induced by glucose.

Total losses of Na^+, K^+, and water have been determined in two types of studies:

1. balance studies in patients who had insulin therapy withheld;
2. retrospective studies involving quantification of the net amounts of Na^+, K^+, and water that were retained during therapy in patients with diabetic ketoacidosis (DKA).

Each of these types of studies has its own limitations. The average losses of electrolytes and water to anticipate during severe hyperglycemia are discussed in a quantitative fashion below (Table 12·6).

Table 12·6
TYPICAL DEFICITS IN A PATIENT WITH DKA

	Quantity	Comment	Danger
• Na^+	• 5–10 mmol/kg	• Restore quickly	• Too rapid a rise in the [Na^+] • Hypernatremia
• K^+	• 5–10 mmol/kg	• Must wait for insulin to shift K^+ into cells if hyperkalemic • Look at the excretion of K^+ when deciding the rate of infusion of K^+	• Hyperkalemia initially • Hypokalemia 2 hours later • Half of the K^+ were excreted with phosphate (see the margin)
• H_2O	• Usually many liters	• Half ICF, half ECF	• Too rapid a repair of water deficit
• HCO_3^-	• Can be > 500 mmol of H^+ because of buffering in the ICF	• If plasma anion gap is increased, do not give HCO_3^- unless acidosis is very severe	• A rapid fall in plasma [K^+] (strong opinions held, but not backed up with clean data)

SODIUM

During an osmotic diuresis, the [Na^+] in the urine at presentation is usually close to 50 mmol/l (Table 12·4). Quantitatively, the net loss of Na^+ at the time of presentation in patients with DKA ranges between 3–9 mmol/kg body weight; an average loss of 7 mmol/kg is quite common. This value represents a net loss that is equivalent to 25% of the total body Na^+ (close to 500 of 2000 mmol in a 70-kg subject); it is detected clinically by a markedly contracted ECF volume, which results in an impaired circulating volume and tissue perfusion.

POTASSIUM

Although the net loss of K^+ is close to 5 mmol/kg of body weight, it is difficult to determine the exact impact this loss has on the ECF and ICF volumes. Close to half of the K^+ lost from the ICF is in conjunction with phosphate (typically about 2 mmol/kg). Because the net charge on intracellular phosphodiesters is in effect close to –1, an equal amount of K^+ should leave the cells with phosphate.

The remaining half of K^+ loss will largely represent ECF volume loss because most of the K^+ will be shifted in exchange for Na^+; this shift equates to at most 1.5 liters of ECF fluid. As shown in Figure 10·1, page 354, some of this loss of K^+ results from an "exchange" of K^+ for H^+.

K^+ and phosphate vs K^+ and Na^+:

- Phosphate in cells is in a macromolecular form. Hence, one needs to lose almost 300 mmol of K^+ with phosphate to lose 1 liter of ICF.
- For every 150 mmol of Na^+ that enter cells in exchange for K^+ that are excreted with Cl^-, the ECF loses 1 liter if there is no change in osmolality. The K^+ loss that occurs in conjunction with entry of Na^+ into cells contracts the ECF volume but does not change the ICF volume (no particle loss in the ICF).

WATER

The usual electrolyte-free water deficit on presentation is said to be 2–3 liters in the adult with severe hyperglycemia. This deficit will be quite variable depending on what the intake of water was during the illness. Because patients do not present with a record of their recent balances, it is only possible to calculate a component of free water deficit from the [Na^+] in plasma, the body weight, and an estimation of the ECF volume.

Metabolic Cost of Glucosuria

If one liter of osmotic diuresis contains 300 mmol (54 g) of glucose, this value could represent the breakdown of 0.5 kg of lean body mass if the source of glucose was gluconeogenesis from endogenous proteins (see the calculation in the margin). Hence, there is a large metabolic cost to excrete this urine if its source is endogenous compounds. Further, because this glucose was derived from 90 g of protein, the urine should contain 500–600 mmol of urea. If the cells containing this protein were destroyed, they would release all of their K^+ and phosphate (close to 45 mmol).

Calculation:

- 1 lb = 454 g.
- Muscle is 80% water, so each pound of muscle contains close to 90 g of protein.
- 60% of protein can be converted to glucose.
  0.6×90 g = 54 g.
- Therefore, each liter of osmotic diuresis represents 1 lb (0.5 kg) of muscle in a diabetic in poor control who has not consumed food.
- 90 g of protein will yield 500 mmol of urea.

PART C
DESIGN OF INTRAVENOUS THERAPY FOR HYPERGLYCEMIA

The major issues with regard to intravenous therapy in patients with a severe degree of hyperglycemia are summarized in Table 12·7. They include the following:

1. Replace the majority of the large deficit of ECF particles and volume as rapidly as needed. There should be a further decline in ECF volume because of the expected shift of water from the ECF to the ICF as the level of glycemia declines.
2. Replace ongoing losses of electrolytes and water in urine.
3. Replace the deficit of water and particles in the ICF slowly. Replace the electrolyte-free water deficit associated with resynthesis of glycogen and proteins slowly because these processes take considerable time to occur.

Therapy must be tailored to the individual patient. It is also important to note that the time over which each of these issues must be addressed differs from one patient to another and depends on the etiology of the hyperglycemia, the magnitude of the deficits, and the underlying cardiovascular status of the patient. Each of these issues will now be considered in more detail.

Note:
When glucose disappears from the ECF, water enters the ICF, and the $[Na^+]$ in plasma rises. Now the ECF volume is lower, but the $[Na^+]$ is close to normal. Hence, the ECF needs reexpansion with fluid restricted to the ECF (isotonic saline). This calculation implies the expected degree of hyponatremia for a given degree of hyperglycemia.

Table 12·7
SUMMARY OF THE GENERAL STRATEGY FOR INTRAVENOUS THERAPY IN A PATIENT WITH A SEVERE DEGREE OF HYPERGLYCEMIA

Compartment	Speed	Tonicity of IV	Comments
ECF			
• Replace deficit	• Fast	• The osmolality of the IV should be equal to the glucose + 2 $[Na^+]$ in plasma	• No shift of water into or out of the ICF
• Replace volume of ECF that shifted into ICF	• Slow	• Isotonic (see the margin)	• Depends on decline in glycemia
Ongoing losses via urine	• Depends on urine output	• Infuse at $[Na^+]$ and $[K^+]$ in urine	• Fast in "drinker" • May rise during therapy in "prune"
ICF			
• Encourage gain of ICF particles	• Slow, many hours	• Water without Na^+	• Replace water when particles are replaced (K^+ and $H_2PO_4^-$)

Clinical pearls:
- "Drinker" is our designation for a patient with severe hyperglycemia caused by an excessive intake of fruit juices or sweetened soft drinks. In the "drinker," the deficit in ECF volume may be mild and could be replaced more slowly.
- "Prune" is our designation for a patient with severe hyperglycemia (caused by a low excretion of glucose from a reduced GFR) and clinical evidence of severe ECF volume depletion. In the "prune," replace the ECF volume aggressively.

Replacing the Deficit of ECF Volume

The deficit of ECF volume and the rate at which this deficit must be replaced must be estimated on clinical grounds. If the patient is hemodynamically unstable, the ECF volume must be reexpanded very quickly with a solution that is isotonic to the patient. Isotonic saline (154 mmol/l NaCl and therefore about 300 mosmoles/liter) is the best choice because its osmolality approximates the patient's effective osmolality.

In addition to the clinical impression of the deficit of ECF volume on admission, two other factors must be considered. First, the decline of hyperglycemia will cause water to enter cells. The resulting deficit of ECF volume must be replaced with isotonic saline. Second, some of the K^+ given will enter cells and lead to a shift of Na^+ from the ICF back into the ECF; this shift will expand the ECF volume. Hence, give one less millimole of NaCl for every millimole of KCl infused.

Calculation:

- Assume a blood glucose of 50 mmol/l.
- Glucose was distributed in 15 liters on admission.
  Total glucose = 750 mmol.
- After therapy, the concentration of glucose will be 10 mmol/l, and glucose will distribute in 19 liters.
  Total = 190 mmol.
- With no other input or output of glucose, 560 mmol of glucose will be excreted and the osmotic diuresis will be close to 2 liters.
- Calculate the rate of appearance of urea to reflect net production of glucose (1.72 mmol of urea for each mmol of glucose).

Replacing Ongoing Losses of Electrolytes and Water

Losses of electrolytes and water are easy to estimate from the urine volume and the approximate composition of the urine during an osmotic diuresis (Table 12·4, page 435); these values can be measured, if necessary (urine electrolytes). Solutions used to replace ongoing losses will thus be hypotonic and will reflect the composition of electrolytes in the urine.

During treatment, most of the glucose in the body may be excreted in the urine once the GFR rises appreciably (Table 12·3, page 434). If each liter of osmotic diuresis contains close to 300 mmol of glucose, there is the potential for the additional loss of several liters by osmotic diuresis. These ongoing losses of Na^+, K^+, and water need to be replaced during therapy. The degree of osmotic diuresis may be even higher in the patient who has consumed a large quantity of glucose because the pool of glucose may be larger (smaller degree of contraction of ECF volume); also, the gastrointestinal tract may contain a hidden pool of glucose when gastric emptying is delayed in response to gastroparesis (commonly seen in these patients).

Restoring the Deficit of Water and Particles in the ICF

The general principle for therapy of the ICF is to replace this deficit slowly. Consider separately the three components of the loss of ICF volume.

DEFICIT OF ELECTROLYTE-FREE WATER

If the plasma [Na^+] fell more or less than 1.5 mmol/l for every 100 mg/dl (5.5 mmol/l) rise in glycemia (see the margin), there is an additional imbalance in electrolyte-free water that must be adjusted during therapy. It is not clear whether a rapid increase in the concentration of Na^+ in the plasma of patients with severe hyperglycemia poses the same risk for the development of osmotic demyelination as do other settings of hyponatremia. During therapy for severe hyperglycemia, cerebral edema can develop and is a common observation, as judged indirectly from radiologic evidence.

Fall in [Na^+] when plasma concentration of glucose rises:

- Background from page 438:
  - Add 1000 mmol of glucose.
    - Concentration of glucose rose by 47.8 mmol/l.
    - ECF volume = 16.5 vs 15 liters.
- Calculation:
  - Original content of Na^+ in the ECF is 2100 mmol (140 mmol/l × 15 liters).
  - [Na^+] in plasma is now 127 mmol/l (2100 mmol/16.5 liters).
- Relationships in plasma:
  - [Na^+] fell 13 mmol/l (140 to 127 mmol/l).
  - Concentration of glucose rose 47.8 mmol/l.
  - [Na^+] fell 1.5 mmol/l/5.5 mmol/l (100 mg/dl) rise in glycemia.

SHIFT OF WATER CAUSED BY A FALL IN GLYCEMIA

The concentration of glucose will usually decline to the 10–15 mmol/l (180–270 mg/dl) range in 6–8 hours. In a patient with a blood sugar of 50 mmol/l (900 mg/dl), this decline will cause the [Na^+] in the ECF to rise by 7% because the ECF lost 1 of its 14 liters (see the calculation on page 442).

LOSS OF ICF PARTICLES

One need not replace ICF water associated with the loss of K^+ in the acute management of a patient with severe hyperglycemia because it will take a considerable period of time to replace the anions (phosphate) lost with K^+. Obviously, the part of the deficit of K^+ that is due to a shift of cations (largely Na^+) must be replaced acutely to prevent the expected degree of hypokalemia caused by the actions of insulin; this replacement of intracellular Na^+ with K^+ will not change the ICF volume.

PART D
CLINICAL ASPECTS OF A SEVERE DEGREE OF HYPERGLYCEMIA

The hyperglycemic hyperosmolar syndrome (HHS) includes a heterogeneous population of patients who have in common the hallmarks of this diagnosis: a marked degree of hyperglycemia (most often much greater than 27.5 mmol/l, or 500 mg/dl) and hyperosmolality (usually more than 320 mosm/kg H_2O). The more common name is *hyperglycemic hyperosmolar nonketotic coma*. Because coma is rare and a certain degree of ketoacidosis is present (sufficient to suppress the oxidation of glucose by the brain), this name is misleading and we prefer not to use it.

Hyperglycemic hyperosmolar nonketotic coma (HHNC):
This term is commonly used to define a clinical syndrome in which hyperglycemia, but not ketoacidosis, predominates. It is a misnomer because ketoacids are almost always present and coma is rare.

Based on the underlying pathophysiology for the development of hyperglycemia, four subtypes of patients emerge: the "drinker," the patient in a catabolic state, the "prune," and the patient with renal failure. These divisions are somewhat arbitrary because they can represent a continuum, and each patient may have a mixed etiology for hyperglycemia. It is useful, however, to categorize individual patients because each subtype has unique dangers and requires a different approach to therapy; these principles also apply to patients with DKA and a significant degree of hyperglycemia.

We shall now address the following issues for each subtype: diagnosis, therapeutic issues, and dangers to anticipate.

Diagnosis

Distinguishing between excessive input and low output of glucose as the major determinants of the severity of hyperglycemia provides the basis for this classification (Figure 12·6). The three questions listed below should be answered.

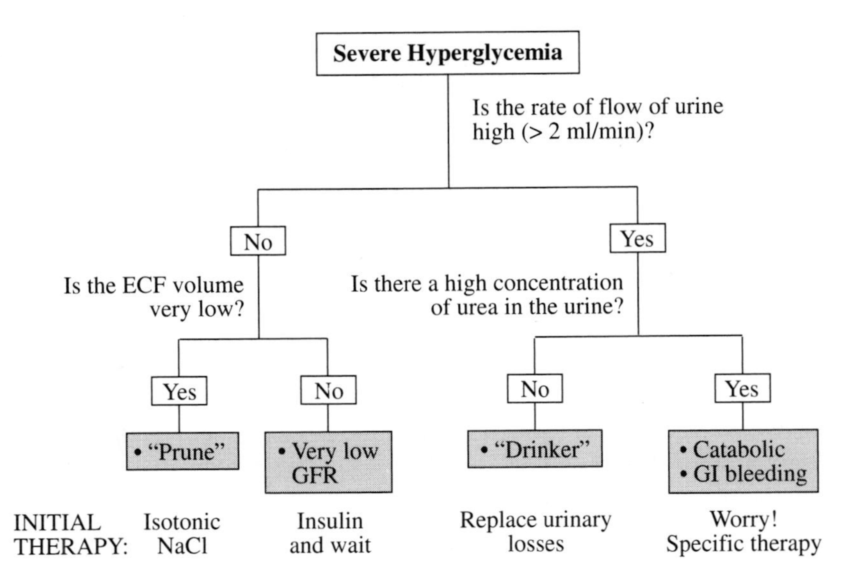

Figure 12·6
Clinical approach to the patient with hyperglycemia
The purpose of this approach is to identify those patients whose primary reason for hyperglycemia is high input of glucose vs those with a low output of glucose. Obviously, most patients will have some degree of excessive intake of glucose and a degree of reduction of the GFR. Nevertheless, if polyuria is present, it should be emphasized because it will help with decision-making concerning intravenous fluid administration.

Question 1: What is the rate of excretion of glucose?

Patients with a high rate of excretion of glucose will have a rate of urine flow of at least 2–3 ml/min. This category includes those with an exogenous source of glucose (the "drinker") and those with a very high rate of production of glucose from endogenous sources (upper gastrointestinal bleeding and/or a very high rate of catabolism). This latter distinction requires the answer to a second question.

Question 2: What is the rate of appearance of urea relative to that of glucose?

Patients with an endogenous protein source for glucose will have a high rate of appearance of urea, and the quantity of glucose formed from protein sources can be quantitated via the *urea:glucose ratio*. Each millimole of glucose is made with 1.7 mmol of urea during the conversion of amino acids to glucose. Therefore, one must calculate the appearance of new urea in the body (the concentration of urea in plasma × total body water) plus the quantity of urea excreted in that time interval. The appearance of glucose is its concentration in plasma × one-half total body water plus the quantity of glucose excreted.

Urea:glucose ratio:
When the source of glucose production is protein, urea is made with glucose and the expected ratio of urea to glucose is 1.7:1. The appearance of glucose without urea implies that glycogen or dietary glucose was the source of this glucose.

In patients with a low rate of excretion of glucose, a renal mechanism is present that permits continued severe hyperglycemia. These patients can be subdivided based on ECF volume.

Question 3: In a patient with a low excretion of glucose, is the ECF volume very low?

Patients with chronic renal insufficiency who were unable to mount an osmotic diuresis will have an expanded ECF volume caused by a shift of water from the ICF. Patients with prerenal failure will have very low GFR because of severe contraction of the ECF volume.

Therapeutic Issues

The issues to consider are summarized in Table 12·8. One must replace the deficits of Na^+, K^+, and water, and the ongoing losses of electrolytes and water. There is some controversy concerning the immediate need for insulin (see the margin). Each of these issues will be considered below in the common clinical settings.

Note:
Insulin is needed to:
- treat severe hyperkalemia,
- prevent severe ketoacidosis.

Insulin is not needed acutely to lower blood sugar (dilution and renal excretion will do that).

THE "DRINKER"

The main issue is that patients with high input and output of glucose are not usually in great danger. The priorities in treating these patients include removing the source of glucose and then replacing ongoing losses of electrolytes and water (see the margin). Any initial deficits should be replaced as well (recognize the degree of ECF volume contraction, an inappropriate $[Na^+]$ for the degree of hyperglycemia, and the degree of hyperkalemia). As a result of kaliuresis during the early osmotic diuresis, a large deficit of K^+ is likely at presentation. The usual values to anticipate are a $[K^+]$ in plasma of close to 5.5 mmol/l

Losses in urine:
The anticipated losses depend on the quantity of glucose to excrete and the composition of the urine.
- One liter of urine contains 300 mmol of glucose.
- Glucose distributes in close to half of body water.
- Therefore, for every 15 mmol/l (270 mg/dl) rise in the concentration of glucose in plasma, expect 1 liter of urine.

and a deficit of K^+ of close to 5 mmol/kg of body weight (Table 12·6, page 438).

Therapy should be tailored to measured serum and urinary electrolytes, but, generally, hypotonic saline supplemented with KCl is required. Insulin is not required acutely if there is not an excessive degree of hyperkalemia or catabolism or if there is no imminent threat of severe ketoacidosis.

Table 12·8
ISSUES DURING DIAGNOSIS AND THERAPY OF HHS

Subgroup of HHS	Diagnostic feature	Unique threat	Therapeutic emphasis
A) Polyuric group			
1. The "drinker"	• Urine > 3 ml/min • Ingestion of glucose • Low urea excretion	• Delayed hypernatremia	• Stop glucose intake • Replace ECF volume • Give free water • Match IV to urine
2. The patient in a catabolic state	• Urine > 3 ml/min • High urea in blood and/or urine	• Loss of lean body mass or GI blood loss	• Give insulin • Defend blood volume if there is GI bleeding • Give free water
B) Oliguric group			
3. The "prune"	• Oliguria • Very low ECF volume	• Shock • More severe hyperkalemia	• Reexpand ECF volume rapidly at first
4. The patient with chronic renal insufficiency	• Expanded ECF volume	• Congestive heart failure • Hyperkalemia	• With insulin, the fall in glucose may be slow • Hypoglycemia occurs after 6–8 hours

THE PATIENT IN A CATABOLIC STATE

Rarely, patients may have excessive breakdown of body proteins. Catabolism of protein yields glucose and urea. If a patient loses 1 kg of lean body mass or digests 1 liter of blood, 667 mmol of glucose and close to 1000 mmol of urea will be produced. This production might aggravate the degree of hyperglycemia in a patient with low net insulin. The clinical settings are GI bleeding or excessive levels of glucocorticoids (usually given exogenously). In these settings, insulin may be valuable to decrease the rate of oxidation of amino acids.

Calculation:
- 1 liter of blood and 1 kg of lean body mass contain 200 g of protein.
- 60% of protein → glucose, so 200 g of protein → 120 g of glucose.
- Molecular weight of glucose is 180, so 120 g of glucose is 667 mmol.

THE "PRUNE"

The first therapeutic priority for this subgroup of patients is administration of isotonic saline to restore effective circulating volume. Once the ECF volume is reexpanded and/or the urine output rises, switch to hypotonic saline.

These patients may have a low GFR and enhanced reabsorption of Na^+ for a prolonged period after the ECF volume is restored so that the extra Na^+ administered will be retained. Replacement of K^+ and the use of insulin are the same as outlined above in the "drinker."

THE PATIENT WITH CHRONIC RENAL INSUFFICIENCY

Severe hyperglycemia in this group of patients cannot be treated by dilution or inducing diuresis; insulin is therefore required to lower the plasma glucose concentration. Administration of K^+ should be avoided unless the patient is hypokalemic on presentation. Because insulin may take hours to lower the level of glucose, a delay in resolution of hyperglycemia should be anticipated, and hypoglycemia should be avoided once the concentration of glucose has begun to fall appreciably. Frequent monitoring of the blood glucose is the best way to avoid hypoglycemia.

Dangers to Anticipate

In all patients, there is the threat of the underlying illness and the general problems related to a slower circulation (e.g., thrombotic events). These threats will not be discussed. Specific items are considered below.

1. **The "drinker":**
   Delayed hypernatremia should be avoided by assuring that electrolyte-free water intake matches its exit from the ECF.

2. **The patient in a catabolic state:**
   Loss of lean body mass should be reduced by avoiding catabolic hormones (glucocorticoids, catecholamines) and by using insulin. Delayed hypernatremia is a risk, as with the drinker. Gastrointestinal blood loss should also be ruled out as a source of protein for gluconeogenesis. With cell lysis, hyperkalemia is also a potential threat.

3. **The "prune":**
   Hypotension should be treated with normal saline to restore tissue perfusion and gradually improve renal function so that glucose can be excreted.

4. **The patient with chronic renal insufficiency:**
   Patients with this background may present with pulmonary edema caused by a shift of water from ICF into ECF. Pulmonary edema may be poorly tolerated if underlying cardiac disease is present.

PART E
REVIEW

Discussion of Introductory Case
Terry Is Confused

(Case presented on page 428)

What is the basis for her severe degree of hyperglycemia?

A severe degree of hyperglycemia has two components. First, there must be "hormonal permission" to have hyperglycemia (i.e., a low level of insulin or a resistance to its actions). Second, there must be a low rate of removal or a large exogenous source of glucose.

Hormonal permission: Terry had several possible reasons for developing relative insulinopenia (NIDDM, α-adrenergic response to ECF volume contraction, drug actions—thiazides—K^+ deficiency or advanced age). Terry also had reasons for a high level of adrenaline, a hormone that opposes the actions of insulin. The concentration of adrenaline should be high because of ECF volume contraction. Adrenaline will stimulate the release of fatty acids by activating hormone sensitive lipase and will also inhibit the release of insulin from her β cells.

Reduced output of glucose: The major source of output of glucose should be via renal excretion when hyperglycemia is severe (Table 12·3, page 434). Terry's GFR is already low as a result of her age and her longstanding NIDDM. She has a further reduction in her GFR because of the ECF volume contraction.

Terry will not be able to oxidize glucose at an appreciable rate because most of her organs will be oxidizing fatty acids and her brain will be oxidizing ketoacids in this setting.

Increased input of glucose: Terry drank fruit juice, which has a high concentration of glucose (Table 12·5, page 436).

Summary: Although an excessive intake of carbohydrates was probably important one week prior to admission, a reduced degree of glucosuria was the critical factor in sustaining the severe degree of hyperglycemia. This low glucosuria was the expected response to a fall in her GFR as a result of ECF volume contraction.

Why does Terry have hyponatremia?

There are several factors operating. She has a loss of Na^+ in the osmotic and drug-induced diuresis and an increased desire to drink water in the face of ADH release. Of more importance, Terry is severely hyperglycemic.

The degree of hyponatremia at presentation was as expected for the level of glycemia (i.e., there was no evidence of an additional free water deficit).

What is the appropriate therapy?

The major threats to Terry are cardiovascular collapse and possibly acute tubular necrosis from the marked ECF volume contraction. The other early danger to anticipate is severe hypokalemia once insulin acts because she had an unusually low [K^+] in plasma on admission (4.1 mmol/l vs the expected 5.5 mmol/l). This low [K^+] was probably the result of the excessive kaliuresis (osmotic and pharmacologic diuretics).

Isotonic saline should be infused to restore her ECF volume. Because the clinical estimate of the deficit of ECF volume was 3 liters, most of this volume should be replaced rapidly using a tonicity adjusted to Terry. Further IV therapy with 150 mmol/l saline should be given to replace the water that shifts into cells as hyperglycemia abates. The volume of urine and electrolytes in urine should also be replaced when the urine flow increases. The deficit of intracellular water is not an immediate concern; phosphate need not be administered acutely since it takes time to resynthesize intracellular macromolecules.

Note:
For tonicity considerations, the KCl infused should be considered equivalent to NaCl in the IV.

The deficit of K^+ is larger than expected, so K^+ should be added to the initial IV solutions (at least 10 mmol/l). If insulin becomes part of the treatment (or is released from her β cells), a fall in plasma [K^+] will be likely, especially since she had an unusually low [K^+] on admission. The anticipated fall in the concentration of K^+ in plasma might be close to 1 mmol/l when insulin acts.

Why will the concentration of glucose fall during therapy?

The concentration of glucose should fall as a result of dilution (infusion of saline), glucosuria (300 mmol/l of urine), and metabolism (which is slow initially, even if insulin is given). The rate of fall in the concentration of glucose should be 100 mg/dl (5.5 mmol/l) per hour until it approaches 270 mg/dl (15 mmol/l). Because there is no major threat of acidosis and the [K^+] in plasma is lower than expected, insulin should be withheld during the first several hours of therapy.

Cases for Review

Case 12·1 An Apple Juice Overdose

(Case discussed on pages 449–50)

A 40-year-old female had a one-week history of extreme thirst and polyuria. She drank large volumes of fruit juice. She did not have a history of diabetes mellitus but did have acute pancreatitis one year ago. The only abnormalities on physical examination were related to a modest degree of contraction of her ECF volume. There was no obvious disturbance in CNS function, and acetone was not detected on her breath. The urine flow was brisk, but the rate of excretion of urea was not high. Laboratory results on admission are summarized in Table 12·9.

Why was the hyperglycemia so severe?
What is the appropriate therapy?

Case 12·2
A Sudden Shift in Emphasis

(Case discussed on pages 450–51)

A 30-year-old male with longstanding IDDM has renal and heart failure. He ingested a carbohydrate-rich meal after stopping his usual administration of insulin because of episodes of hypoglycemia. Laboratory data are provided in Table 12·9.

What do you expect to happen to his degree of congestive heart failure? What is the appropriate therapy?

Table 12·9
BLOOD VALUES ON ADMISSION IN CASES 12·1 AND 12·2

Blood or plasma		Apple juice overdose	Shift in emphasis
Glucose	mmol/l (mg/dl)	90 (1620)	50 (900)
Na^+	mmol/l	116	126
K^+	mmol/l	6.9	5.9
Creatinine	μmol/l (mg/dl)	115 (1.0)	1150 (10)
Urea	mmol/l (mg/dl)	4.0 (11)	30 (84)

Discussion of Cases

Discussion of Case 12·1
An Apple Juice Overdose

(Case presented on page 448)

Why was the hyperglycemia so severe?

Hyperglycemia develops primarily as a result of a low net level of insulin. This hormonal change could reflect destruction of β cells in the pancreas (pancreatitis) and an α-adrenergic response to ECF volume contraction.

Her severe degree of hyperglycemia and high rate of excretion of glucose (brisk urine flow rate) imply a high intake of glucose (history). There could also be a high production of glucose from endogenous sources. If glycogen breakdown in muscle was the source of glucose, there would be muscular contraction and a high rate of release of L-lactate (not present) or high protein catabolism (urea excretion was not high). Hence, this is a "drinker" type of hyperglycemia. Given the twofold higher concentration of glucose in fruit juice compared with that in urine during an osmotic diuresis (Table 12·5, page 436, and Table 12·4, page 435), the excretion of glucose following the intake of one liter of apple juice will oblige the excretion of 2 liters of water plus close to 100 mmol of Na^+ and close to 50 mmol of K^+ (Table 12·5). With time, the ECF volume will decline somewhat if the intake of Na^+ is inadequate to match these losses. This decline will lead to ECF volume contraction, to a reduction in the GFR, and to a more severe degree of hyperglycemia.

What is the appropriate therapy?

K^+ deficit:
Replace only that portion associated with a gain of Na^+ or H^+ acutely. Replacing the K^+ plus phosphate deficit requires time for anabolic events to occur.

Avoid a cardiac arrhythmia: Because hyperkalemia was present (chance of cardiac arrhythmias) and the ECG displayed a pattern indicating hyperkalemia, Ca^{2+} and insulin should be given. The expected loss of K^+ in the urine in the first hour will be small because the $[K^+]$ in the urine in HHS or DKA is only about 20 mmol/l (see the margin, page 436). Given a urine volume of 200 ml in one hour (3 ml/min), only 4 mmol of K^+ will be excreted in this hour. Further, one cannot raise this urinary $[K^+]$ by giving aldosterone. Attempts to replace the anticipated deficit of K^+ should not be initiated until the $[K^+]$ in plasma approaches 4 mmol/l (see the margin).

Lower the concentration of glucose in plasma: One aim of therapy is to lower the concentration of glucose in plasma while preventing contraction of the ECF and ICF volumes. Stopping the intake of glucose should lead to a prompt fall in blood sugar as a result of glucosuria; minor factors that cause a decline in glycemia will be the dilution of glucose (ECF volume reexpansion) and the metabolism of glucose. If the patient has a large dilated stomach containing carbohydrate, there may be a much slower fall in glycemia.

Maintain a normal ECF volume: Clinically, a deficit of about 1 liter of ECF volume was estimated and should be replaced with 150 mmol/l saline, which has almost the same "effective osmolality" as the patient. Over the first six hours, an additional 1.5 liters of isotonic saline will be required to prevent contraction of the ECF volume when hyperglycemia declines and water shifts into myocytes.

Replace losses in the urine: The urinary excretion of water and Na^+ should be replaced stoichiometrically. Approximately 3.5 liters of osmotic diuresis is likely to be excreted (glucose surplus of 1000 mmol), so a total of almost 200 mmol of Na^+ will need to be replaced. Hypernatremia should be avoided by adjusting the osmolality of IV fluids for the ongoing losses of Na^+ and water; also take into consideration the degree of hyperglycemia.

In summary, the main purpose of therapy is to correct the unusual degree of hyperkalemia with an administration of insulin. Restoration of the ECF volume will remove the inhibitory signals to insulin secretion and raise the GFR, which, in turn, will lead to a prompt fall in glycemia. One should anticipate ongoing renal water losses from the osmotic diuresis and replace these losses.

Discussion of Case 12·2 A Sudden Shift in Emphasis

(Case presented on page 449)

What do you expect to happen to his degree of congestive heart failure?

In the setting of insulinopenia and and very low GFR, the glucose ingested will lead to a severe hyperglycemia, which, in turn, will lead to a shift of water from the ICF to the ECF volume. Because there is no major osmotic diuresis, an extra liter or so of water will be retained in the ECF and aggravate the degree of congestive heart failure.

What is the appropriate therapy?

The best therapy is the administration of insulin. Because excretion of glucose is not possible, insulin is the only option for lessening the degree of hyperglycemia (and the shift of water out of the ECF). Fat-derived fuels were most likely present in the setting of relative insulinopenia (allowing this degree of hyperglycemia), so one could not expect a maximal rate of oxidation of glucose (25 g/hr) early in therapy.

Hyperkalemia is also a concern in this patient. Its degree will be diminished by the insulin therapy.

If the degree of heart failure becomes critical, the blood volume could be lowered promptly by phlebotomy. Alternatively, hemodialysis could be used to control the volume of the ECF and lessen the degree of hyperglycemia.

Summary of Main Points

When outlining the appropriate therapy for patients with a severe degree of hyperglycemia, identify the subgroup to which the patient belongs (Figure 12·6, page 443) and address the following issues:

- **Concerning Na^+:** Is the degree of contraction of the ECF volume a major threat? How large is the deficit of Na^+? Will it increase or decrease in the next several hours? How large was the shift of Na^+ into the ICF?
- **Concerning water:** How large is the deficit of water (compare expected vs observed [Na^+] in plasma)? How much water was lost from the ICF and the ECF? Is the deficit different in individual organs? How much more water will be lost over the first 5–6 hours of therapy? Will a rapid rise in [Na^+] cause osmotic demyelination?
- **Concerning glucose:** How quickly will the concentration of glucose fall over the next 5–6 hours? How much will be excreted, metabolized, and produced? Does it matter what the source of this glucose is? Does it matter how quickly the concentration of glucose will fall?
- **Concerning K^+:** There is a large deficit of K^+ despite hyperkalemia. Expect a large fall in [K^+] (about 1 mmol/l) once insulin acts. This fall will be increase with the administration of HCO_3^-.

It is important to emphasize that therapy must reflect the underlying pathophysiology and complications.

Discussion of Questions

12·1 In a patient with hypoglycemia, glycogen in muscle breaks down. Why?

There are two major stimuli for the breakdown of glycogen in muscle. First, with local anaerobiosis (a sprint), the need to regenerate ATP rapidly leads to glycogenolysis. Second, glycogenolysis is augmented when adrenaline levels rise. The adrenergic response to hypoglycemia should lead to the breakdown of glycogen in muscle. The product of this glycogenolysis is L-lactic acid, which may be released from muscle cells.

How might this pathway be activated in a patient with hyperglycemia?

The key to the control of glycogenolysis in muscle is adrenaline. The release of this hormone is augmented by a contracted ECF volume, a feature common to most patients with hyperglycemia. In addition, adrenaline can be released in response to stress or an underlying illness that precipitated the hyperglycemia.

12·2 In, what organ(s) might glucose be oxidized to CO_2 at an appreciable rate in a patient with chronic hyperglycemia?

In chronic hyperglycemia caused by a relative lack of insulin, more fatty acids are released from adipose tissue. Because fatty acid oxidation inhibits the oxidation of glucose, high levels of fatty acids will limit the oxidation of glucose in all organs that can take them up. The blood-brain barrier prevents fatty acids from entering brain cells at an appreciable rate; therefore, in the absence of ketoacid accumulation, the brain is the only organ that oxidizes glucose to CO_2 during hyperglycemia.

12·3 A patient with NIDDM has hyperglycemia (50 mmol/l, 900 mg/dl). Assume that the GFR is normal. What are the implications for the intake of glucose and the electrolyte balance over the next 24 hours?

It is extremely rare that a very high level of glucose can be maintained in the circulation when the GFR is normal because the kidney has an enormous capacity to excrete glucose in this setting (Table 12·3, page 434). This patient therefore has an enormous intake of glucose or a very unusual renal lesion of hyper-reabsorption of glucose. Only in the former setting will there be very large losses of water, Na^+, and K^+. Judging from Table 12·3, one might expect an excretion of close to 1500 g of glucose. At a glucose concentration of 50 g/liter of urine, the urine volume could conceivably be 30 liters per day, and a loss of 1500 mmol of Na^+ could be anticipated (Table 12·4, page 435). It is obvious from these numbers that one of the assumptions in the question is not valid—i.e., the GFR is much lower than predicted or the values do not represent a steady state.

SUGGESTED READINGS

ACID-BASE

The following four sources provide background in the area of acid-base balance.

Hochachka, P., and T. Mommsen. 1983. Protons and anaerobiosis. *Science* 219:1391–97.

Kamel, K. S., and M. L. Halperin. 1992. Metabolic aspects of metabolic acidosis. In *Clinical disorders of fluid and electrolyte metabolism.* 5th ed. Ed. M. H. Maxwell, C. R. Kleeman, and R. G. Narins, 911–31. New York: McGraw Hill.

Rahn H. 1979. Acid-base balance and the "milieu interieur." In *Claude Bernard and the internal environment: A memorial symposium*, ed. E. Robin, 179–90. New York: Marcel Dekker, Inc.

Relman, A., E. Lennon, and J. Lemann, Jr. 1961, vol. 6. Endogenous production of fixed acid and the measurement of the net balance of acid in normal subjects. *J. Clin. Invest.* 1621–30.

The following two articles provide background in the area of intracellular pH.

Aw, T. Y., and D. P. Jones. 1989. Heterogeneity of pH in the aqueous cytoplasm of renal proximal tubule cells. *FASEB J* 3:52–58.

Boron, W. F. 1992. Control of intracellular pH. In *The kidney: Physiology and pathophysiology.* 2d ed. Ed. D. W. Seldin and G. Giebisch, 219–64. New York: Raven Press Ltd.

The following five articles provide background in the area of buffering.

Fernandez, P., R. Cohen, and G. Feldman. 1989. The concept of bicarbonate distribution space: The crucial role of body buffers. *Kid. Internat.* 36:747–52.

Green, J., and C. Kleeman. 1991. Role of bone in regulation of systemic acid-base balance. *Kid. Internat.* 39:9–26.

Oh, M. 1991. Irrelevance of bone buffering to acid-base homeostasis in chronic metabolic acidosis. *Nephron* 59:7–10.

Swan, R. C., and R. F. Pitts. 1955. Neutralization of infused acid by nephrectomized dogs. *J. Clin. Invest.* 34:205–12.

Vasuvattakul, S., L. Warner, and M. L. Halperin. 1992. Quantitative role of the intracellular bicarbonate buffer system in response to an acute acid load. *Am. J. Physiol.* 262:R305–R309.

The following five articles provide background in the area of renal influences on bicarbonate balance.

Alpern, R. J. 1990. Cell mechanisms of proximal tubule acidification. *Physiol. Rev.* 70:79–114.

Edelman, C. M., et al. 1967. Renal bicarbonate reabsorption and hydrogen ion excretion in normal infants. *J. Clin. Invest.* 46:1309–17.

Halperin, M. L., et al. 1992. Biochemistry and physiology of ammonium excretion. In *The kidney: Physiology and pathophysiology.* 2d ed. Ed. D. W. Seldin and G. Giebisch, 1471–89. New York: Raven Press Ltd.

Kamel, K., et al. 1990. The removal of an inorganic acid load in subjects with ketoacidosis of chronic fasting: The role of the kidney. *Kid. Internat.* 38:507–11.

Knepper, M., R. Packer, and D. Good. 1989. Ammonium transport in the kidney. *Physiol. Rev.* 69:179–249.

The following six articles provide background on diagnostic tests in patients with acid-base disorders.

Dyck, R., et al. 1990. A modification of the urine osmolal gap: An improved method for estimating urine ammonium. *Am. J. Nephrol.* 10:359–62.

Goldstein, M., et al. 1986. The urine anion gap: A clinically useful index of ammonium excretion. *Am. J. Med. Sci.* 29:198–202.

Halperin, M. L., et al. 1988. The urine osmolal gap: A clue to estimate urine ammonium in 'hybrid' types of metabolic acidosis. *Clin. Invest. Med.* 11:198–202.

Oh, M., and H. Carroll. 1977. The anion gap. *NEJM* 297:814–17.

Stewart, P. A. 1983. Modern quantitative acid-base chemistry. *Can. J. Physiol. Pharmacol.* 61:1444–61.

Van Leeuwen, A. M. 1964. Net cation equivalency (base-binding power) of the plasma proteins. *Acta. Med. Scand.* 176:36–57.

The following six articles provide background on ketoacidosis.

Atchley, D., et al. 1933. On diabetic acidosis: A detailed study of electrolyte balances following withdrawal and reestablishment of insulin therapy. *J. Clin. Invest.* 12:297–325.

Flatt, J. 1972. On the maximal possible rate of ketogenesis. *Diabetes* 21:50–53.

Halperin, M. L., K. S. Kamel, and S. Cheema-Dhadli. 1992. Lactic acidosis, ketoacidosis, and energy turnover: "Figure" you made the correct diagnosis only when you have "counted" on it—Quantitative analysis based on principles of metabolism. *Mt. Sinai J. Med.* 59:1–12.

Halperin, M. L., and F. S. Rolleston. 1993. *Clinical detective stories: A problem-based approach to clinical cases in energy and acid-base metabolism.* London: Portland Press.

Kamel, K. S., et al. 1993. Rate of production of carbon dioxide in patients with a severe degree of metabolic acidosis. *Nephron* 64:514–17.

Schreiber, M., et al. 1994 (in press). Ketoacidosis: An integrative view. *Diabetes Rev.* 2:000–000.

The following three articles provide background on the ketoacidosis of fasting.

Hannaford, M. C., et al. 1982. Protein wasting due to the acidosis of prolonged fasting. *Am. J. Physiol.* 243:E251–E256.

Owen, O. E., S. Caprio, and G. A. Reichard, Jr. 1983. Ketosis of starvation: A revisit and new perspectives. *Clin. Endocrinol. Metab.* 12:359–79.

McGarry J. D., et al. 1994 (in press). Regulation of ketogenesis and the renaissance of carnitine palmitoyltransferase. *Diabetes Rev.* 2: 000–000.

The following article provides background on alcoholic ketoacidosis.

Halperin, M. L., M. Hammeke, and R. G. Josse. 1983. Metabolic acidosis in the alcoholic: A pathophysiologic approach. *Metabolism* 32:308–15.

The following two articles provide quantitative data about the production of L-lactic acid during exercise.

Cheetham, M. E., et al. 1986. Human muscle metabolism during sprint running. *J. Appl. Physiol.* 61:54–60.

Osnes, J. B., and L. Hermansen. 1972. Acid-base balance after maximal exercise of short duration. *J. Appl. Physiol.* 32:59–63.

The following three articles provide a general review of L-lactic acidosis.

Cohen, R., and H. Woods. 1983. Lactic acidosis revisited. *Diabetes* 32:181–91.

Halperin, M. L., K. S. Kamel, and S. Cheema-Dhadli. 1992. Lactic acidosis, ketoacidosis, and energy turnover: "Figure" you made the correct diagnosis only when you have "counted" on it—Quantitative analysis based on principles of metabolism. *Mt. Sinai J. Med.* 59:1–12.

Robinson, B. 1989. Lactic acidemia. In *Metabolic basis of inherited disease*, ed., C. Scriver, A. Beaudet, W. Sly, and D. Valle, 869–88. New York: McGraw-Hill.

The following three articles address therapy of metabolic acidosis with $NaHCO_3$.

Arieff, A. I. 1991. Indications for use of bicarbonate in patients with metabolic acidosis. *Br. J. Anes.* 67:165–77.

Halperin, M. L. 1994 (in press). Rationale for the use of sodium bicarbonate in a patient with lactic acidosis due to a poor cardiac output. *Nephron* 65:000–000.

Narins, R. G. 1994 (in press). Bicarbonate therapy in lactic and ketoacidosis. *Diabetes Rev.* 2:000–000.

The following two articles provide background for changing the emphasis of the classification of metabolic acidosis.

Carlisle, E., et al. 1991. Glue-sniffing and distal renal tubular acidosis: Sticking to the facts. *J. Am. Soc. Nephrol.* 1:1019–27.

Halperin, M. L., S. Vasuvattakul, and A. Bayoumi. 1991. A modified classification of metabolic acidosis: A pathophysiologic approach. *Nephron* 60:129–33.

RENAL TUBULAR ACIDOSIS

The following articles are of interest from an historic perspective.

Albright, F., et al. 1946. Osteomalacia and late rickets: Various etiologies met in United States with emphasis on that resulting from specific form of renal acidosis, therapeutic indications for each etiological subgroup, and relationship between osteomalacia and Milkman's syndrome. *Medicine* 25:399–479.

Morris, R. C. J. 1969. Renal tubular acidosis: Mechanisms, classification and implications. *NEJM* 281:1405–13.

Soriano, J., et al. 1967. Proximal renal tubular acidosis:A defect in bicarbonate reabsorption with normal urinary acidification. *Pediat. Res.* 1:81–98.

Wrong, O., and H. E. F. Davies. 1959. The excretion of acid in renal disease. *Quart. J. Med.* 28:259–313.

The following six sources provide background on the pathophysiology of RTA.

Cohen, E. P., et al. 1992. Absence of H^+ATPase in cortical collecting tubules of a patient with Sjogren's syndrome and distal renal tubular acidosis. *JASN* 3:264–71.

Donnelly, S. M., et al. 1992. Might distal renal tubular acidosis be a proximal disorder? *Am. J. Kid. Dis.* 19: 272–81.

Halperin, M. L., et al. 1974. Studies on the pathogenesis of type I (distal) renal tubular acidosis as revealed by the urinary Pco_2 tensions. *J. Clin. Invest.* 53:669–77.

Halperin, M. L., et al. 1992. In *Clinical disorders of fluid and electrolyte metabolism.* 5th ed. Ed. M. H. Maxwell, C. R. Kleeman, and R. G. Narins, 910–31. New York: McGraw Hill.

Vasuvattakul, S., et al. 1992. Should the urine Pco_2 or the rate of excretion of NH_4^+ be the gold standard to diagnose distal renal tubular acidosis? *Am. J. Kid. Dis.* 19:72–75.

Wrong, O. 1991. Distal renal tubular acidosis: The value of urinary pH, Pco_2 and NH_4^+ measurements. *Pediatr. Nephrol.* 5:249–55.

The following articles deal with the pathophysiology of metabolic alkalosis.

Galla, J. H., et al. 1991. Adaptations to chloride-depletion alkalosis. *Am. J. Physiol.* 261:R771–R781.

Galla, J. H., and R. G. Luke. 1987. Pathophysiology of metabolic alkalosis. *Hosp. Practice* 22:123–46.

Kassirer, J. P., and W. B. Schwartz. 1966. The response of normal man to selective depletion of hydrochloric acid. *Am. J. Med.* 40:10–18.

The following two articles deal with the reabsorption of bicarbonate during metabolic alkalosis.

Maddox, D., and F. Gennari. 1986. Load dependence of proximal tubular bicarbonate reabsorption in chronic metabolic alkalosis in the rat. *J. Clin. Invest.* 77:709–16.

Wesson, D. 1989. Augmented bicarbonate reabsorption by both the proximal and distal nephron maintains chloride-deplete metabolic alkalosis in rats. *J. Clin. Invest.* 84:1460–69.

The following two articles deal with the treatment of metabolic alkalosis.

Halperin, M. L., and A. Scheich. 1994 (in press). Why do we continue to recommend that NaCl be used to treat a deficit of KCl? *Nephron* 65:000–000.

Kassirer, J. P., and W. B. Schwartz. 1966. Correction of metabolic alkalosis in man without repair of potassium deficiency. *Am. J. Med.* 40:19–26

The following seven articles provide information about respiratory acid-base disorders.

Arbus, G.S. 1973. An in vivo acid-base nomogram for clinical use. *CMAJ* 109:291–92.

Bercovici, M., et al. 1983. Effect of acute changes in the P_aco_2 on acid-base parameters in normal dogs and dogs with metabolic acidosis or alkalosis. *Can. J. Physiol. and Pharmacol.* 61:166–73.

Brackett, N. C., J. J. Cohen, and W. B. Schwartz. 1965. Carbon dioxide titration curve of normal man. *NEJM* 272:6–12.

Kamel, K. S., et al. 1993. Rate of production of carbon dioxide in patients with a severe degree of metabolic acidosis. *Nephron* 64:514–17.

Krapf, R., et al. 1991. Chronic respiratory alkalosis: The effect of sustained hyperventilation on renal regulation of acid-base equilibrium. *NEJM* 324:1394–1401.

Schwartz, W. B., N. C. Brackett, and J. J. Cohen. 1965. The response of extracellular hydrogen ion concentration to graded degrees of chronic hypercapnia: The physiologic limits of the defense of pH. *J. Clin. Invest.* 44:291–301.

Weinberger, S. E., R. M. Schwartzstein, and J. W. Weiss. 1991. Hypercapnia. *NEJM* 321:1223–31.

The following two articles provide information about mixed acid-base disturbances.

Bear, R., et al. 1977. Effect of metabolic alkalosis on respiratory function in patients with chronic obstructive lung disease. *CMAJ* 117:900–03.

Narins, R., and M. Emmett. 1980. Simple and mixed acid-base disorders: A practical approach. *Medicine* 59:161–87.

SODIUM AND WATER

The following four review articles provide background concerning salt and water balance.

Baylis, P. H., and C. J. Thompson. 1988. Osmoregulation of vasopressin secretion and thirst in health and disease. *Clin. Endocrinol.* 29:549–76.

Feig, P. U., and D. K. McCurdy. 1977. The hypertonic state. *NEJM* 297:1444–54.

Schrier, R. W. 1992. An odyssey into the milieu interieur: Pondering the enigmas. *J. Am. Soc. Nephrol.* 2:1549–59.

Sonnenberg, H. 1990. Renal regulation of salt balance: A primer for non-purists. *Pediatr. Nephrol.* 4:354–57.

The following two articles provide a reference source on Na^+ and water balance.

Fitzsimons, J. T. 1993. Physiology and pathophysiology of thirst and sodium appetite. In *Clinical disturbances of water metabolism*, ed. D. W. Seldin and G. Giebisch, 65–97. New York: Raven Press Ltd.

Robertson, G.L. 1993. Regulation of vasopressin secretion. In *Clinical disturbances of water metabolism*, ed. D. W. Seldin and G. Giebisch, 99–118. New York: Raven Press Ltd.

The following two articles deal with the control of cell volume.

Grinstein, S., W. Furuya, and L. Bianchini. 1992. Protein kinases, phosphatases, and the control of cell volume. *News in Physiol. Sci.* 7:232–37.

Gullans, S. R., and J. G. Verbalis. 1993. Control of brain volume during hyperosmolar and hypoosmolar conditions. *Annu. Rev. Med.* 44:289–301.

The following four articles deal with the pathophysiology of hyponatremia.

Graber, M., and D. Corish. 1991. The electrolytes in hyponatremia. *Am. J. Kid. Dis.* 18:527–45.

Nelson, P. B., et al. 1981. Hyponatremia in intracranial disease: Perhaps not the syndrome of inappropriate secretion of antidiuretic hormone (SIADH). *J. Neurosurg.* 55:938–41.

Verbalis, J. G. 1991. Hyponatremia: Answered and unanswered questions. *Am. J. Kid. Dis.* 18:546–52.

Weisberg, L. S. 1989. Pseudohyponatremia: A reappraisal. *Am. J. Med.* 86:315–18.

The following five articles focus on the controversy about the rate of treatment of hyponatremia.

Arieff, A. I., and J. C. Ayus. 1991. Treatment of symptomatic hyponatremia: Neither haste nor waste. *Crit. Care Med.* 19:748–51.

Berl, T. 1990. Treating hyponatremia: Damned if we do and damned if we don't. *Kid. Internat.* 37:1006–18.

Decaux, G., et al. 1981. Treatment of the syndrome of inappropriate secretion of antidiuretic hormone with furosemide. *NEJM* 304:329–30.

Laureno, R., and B. I. Karp. 1988. Pontine and extrapontine myelinolysis following rapid correction of hyponatremia. *Lancet* 1:1439–41.

Sterns, R. H., E. C. Clark, and S. M. Silver. 1993. Clinical consequences of hyponatremia and its correction. In *Clinical disturbances of water metabolism*, ed. D. W. Seldin and G. Giebisch, 225–36. New York: Raven Press Ltd.

The following three articles deal with aspects of hypernatremia.

Marsden, P. A., and M. L. Halperin. 1985. Pathophysiologic approach to patients presenting with hypernatremia. *Am. J. Nephrol.* 5:229–35.

Perez, G. O., J. R. Oster, and G. L. Robertson. 1989. Severe hypernatremia with impaired thirst. *Am. J. Nephrol.* 9:421–34.

Star, R. A. 1993. Clinical consequences of hypernatremia and its correction. In *Clinical disturbances of water metabolism,* ed. D. W. Seldin and G. Giebisch, 237–47. New York: Raven Press Ltd.

The following four articles deal with polyuria.

Magner, P.O., and M. L. Halperin. 1987. Polyuria—a pathophysiological approach. *Med. N. Am.* 15:2971–78.

Narins and Riley. 1991. Polyuria: Simple and mixed disturbances. *Am. J. Kid. Dis.* 17:237–41.

Robertson, G. L. 1988. Differential diagnosis of polyuria. *Annu. Rev. Med.* 39:425–42.

Star, R. A. 1993. Pathogenesis of diabetes insipidus and other polyuric states. In *Clinical disturbances of water metabolism,* ed. D. W. Seldin and G. Giebisch, 211–24. New York: Raven Press Ltd.

The following three articles deal with the clinical interpretation of urine electrolytes.

Halperin, M. L., and K. L. Skorecki. 1986. Interpretation of the urine electrolytes and osmolality in the regulation of body fluid tonicity. *Am. J. Nephrol.* 6:241–45.

Kamel, K., et al. 1990. Urine electrolytes and osmolality: When and how to use them. *Am. J. Nephrol.* 10:89–102.

Rose, B. D. 1986. New approach to disturbances in the plasma sodium concentration. *Am. J. Med.* 81:1033–40.

The following two review articles deal with hyperglycemia and hyponatremia.

Halperin, M.L., et al. 1993. Clinical consequences of hyperglycemia and its correction. In *Clinical disturbances of water metabolism,* ed. D. W. Seldin and G. Giebisch, 249–72. New York: Raven Press Ltd.

Roscoe, J., et al. 1975. Hyperglycemia-induced hyponatremia: Metabolic considerations in calculation of serum sodium depression. *CMAJ* 112:452–53.

The following two articles deal with diuretics.

Brater, D. C. 1993. Resistance to diuretics: Mechanisms and clinical implications. *Adv. in Nephrol.* 22:349–69.

Rose, B. D. 1991. Diuretics (clinical conference). *Kid. Internat.* 39:336–52.

POTASSIUM

The following seven articles provide background on the physiology of the excretion of K^+.

Carlisle, E. J. .F., et al. 1991. Modulation of the secretion on potassium by accompanying anions in humans. *Kid. Internat.* 39:1206–12.

Lang, F., and W. Rehwald. 1992. Potassium channels in renal epithelial transport regulation. *Physiol. Rev.* 72:1–32.

Schuster, V., and J. Stokes. 1987. Chloride transport by the cortical and outer medullary collecting duct. *Am. J. Physiol.* 253:F208–F212.

Vasuvattakul, S., et al. 1993. Kaliuretic response to aldosterone: Influence of the content of potassium in the diet. *Am. J. Kid. Dis.* 21:152–60.

Velazquez, H., F. S. Wright, and D. W. Good. 1982. Luminal influences on potassium secretion: Chloride replacement with sulfate. *Am. J. Physiol.* 242:F46–F55.

Wright, F. S. and G. Giebisch. 1992. Regulation of potassium excretion. In *The kidney: Physiology and pathophysiology.* 2d ed. Ed. D. W. Seldin and G. Giebisch, 2209–47. New York: Raven Press Ltd.

Young, D. B. 1988. Quantitative analysis of aldosterone's role in potassium regulation. *Am. J. Physiol.* 255:F811–F822.

The following eight articles deal with movement of K^+ across cell membranes.

Adrogue, H. J., and N. E. Madias. 1981. Changes in plasma potassium concentration during acute acid-base disturbances. *Am. J. Med.* 71:456–67.

Akaike, N. 1988. Regulation of sodium and potassium in muscle of potassium-deficient rats. *News in Physiol. Sci.* 3:25–27.

Brown, A. M. 1992. Ion channels in action potential generation. *Hosp. Practice* 27: 125–32.

Brown, R. S. 1986. External potassium homeostasis. *Kid. Internat.* 30:116–27.

Magner, P. O., et al. 1988. The plasma potassium concentration in metabolic acidosis: A re-evaluation. *Am. J. Kid. Dis.* 11:220–24.

Moore, R. D. 1983. Effects of insulin upon ion transport. *Biochem. Biophys. Acta.* 737:1–49.

Rosa, R. M., M. E. Williams, and F. H. Epstein. 1992. Extrarenal potassium metabolism. In *The kidney: Physiology and pathophysiology.* 2d ed. Ed. D. W. Seldin and G. Giebisch, 2165–90. New York: Raven Press Ltd.

Swan, R. C., and R. F. Pitts. 1955. Neutralization of infused acid by nephrectomized dogs. *J. Clin. Invest.* 34:205–12.

The following two articles provide background on diurnal excretion of K^+.

Moore-Ede, M. C. 1986. Physiology of the circadian timing system: Predictive versus reactive homeostasis. *Am. J. Physiol.* 250:R735–R752.

Steele, A., et al. 1994 (in press). What is responsible for the diurnal variation in potassium excretion? *Am. J. Physiol.*, 258:000–000.

The following four articles provide information on drugs and hyperkalemia.

Choi, M.J., et al. 1993. Trimethoprim induced hyperkalemia in rats. *NEJM* 328:703–06.

Kamel, K.S., et al. 1992. Studies to determine the basis for hyperkalemia in recipients of a renal transplant who are treated with cyclosporine. *JASN* 2:1279–84.

Ponce, S., et al. 1985. Drug-induced hyperkalemia. *Medicine* 64:357–70.

Velazquez, H., et al. 1993. Renal mechanism of trimethoprim-induced hyperkalemia. *Ann. Intern. Med.* 119:295–301.

The following articles provide background on some of the rarer disorders of K^+ homeostasis.

Cannon, S. C., R. H. Brown, Jr., and D. P. Corey. 1991. A sodium channel defect in hyperkalemic periodic paralysis: Potassium-induced failure of inactivation. *Neuron* 6:619–26.

Clore, J., A. Schoolwerth, and C. O. Watlington. 1992. When is cortisol a mineralocorticoid? *Kid. Internat.* 42:1297–1308.

Gitelman, H. J. 1992. Hypokalemia, hypomagnesemia, and alkalosis: A rose is a rose—or is it? *J. Pediatr.* 120:79–80.

Gordon, R. D. 1986. Syndrome of hypertension and hyperkalemia with normal glomerular filtration rate. *Hypertension* 8:93–102.

Griggs, R. C., and L. J. Ptacek. 1992. The periodic paralyses. *Hosp. Practice* 27: 123–37.

Lehmann-Horn, F., et al. 1991. Altered gating and conductance of Na^+ channels in hyperkalemic periodic paralysis. *Pflugers Arch.* 418:297–99.

Lifton, R. P., et al. 1992. A chimeric 11b-hydroxylase/aldosterone synthetase gene causes glucocorticoid-remediable aldosteronism and human hypertension. *Nature* 355:262–65.

Monder, C. 1991. Corticosteroids, receptors, and the organ-specific functions of 11b-hydroxysteroid dehydrogenase. *FASEB J* 5:3047–54.

Schambelan, M., A. Sebastian, and F. C. Rector, Jr. 1981. Mineralocorticoid-resistant renal hyperkalemia without salt wasting (type II pseudohypoaldosteronism): Role of increased renal chloride reabsorption. *Kid. Internat.* 19:716–27.

Warnock, D.G. 1993. Liddle's syndrome: 30 years later. *J. Nephrol.* 6:142–48.

The following two articles provide information about the utility of the TTKG.

Ethier, J. H., et al. 1990. The transtubular potassium concentration in patients with hypokalemia and hyperkalemia. *Am. J. Kid. Dis.* 15:309–15.

West, M. L., P. O. Magner, and R. M. A. Richardson. 1988. A renal mechanism limiting the degree of potassium loss in severely hyperglycemic patients. *Am. J. Nephrol.* 8:373–78.

The following three articles provide information about the treatment of hypokalemia and hyperkalemia.

Donnelly, S. M., et al. 1991. Hypokalemia. In *Current therapy in nephrology and hypertension*. 3rd ed. Ed. J. P. Kassirer, 1015–23. Philadelphia: B.C. Decker, Inc.

Kurtz, I., and L. G. Fine. 1991. Hyperkalemia. In *Current therapy in nephrology and hypertension*. 3rd ed. Ed. J. P. Kassirer, 1023–28. Philadelphia: B.C. Decker, Inc.

Tannen, R. 1985. Diuretic-induced hypokalemia. *Kid. Internat.* 28:988–1000.

HYPERGLYCEMIA

The following four articles provide background on aspects of carbohydrate metabolism.

Bjorntorp, P., and L. Sjostrom. 1978. Carbohydrate storage in man: Speculations and some quantitative considerations. *Metabolism* 27:1853–85.

Jungas, R. L., M. L. Halperin, and J. T. Brosnan. 1992. Lessons learnt from a quantitative analysis of amino acid oxidation and related gluconeogenesis in man. *Physiol. Rev.* 72:419–48.

Martin, B. J. 1990. Gut transit with exercise. *Gastroenterology* 99:290.

Randle, P. 1986. Fuel selection in animals. *Biochem. Soc. Trans.* 14:799–806.

We recommend the following textbooks.

Halperin, M. L., and F. S. Rolleston. 1993. Clinical detective stories: A problem-based approach to clinical cases in energy and acid-base metabolism. London: Portland Press.

Voet, D. and J. G. Voet. 1990. *Biochemistry*. New York: John Wiley and Sons.

The following five articles provide information on the renal handling of glucose.

Deetjen, P., H. V. Baeyer, and H. Drexel. 1992. Renal glucose transport. In *The kidney: Physiology and pathophysiology*. 2d ed. Ed. D. W. Seldin and G. Giebisch, 249–72. New York: Raven Press Ltd.

Halperin, M. L., et al. 1980. Quantitative aspects of hyperglycemia in the diabetic: A theoretical approach. *Clin. Invest. Med.* 2:127–30.

Kurtzman, N., and V. Pillay. 1973. Renal reabsorption of glucose in health and disease. *Arch. Intern. Med.* 131:901–04.

Magner, P. O., and M. L. Halperin. 1990. Effect of metabolic acidosis on glucose reabsorption in rats with acute hyperglycemia. *Can. J. Physiol. Pharmacol.* 68:79–83.

Mogensen, C. 1971. Maximum tubular reabsorption capacity for glucose and renal hemodynamics during rapid hypertonic glucose infusion in normal and diabetic subjects. *Scand. J. Clin. Lab. Invest.* 28:101–09.

The following two articles help define the deficits of electrolytes in patients with a severe degree of hyperglycemia.

Danowski, T. S., et al. 1949. Studies in diabetic acidosis and coma, with particular emphasis on the retention of administered potassium. *J. Clin. Invest.* 28:1–9.

Nabarro, J,. A. Spencer, and J. Stowers. 1952. Metabolic studies in severe diabetic ketosis. *Quart. J. Med.* 82:225–43.

The following two articles deal with clinical aspects of hyperglycemia.

Arieff, A. I., and H. J. Carroll. 1971. Hyperosmolar nonketotic coma with hyperglycemia: Abnormalities of lipid and carbohydrate metabolism. *Metabolism* 20:529.

West, M., et al. 1986. A quantitative analysis of glucose loss during acute therapy for the hyperglycemia hyperosmolar syndrome. *Diabetes Care* 9:465–71.

The following two articles deal with hyperglycemia and hyponatremia.

Halperin, M. L., et al. 1993. Clinical consequences of hyperglycemia and its correction. In *Clinical disturbances of water metabolism*, ed. D. W. Seldin and G. Giebisch, 249–72. New York: Raven Press Ltd.

Roscoe, J., et al. 1975. Hyperglycemia-induced hyponatremia: Metabolic considerations in calculation of serum sodium depression. *CMAJ* 112:452–53.

INDEX